AF556435

Potentials of Living Resources

POTENTIALS
OF
LIVING RESOURCES

Editors

Dr. G. Tripathi

&

Dr. A. Kumar

DISCOVERY PUBLISHING HOUSE

NEW DELHI-110002

First Published-2003

ISBN 81-7141-718-3

Published by

DISCOVERY PUBLISHING HOUSE
4831/24, Ansari Road, Prahlad Street,
Darya Ganj, New Delhi-110002 (India)
Phone: 23279245 • Fax: 91-11-23253475
E-mail:dphtemp@indiatimes.com

Printed at: Tarun Offset Printers, Delhi-53

PREFACE

There is enormous potential value of natural life to fulfill human needs on a sustainable basis without deteriorating environment. Peoples are now fade up with chemical world. They are in search of an alternative and relying on biological potentials. Genetically modified organisms may have a very serious impact on ecosystem and environment in the days to come. So it becomes wise and imperative to explore and use natural potentials of living resources for welfare of mankind and biospheric peace. There are about 10^9 genes in the world biota. One can earn millions of dollar from product of a single important and useful gene. We have gained the knowledge of intellectual property right (IPR) and are moving towards gene economics. An integrated approach involving traditional knowledges, bioresource survey and application of modern techniques will strengthen the world's economy particularly in biodiversity-rich areas. Today 80 per cent of human kind depends entirely on plant resources for medicinal or preventive measures. About 10,000 toxins have been found in plants and many of them are potential candidates for drugs. The contribution of microbial world are not less than plants. A whole range or enzymes have been found in microorganisms that are useful for biocatalytic processes. Microbes also produces antibiotics, antifungals and antiviral proteins. There are manifold applications of yeasts to food processing. Undoubtedly, billions of proteins are yet to be screened and evaluated for possible use in future.

Animal world also contributes a lot to our welfare, although only indigenous peoples use them as drug reservoirs such as therapeutic uses of earthworms etc. There is a rich biological potentials in the oceans which need exploitation. Slugs, sponge, worms, crabs and other marine creatures may reveal their secrets. Invertebrates have long enriched human manufacturing such as production of silk, wax, chitin and toxins for pharmaceutical purposes. Conservation and identification of potentials of animal, plant and microbial resources may be of unparallel importance to Man and Environment. Looking into the importance of bioresources, we decided to synthesize a compilation on "Potentials of Living Resources" for those who believed in ecofriendly economic growth with environmental sustainability.

We are grateful to the contributors of different chapters who presented their expert views and made the compilation possible. We are also thankful to the Discovery Publishing House, New Delhi, for publication of this important and interesting book.

Editors

Contents

List of Contributors

Y.K. Jaiswal, School of Studies in Biochemistry, Jiwaji University, Gwalior-474011.

K. Deb, School of Studies in Biochemistry, Jiwaji University, Gwalior-474011.

A. Gupta, Department of Zoology, Dr. H.S. Gour Vishwavidyalaya, Sagar-470003.

M. Bhide, Department of Zoology, Dr. H.S. Gour Vishwavidyalaya, Sagar-470003.

S. Modi, Department of Zoology, Dr. H.S. Gour Vishwavidyalaya, Sagar-470003.

K.K. Sharma, Indian Lac Research Institute, Namkum, Ranchi-470010.

K.K. Kumar, Indian Lac Research Institute, Namkum, Ranchi-470010.

D. Rajagopal, Department of Apiculture, University of Agriculture Sciences, GKVK, Bangalore-560065.

N. Nagaraja, Department of Apiculture, University of Agriculture Sciences, GKVK, Bangalore-560065.

S.K. Verma, Central Soil and Water Cons. Res. & Training Institute, 218-Kaulagarh Road, Dehradoon-248 195.

K.S. Dadhwal, Central Soil and Water Cons. Res. & Training Institute, 218-Kaulagarh Road, Dehradoon-248195.

T.K. Ghosh, NEERI, Nehru Marg, Nagpur-440020.

L.N. Sangolkar, NEERI, Nehru Marg, Nagpur-440020.

A. Saxena, Department of Fishery Biology, College of Fishery, GBPUAT, Pantnagar-263145.

Y.C. Tripathi, Rain Forest Research Institute, P.O. Box 136, Deovan, Jorhat-785001.

S. Singh, Non-Wood Forest Products Division, Arid Forest Research Institute, Jodhpur-342008.

P.K. Mishra, Department of Botany, R.S. More College, Govindpur-828109.

G.Singh, Division of Forest Ecology & Desert Development, Arid Forest Research Institute, Jodhpur-342005.

S. Mutha, Division of Forest Ecology & Desert Development, Arid Forest Research Institute, Jodhpur-342005.

P. Hazarika, Rain Forest Res. Institute, Deovan, Jorhat-785001.

G.P. Juyal, Central Soil and Water Cons. Res. & Training Institute, 218, Kaulagarh Road, Dehradoon-248195.

S.C. Deb, SIES-Indian Instt. Environ. Manag., Sri C.S.S. Vidhyapuram Sector-V, Plot-1E, Nerul, Navi Mumbai-400706.

S. Arun, Department of Animal Science, Bharathi Dasan University, Tiruchirapalli-620024.

P. Subramaniam, Department of Animal Science, Bharathi Dasan University, Tiruchirapalli-620024.

A. Dutta, Environ. Res. Lab., Ranchi University, Ranchi-834010.

S. Datta, Environ. Res. Lab., Ranchi University, Ranchi-834010.

P.N. Pandey, Environ. Res. Lab., Ranchi University, Ranchi-834010.

S. Kaviraj, S.SLN.T, Women's College, Dhanbad.

G. Tripathi, Department of Zoology, J.N.V. University, Jodhpur-342001.

P. Bhardawaj, Department of Zoology, J.N.V. University, Jodhpur-342001.

Omkar, Department of Zoology, University of Lucknow, Lucknow-226007.

G. Mishra, Department of Zoology, University of Lucknow, Lucknow-226007.

S. Ram, Department of Zoology, J.N.V. University, Jodhpur-342001.

B.M. Sharma, Department of Zoology, J.N.V. University, Jodhpur-342001.

1

Antifertility Potential of Gram-Negative Bacterial Endotoxins

Y.K. Jaiswal and *Kaushik Deb*

Gram-negative bacterial infections lead to a lethal condition known as gram-negative septic shock which causes about 1.5 million deaths worldwide (Parrilo, 1993; Stone, 1994). The pathological effect is believed to be due to the fever-inducing substances known as pyrogens. Pyrogens have been divided into two classes i.e. exogenous and endogenous pyrogens. Exogenous pyrogens are those that originate outside the body, and elevate temperature when injected to humans and animals. These generally include microbial components of gram-negative bacteria and some non-microbial pyrogens e.g. drugs and steroids. Of all the known components of gram-negative bacteria, endotoxins are by far the most ubiquitous and important exogenous type of pyrogens (Glauser *et al.,* 1991). However, endogenous pyrogens are produced in response to the stimulus from various exogenous pyrogens internally by the host. The proinflammatory type of cytokines like TNF, interleukin (IL-1), IFN etc. belong to this category and are known to be the primary mediators of fever (Raetz *et al.,* 1991).

Endotoxin is the name given to a group of heat stable lipopolysaccharide (LPS) molecules present in the cell walls of gram-negative bacteria that possess toxic effect (Burell, 1994). These toxins although intimately associated with the bacterial cell membrane are constantly shed into the environment of the bacterium much like the daily shedding of the superficial layers of the human skin. When bacterium lyse disintegrate all endotoxin is released from the cell. Once endotoxins are free to act, they recruit particular host cells to secrete mediator molecules. These mediators act locally or float through the blood or both to elicit a wide variety of responses. In higher doses endotoxins have the capacity to activate the coagulation system, alter carbohydrate and lipid metabolism, modify hemodynamics, cause platelet aggregation, induce disseminated intravascular coagulation leading to shock and ultimately death (Kielian and Blecha, 1995). Endotoxins derived from gram-negative bacteria have been found to extensively damage preimplantation stage embryos (Dostal *et al.*, 1996). Endotoxins decrease embryo cleavage rate, blastocyst formation and pregnancy rates in human. Endotoxins also cause fragmentation and degeneration of preimplantation embryos in vivo and in vitro. The presence of endotoxins during IVF of oocytes results in a high rate of polyspermy in human and bovine species (Berger *et al.*, 1982). Microorganisms such as *E.coli*, isolated from the uterine cervix, decrease sperm motility and possess spermicidal activity (Diemer *et al.*, 2000).

As a crucial step towards understanding how endotoxins might achieve diverse pathological effects, research teams have concentrated on elucidating their chemical and three-dimensional structure (Rietschel *et al.*, 1994). Endotoxins, isolated from the outer membrane of gram-negative bacteria, consist of a polysaccharide or long chain of sugars and a fat called Lipid A. Studies indicate that the sugars, the phosphates and the fatty acids combine into a three-dimensional shape that best facilitates interactions with the cells of the host. Recognition of LPS triggers gene induction by myeloid and nonmyeloid lineage cells. These inducible genes encode proteins that induce cytokines, adhesive proteins, and enzymes that produce low molecular weight proinflammatory mediators, which upregulate host defense systems that participate in eliminating the bacterial infection (Manthey and Vogel, 1994).

Many of the biological effects of LPS are mediated by the proinflammatory cytokines like IL-1, IL-6, TNF, etc. Several of these cytokines have been implicated in delicate immune system balances that exist within the feto-maternal interface (Kauma 2000). The presence of cytokines and growth factors in the endometrium and or embryos and the expression of their corresponding receptors on the implanting blastocyst and or endometrium, suggest that they may play important role in the process of development and blastocyst invasion into the endometrium. Uterine secretions of many of these cytokines are under the control of ovarian steroids, confirming their close association with a distinct regulatory network, that may overlap or complement the immune system represented by the lympho-hematopoietic cells residing in the uterine endometrium (Paria *et al.,* 2000). This may be a general mechanism for the regulation of early embryonic development and for maintaining a synchrony between the embryo and uterine preparation for implantation. Any disturbance of this delicate immune balance within the maternal-fetal interface may result in pregnancy loss or other perinatal complications. It is likely that LPS brings about its effects by modulating the synthesis and secretion of the cytokines and growth factors from the lymphohaematopoetic and uterine cells. These cytokines in turn could alter the embryonic expression of growth factors leading to preimplantation embryonic loss, implantation failure, or abortions. The major goal of our laboratory is to define the cytokine basis of embryo-uterine interactions during implantation and the effect of gram-negative bacterial endotoxins on it. In the discussion below, we have made an attempt to illucidate a probable mechanism of LPS/ endotoxin induced pregnancy loss/infertility, primarily focusing on the uterine and embryonic development.

STRUCTURE AND OVERVIEW OF ENDOTOXIN PHARMACOLOGY

Cox *et al.* (1988) have reported that gram-negative bacteria (aerobic and anaerobic) are involved in colonization and infection of the genitourinary tract in pregnant woman and is associated with preterm labor/birth and other perinatal complications. All gram-negative bacteria have LPS as a complex glycolipid present on the outer membrane (see Fig.1). These cell associated toxins of gram-negative bacteria (endotoxins) consist of carbohydrates and lipids and are composed of two chemically dissimilar structural regions: the hydrophilic repeating, polysaccharides of the core and O-antigenic structure and a hydrophobic domain

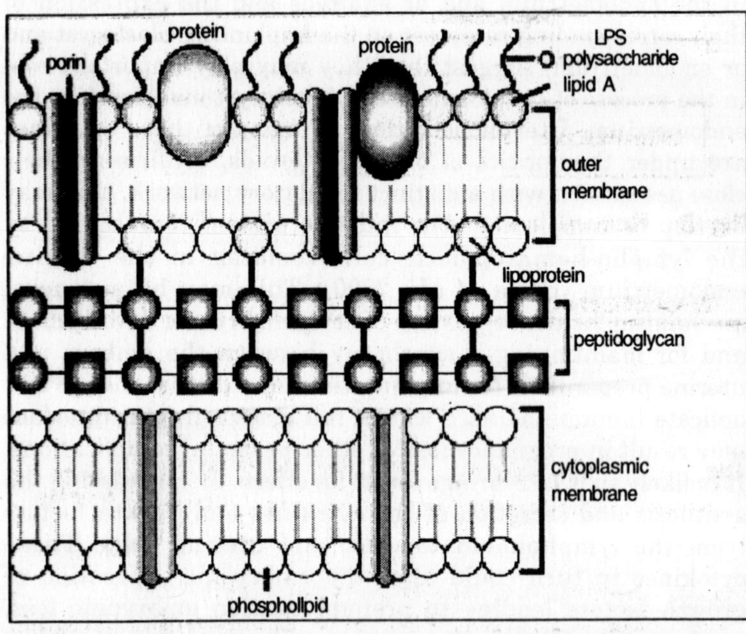

Fig. 1: Structure of the cell surface of a gram-negative bacterium

known as lipid-A (Fig.2). The O-specific chain contains repeating oligosaccharides of up to five repeating sugar residues. Different gram-negative bacteria show considerable structural diversity among the O-specific chain structures. However, the structure of the core region is substantially more conserved (Fig.3). The chemical nature of LPS was deciphered as lipid-A domain was synthesised chemically. Raetz (1993) has shown that lipid A component possesses the biological activities of endotoxins. Moreover, all of the LPS-induced biological responses are lipid-A dependent. Thus recognition of the lipid-A component of LPS by cells will be the initial step in LPS induced cellular response (Rietschel *et al.*, 1994).

The general chemical structure of Lipid A from diverse gram-negative bacteria is highly conserved (Fig.4). The Lipid A of *Escherichia coli* or *Salmonella* has β-1,6-linked D-glucosaminedisaccharide phosphorylated in position 1 and 4. Two KDO (2-keto-3 deoxyoctonic acid) moieties are attached

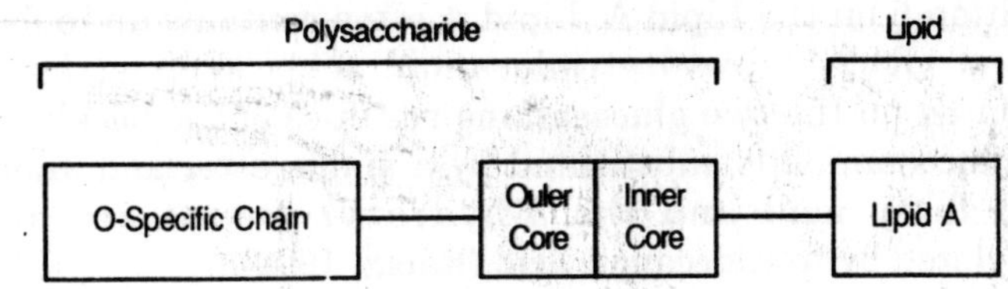

Fig. 2: General architecture of Lipopolysaccharide

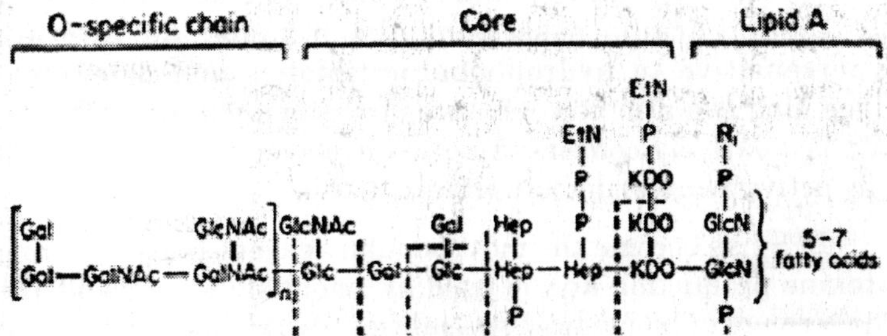

Fig. 3: General structure of *Samonella* LPS
Glc = glucose; GlcNac = N-acetyl-glucosamine;
Gal = galactose; KDO (2-keto-3 deoxyoctonic acid)
H = heptose; P = phosphate; Etn = ethanolamine

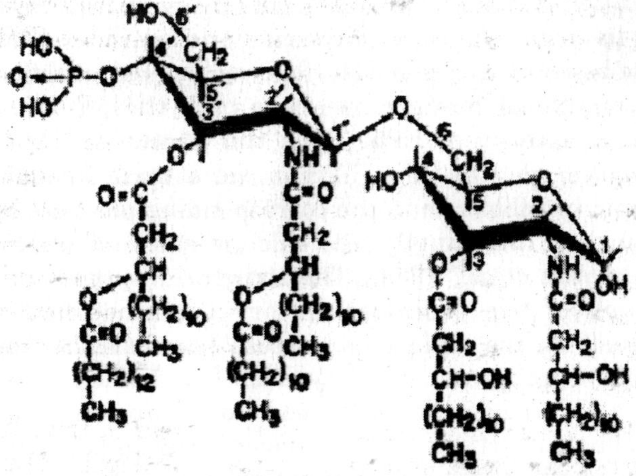

Fig. 4: Complete structure of the Lipid A moiety of LPS of *S. typhimurium, S. minnesota,* and *E. coli*

at position 6 on the Lipid A. Lipid A often contains up to four moles of (R)-3-hydroxytetradecanoic acids symmetrically distributed on the two glucosamine residues of the backbone. Each glucosamine is substituted by a single ester and amide linked fatty acid and the 3-hydroxyl groups are also 3-O-acylated by tetradecanoic acid (Raetz, 1990).

The minimal LPS structure required for the growth of *Escerichia coli* and other gram-negative (termed Re LPS) bacteria consist of Lipid A and two KDO moieties (Rietschel *et al.*, 1988). Strains possessing the minimal Re LPS are hypersensitive to hydrophobic antibiotics and detergents suggesting evolutionary advances to a complete LPS. The Re LPS is relatively conserved between bacterial species and is fully active as an endotoxin (Raetz, 1990).

LPS has become the most popular reagent for stimulating cytokine production and is used by scientists throughout the whole range of biomedical specialities. According to Hitchcock *et al.*, (1986) the term LPS "should be reserved for purified bacterial extracts which are reasonably free from detectable contaminants particularly proteins". The physical state of LPS can markedly affect its biological activities. This has been clearly demonstrated with the LPS extracted from the *Salmonella sp.* which exist as a negatively charged molecule bound to amines such as putrescine and spermine. Removal of these bases by electrodialysis followed by neutralization with triethylamine or metal ions alters both the solubility and biological activity of LPS. Thus the triethylamine salt of *Salmonella abortus equi* LPS is more toxic in mice and pyrogenic in rabbits than the corresponding partially soluble calcium salt. Intriguingly, the calcium salt is more toxic in rats (Galanos *et al.*, 1984). The structural requirements for induction of the proinflammatory cytokines have been examined. Active molecules possess the disaccharide, two phosphoryl groups and six fatty acids in a defined location as found in *E.Coli* LPS. The fatty acid side chains are important determinants of activity with the major variables being number, location, chain length and stereochemistry. *E. coli* lipidA with five or four fatty acid residues are, respectively, 10 and 100-fold less potent at inducing proinflammatory cytokines than is the hexaacyl form of *E.coli* lipid A (Almer *et al.*, 1992) . Cavaillon *et al.*, (1990) have reported that lipid A must be linked to at least one KDO residues in order to induce monocytes/macrophages to synthesize IL-1. It has also been reported that the heptose-heptose-KDO from a *Salmonella minnesota* mutant and the core oligosacccharides from this organism are able to release IL-1 from human monocytes. Feist *et al.*, (1989) demonstrated that R-type LPSs from *Salmonella minnesota* were more potent than were S-type LPSs in stimulating TNF-release.

In humans and experimental animals the presence of bacterial lipopolysaccharides (LPS) signals the presence of gram-negative bacteria. Considerable progress has been made recently in defining the mechanisms by which endotoxins interact with animal cells. When gram-negative bacteria are ingested by phagocytes and degraded in vacuoles, the LPS is released. When LPS enters the circulation, it is bound either to lipoproteins or LPS-binding proteins (LBP), which may stimulate macrophages by binding to a glycoprotein receptor (mCD14) present on the cell surface (Saito *et al.*, 2000). CD14 can also exist in a soluble form in the circulation (sCD14) (Tapping and Tobias 1997) which can facilitate LBP mediated LPS transfer to high density lipoproteins (HDL) (Vesy *et al.*, 2000) and enables cells which do not express mCD14 to respond to low doses of LPS (Frey *et al.*, 1992). An understanding of the mechanism of action of LPS was greatly expedited by the identification of Toll-like receptor (TLR) family and the parallels between recognition of bacterial stumuli in mammals and drosophila. Recent discoveries suggest that CD14 interacts with toll-like receptors to transmit a signal (Chow *et al.*, 1999; and Kirschning *et al.*, 1998). Toll-like receptors comprise a family of innate immune signaling receptors that are related to the *Drosophila* Toll protein, a molecule that is implicated in defence against fungal infection in the fly (Lemaitre *et al.*, 1997). It was found that mice with mutations in the TLR_4 genes were hypo-sensitive to LPS (Hoshino *et al.*, 1999, Poltorak *et al.*, 1998). Activation of TLR_4 on the cell surface which requires cooperation with at least CD14 and MD_2 (Nomura *et al.*, 2000) leads to recruitment of adopter proteins

(MyD88, IRAK, TRAF6) that ultimately couple the recognition of the microbial LPS to activation of the transcription factor complex NF-kB (Aderem and Ulevitch, 2000). Translocation of this transcription factor complex in turn contributes to activation of the promoters of many inducible genes in macrophages (Sweet and Hume, 1996) (Fig.5).

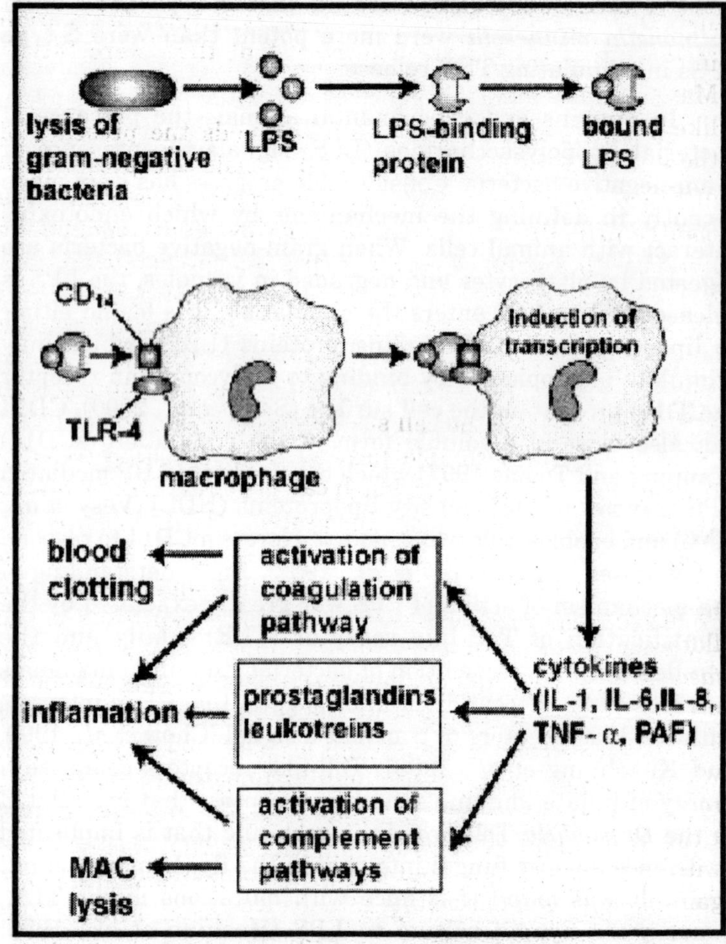

Fig. 5: General physiological action of gram-negative bacterial LPS, through the activation of macrophages

These stimulated macrophages in turn generate cytokines, which include the principal host mediators of clinical manifestations associated with endotoxemia, IL-1 and TNFα (Dinarello, 1994, and Raabe *et al.,* 1998). These cytokines modulate the expression of cell surface adhesion proteins or modules on leukocytes and endothelial cells, such as selectin and integrins (CD11/CD18) and their corresponding ligands (ICAM-1) on endothelial cells (Thomson *et al.,* 1999). This initiates the recruitment of leukocytes at the site of infection. Many of these inducible genes are proinflammatory cytokines like TNF, IL-1, IL-12, IL-6, M-CSF etc., which not only affect the preimplantation embryonic growth and development, but are also involved in regulating the delicate inflammatory reactions at the site of blastocyst implantation. Interleukin-1 (an endogenous pyrogen) is carried to the hypothalamus by the blood. This causes the hypothalamus to release lipids (prostaglandins) that reset the body's temperature to a higher setting, causing fever (Dinarello 1989). Excessive production of these proteins is responsible for the symptoms of endotoxin-induced shock.

THE VAGINAL MICROBIAL ECOSYSTEM AND INFERTILITY

Normal flora of the urinogenital tract

The flora of the anterior urethra, as indicated principally by urine cultures, suggests that a relatively consistent normal flora consisting of *Staphylococcus epidermidis, Enterococcus faecalis* and some alpha-hemolytic streptococci may inhabit the area. However, their numbers are not plentiful. In addition, some enteric bacteria (e.g. *E. coli, Proteus*) and *corynebacteria,* which are probably contaminants from the skin, vulva or rectum, may occasionally be found at the anterior urethra. Urine is normally sterile, and since the urinary tract is flushed with urine every few hours, microorganisms have problems gaining access and becoming established.

The vagina becomes colonized soon after birth with *corynebacteria, staphylococci,* nonpyrogenic *streptococci, E. coli,* and a lactic acid bacterium historically named "Doderlein's bacillus" (*Lactobacillus acidophilus*). During reproductive life, from puberty to menopause, the vaginal epithelium contains glycogen due to the actions of circulating estrogens. Doderlein's bacillus predominates, being able to metabolize the glycogen to lactic acid. The lactic acid and other products of metabolism inhibit colonization of all except Doderlein's bacillus and a selected number of lactic acid bacteria. The resulting low pH of the vaginal epithelium prevents establishment of most of the bacteria and also the potentially pathogenic yeast, *Candida albicans*. In an in vitro study it has been demonstrated that Lactobacillus strains taken from normal vaginal flora show antagonistic activity against a variety of bacteria related to vaginal and urinary tract infections (Strus *et al.*, 2002). Thus the normal bacterial flora possess a protective effect for their human host. Any disturbances in this delicate vaginal microbial ecosystem lead to the colonization of the genitourinary tract with other harmful/pathogenic microbes leading to vaginitis and Urinary Tract Infections.

Pathogenic flora of genitourinary tract infections

Genital tract infections affect primarily the cervix, uterus, or fallopian tubes and during severe infections one or both ovaries may also get affected. Genital tract infection/vaginitis is one of the major causes of gynecologic morbidity, such as infertility, ectopic pregnancy, preterm labor and chronic pelvic pain (Cram *et al.,* 2002). The normal urinary tract is sterile and very resistant to bacterial colonization. However, urinary tract infections (UTI) in human is the most common bacterial infection in all age groups. Gram-negative aerobic bacteria (Table-1) cause most bacterial UTIs. A few UTIs are acquired hematogenously but about 95% occur when bacteria ascend from a colonized vaginal introitus and urethra to the bladder and in the case of acute uncomplicated pyelonephritis, up the ureter to the kidney. *Escherichia coli* is the most common bacterium isolated and accounts for about 80% of community-acquired infections, and *Staphylococcus saprophyticus* for about 10%. UTI, asymptomatic bacteriuria, and pyelonephritis are associated with an increased incidence of preterm labor and premature rupture of the membranes in hospitalized patients, *E. coli* accounts for about 50% of cases; the gram-negative species *Klebsiella, Proteus, Enterobacter,* and *Serratia* for about

Table 1. Normal vaginal flora

Type	*Sub-type*	*Name of the organism*
Bacteria	Gram-negative aerobic	Escherichia coli Proteus vulgaris Klebsiella Enterobacter
	Gram-negative anaerobes	Bacteroides fragilis Bacteroides urolyticus Pervotella Mombiluncus spp.
	Gram-positive aerobic	Enterococcus faecalis
	Gram-positive anaerobic	Streptococcus faecalis Staphylococcus epidermidis Lactobacillus Clostridium
Unusual organisms	Parasites	
	Protozoa Yeast Acid fast organism Virus	

40%; and the gram-positive bacterial cocci *Enterococcus faecalis* and *Staphylococcus sp* (*saprophyticus, aureus*) for the remainder.

Colonization of the vagina by *Lactobacillus sp* keeps the pH normal (3.8 to 4.2), preventing the overgrowth of bacteria and yeast (Di Renzo and De Domenico, 2000). Menstrual blood, certain infections, or semen often makes the vaginal pH alkaline. Among nuns bacteremia is significantly less frequent (0.4 to 1.6% from 15 to 54 yr of age) than among sexually active women, suggesting a probable role for sexual intercourse in the development of acute uncomplicated UTI in women. Use of spermicide with a diaphragm is linked to an increased risk of UTI in women probably because spermicide-induced alterations in vaginal flora permit overgrowth of *E. coli*. Silent (subclinical) microbial infection in the genitourinary tract leads to preimplantation embryonic loss and is one of the major causes of reduced reproductive efficiency in most mammals. Various (aerobic and anaerobic) microorganisms (Table 1) found in the vagina have been associated with increased risk for early pregnancy loss (Donder *et al.*, 2002) and or preterm labour but no single organism could be identified as having a causal relationship (Gibbs *et al.*, 1992). In the case of gram-negative bacterial infections, fetal loss may result in part from host immune responses to the main antigenic component of the bacterial cell wall, endotoxin/lipopolysaccharide (LPS) (Satoh *et al.*, 1994).

Vaginitis

Vaginitis is characterized by vaginal discharge or vulvar itching and irritation with a vaginal odor. The following three diseases most frequently associated with vaginal discharge:

1. **Trichomoniasis:** caused by *Tricomoniasis vaginalis*, a protozoa.
2. **Bacterial Vaginosis (BV):** caused by a replacement of the normal vaginal flora by an overgrowth of anaerobic gram-negative microorganisms like *Gardnerella vaginalis*; and some gram positive facultative anaerobe like *Mycoplasma hominis*.
3. **Candidiasis:** usually caused by *Candida albicans*, a yeast.

Other common causes of vaginal infections include such bacteria as *N. gonorrhoeae, Chlamydia trachomatis, Mycoplasma hominis,* streptococci, and staphylococci; foreign bodies; some viruses (eg, herpes simplex); fistulas; radiation and genital tract neoplasms. BV/Bacterial vaginosis represents 60% of all vaginal infections. Women with bacterial vaginosis also have more chance of getting pelvic inflammatory diseases (PID). PID can damage reproductive organs, cause infertility, pelvic pain and tubal pregnancy.

Bacterial vaginosis

Bacterial vaginosis (BV) is the most common vaginal infection in women who have not gone through menopause.

At least one-third of all vaginal infections are due to bacterial vaginosis.

Bacterial vaginosis was previously called nonspecific vaginitis, Haemophilus vaginitis, or Gardnerella vaginitis. Vaginitis is an inflammation of the vagina and vulva that usually causes itching, pain, and vaginal discharge. Bacterial vaginosis is not usually accompanied by inflammation. Nearly half of all women with BV have no symptoms or irritating or itchy vaginal discharge. The most distinctive symptom of BV, which helps distinguish it from vaginitis, is a thick, milky vaginal discharge with a strong fishy odor.

BV is an imbalance among the microorganisms that normally exist in the vagina. Those involved in BV include gram-negative facultative anaerobes like *Gardnerella vaginalis*, *Chlamydia trachomatis* and other types of bacteria (*Mobiluncus*, *Bacteroides*, and *Mycoplasma*). For reasons not well understood, the numbers of these organisms increase with BV while the number of lactobacillus organisms decreases.

Most cases of bacterial vaginosis occur in sexually active women between the ages of 15 and 44. Pregnant women and women with a sexually transmitted disease (STD) are especially at risk for getting this infection. Bacterial vaginosis does not usually affect women after menopause. Pregnant women with BV are at risk for developing an infection of the amniotic fluid and placenta and are more likely to have a preterm labor and birth (Hay *et al.*, 1994).

Bacterial vaginosis (BV) is, therefore, a change in the normal vaginal microbial ecosystem (Table 1) with three important features:

1. **An overgrowth of anaerobic bacteria** and the facultatives, *Gardnerella vaginalis* and *Mycoplasma hominis* (Table 2). These bacteria are frequently present in normal vaginal flora but quantitative culture demonstrates one to two log increases in BV. Anaerobes commonly identified in BV include the Gram-negative rods *Prevotella spp*. (especially *P.bivia*),

Table 2. Microbes associated with genitourinary tract infections and infertility

Type	*Sub-type*	*Name of the organism*
Bacteria	Gram-negative aerobic	Escherichia coli
		Proteus vulgaris
		Klebsiella
		Acinetobacter calcoaceitus
		Enterobacter
		Pseudomonas aeruginesa
		Serratia
		Neisseria gonorrhoeae
	Gram-negative anaerobes	Bacteroides fragilis
		Bacteroides urolyticus
		Pervotella
		Porphyromonas
		Mombiluncus spp.
		Chlamydia trachomatis
		Gardnerella vaginalis
	Gram-positive aerobic	Enterococci
		Straphylococcus saproplyticus
		Staphylococcus aureus
	Gram-positive anaerobic	Peptostreptococcus
		Mycoplasma hominis
Unusual organisms	Parasites	Echinococcus
		Schistosoma haematobium
		Schistosoma mansoni
	Protozoa	Trichomoniasis vaginalis
	Yeast	Candida albicans
		Blastomyces
		Coccidioides immitis
	Acid fast organism	Mycobacterium tuberculosis.
	Virus	Herpes simplex virus

Porphyromonas and *Bacteroides ureolyticus*, the curved *Mobiluncus spp.* and Gram-positive *Peptostreptococci*. Several observations demonstrate interdependence between these organisms (Pybus and Onderdonk, 1999). For example treating BV with metronidazole frequently leads to the disappearance of *M. hominis* despite its resistance to metronidazole, suggesting it requires the presence of anaerobes. *In vitro* experiments demonstrate nutritional relationships between *Prevotella bivia* and other organisms. *P. bivia* produces ammonia which stimulates the growth of *G. vaginalis*, and amino acids which similarly promote growth of *Peptostreptococcus* and probably *G. vaginalis*.

2. **Biochemical changes** (Pybus and Onderdonk, 1999): The normal pH of 3.8 to 4.2 (acidic) is maintained by the *Lactobacillus sp.* in the vaginal flora. In bacterial vaginosis the vaginal pH becomes more than 4.5 (i.e. loss of normal acidity). Production of amines by the anaerobes (trimethylamine, putrescine, cadaverine) gives a fishy smell to the discharge, which is often a source of complaint. The succinate:lactate ratio increases, as the succinate-producing anaerobic gram-negative rods outnumber lactobacilli. A number of enzymes, including sialidases and proteases, are produced which may interfere with vaginal defences against pathogens (discussed below). The amines and proteases may act as tissue irritants, causing sloughing of epithelial cells and transudation, leading to discharge.

3. **Altered/absent lactobacilli:** Lactobacilli produce lactic acid, and many isolates produce hydrogen peroxide, both thought to inhibit the growth of some pathogens (Hillier, 1998). BV is most common in women without lactobacilli. Moreover, among those women with lactobacilli, it is more common in those whose lactobacilli produce no hydrogen peroxide. However, lactobacilli are frequently present in BV, but are greatly outnumbered by the overgrowth of

BV organisms. Lactobacilli numbers may not be reduced (Hillier and Holmes, 1999), however, there is an inverse relationship between the presence of hydrogen peroxide-producers and BV.

Complications due to Bacterial Vaginosis (BV)

Complications due to bacterial vaginosis include:

- post-partum endometritis (particularly following Cesareans),
- post-hysterectomy vaginal-cuff cellulitis,
- PID following termination of pregnancy and there is weaker evidence linking BV with non-surgical PID.

BV in pregnancy is strongly associated with:

- chorioamnionitis and amniotic fluid infection,
- increased fetal loss at all stages of pregnancy - Ralph *et al.*, (1999) have confirmed this for first trimester miscarriage by studying women having IVF: those with BV had the same conception rate but roughly double the relative risk of miscarriage compared to those with normal flora.
- premature rupture of membranes,
- low birth weight and
- preterm birth – a body data demostrate that the risk is higher if BV is diagnosed before 20 weeks' gestation, or even earlier.

Diagnosis for bacterial vaginosis (BV)

Bacterial vaginosis can be distinguished from vaginitis by the following methods

Wet mount. A sample of vaginal discharge is mixed with a salt solution after placing it on a microscope slide. The prepared slide is examined to identify the bacteria present, to look for white blood cells that indicate an infection, and to look for unusual cells called clue cells. Clue cells are cells covered with Gardnerella bacteria. The bacteria give clue cells a characteristic granular look. Clue cells can flake off of the walls of the vagina and may be found in a vaginal smear.

Chemicals released by the bacteria that cause bacterial vaginosis (BV) may damage vaginal wall cells, causing them to be flake off in greater numbers than usual. The presence of clue cells is the most reliable indicator of BV (Mota *et al.*, 2000).

Whiff test. A strong fishy odor is produced after addition of several drops of a potassium hydroxide (KOH) solution to a sample of vaginal discharge. A fishy odor on the whiff test suggests BV.

Vaginal pH. The normal vaginal pH is between 3.8 to 4.5. Bacterial vaginosis often raises the vaginal pH above 4.5.

Gram stain. A sample of the vaginal discharge is placed on a microscope slide. A special dye (Crystal violet and Iodine as moderant) is applied to the slide, causing certain types of bacteria ("gram-positive" bacteria) to turn a shade of purple while coloring others ("gram-negative" bacteria) pink by counter staining with Saffranin. In bacterial vaginosis, gram-negative bacteria, especially Gardnerella vaginalis, are most common.

Oligonucleotide probes. This test detects the genetic material (DNA) of BV bacteria. Oligonucleotide probe testing is very accurate but is not routinely available in most labs.

Pap test. The Pap test (Papanicolaou test or Pap smear) is used to screen women for cancerous and precancerous cells of the opening of the uterus (cervix). The Pap test is very reliable for detecting early abnormal cell changes that could lead to cervical cancer. A Pap test may also detect infection (such as yeast infection), but this is not the main purpose of the test. During a Pap test, a small sample of cells from the surface of the cervix is collected by a health professional. The sample is then spread on a slide and sent to a lab for examination under a microscope. The cells are examined for abnormalities that may indicate cancer or changes that could lead to cancer. Any abnormal cells are classified according to their "degree" of abnormality.

BV may be detected during routine Pap testing. However, Pap testing is not a standard test to diagnose BV.

Culture. A culture of vaginal discharge may show heavy growth of *Gardnerella vaginalis*, which can be associated with BV (Lurie *et al.*, 2001).

The presence of clue cells, an increased vaginal pH, and a positive whiff test usually are enough evidence to treat for bacterial vaginosis (Mikamo *et al.*, 2000). The diagnosis can even be made without inserting a speculum, as there is good agreement between diagnoses made from wet preps from blindly-inserted swabs and those made with the aid of a speculum (Blake *et al.*, 1998).

Treatment for bacterial vaginosis

The principal goal of therapy for BV is to relieve vaginal symptoms and signs of infection. All women who have symptomatic disease require treatment, regardless of pregnancy status. BV during pregnancy is associated with adverse pregnancy outcomes. The results of several investigations indicate that treatment of pregnant women who have BV and who are at high risk for preterm delivery (i.e., those who previously delivered a premature infant) might reduce the risk for prematurity (Blockehurst, 1999). Therefore, high-risk pregnant women who do not have symptoms of BV may be evaluated for treatment.

Although some experts recommend treatment for high-risk pregnant women who have asymptomatic BV, others believe more information is needed before such a recommendation is made. The bacterial flora that characterizes BV has been recovered from the endometria and salpinges of women who have pelvic inflammatory diseases (PID). BV has been associated with endometritis, PID and vaginal cuff cellulitis after invasive procedures such as endometrial biopsy, hysterectomy, hysterosalpingography, placement of an intrauterine device, cesarean section, and uterine curettage. The results of one randomized controlled trial indicated that treatment of BV with metronidazole substantially reduced postabortion PID. On the basis of these data, consideration should be given to treatment of women who have symptomatic or asymptomatic BV before surgical abortion procedures are performed. However, more information is needed before recommending whether patients who have asymptomatic BV should be treated before other invasive procedures are performed.

Recommended Regimens for Nonpregnant Women

Metronidazole 500 mg orally twice a day for 7 days,

OR

Clindamycin cream 2%, one full applicator (5 g) intravaginally at bedtime for 7 days,

OR

Metronidazole gel 0.75%, one full applicator (5 g) intravaginally twice a day for 5 days.

Note: Patients should be advised to avoid consuming alcohol during treatment with metronidazole and for 24 hours thereafter. Clindamycin cream is oil-based and might weaken latex condoms and diaphragms.

Alternative Regimens

Metronidazole 2 g orally in a single dose,

OR

Clindamycin 300 mg orally twice a day for 7 days.

Metronidazole 2-g single-dose therapy is an alternative regimen because of its lower efficacy for BV. Oral metronidazole (500 mg twice a day) is efficacious for the treatment of BV, resulting in relief of symptoms and improvement in clinical course and flora disturbances. Based on efficacy data from four randomized controlled trials, overall cure rates 4 weeks after completion of treatment did not differ significantly between the 7-day regimen of oral metronidazole and the clindamycin vaginal cream (78% vs. 82%, respectively). Similarly, the results of another randomized controlled trial indicated that cure rates 7-10 days after completion of treatment did not differ significantly between the 7-day regimen of oral metronidazole and the metronidazole vaginal gel (84% vs. 75%, respectively). FDA has approved Flagyl ER™ (750 mg) once daily for 7 days for treatment of BV.

Some health-care providers remain concerned about the possible teratogenicity of metronidazole, which has been suggested by experiments using extremely high and prolonged doses in animals. However, a recent meta-analysis does not indicate teratogenicity in humans. Some health-care providers prefer the intravaginal route because of a lack of systemic side effects (e.g., mild-to-moderate gastrointestinal disturbance and unpleasant taste). Mean peak serum concentrations of metronidazole after intravaginal administration are less than 2% the levels of standard 500 mg oral doses and the mean bioavailability of clindamycin cream is approximately 4%.

Follow-Up

Follow-up visits are unnecessary if symptoms resolve. Recurrence of BV is not unusual. Because treatment of BV in high-risk pregnant women who are asymptomatic might prevent adverse pregnancy outcomes, a follow-up evaluation, at 1 month after completion of treatment, should be considered to evaluate whether therapy was successful. The alternative BV treatment regimens may be used to treat recurrent disease. No long-term maintenance regimen with any therapeutic agent is recommended.

Management of Sex Partners

The results of clinical trials indicate that a woman's response to therapy and the likelihood of relapse or recurrence are not affected by treatment of her sex partner(s). Therefore, routine treatment of sex partners is not recommended.

Special Considerations

Allergy or Intolerance to the Recommended Therapy

Clindamycin cream is preferred in case of allergy or intolerance to metronidazole. Metronidazole gel can be considered for patients who do not tolerate systemic metronidazole but patients allergic to oral metronidazole should not be administered metronidazole vaginally.

Treatment during pregnancy

BV has been associated with adverse pregnancy outcomes (e.g., premature rupture of the membranes, preterm labor, and preterm birth) and the organisms found in increased concentration in BV also are frequently present in postpartum or postcesarean endometritis. Because treatment of BV in high-risk pregnant women (i.e., those who have previously delivered a premature infant) who are asymptomatic might reduce preterm delivery, such women may be screened and those with BV can be treated (Lamont, 2000). The screening and treatment should be conducted at the earliest part of the second trimester of pregnancy. The recommended regimen is metronidazole 250 mg orally three times a day for 7 days (Carey 2000). The alternative regimens are a) metronidazole 2 g orally in a single dose or b) clindamycin 300 mg orally twice a day for 7 days.

Low-risk pregnant women (i.e., women who previously have not had a premature delivery) who have symptomatic BV should be treated to relieve symptoms. The recommended regimen is metronidazole 250 mg orally three times a day for 7 days. The alternative regimens are a) metronidazole 2 g orally in a single dose; b) clindamycin 300 mg orally twice a day for 7 days; or c) metronidazole gel, 0.75%, one full applicator (5 g) intravaginally, twice a day for 5 days. Some experts prefer the use of systemic therapy for low-risk pregnant women to treat possible subclinical upper genital tract infections.

The lower doses of medication are recommended for pregnant women to minimize exposure to the developing fetus. Data are limited concerning the use of metronidazole vaginal gel during pregnancy. The use of clindamycin vaginal cream during pregnancy is not recommended because two randomized trials indicated an increase in the number of preterm deliveries among pregnant women who were treated with this medication.

GRAM-NEGATIVE BACTERIAL ENDOTOXINS (LPS): EFFECT ON EMBRYONIC DEVELOPMENT AND PREGNANCY

Endotoxin-induced pregnancy losses associated with embryonic death, abortion, preterm labor and delivery and stillbirth have been documented in a number of species (Schlafer *et al.*, 1994, Edqvist *et al.*, 1984, Fredriksson *et al.*, 1986, Giri *et al.*, 1990). Approximately 7% of pregnant women deliver preterm. Clinical studies show intra-amnionic infections in 1-13.5% of cases of preterm labor. Intra-amnionic infections are frequently caused by gram-negative organisms that release endotoxins found on the outer surface of their cell wall.

The effect of gram-negative bacterial endotoxin on gamet function, fertilization and embryo development has been widely studied (Dumoulin *et al.,* 1991). Endotoxins decrease embryo cleavage rate, blastocyst formation and pregnancy rates in human (Dumoulin *et al.,* 1991). The presence of endotoxins during *in vitro* fertilization (IVF) results in a high rate of polyspermy in humans and bovine species (Berger *et al.,* 1982, and Diemer *et al.,* 2000). Such polyspermic fertilizations lead to chromosomal aneuploidy resulting in embryonic loss. Local and systemic presence of endotoxins also induce degeneration, fragmentation and developmental arrest of the preimplantation embryo, and is one of the significant causes of reduced reproductive efficiency in most mammal (Dostal *et al.,* 1996). A large body of *in vivo* and *in vitro* evidence suggests that many of the biological effects of LPS are indirectly mediated by the pro-inflammatory cytokines like interleukin-1alpha (IL-1α), interleukin-1beta (IL-1β), tumor necrosis factor alpha (TNF-α) etc. since exogenously administered TNF-a can cause fetal loss in mice and rats. TNF and LPS show synergistic detrimental effects on the growth of preimplantation embryos *in vitro* (Randall *et al.,* 1991). Moreover, high levels of TNF-α and other cytokines appear in the amniotic fluid of LPS treated mice suggesting their probable role in abortions or fetal deaths (Wang and Zhang, 1998). These proinflammatory cytokines are produced primarily by activated monocytes and a variety of other cell types including gestational tissues and embryos.

MOLECULAR MECHANISMS OF LPS INDUCED ABORTIONS

Mechanism of Endotoxin-induced Preimplantation Embryonic loss

Pregnancy is a natural condition in which the receipt and tolerance of a foreign tissue, the fetus, must occur for a successful outcome. Any disturbance of the delicate immune balance within the maternal-fetal interface may result in pregnancy loss or other perinatal complications. Several cytokines have been implicated in this immune system balance and may influence placental and fetal growth, eg. high levels of interleukin-1 (IL-1α and IL-1β) and TNF-α are linked to fetal deaths (Silver *et al.*, 1994, and Silver *et al.*, 1997) and colony stimulating factor-1 (CSF-1) in the amniotic fluid is linked to positive fetal growth (Wegmann *et al.*, 1989). It has been reported that lymphocytes and macrophages are present in the uterine endometrium throughout the pregnancy (McMaster *et al.*, 1993) and that they are present in a higher number in the case of uterine infections (Kurman, 1982). Several studies have shown that uterine leukocytosis and leukocyte products (like cytokines) are toxic to the preimplantation embryos.

Many of the cytokines and growth factors like IL-1, IL-6 TNF, CSF-1, TGF-α1 and LIF exist in both soluble and membrane associated or matrix bound forms, and are expressed in the uterus during pregnancy (De M *et al.*, 1993) Receptors for many of these growth factors (like IL-1, TNF, CSF-1, PDGF, EGF etc.) are expressed on preimplantation embryos and are mostly transmembrane tyrosine kinases that are activated by ligand binding (Geng *et al.*, 1993). The presence of ligands for these receptors in the uterine environment suggests that signal transduction cascades involving complex phosphorylation patters might be activated in preimplantation embryos leading to proliferation and differentiation. Moreover, many of these cytokines like IL-1α, IL-1β, PDGF-A, TGF, TNF, CSF-1 etc. are expressed in a developmentally regulated manner in the preimplantation stage blastocyst (Gerwin *et al.*, 1995; Deb *et al.*, 2000; Bischof and Campana, 2000). This may be a general mechanism for the regulation of early embryonic development and for maintaining a synchrony between the embryo and uterine preparation for implantation. Endotoxins activate the macrophages residing in the female reproductive tract, which leads to the production of several cytokines, that are targeted towards the embryo, and can adversely affect early embryonic development (Andrews *et al.*, 1994; Jaiswal *et al.*, 2001a) trophoblast proliferation, and fetal growth during later stages of gestation.

Reverse transcription –polymerase chain reaction (RT-PCR) experiments carried out in our laboratory suggests that the pattern of expression of proinflammatory cytokines like IL-1α, IL-1β, TNF-α, and CSF-1 at the various developmental stages of the mouse preimplantation blastocyst, get altered dramatically in presence of LPS *in vivo* (unpublished observations). We have seen that the pattern of positive mRNA expression for IL-1α was found from day 1 to day-4.125 of pregnancy in the preimplantation mouse blastocyst whereas IL-1β expression started from day-4 and continued till implantation. Positive signals for TNF-α were found from day-4 of pregnancy and continued till implantation, and CSF-1 was found in the two-cell stage to the morula stage embryos (Deb *et al.,* 2000). However, after administration of LPS we observed an early expression of IL-1β and TNF-α (i.e. from day 1.5 of pregnancy onwards and IL-1α was found from day 1.5 of pregnancy and continued till implantation (unpublished observations). IL-1α is responsible for the regulation of the various matrix metaloprotinases involved in tissue remodeling during the implantation event (Sharkey *et al.,* 1996). An observed change in the expression of this cytokine could, therefore, be responsible for implantation failure. Moreover, morphological studies of the preimplantation embryos recovered from LPS treated pregnant mice show developmental arrest, degeneration, and fragmentation of the blastocysts (Jaiswal *et al.,* 2001). This observed asynchronous cleavage and degeneration of the preimplantation embryos could be due to apoptosis triggered by the early expressions of TNF-α and IL-1β. This is further supported by the fact that only the TNF-RI/p60 form of the TNF receptor is found in the mouse blastocyst, which has the death domain for triggering apoptotic pathway (Pampfer *et al.,* 1994).

Uterine secretion of many of these cytokines are under the control of ovarian steroids, confirming their close association with a distinct regulatory network that may overlap or complement the immune system represented by the lympho-hematopoetic cells residing in the uterine endometrium. It has been shown that high levels of circulating estrogen can also inhibit normal embryonic development (Cline *et al.*, 1977). A reduction in the plasma progesterone (P): estradiol 17b (E) ratio decreases the number of normal blastocysts formed in rats. High level of progesterone in the circulation, during pregnancy has been known to have an immunosuppressive effect, which suppresses the production of IL-1, IL-6 etc. (Kelly *et al.*, 2001). However, LPS has been known to inhibit steroidogenesis in steroidogenic tissues (Telleria *et al.*, 1998). LPS is known to activate the macrophages and/or the corpora luteal cells in the corpus luteum of the ovaries (Taylor and Terranova,1996). These activated macrophages and/or corpora luteal cells produce a number of cytotoxic proinflammatory cytokines like TNF, IL-1, IL-6 etc. (Ault and Faro, 1993; Gorospe *et al.*, 1992). These cytokines, exert autocrine and paracrine effects causing regression of the ovary, resulting in reduced steroidogenesis (Bagavandoss *et al.*, 1990). Expression for IL-6 has been found in the luteal cells of the corpora lutea, which may account for pregnancy loss, through the inhibition of progesterone production, during pelvic bacterial infections (Terranova and Mongomery–Rice 1997). Thus the modulation in the progesterone/estrogen ratio induced by LPS, could lead to early embryonic losses/abortions (LaPolt *et al.*, 1990). Thus in woman with immunologic reproductive failure the embryo may be killed as an innocent bystander to leukocyte activation in response to bacterial antigens.

LPS induced alterations in the uterine receptivity for implantation

The major uterine cell-types respond uniquely to progesterone (P_4) and/or estrogen. In mouse on days 1 and 2 of pregnancy, preovulatory ovarian estrogen stimulates epithelial cell proliferation. On day 3 of pregnancy, P_4 from newly formed corpora lutea initiates stromal cell proliferation, which is further potentiated by ovarian estrogen secretion on day 4 of pregnancy. In contrast the epithelial cells cease to proliferate and become differentiated on day 4. The uterus becomes receptive on day 4 of pregnancy in the mouse and a 'cross-talk' between the blastocyst and uterus ensues to initiate implantation (Paria *et al.*, 2000). The receptive state of the uterus is defined as the limited time when the uterine environment is conducive to blastocyst acceptance and implantation. In the absence of blastocyst the receptive uterus enters into the non-receptive phase. Thus the receptive state of the uterus is defined as the 'window' when the uterus is conducive to blastocyst for implantation (Paria *et al.*, 1993).

Cytokines within the endometrium participate in both menstruation and implantation but also contribute to the defence mechanisms of the mucosal epithelium. Endometrium is under the control of steroid hormones, particularly progesterone and thus control of cytokines by this steroid is very important (Kelly *et al.*, 2001). Although appreciable numbers of progesterone receptors are not found in endometrial leucocytes, however, it can modulate the expression of cytokines by acting on the uterine cells expressing its receptor. It has been reported that failure of implantation after IVF is sometimes associated with elevated estradiol: progesterone ratios. This elevated estrogen: progesterone ratio creates an uterine luminal milieu that suppresses embryonic metabolism (Safro *et al.*, 1990). These effects could be reversed and implantation re-established by the correction of this ratio by exogenous progesterone. Studies are currently being carried out in our laboratory to find if exogenous progesterone administration could improve fertility rates in LPS treated mice.

The spatiotemporal expression of the cytokines and growth factors like EGF, TGF, TNF, IL-1, CSF-1,GM-CSF, PDGF-A etc and their receptors in the uterus and embryo during the preimplantation period suggests that these cytokines/growth factors serve as local mediators of embryo-uterine interactions during implantation (Kauma 2000). Other molecules like the cell adhesion molecules, endometrial epithelial β3 integrins, matrix metalloproteinases (Bischof *et al.* 2000), prostaglandins and the inducible prostaglandin synthesis enzyme COX-2 are important for successful implantation (Carson *et al.*, 2000). They play a primary role in decidual tissue remodeling and regulation of trophoblast invation. Expression of these molecules in the trophoblast and the uterine cells strongly regulated by physiologically relevant cytokines and growth factors (eg. IGF, TGF, IL-1, IL-6, TNF, etc.)

Gram-negative bacterial endotoxins/LPS has been known to modulate the synthesis of proinflammatory cytokines and growth factors, and this in turn can potentially alter the synthesis/expressions of other cytokines, growth factors, their receptors, steroid hormones, prostaglandins, COX-2, cell adhesion molecules, integrins etc. Proinflammatory cytokines like IL-1α, IL-1β, and TNFα are strongly expressed by the lymphohematopoetic and uterine tissues after LPS administration. Embryonic IL-1α and β, and their receptors on the uterine endometrium, are known to trigger the β3 integrin expression in the endometrium during implantation (Simon *et al.*, 1998). Endotoxin induced alteration in the levels of these implantation relevant cytokines can lead to implantation failure. TNF-α has been shown to regulate the window of implantation in human (Tabibzadeh, 1998). The stimulation of TNF-α in response to endotoxins is well established and therefore high levels of TNF-α may possibly disturb the implantation window, leading to infertility. The recent findings of our laboratory suggest that IL-1α, IL-1β, TNF-α and CSF-1, by semiquantitative RT-PCR in the uterus of pregnant mouse treated with LPS increases as compared to that of the control (unpublished data). The overexpression of the developmentally relevant cytokines in the uterine tissue may be responsible for embryonic loss and or implantation failure. Histopathological studies of the female reproductive organs in LPS treated mouse show marked morphological and physiological alterations in the uterus, ovaries and oviducts as compared to that of the normal. Degenetation of the luminal and glandular epithelium, reduction in the number of glands and heavy infiltration of macrophages in the stroma and hyperplasia were observed in the uterus of LPS-treated animals. The corpus luteum showed signs of regression in response to LPS and macrophage infiltration (Jaiswal *et al.*, 2001). Therefore, it has been proposed that LPS induced histopathological alterations in the various reproductive organs of the pregnant mice would be one of the causes of embryonic loss during early stages of pregnancy in mouse.

The NF kappa B (NF-kB) pathway is important in the regulation of cytokine synthesis and can modulate the production of chemokines, matrix metallopoteinases and COX-2. Kelly *et al.* (2001) has reported that LPS/ endotoxins strongly induces NF-kB activation and its activity can be inhibited by progesterone by either stimulating I kappa B; the molecule that sequester in the cytosol, or after binding to the nuclear receptor, competing with NF-kB for recognition sites on the relevant genes. Therefore, LPS induced infertility appears to operate mainly through a NF-kB mediated pathway. However, NF-kB is not the only transcription factor involved in the intracellular signaling events induced by LPS or the cytokines. It has been observed that TNF-α and IL-1 activate NF-kB whereas others like IFN-α, EGF and CSF-1 activate p91 (a subunit of interferon-stimulated gene factor-3) through its phosphorylation directly by a receptor-associated tyrosine kinase (JAK-STAT pathway). These observations indicate that more than one pathway may exist for transduction of signals to the nucleus induced by the different cytokines and LPS and ultimately leading to an alteration in the uterine receptivity for implantation of blastocysts.

Mechanism of Endotoxin-induced Preterm birth

The mechanisms through which endotoxins cause abortions are poorly understood and appear to differ between groups of animals. For example, abortion in pregnant rodents following endotoxin administration is accompanied by severe damage to the placenta (Silen *et al.,* 1989) and endometrium (Shalaby *et al.,* 1989), however, in other species no recognized lesions accompany abortion or preterm delivery has been observed. In the goat (Edqvist *et al.,* 1984), pig (Wrathall *et al.,* 1978), horse (Fredricksson, 1986), and cow (Giri *et al.,* 1990, and Foley *et al.,* 1993) pregnancy loss is attributed to altered maternal endocrine function, most notably, regression of the corpus luteum followed by abortion or onset of preterm labor. In sheep and primates maintenance of pregnancy towards the end of gestation is dependent on the placental progesteron and thus a functional corpus luteum is not required. In sheep the duration of pregnancy is determined by the activity of the fetal hypothalamo-hypophyseal-adrenal-axis (HHAA) (Liggins, 1990). In nonhuman primates, the dramatic increase of fetal plasma androgens in the last 2 wk of pregnancy also suggests that the fetal adrenal plays a role in the initiation of labor in these animals. LPS-infused pregnant sheep experienced marked stimulation of HHAA as evidenced by the increase in adreno corticotrophic hormone (ACTH) and cortisol concentrations in both maternal and fetal circulation (Schlafer *et al.*, 1994). Endotoxin administration to pregnant cattle and ruminants also stimulate ACTH and cortisol production.

TNF-α secreted by the macrophages in response to LPS bind to many tissues and alters their metabolism. It can damage blood capillaries and increase their permeability resulting in a drop in blood pressure that results in shock. TNF-α can stimulate IL-1 production (Vassalli, 1992) and this cytokine is known to cause fetal death in a dose-dependent fashion (Silen *et al.*, 1989). TNF-α can cause microvascular injury and thrombosis via the elaboration of numerous factors leading to vasoconstriction, platelet aggregation, and clotting. These events could potentially lead to ischemia and hemorrhagic necrosis in deciduas and trophoblast resulting in fetal death. Experimental i.v. administration of TNF-α causes plasma levels of cortisol and corticosterone to rise rapidly, however, it is unclear whether this rise is due to release of ACTH or to direct stimulation of the adrenal (Jattela, 1990).

Another possible mechanism of TNFα-induced fetal death is via the production of prostaglandins (PG). Eicosanoids are known abortifacients and are produced by gestational tissues in response to TNF-α *in vitro*. Inhibition of prostaglandin production with indomethacin can decrease LPS induced fetal death (Silver *et al.*, 1997). The largest part of pregnancy is characterized by a quiescent state of the uterus. The most important substrates responsible for uterine quiescence are progesterone and prostacyclin (PgI_2) and both have an inhibitory effect on uterine contractions (Keelan *et al.*, 1997). PGs are generated via cyclooxygenase (COX), the rate limiting enzyme for the conversion of arachidonic acid into PgH_2, the common substrate for various PGs. It has recently been shown that COX-2 derived PgI_2 is the primary PG that is essential for implantation and decidualization (Lim *et al.*, 1999):

Beside fetal stress and fetal hypoxia, maternal stress can also stimulate the fetal HHAA and induce secretion of corticotrophin-releasing-hormone (CRH) leading to prostaglandin E_2 (PgE_2) production. PgE_2 changes the quiescent state of the uterus and induce softening of the cervix and the occurrence of synchronous contractions of the myometrium. A prerequisite for myometrial contractions is the formation of gap-junctions and the formation of receptors for prostaglandins and oxytocin (Keelan *et al.*, 1997). The degradation of the cervical collagen is necessary for the softening and dilatation of the cervix. The increase in PgE_2 levels results in gap-junction formation and increased contractility. Microorganisms found in infections of the uterine cavity exert phospholipase A_2 (PIA_2) activity (Bejar *et al.*, 1981). Phospholipase A_2 is an enzyme necessary for the release of arachidonic acid from phospholipids. PIA_2 levels have been found to increase near the end of the term (Mitchell *et al.*, 1995). PIA_2 thus increases the level of arachidonic acid, which is converted to PgE_2 by COX-2. Also the production of prostaglandin dehydrogenase (PGDH), the enzyme which catabolises PgE_2 decreases near term, intriguingly, the production of this enzyme is regulated by the local estrogen: progesterone ratio (Van-Meir *et al.*, 1997).

The other pathway, through which LPS-induced preterm labor can be initiated, is an inflammatory reaction at the chorio-decidual interface. This inflammatory reaction of the decidua leads to a release of cytokines, mostly IL-1 and TNF. Decidual macrophages and polymorphonuclear leukocytes (PMN) can react to this inflammatory reaction and migrate to the chorio-decidual interface. IL-1 and TNF initiate production and secretion of IL-6, IL-8, and PgE_2 from these decidual macrophages (Mitchell *et al.*, 1995). IL-8 stimulates migration of PMN (Osmer *et al.*, 1995) and these cells can release cytokines, collagenases and proteases. These enzymes are capable of weakening the membranes resulting in premature rupture of membranes or softening of the cervix (Challis and Mitchell, 1994).

CONCLUSION AND FUTURE DIRECTIONS

Allevation of infertility on one hand and development of improved methods of contraception on the other are global concerns to woman's health. The molecular signals that regulate implantation are of clinical relevance since understanding the nature of these signals may lead to strategies to correct implantation failure and to develop novel contraceptive approaches. The other pressing concern is the poor pregnancy rate resulting from *in vitro* fertilization (IVF). The pregnancy rate in IVF programs remains about 20-30 % in spite of the high rate of successful fertilization. This has led to the proposition that additional uterine factors, critical for the implantation process, must be limiting. Identification of such parameters could help in determining the appropriate physiological state of the uterus for embryo transfer. Several factors are known to have a direct or indirect impact on the ability of the uterus to develop to a functionally receptive state. LPS induced cytokines and growth factors, for instance, can significantly accelerate or delay the uterine transition to the receptive state. This would disrupt the normal coordination between embryonic and uterine development even though all molecular players may seem otherwise normal.

Preterm labor and delivery continues to be the most important unsolved problem in obstetrics till today. Perhaps the primary reason for the little progress in the prevention and treatment of preterm labor is the lack of understanding of its pathophysiology. 10-30% of women with preterm labor have clinically evident or subclinical intrauterine infections. Administration of LPS/endotoxins mimics the pathophysiology of infection induced preterm labor and embryonic loss. The sequels of fetal growth retardation/death by LPS may not only be similar to clinical intra-uterine growth retardations (IUGR) caused due to gram-negative bacterial infections but also to an extended clinical spectrum ranging from premature labor to fetal demise or still birth. LPS/endotoxins agument the production of immune and inflammatory mediators, like IL-1, and TNF and leads to a concentration dependent increase in the production of PGE_2 and IL-6. All these substances have been detected in the amniotic fluid of woman with infection associated preterm labor. Thus it underscores the importance of these mediators as determinants of a successful pregnancy.

Maternal TNF production increases markedly after administration of LPS. Moreover, administration of recombinant TNF-α mimics endotoxemia. However, it is not clear whether TNF-α can cross the placenta although mRNA and protein for TNF-α and its receptors have been found in many tissues including the placenta and deciduas. Fetal and placental tissues can both generate and respond to TNF-α. Clinical data suggests that TNF-α may act as a mediator in parturition associated with infection of amniotic fluid in woman, but is not clear whether TNF released into the systemic circulation affects pregnancy or not.

High levels of endotoxin in the vagina are an obvious consequence of high frequency of occurance (~40%) of gram-negative bacteria. Presence of commonly occurring non-pathogenic gram-negative bacteria like *E. coli* and *S. faecalis* in the uterine environment has not been found to cause developmental abnormalities in experimental animals. However, when endotoxin level in the uterus and oviduct high it is associated with highly degenerate and fragmented embryos. High level of endotoxins in the uterine environment could be a result of gram-negative pathogenic bacterial vaginosis. This indicates that preimplantation embryos are susceptible to direct exposure to endotoxin but the influence may depend on the concentration of the endotoxins.

There are numerous mechanisms through which LPS may cause fetal death, implantation failure or embryonic loss. However, it is not known whether endotoxins in the maternal system reach the fetus since there is little information available concerning movement of endotoxin across the placenta in any species. ^{32}P labeled *E. coli* endotoxin was shown to cross the placenta in the rabbit but *in vitro* studies using human chorioamniotic membranes failed to detect moment of LPS across fetal membrane. It has also been shown that the fetus is generally less sensitive to LPS than is the adult.

Since fetal death caused by LPS administration appears predominantly to be due to a maternal response as opposed to a direct fetal effect, it is suggested that suppression of the maternal immune response, particularly the actions of proinflammatory cytokines, may improve fetal outcome in cases of maternal bacterial infection. TNF-α inhibition has been used to improve survival outcomes in several clinical settings such as septic shock due to gram-negative bacterial sepsis. In view of the current status concerning the interaction between the immune and reproductive systems that result from sharing the lymphohaematopoetic cytokines and their receptors, we hypothesize that swiping out of the LPS-induced embryotoxic TNF-α from maternal circulation during the preimplantation period could improve fertility rates. Trials in this direction using anti-TNF-α monoclonal antibodies and or recombinant chimeric p75-TNF-α-receptor-Ig proteins to block LPS-induced circulating TNF-α is currently being carried out at our laboratory. However, in addition to TNF-α blockade the inhibition of other proinflammatory mediators/cytokines known to be triggered by LPS, eg. IL-1, should also be investigated during pregnancy.

ACKNOWLEDGMENTS

We thank the All India Council for Technical Education (AICTE), New Delhi; the Rockefeller Foundation, New York, USA, for financial support and the Council for Scientific and Industrial Research (CSIR), New Delhi for a Senior Research Fellowship to KD. We are thankful to Dr. M.M. Chaturvedi, CAS in Zoology, Banaras Hindu University, Varanasi for his valuable technical help.

REFERENCES

Aderem A, Ulevitch RJ. (2000). Toll like receptors in the induction of the innate immune response. Nature. 406: 782-787.

Almer AJ, Feist W, Hein H, Kirekae T, Kusumoto S, Kusama T, Brade H, Schade U, Rietschell ET, Flad HD. (1992). Modulation of endotoxin-induced monokine release in human monocytes by Lipid A partial structures that inhibit binding of ^{125}I-lipopolysaccharide. Infect Immun. 60: 5145-5152.

Ault KA, Faro S. (1993). Current diagnostic criteria and treatment guidelines. *Postgrad Med.* 93; 89-91.

Andrews GK, Mc Master MT, Liang L, Kover K, Dey SK. (1995). Cytokine gene expression and distribution of inflammatory leukocytes in the periimplantation mouse uterus. (Dey SK. Ed.) In *Molecular and cellular aspects of preimplantation processes.* Serono Symposia USA. Norwell Massachussets. Springer Verlag New York. 13; pp. 205 –230.

Bagavandoss P, Wiggins RC, Kunkel SL, Remick DG, Keyes PL. (1990). Tumor necrosis factor production and accumulation of inflammatory cells in the corpus luteum of pseudopregnancy and pregnancy in rabbits. *Biol Reprod.* 42, 367-376.

Berger RE, Karp LE, Williamson RA, Koehler J, Moore DE, Holmes KK. (1982). The relationship of pyospermia and seminal fluid bacteriology to sperm function as reflected in the sperm penetration assay. *Fertil Steril.* 37(4): 557-64.

Bischof P, Campana A. (2000). Molecular mediators of implantation. *Baillers Best Pract Res Clin Obest Gyneacol.*14 (5): 801-814.

Bischop P, Messer A, Campana A. (2000). Mechanisms of endometrial control of trophoblast invation. *J Reprod Fertility Suppl* 55: 65-71.

Blake DR, Duggan A, Quinn T, Zenilwan J, Joffe A. (1998). Evaluation of vaginal infections in adolescent women: can it be done without a speculum? *Pediatrics.*102:939-44.

Brocklehurst P. (1999). Interventions for treating bacterial vaginosis in pregnancy. Cochrane Library, Issue 3.

Burell R. (1994). Human responses to bacterial endotoxins. *Circ. Shock* 43: 137-153.

Carey CJ. (2000). Metronidazole to prevent preterm delivery in pregnant women with aymptomatic bacterial vaginosis. *N Engl J Med.* 342: 534-40.

Carson DD, Bagchi I, Dey SK, Enders AC, Fazlebas AT, Lessey BA, Yoshinaga K. (2000). Review: Embryo implantation. *Dev. Biol.* 223: 217-237.

Cavallion JM, Haeffner-Cavallion N. (1990). Signals involved in interleukin–1 sysnthesis and release by lipopolysaccharide-induced monocytes, macrophages. *Cytokine.* 2: 313-329.

Chow JC, Young DW, Golenbock DT, Christ WJ, and Gusovsky F. (1999). Toll-like receptor-4 mediates lipopolysaccharide-induced signal transduction. *J. Biol.Chem.* 274, 10689-10692.

Cline EM, Randall PA, Oliphant G. (1977). Hormone-mediated oviductal influence on mouse embryo development. *Fertil Steril.* 28: 766-771.

Cox SM, Mac Donald PC, Casey ML. (1988). Assay of bacterial endotoxins in human amniotic fluid: potential usefulness in diagnosis and treatment of preterm labor. *Am. J. Obest. Gyneacol.* 159: 99-106.

Cram LF, Zapata MI, Toy EC, Baker B 3rd. (2002). Genitourinary infections and their association with preterm labor. *Am Fam Physician.* 15; 65(2): 241-8.

Deb K, Chatturvedi MM, Jaiswal YK. (2000). Developmental expression of IL-1α, IL-1β, TNF-α and CSF-1 in preimplantation mouse embryos. In *The 69th Annual Meeting of the Society of Biological Chemists, India*. pp.-J 2.

De M, Sanford TR, Wood GW. (1993). Expression of interleukin-1, interleukin –6, and tumor necrosis factor alpha in mouse uterus during the peri-implantation period of pregnancy. *J. Reprod Fertil*. 97(1): 83-89.

Diemer T, Huwe P, Michelmann HW, Mayer F, Schiefer HG, Weidner W. (2000). Escherichia coli-induced alterations of human spermatozoa. An electron microscopy analysis. *Int J Androl.* 23(3): 178-86

Dinarello CA. (1989). Fever. In *Interleukin –1 inflammation and disease*. (Bomford R and Henderson B eds.) pp. 175-190. Elsevier/ North-Holland Publishing Co., Amsterdam.

Dinarrelo CA. (1994). Interleukin –1 Adv.Pharmacol. 25: 21-51.

Di Renzo, De Domenico. (2000). Cervico-inguinal microbiology, vaginal pH, infections, and premature labor. *Acta Biomed Ateneo Parmense*; 71(1): 513-7.

Donder GG, Vereecken A, Bosmans E, Dekeersmaecker A, Salembier G, Spitz B. (2002). Definition of a type of abnormal vaginal flora that is distinct from bacterial vaginosis: aerobic vaginitis. *BJOG*. 109(1): 34-43

Dostal J, Oborna I, Talas M, Chrastinova L, Machovska K. (1996). Infectious agents and treatment of infertility with IVF and ET. *Ceska Gynekol.* 61(3): 144-7.

Dumoulin JC, Menheere PP, Evers JL, Kleukers AP, Pieters MH, Bras M, Geraedts JP. (1991). The effects of endotoxins on gametes and preimplantation embryos cultured in vitro. *Hum Reprod*. 6(5): 730-4.

Edqvist L, Fredriksson G, Kindahl H. (1984). Some aspects of endotoxins and corpus luteum function in ruminants. In: Panel proceedings of nuclear techniques in tropical animal diseases, International Atomic Energy Agency; Vienna, Australia. pp. 57-68

Feist W, Ulmer AJ, Musehold J, Brade S, Kusumoto S, Flad HD. (1989). Induction of tumor necrosis factor alpha, release by lipopolysaccharide and defined lipopolysaccharide partial structures. *Immunobiology*. 179: 293-307.

Foley GL, Schlafer DH, Elsasser TH, Mitchell M. (1993). Endotoxemia in pregnant cows: comparison of maternal and fetal effects utilizing the chronically catheterized fetus. *Threiogenology*. 39: 739-762.

Fredriksson G, Kindahl H, Stabenfeldt G. (1986). Endotoxin-induced and prostaglandin mediated effects on the corpus luteum in the mare. *Theriogenology* .25:309-316.

Frey EA, Miller DS, John TSI, Sundam A, bazil V, Espevik T, Finlay BB, Wright SD. (1992). Soluble CD14 participates in the response of cells to lipopolysaccharide. *J. Exp. Med.* 176: 1665-67.

Galanos C, Luderitz O, Westphal O. (1984). Isolation and purification of a standard lipopoysaccharide from *Salmonella abortus-equi*. In *Endotoxin: chemical biological and clinical aspects.* (Homma JY, Kanegasaki S, Luderitz O, Westphal O. eds.) pp. 409-420.

Geng Y, Zhang B, Lotz M. (1993). Protein Tyrosine Kinase activation is required for Lipopoysaccharide induction of cytokines in Human Blood Monocytes. *Immunol* 151: 6692-6700.

Gerwin N, Jia GQ, Kulbacki R, Gutierrez-Ramos JC. (1995). Interleukin gene expression in mouse preimplantation development. *Dev. Immunol.* 4(3): 169-79.

Gibbs RS, Romero R, Hiller SL, Eschenbach Da, Sweet RL. (1992). A review of premature birth and subclinical infection. *Am J. Obestet Gynecol.* 166(%): 1515-28.

Giri SN, Stabenfeldt GH, Bruss MI, BonDurant RH, Osburn BI. (1990). Effect of endotoxins on circulating levels of eicosanoids, progesterone, cortisol, glucose, and lactic acid, and abortion in pregnant cows. *Vet. Microbiol.* 21: 211-231.

Glauser MP, Zannetti G, Baumgartner JD, Cohen J. (1991). Septic shock: pathogenesis. *Lancet* 338: 732-36.

Gorospe WC, Hughes FM, Spangelo BL, (1992). Inlerleukin-6: effects on and production by rat granulose cells in vitro. Endocrinology. 130; 1750-1752.

Hay PE, Lamont RF, Taylor-Robinson D, Morgan DJ, Ison C, Pearson J. (1994). Abnormal bacterial colonization of the genital tract and subsequent preterm delivery and late miscarriage. *BMJ*. 308(6924):295-8.

Hillier S. (1998). The vaginal microbial ecosystem and resistance to HIV. *AIDS Research and Human Retroviruses*. 14, S17-S21.

Hillier S and Holmes KK. (1999). Bacterial vaginosis. In *Sexually Transmitted Diseases*, 3rd ed, pp. 563-586, McGraw Hill.

Hill JA, Haimovici F, Anderson DJ (1987). Products of activated lymphocytes and macrophages inhibit mouse embryo development in vitro. *J. Immunol.* 139: 2250-2259.

Hitchock PJ, Leive L, Makela PH, Reitschell ET, Strittmarter W, Morrison DC. (1986). Lipopoysaccharide nomenclature-past, present and future. *J . Bacteriol.* 166: 699-701.

Hoshino K, Takeuchi O, Kawai T, Sanjo H, Ogawa T, Takeda Y, Takeda K, Akira S. (1999). Cutting edge: Toll-like receptor 4 (TLR4)-deficient mice are hyporesponsive to lipopolysaccharide: evidence for TLR4 as the LPS gene product. J Immunol. 162: 3749-3752.

Jaattela M, Ilvesmaki V, Voutilainen R, Stenman U, Saksela E. (1990). Tumor necrosis factor alpha is a potent inhibitor of adrenocorticotrophin-induced cortisol production and steroidogenic P450 enzyme gene expression in cultured human fetal adrenal cells. *Endocrinology*. 128: 623-629.

Jaiswal YK, Deb K, Chatturvedi MM. (2001a). Lipopoysaccharide induced histopathological alteraions in murine reproductive organs: A probable cause of failure of Blastocyst implantation. In The 1st Conference of Biotechnology Society of India (Biotechcon-2001) pp. 22-23.

Jaiswal YK, Deb K, Chatturvedi MM. (2001). Effect of LPS on the development of in-vbivo produced preimplantation mouse embryos. In *The 70th Annual Meeting of the Society of Biological Chemists, India.* p. 197.

Kauma SW. (2000). Cytokines in implantation. *J Reprod Fertil.* 55: 31-42.

Keelan JA, Coleman M, Mitchell MD. (1997). The molecular mechanisms of term and preterm labor: recent progress and clinical implications. *Clin Obest Gyneacol.* 40(3): 460-78.

Kelly RW, King AE, Critchley HO. (2001). Cytokine control in human endometrium. *Reproduction.* 121(1): 3-19.

Kurman RJ. (1982). Benign diseases of the endometrium. In *Pathology of the Female Genital Tract.* (A. Blaustein ed.) pp. 279-284. Springer Verlag. New York.

Kielian TL, Blecha F. (1995). CD14 and other recognition molecules for lipopoysaccharides: a review. *Immunopharmacology* 29: 187-205.

Kirschning CJ, Wesche H, Merrill Ayres T, Rothe M. (1998). Human toll like receptor 2 confers responsiveness to bacterial lipopolysaccharide. 188. 2091-2097.

Lamont RF. (2000). Antibiotics for the prevention of preterm birth. *N Engl J Med.* 342:581-3.

LaPolt PS, Day JR, Lu JK. (1990). Effects of estradiol and progesterone on early embryonic development in aging rats. *Biol Reprod*. 43 (5):843- 50

Lemaitre B, Reichhart JM, Hoffmann JA. (1997). Drosophila host defense: differential induction of antimicrobial peptide genes after infection by various classes of microorganisms. *Proc Natl Acad Sci USA*. 94, 14614-14619.

Lim H, Gupta RA, Ma WG, Paria BC, Moller DE, Morrow JD, Du Bois RN, Trzaskos M, Dey SK. (1999). Cyclooxygenase –2 derived prostacyclin mediates embryo implantation in the mouse via PPARδ. *Genes Dev*. 13:1561-74.

Liggins GC. (1990). The foetal role in the initiation of parturition in the ewe. In: Wolstenhome GEW, O'Connor M (eds.), Foetal Autonomy (CIBA Foundation Symposium). London: Churchill; pp. 218.

Lurie S, Woliovitch I, Rotmensch S, Sadan O, Glezerman M. (2001). Value of vaginal culture in management of acute vaginitis. *Arch Gynecol Obstet*. 265(4):187-9.

McMaster MT, Dey SK, Andrews GK. (1993). Association of monocytes and neutrophils with early events of blastocyst implantation in mice. *J. Reprod. Fertile.*99(2):561-9.

Manthey CL, Vogel SN. (1994). Interactions of lipoplysaccharides with macrophages. *Immunol. Ser*. 60:63-81.

Mikamo H, Sato Y, Hayasaki Y, Hua YX, Tamaya T. (2000). Vaginal microflora in healthy women with Gardnerella vaginalis. *J Infect Chemother*. 6(3):173-7

Mitchell MD, Romero RJ, Edwin SS, Trautman MS. (1995). Prostaglandins and parturition. Reprod Fert Dev. 7(3):623-32.

Mota A, Prieto E, Carnall V, Exposto F. (2000). Evaluation of microscopy methods for the diagnosis of bacterial vaginosis. Acta Med Port. 13(3):77-80.

Nomura F, Akashi S, Sakao Y, Sato S, Kawai T, Matsumoto M, Nkanishi K, Kimotio M, Miyake K, Takeda K, Akira S. (2000). Cutting edge: endotoxin tolerance in mouse peritoneal macrophages correlates with downrtegulation o surface toll-like receptor4 expression. J Immunol. 164: 3476-3479.

Osmer RG, Blaser J, Kuhn W, Tschesche H. (1995). Interleukin–8 synthesis and the onset of labor. Obest. Gyneacol. 86(2): 223-9.

Pampfer S, Wuu YD, Vanderhyden I, Hertogh R De. (1994). Expression of Tuimor necrosis factor-α receptors and selective effect of TNF-α on the inner cell mass in mouse blastocyst. Endocrinology. 134:206-212.

Paria BC, Huet-Hudson YM, Dey SK. (1993). Blastocysts state of activity determineds the window of implantation in the receptive mouse uterus. Proc. Natl. Acad. Sci. USA. 90:10159-10162.

Paria BC, Lim H, Das SK, Reese J, and Dey SK. (2000). Molecular signaling in uterine receptivity for implantation. Seminars in Cell and Developmental Biology. 11:67-76.

Parrilo JE. (1993). Pathogenetic mechanisms of septic shock. New Engl J Med. 328:1471-77.

Poltorak A, He X, Simimova I, Liu MY, Huffel CV, Du X, Birdwell D, Alejos E, Silva M, Galanos C, Freudenberg M, Ricciardi-Castagnoli P, Layton B, Beutler B: (1998). Defensive LPS signaling in C3H/HeJ and C57BL/10ScCr mice: mutations in TLR 4 gene. Science. 282: 2085-2088.

Pybus V and Onderdonk A. (1999). Microbial interactions in the vaginal ecosystem, with emphasis on the pathogenesis of bacterial vaginosis. Microbes and Infection, 1:285-292.

Raabe T, Bukrinsky M, Currie AR. (1998). Relative contribution of transcription and translation to the induction of Tumor necrosis factor –alpha by lipopolysaccharide. J. Biol. Chem. 273:974-980.

Raetz CRH (1990). Biochemistry of endotoxins. *Annu. Rev. Biochem.* 59:129-170.

Raetz CRH. (1993). Bacterial Endotoxins: Extraordinary lipids that activate eukaryotic signal transduction. *J. Bacteriol.* 175:5745-55.

Raetz CRH, Ulevitch RJ, Wright SD, Sibley CH, Ding A, and Nathna CF. (1991). Gram negative endotoxin: an extracellular lipid with profound effects on eukaryotic signal transduction. *FASEB J.* 5:2660-2692.

Ralph SG, Rutherford AJ, Wilson JD. (1999). Influence of bacterial vaginosis on conception and miscarriage in the first trimester: cohort study. *BMJ.* 39: 220-3.

Randall GW, O'Connor EF, Gantt PA. (1991). Synergy between tumor necrosis factor and endotoxin decreases early embryo development in vitro. *J In Vitro Fert Embryo Transf.* 8(6):304-7

Rietschel ET, Brade L, Schade U, Seydel U, Zaheringer U, Kusumoto S, Brade H. (1988). Bacterial endotoxins: Properties and structure of biologically active domain. In *Surface Structure of Microorganisms and their interactions with the mammalian host.* (Schrinner E, Richmond MH, Seibert G, and Schwarz U, eds.) pp.1-41, VCH, New York.

Rietschel ET, Kirikae T, Schade FU, Mamat U, Schmidt g, Loppnow H, Ulmer AJ, Zahringer U, Seydel U, di Padova F, Schreier M, Brade H. 1994. Bacterial endotoxins: molecular relationships of structure to activity and function. *FASEB J.* 8:217-25.

Safro E, O'Neill C, Saunders DM. (1990). Elevated leutal pohase estradiol: progesterone ratio in mice causes implantation failure by creating a uterine environment that suppresses embryonic metabolism. *Fert. Steril.* 54: 1150-1153.

Saito S, Matsuura M, Tominanga K, Kirikae T, Nakano M. (2000). Important role of membrane associated CD14 in the induction of INF-beta and subsequent nitric oxide production by murine amcrophages in response to bacterial lipopolysaccharide. *Eur. J Biochem.* 267(1):37-45.

Satoh T, Kumamota Y, Hirose T. (1994). Studies on the local immune response in mouse model of experimental Escherichia coli intrauterine infections. *Kansenshogaku Zasshi.* 68 (11): 1381-9.

Schlafer DH, Yuh B, Foley GL, Elssaser TH, Sadowsky D, Nathanielsz PW. 1994. Effect of Salmonella Endotoxin administered to the pregnant sheep at 133-142 day gestation on fetal oxygenation, maternal and fetal adrenocorticotropic hormone and cortisol, and maternal plasma tumor necrosis factor α concentrations. *Biol Repod* 50:1297-1302.

Shalaby MR, Laegreid WW, Amman AJ, Liggitt HD. (1989). Tumor necrosis factor α associated uterine endothelial injury in vivo. Influence of dietary fat. *Lab Invest.* 61:564-570.

Sharkey ME, Adler RR, Brenner CA, Nieder GL. (1996). Matrix metalloprotinase expression during mouse preimplantation development. *Am. J. Reprod. Immunol.* 36(2):72-80.

Silen ML, Firpo A, Morgello S, Lowry SF, Francus T. (1989). Interleukin -1α and tumor necrosis factor alpha cause placental injury in rats . *Am J Pathol.* 135:239-244.

Silver RM, Edwin SS, Umar F, Dubley DJ, Branch DW, Mitchell MD. (1997). Bacterial lipopolysaccharide-mediated murine fetal death: The role of interleukin-1. *Am. J. Obest. Gyneacol.* 176:544-549.

Silver RM, Lohner WS, Dayner RA, Mitchell MD, Branch DW. (1994). Lipopolysaccharide induced fetal death: The role of tumor necrosis factor alpha. *Biol. Reprod* . 50:1108-1112.

Simon C, Moreno C, Remohi J, Pellicer A. (1998). Cytokines and embryo implantation. *J Repod Immunol.* 39:117-131.

Stone R. (1994). Search for sepsis drugs goes on despite past failures. *Science* 64:365-67 .

Strus M, Malinowska M, Heczko PB. (2002). *In vitro* antagonistic effect of Lactobacillus on organisms associated with bacterial vaginosis. *J Reprod Med* 47(1): 41-6.

Sweet MJ, Hume DA. 1996. Endotoxin signal transduction in macrophages. *J. Leukoc. Biol.* 60; 8-26.

Tabibzadeh S. (1998). Molecular control of the implantation window. *Hum. Reprod Update.* 4(5):465-71.

Tapping RI, Tobias PS. (1997). Cellular binding of soluble CD14 requires lipopolysaccharide (LPS) and LPS-binding protein. *J. Biol. Chem.* 272(37):23157-64.

Taylor CC, and Terranova PF. (1996). Lipopolysaccahride inhibits in vitro lutenizing hormone–stimulated rat ovarian granulose cell estradiol but not progesterone secretion. *Biol Reprod.* 54; 1390-1396.

Telleria CM, Ou J, Sugino N, Ferguson S, Gibori G. (1998). The expression of Interleukin-6 in the pregnant rat corpus lutuem and its regulation by progesterone and glucocorticoid. Endocrinol. 139(6): 3597-3605.

Terranova PF, Mongomery-rice V. (1997). Review: Cytokine involvement in ovarian process. Am. J. Reprod. Immunol. 37: 50-63.

Thompson AJ, Greer MR, Young A, Boswell F, Telfer JF, Cameron IT, Norman JE, Compbell S. (1999). Expression of intracellular adhesion molecules ICAM-1 and ICAM-2 in human endometrium: regulation by Interferon-g. Mol Hum Reprod. 5, 64-67.

Van-Meir C, Ramier MM, Matthews SG, Calder AA, Keirse MJ, Challis JR. 1997.Chronic prostaglandin catabolism is decreased in the lower uterine segment with term labour. *Placenta* 18: 109-14.

Vassilli P. (1992). The pathophysiology of tumor necrosis factors. *Annu. Rev. Immunol.* 10: 411-452.

Vesy CJ, Kitchens RL, Wolfbauer G, Albers JJ, Munford RS. (2000). Lipopolysaccharide-binding protein and phospholipid transfer protein release lipopolysaccharides from gram negative bacterial membranes. *Infect Immun*. 68(5): 2410-2417.

Wang L, Zhang W. (1998). Detecting cytokines in patients with preterm premature rupture of membrane. *Chung Hua I Hsueh Tsa Chih.* 78(3): 216-7.

Wegmann TG, Athanassakis I, Guilbert I, Branch D, Dy, M, Menu E. (1989). The role of M-CSF and GM-CSF in fostering placental growth, fetal growth and fetal survival. Transplant. Proc 21: 566-568.

Wrathall AE, Wray C, bailey J, Wells DE. (1978). Experimentally induced bacterial endotoxemia and abortion in pigs. *Br. Vet. J.* 134: 225-230.

2

Antifertility Potential of Carica papaya in Mammalian Reproduction

Angel Gupta, Mangla Bhide and Seema Modi

Rats are very harmful pests of standing crops, stored food grains, fodder, vegetables, fruits and other valuable households. They not only cause damages but also act as vector of so many diseases, so to control their progeny, it is most significant to introduce them with either some antifertility agents of plant origin because synthetic antifertility agents cause a lot of side effects to user. The effect of antifertility agent of plant origin causes adverse effect on the GSI, FFI, UI, inhibits ovarian growth which not only effect the most of the aspects of reproductive performance but also create deviation from normal strain in protein and lipid fractions of gonadal tissue hence directly or indirectly affecting the fertility of experimental animals e.g. in albino rats in the present investigation.

The parameter on GSI plays a very significant role which evaluates the reproductive potency of the concerning animal. Deviation from normal and increase in GSI indicating the oedematous condition or cancerous growth or hyperactivity of gonads or decrease in GSI indicated the poor development, disfunctioning or inactiveness of gonads.

The increase or decrease in GSI is correlated with the increase or decrease in the body weight with respect to the increase or decrease in ovarian weight as observed in rate by Arora *et al.* (1971) after *Embelia ribes* extract treatment; Adhikari *et al.* (1990), Choudhary *et al.* (1991) after *Piper betle* leaf stalk extract treatment; Pandey *et al.* (1992) after *Hibiscus rosa sinensis* flower extract exposure and Sreenivasulu (1992) after CDRI 85/287 compound treatment and stated that reduction in weight of ovaries was correlated with the anti-estrogenic activity.

Normal ovarian cycle showed all types of primary, developing and Graafian follicles with normal ovulation and corpora lutea formation as reported by Robert (1961); Copen (1964); Rough (1968); Craigmyle and Presely (1975)' Bingle and Fransworth(1980); Saxena (1980); Browder (1984).

Degenerative changes in ovarian tissue affected the fertility of affected female rat as reported by Radhakrishnan and Alam (1975) in female albino rats after *Embelia ribes* treatment. Germinal epithelium was irregular but intact as observed by Bahar and Gerg. (1998) with emarginatosides B and C extracted from fruit of *Sapindus emarginatus* fed orally to rats as also observed by Pandey *et al.* (1992) in female rats after *Hibiscus rosa sinensis* flower extract treatment and by Bhide (1999) in albino rats after treatment with *Argimone maxicana* seed extract treatment.

Necrotic changes in different types of ovarian follicles were clearly marked, as reported by Arora *et al.* (1971) in rats after *Embelia ribes* and kholkute and Udupa (1976) in female rat after *Hibiscus rosa sinensis* and by Singh and Singh (1991) in albino rats after treatment with *Abrus precatorius, Daucus carota* and *Embelia ribes* herbal preperation respectively and reported gonadotixic nature of treated agents.

Gradual decrease in number of developing Graafian follicles was reported by Modi (1991) in albino rats. Number of degenerative follicles was increases in such a way that no follicle attains such maturity which is essential for ovulation in normal developing ovaries and instead of ovulation it showed the degeneration and resorption of ovum along with zona pellucida, corna radiata and discus proligerous showed the dose dependent toxic effect as also reported by Singh and Singh (1991) in albino rats after *Abrus precatorius, Daucus carota* and *Embelia ribes* exposure and Pandey *et al.* (1992) in female rats after *Hibiscus rosa sinensis* extract exposure and by Shukla (1995) in rats, mice, guinea pig and golden hamsters after *Pueraria tuberosa butanolic* extract treatment respectively.

Membrana granulosa, theca interna and theca externa of atretic Graafian follicles showed pycnosis as reported by Arora *et al.* (1971) with *Embelia ribes* and Kholkute and Udupa (1976) with the treatment of *Hibiscus rosa sinensis* flower extract to female rats and by Singh and Singh (1991) in albino rat after *Abrus precatorius, Daucus carota* and *Embelia ribes* exposure respectively.

Between the inner and outer structure of follicle, prominent basement membrane delimited the granulosa layer from the outer stromal tissue as observed in guinea pig by Suzuki and Yamaguchi (1980) while bred them on less calcium diet. The basement membrane became irregular, ruptured at places and loosened due to *Argimone maxicana* seed extract treatment [Bhide (1999)].

Degenerative changes were clearly marked in stromal tissue. Pycnosis and clumping of stromal tissue resulted into syncytium at places or giant nucleated cells within stroma as reported by Bhide (1999) in albino rats with *Agrimone maxicana* treatment.

The vacuolization and space formation in stromal tissue were the ultimate consequences due to degenerative action of antifertility agent. The rate of vacuolization is directly proportional to the rate of degeneration of ovarian tissue as reported by Kholkute and Udupa (1976) and Kholkute and Uza (1976) in female rats after *Hibiscus rosa sinensis* flower extract administration, Adhikari *et al.* (1990) and Pandey *et al.* (1992) in the ovaries of rats treated with *Piper betle* leaf stalk extract and by Gupta and Bhide (2000) in albino rat after OVRAL-G treatment.

Dilation of arteries and veins was clearly marked, which were probably formed by the hemorrhage of blood vessels due to intoxication of antifertility agent and their synergistics. There was dose dependent significant increase in the vacuolization of stromal tissue as observed by Singh and Singh (1991) in albino rats after treatment with *Abrus precatorius, Daucus carota* and *Embelia ribes* seed extract exposure, while usual compact cortex and loosely packed medulla was seen in normal, control and placentract treated dams as reported by Robert (1961); Copen (1964); Browder (1984) Cariagmyle and Presley (1975), Gandhi *et al.* (1991) on rats and hamsters after *Ruta graveolens* treatment.

The antifertility agents extracted from plants also act as antagonistic to estrogen as reported by Sreenivasulu *et al.* (1992) after CDRI 85/287 exposure and Dhar *et al.* (1993) in female rats after *Ischnochiton camptus* exposure respectively, while *Carica papaya* seed extract in long term treatment exhibited some antioestrogenic activity which resulted into inhibition of normal estrous cycle in the severely affected female rats as reported by Adhikari *et al.* (1990) in rats after *Piper betle* leaf stalk extract ingestion and have noticed the changes in ovarian steroidogenesis with altered oestrous cycle and degenerative changes in ovarian tissue respectively, while Deans (1952) in albino rats and Hiremath *et al.* (1996) in female albino rats reported weak oestrogenic activity after *Stringa densiflora* extract administration respectively.

Degeneration of Graafian follicle resulted into inhibition of ovulatory activity. Ovulation was totally arrested and was the main potent cause of total absence of corpora lutea as reported by Gupta and Bhide (2000) in albino rat after OVRAL-G treatment while normal, control and placentract treated rats showed progressive development of all types of ovarian follicles as also reported by Sangha and Guraya (1989), Swariz and Gray (1989) and Bhide (1999) in normal ovaries of rats respectively.

Necrotic degenerative changes were well marked in granulosa lutein cells as well as theca lutein cells resulted into multinucleated giant cells and giant nucleated cells which were formed by the fusion of most of the adjacent lutein cells or their nuclei respectively, while Bhide (1999) also reported atretic corpora lutea in albino rats after *Argimone maxicana* seed extract treatment and reported the main cause of antiovulatory activity and such female rats did not conceive even though kept with fertile male resulted into zero per cent fertility. Necrotic changes in ovarian tissue and luteolytic effect with reduction or total loss of libido (sex desire) were well marked in albino rats as reported by Choudhary *et al.* (1991) after treated with *Alstonia scholaris.*

Lerner (1979) in mammals also reported the contraceptive activities of plant extract as also observed by Kalla and Frick (1989) in human beings after gossypol treatment and Gandhi *et al.* (1991) in female rats and hamster after treatment with benzene and chloroform extract of *Ruta Graveolens* respectively.

Degeneration, disintegration and resorption of corpora lutea were clearly observed by Bhide (1999) and Gupta and Bhide (2000) in albino rats after Argimone maxicana and OVRAL-'G' treatment respectively, while Johnsson *et al.* (1979) in mature female rats after ingestion with antagonists and Choudhary *et al.* (1991) in albino rats not only reported the luteolytic activity but also observed the abortificient and anti-implantation activity after administration of leaf extract of *Alstonia scholaries, Cleistanthus collinus, Terminalia bellirica* and *Murraya peniculata* root extract exposure but Bhide (1999) reported the marked inhancement in the development of corpora lutea in the ovaries of albino rats after placentract treatment.

Some plant extracts are contraceptive and acts as gonadotoxic, blastotoxic or luteolytic agents so as to control the population of rodents [as they are prolific breeders] these plant extracts were used as antifertility agents as observed in *Rattus rattus* (Dixit, 1977) with *Malva viscus* conzatti flower extract, male rats (Rao, 1988) after *Solanum xanthocarpus* alcoholic seed extract and in albino rats (Mathur *et al.,* 1990) with *Costus speciousus* rhizome extract while Oluwole and Bolarinwa (1995) studies the leucopenic properties of *Jatropa curacus* extract in rats and Keshri *et al.* (1995) reported the contraceptive activity by the efficacy of *Nigella sativa* seed extract treatment in rat which was directly proportional to the antifertility activity.

Moreover so far no literature was available on the depletion or increase in protein and lipid fractions in the ovaries of female rats treated with any antifertility agents, so the findings regarding this aspect is original and that of author's only.

EXPERIMENTAL APPROACH

Experiments were conducted on the live specimens of albino rats and the details of the same were as follows:

1. Procurement and Rearing

Indian Veterinary Research Institute colony-bred albino rats procured from Jabalpur (M.P.) were used in the present investigation. The rats were acclimatised for 15 days to the normal laboratory conditions prior to experiments according to the methods of Edmond (1950) and were maintained on the balanced diet (Hindustan Lever Ltd.) and water ad libitum.

2. Test Animal

Albino rats of both the sexes averaging between 180g to 200g initial body weight and aged about 90 to 95 days were randomly bred and the male and female rats obtained from first litter were used for experiments. All the normal control and experimental rats were housed into wooden cages at the room temperature under normal laboratory conditions and provided with balanced diet and clean water ad libitum. (Edmond, 1950).

3. Procurement of Carica Papaya Seeds and Extraction of Antifertility Agent

Seeds of *Carica papaya* (belongs to family Caricaceae) were collected locally, washed, shadow dried, powdered mechanically, weighed 100 gm. The antifertility agent was extracted with petroleum ether (B.P.40^0 C-60^0C procured from BDH) from powdered *Carica papaya* seeds by Soxhlet method (after Sharma, 1988). The extract was weighed and 1% stock solution was prepared in distilled water. For experimental purpose further dilution of stock solution with distilled water to prepare 0.1% experimental solution.

4. Experiments with Different Dosages of Antifertility Agents

Each groups of 5 albino rats (male or female) were treated orally with doses of antifertility agents as follows:

(a) Carica papaya Seed Extract Treatment

0.2ml Carica papaya seed extract/100 gm body wt./dose/ rat daily orally ingested upto 90days.

5. Calculation of Gonadosomatic Index

Mean weight of each group and their gonads and Uteri have been recorded and summarized in Table No. 1 (Gonado-somatic Index) was calculated by the following formula:

$$\text{GSI} \,♀\, \text{ovarian Index} = \frac{\text{Wet wt. of the ovary}}{\text{Wet wt. of the female rat}}$$

$$\text{Uterine Index} = \frac{\text{Wet wt. of the Uterus}}{\text{Wet wt. of the female rat}} \times 100$$

6. Histological, Histopathological Studies

(i) Tissue Fexation and Fixatives

Five rats of each group first weighed and then were sacrified under light chloroform anesthesia and the ovaries of control and treated rats were fixed in Carnoy's fluid (6:3:1, Pearse, 1960) after 20, 40, 60 and 90 days.

(ii) Preparation of Paraffin Blocks and Microtomy

After fixation and dehydration paraffin blocks of ovaries were prepared in usual manner and sections were cut at 6 μm thickness.

(iii) Staining Techniques

(a) **Histological and Histopathological Studies:** Delafield's Haematoxylin and Eosin (Pearse, 1960) technique was used for histological and histopathological studies.

7. Detection of Positively and Negatively Charged Protein and Lipid Fractions in the Ovaries of Control and Treated Groups of Albino Rats by Paper Electrophoresis

Method: Procedure were the same as described in manual Supplied by Systronics. Co Ltd Ahmadabad and adapted by Bhide (2000). The data were recorded and summarized in electrophoretograms 1 and 2 and Table 2 and 3.

Electrophoretograms Showing Detection of Positively Charged Protein Fractions in the Overy of Control and Treated Female Albino Rats

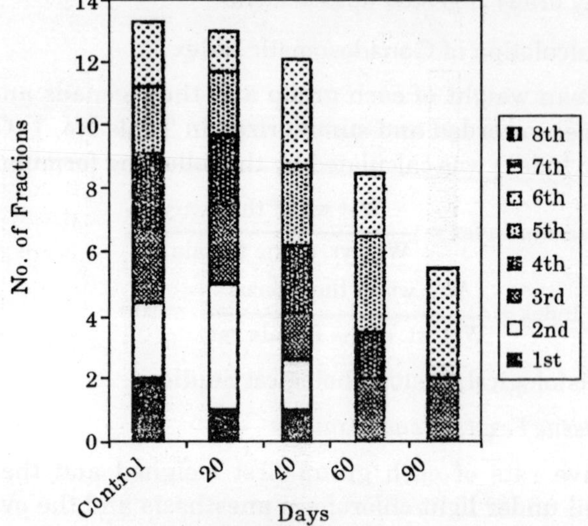

Electrophoretograms Showing Detection of Negatively Charged Protein Fractions in the Overy of Control and Treated Female Albino Rats

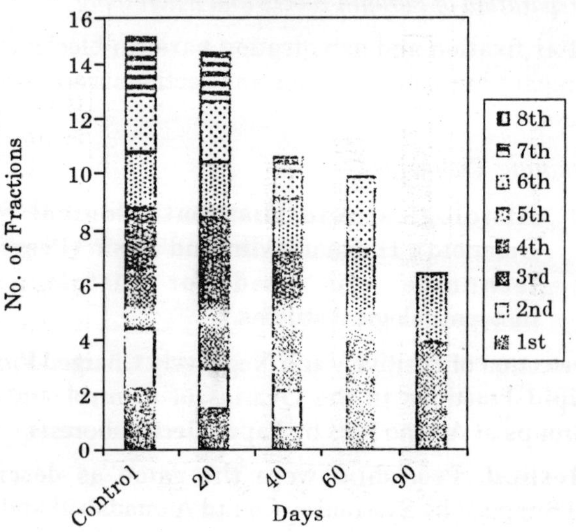

Electrophoretograms 1

Electrophoretograms Showing Detection of Positively Charged Protein Lipid Fractions in the Overy of Control and Treated Female Albino Rats

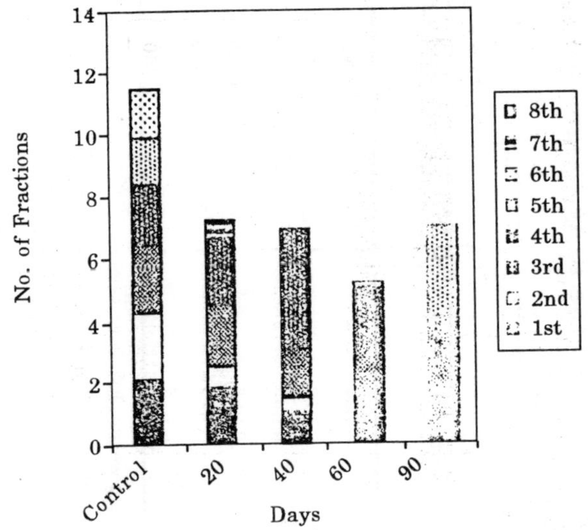

Electrophoretograms Showing Detection of Negatively Charged Protein Lipid Fractions in the Overy of Control and Treated Female Albino Rats

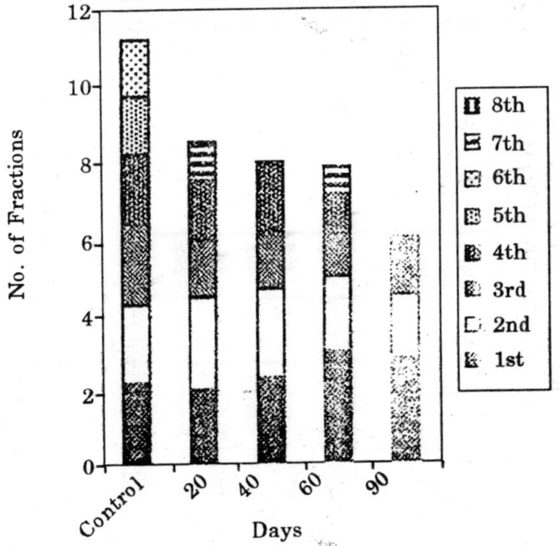

Electrophoretograms 2

Table 1. Effect of *Carica papaya* Seed Extract Treatment on the Body Weight and Genital Organs Weight of Female Albino Rats upto 90 days

Sl. No.	*Experimental Group of female rats*	*Average Body Weight*		*Increase and decrease in body weight*	***Genital organs weight in female*		$GSI = \frac{\text{Wet wt. of overy}}{\text{Wet wt. of female rat}} \times 100$	$GSI = \frac{\text{Wet wt. of uterus}}{\text{Wet wt. of female rat}} \times 100$
		*IW**	*FW**		*Overy*	*Uterus*		
1.	Normal	180g.	240g.	60g	1.25g.	8.20g.	0.520	3.41
2.	Control	180g.	235g.	55g	1.20g.	8.00g.	0.510	3.33
3.	20 Days Treated	180g.	160g.	−20g	0.100g.	0.700g.	0.062	0.437
4.	40 Days Treated	180g.	130g.	−50g	0.90g.	0.690g.	0.069	0.530
5.	60 Days Treated	180g.	120g.	−60g	0.70g.	0.500g.	0.058	0.416
6.	90 Days Treated	180g.	100g.	−80g	0.50g.	0.450g.	0.050	0.450

* Body wt.→Average wt. of 5 female albino rats in each group. IW initial weight, FW final weight.

** Wt. of ovary and uterus→Average wt. of ovary and uterus of 5 female albino Rats in each group.

OBSERVATIONS

Investigation was carried out on the following aspects:-

(a) *Studies on the Effect of Carica Papya Seed Extract Ingestion on the Body Weight, Genital Organs Weight and GSI in albino Rats.*

The data of FFI was recorded and summarized in Table No.1

(b) *Histological and histopathological studies on the ovarian tissue of female albino rats.*

I. **In Control Group:** There was no change in histological architecture of control rats in ovarian structure and it resembled to that of the ovarian structure of normal group (Fig. 1 & 2).

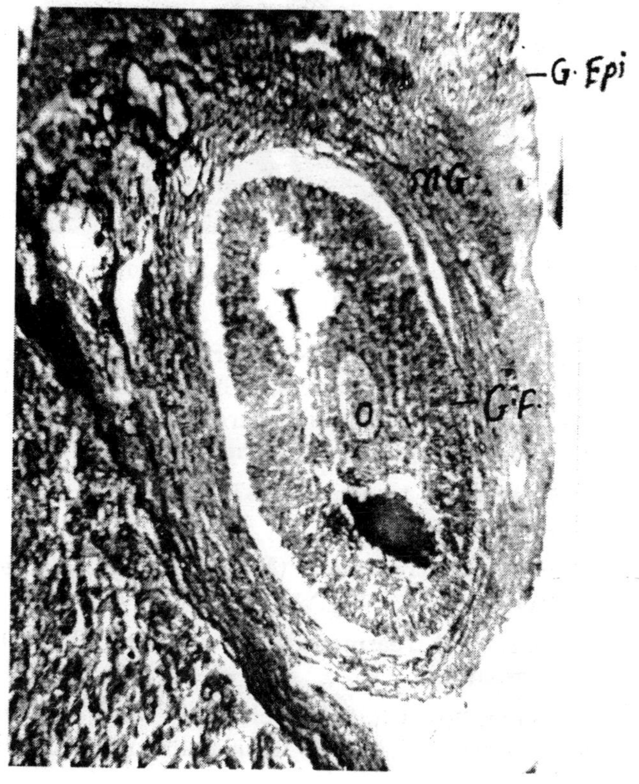

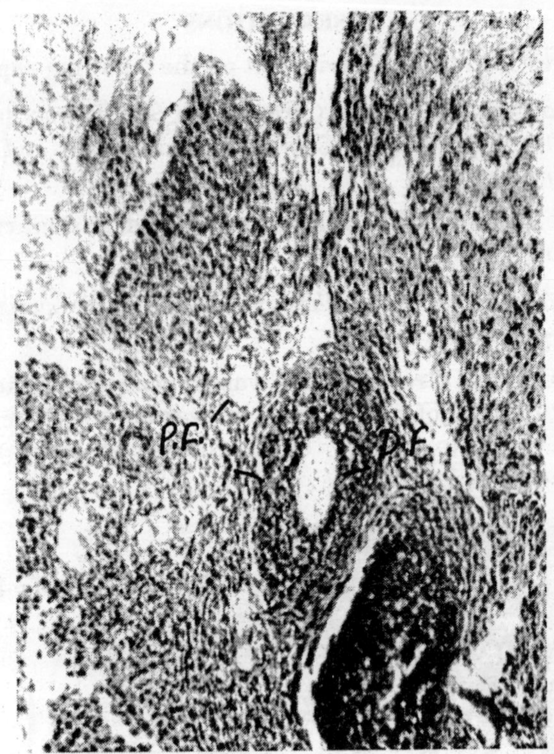

Fig. 1 & 2: Photomicrographs of sections of ovaries of control albino rats (x200).

II. **In Treated Groups:** Treated group revealed dose and duration of treatment dependent histopathological changes in the ovarian tissue architecture. All the treated groups were divided into three categories as follows:-

(i) Short term treatment (STT) i.e. upto 20 days of treatment.

(ii) Mid term treatment (MTT) i.e. upto 40 days of treatment.

(iii) Long term treatment (LTT) i.e. from 60 to 90 days of treatment.

Gradual histopathological changes were observed in ovarian tissue from STT to LTT female albino rats as follows:-

(a) *Histopathological studies in carica papaya seed extract treated female albino rats*

(i) **In Short Term Treatment (STT):** The ovarian architecture did not show any remarkable deviation in comparison to normal and control ovaries. Germinal epithelium was somewhat regular in outline with darkly stained nuclei. Cortex showed the presence of a number of progressively developing follicles. Graafian follicles were a few in number which showed the presence of somewhat degenerated condition in membrana granulosa, theca intena, theca externa and ovum along with zona pellucida and corna radiata. Follicular fluid showed the presence of variable sized vacuoles.

Between the inner and outer structure of follicle prominent basement membrane was situated which delimited the membrana granulosa from the outer stromal tissue was stained darkly. It was irregular in outline. Ovulation was evidenced by the presence of Corpora lutea. Stromal tissue showed somewhat degenerative changes, which were not so prominent. Blood vessels and some spaces were observed in stroma. Interstitial tissues were normal in appearance. A few degenerated corpora lutea were also observed, indicated the ovulataion was not followed by successful fertilization or perfect blastocyst implantation (Fig. 3 & 4).

(ii) **In Mid Term Treatment (MTT):** Ovarian architecture of MTT female albino rats exhibited remarkable changes in comparison to STT. Ovaries of treated female albino rats showed the presence of reduced number of developing ovarian follicles. A number of them became atretic due to necrosis in the corresponding tissue. Graafian follicles with highly vacuolated follicular fluid were showed significant reduction in number with degenerated ovum along with zona pellucida and corna radiata due to active degeneration, disintegration and pycnosis in discus proligerous and membrana granulosa. Some giant nucleated cells were observed in membrana granulosa as well as in stromal tissue.

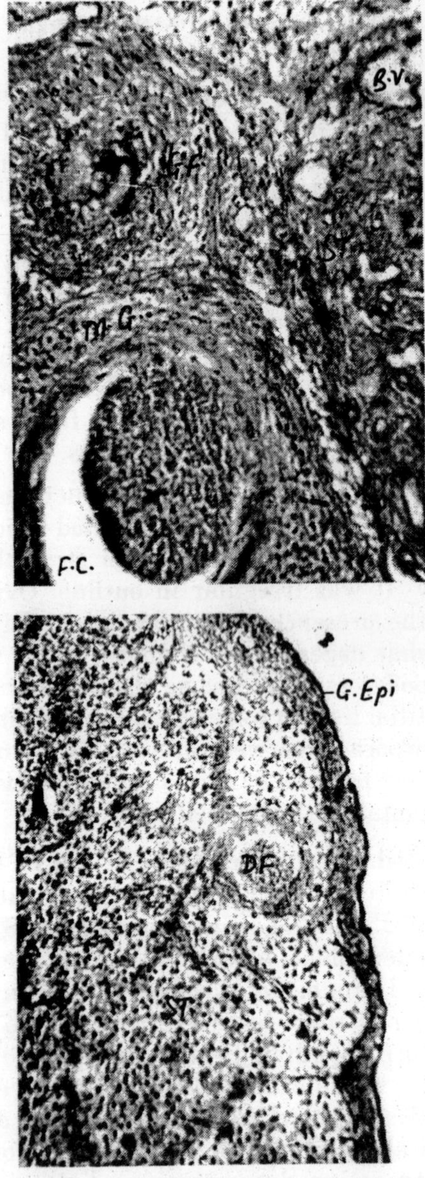

Fig. 3 & 4: Photomicrographs of sections of ovaries of STT female albino rats treated with ***Carica papaya*** seed extract. (x 200).

Basement membrane was irregular in outline while germinal epithelium was intact but ruptured at some places showed some toxicity of the treated antifertility agent e.g. *Carica papaya* seed extract treatment. A few atretic follicles and corpora lutea were observed. Stroma showed the remarkable sign of degeneration, disintegration and resorption evidenced by the presence of a long number of vacuoles and blood spaces due to the haemorrhage in number of blood vessels. Giant nucleated cells were observed in atretic follicles and stroma indicated the fusion of number of neighbouring cells with each other (Fig. 5, 6, 7 & 8).

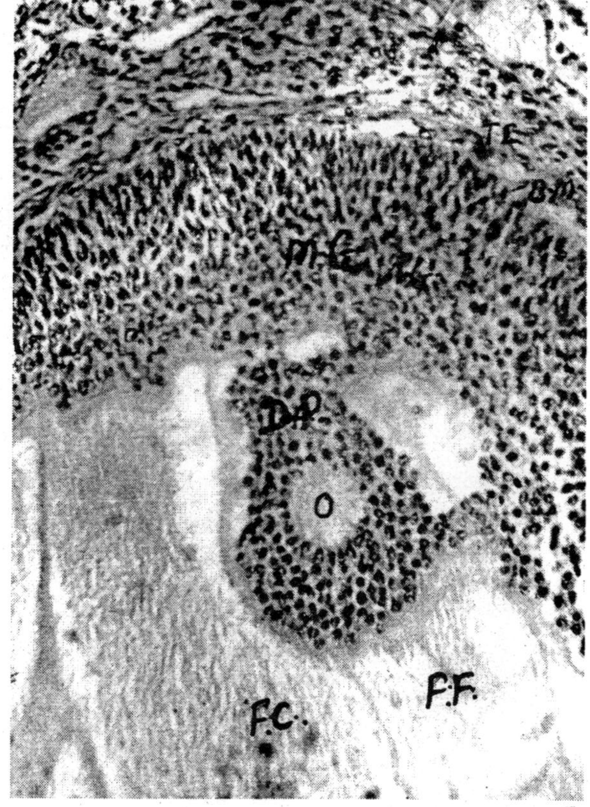

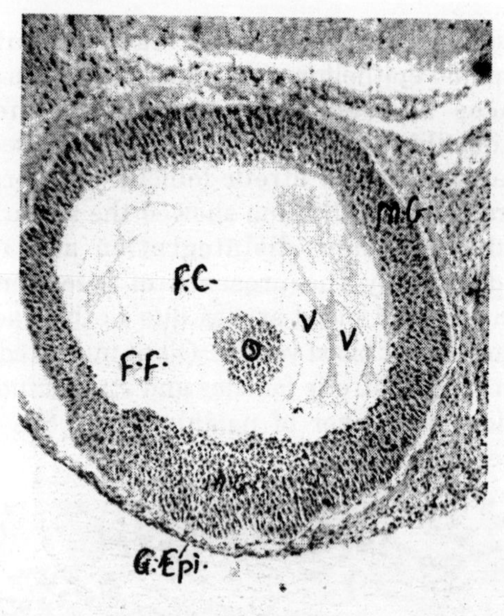
M.G
F.C.
V
V
O
F.F.
M.G.
G.Epi.

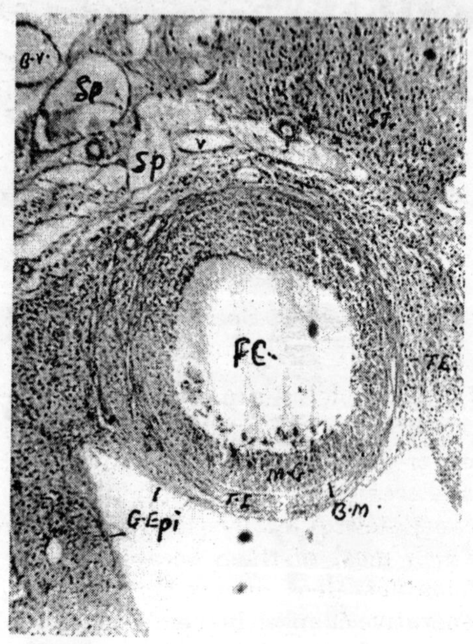
B.V.
Sp
Sp
V
F.C.
M.G.
G.Epi
B.M.

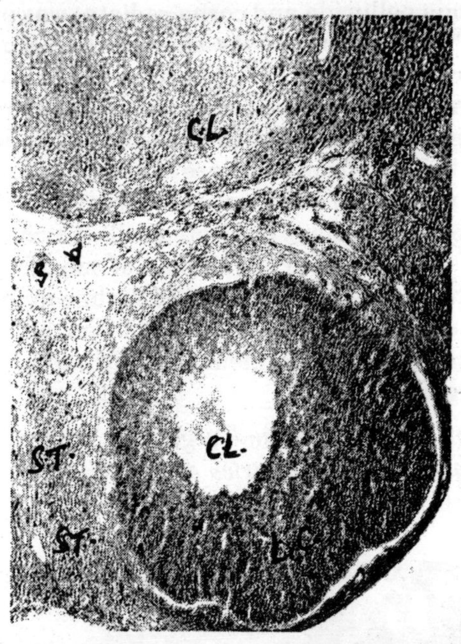

Fig. 5, 6, 7 & 8: Photomicrographs of sections of ovaries of MTT female albino rats treated with ***Carica papaya*** seed extract. (r200). B.M.-Basement membrane; B.V.-Blood vessel; C.L.-Corpus luteum; D.F.-Degenerating follicle; D.P.-Discus proligerous; F.C.-Follicularcavity; F.F.-Follicular fluid; G.E.-Germinal epithelium; G.F.-Graafian follicle; L.C.-Lutein cells; M.G.-Membrana granulosa; O.-Oocyte; S.P.-Space; S.T.-Stroma; T.E.-Tunica externa; T.I.-Tunica interna; V.-Vacuole.

(iii) **In Long Term Treatment (LTT):** Ovaries of LTT female albino rats showed the presence of regular (60 days of treatment) to irregular, ruptured at places (90 days of treatment) germinal epithelium. Oogenesis was very slow. A few ovarian follicles were seen out of which most of them became atretic. Graafian follicles very less in number showed significant degenerative changes in membrana granulosa. Total resorption of ovum along with surrounding envelops.

(Zona pellucida and corna radiata) was prominent in LTT. Follicular cavity showed the accumulation of vacuolated fluid. Discus proligerous also showed the sign of disintegration. Stromal tissue showed the active degeneration, disintegration and resorption indicated by the significant increase in vacuolisation and exceed in number of blood spaces. Stroma also showed the presence of giant nucleated cells which were formed by the fusion of a number of neighbouring cells. Interstitial cells also showed decline in their number. The ovaries of LTT rats showed antioestrogenic activity by the treatment of *Carica papaya* seed extract (Fig. 9, 10, 11, 12).

(c) Detection of positively and negatively charged protein and lipid fractions in the ovaries of control and treated female albino rats by paper electrophoresis

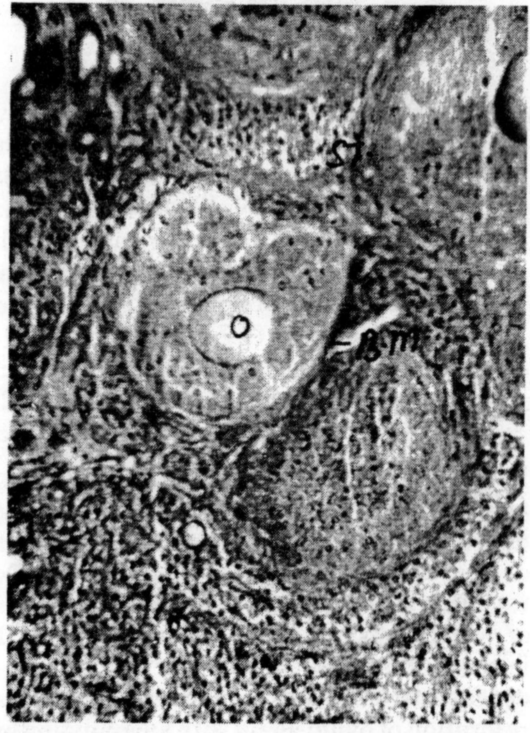

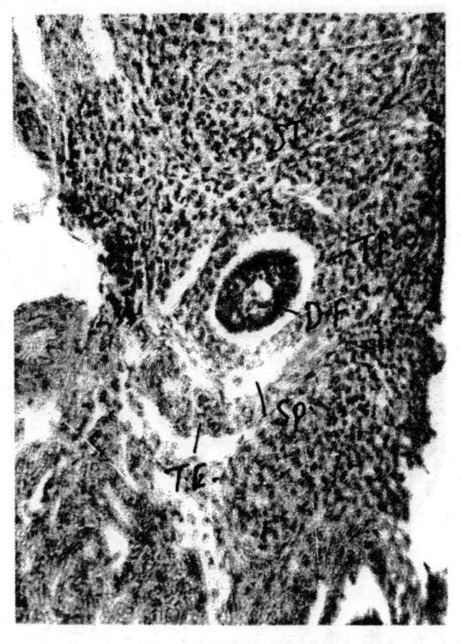
ST
D.F
Sp
TE

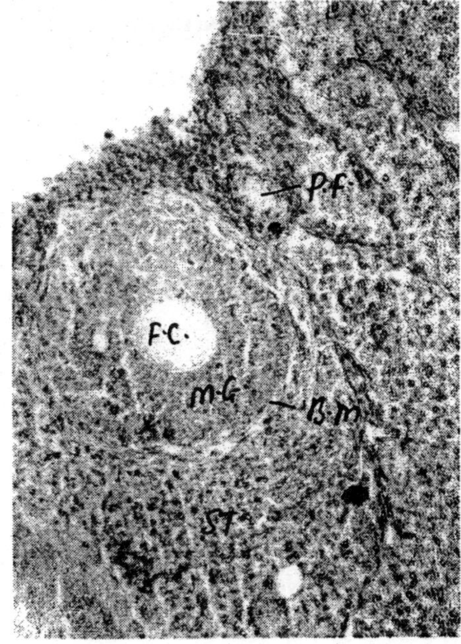
P.f
F.C.
M.G
B.M
ST

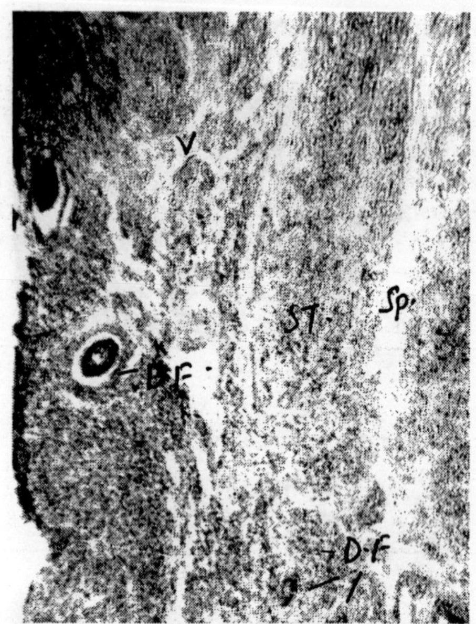

Fig. 9, 10, 11 & 12: Photomicrographs of sections of ovaries of LTT female albino rats treated with ***Carica papaya*** seed extract. (x200)

Samples of ovaries of control and 20, 40, 60 and 90 days treated female and albino rats containing Protein and lipid fraction were analysed separately (in triplicate) by paper electrophoresis and the results are summerised graphically on the basis of coloured bands obtained on the paper strips and the data of the same is summerized and analysed as follows:

IN CONTROL (FEMALE) GROUP →

(i) 6 +vely charged P.F.

(ii) 7 -vely charged P.F.

(iii) 6 +vely charged L.F.

(iv) 6 -vely charged L.F.

INTERPRETATIONS

In the present investigation body weight of female showed gradual decrease (Table No.1) which was dose and duration of treatment dependent, while GSI in female rats showed gradual

Table 2. Detection of P.F. in ovaries of C.P. Treated Female Albino Rats C.P.- Carica papaya; G.T.- Table of Graph; H.I.-High intensity; L.F.- lipid Fractions; L.I.- Low intensity; M.I.- Moderate intensity; P.F.- Protein Fractions; T.A.- Totally absent; VHI- Very high intensity; VLI- Very low intensity; VVHI.- Very Very high intensity. +vely- Positively & - vely - negatively.

Sl. No.	*Ingestion for No. of Days*	*Intensities of+Vely Charged P.F.*		*Intensities of-vely Charged P.F.*	
		6 P.F.		**7 P.F.**	
		No. 2	VHI	No. 4	VHI
1.	20 Days Treatment	No. 3,4,5	HI	No. 3,5,6	HI
		No. 1.6	MI	No. 1,2	MI
		No. 7,8	TA	No. 7	LI
				No. 8	TA
		6 P.F.		**7 P.F.**	
		No. 3,4,5,6	HI	No. 3,4,5,6	HI
2.	40 Days Treatment	No. 1.2	MI	No. 1,2	LI
		No. 7,8	TA	No. 7	VLI
				No. 8	TA
		4 P.F.		**4 P.F.**	
		No. 5	VHI	No. 3	VHI
3.	60 Days Treatment	No. 3,6	HI	No. 5	HI
		No. 4	MI	No. 4,6	MI
		No. 1,2,7,8	TA	No. 1,2,7,8	TA
		2 P.F.		**2 P.F.**	
		No. 6	VHI	No. 3,5	VHI
4.	90 Days Treatment	No. 3	HI	No. 1,2,4,6,7,8	AT
		No. 1,2,4,5,7,8	TA		

Analysis

1. Maximum 7 -vely charged P.F.
2. Manimum 2 +vely and -vely charged P.F. present.
3. No. 7,8 P.F. were totally absent.
4. In 90 days treated female albino rats only 2 P.F. were detected.

Table 3. Detection of L.F. in ovaries of C.P. Treated Female Albino Rats

Sl. No.	*Ingestion for No. of Days*	*Intensities of+Vely Charged L.F.*		*Intensities of-vely Charged L.F.*	
		5 L.F.		**4 L.F.**	
		No. 4	HI	No. 1, 2	HI
1.	20 Days Treatment	No. 1,3	MI	No. 3,4	MI
		No. 2,7	VLI	No. 5,6,7,8	TA
		No. 5,6,8	TA		
		6 L.F.		**7 L. F.**	
		No. 4	VHI	No. 1	VHI
2.	40 Days Treatment	No. 1,3	LI	No. 2	HI
		No. 2	VLI	No. 3,4	MI
		No. 5,6,7,8	TA	No. 5,6,7,8	TA
		4 L. F.		**4 L. F.**	
		No. 5	HI	No. 1	VHI
3.	60 Days Treatment	No. 1,3,7	LI	No. 2	MI
		No. 1,2,7,8	TA	No. 3,4	VLI
				No. 5,6,7,8	TA
		2 L.F.		**2 L.F.**	
		No. 4	HI	No. 1	VHI
4.	90 Days Treatment	No. 1	MI	No. 3	MI
		No. 3	VLI	No. 2	LI
		No. 2,5,6,7,8	TA	No. 4,5,6,7,8	AT

Analysis

1. Maximum 5 +vely charged L.F.
2. Manimum 3 +vely and -vely charged L.F
3. No. 5, 6, 7, 8, L.F. missing from 20 to 90 days treated female.
4. Gradual decrease in No. and intensities of LF.

decrease from day 20th to 90th days of *Carica papaya* treatment (Table No.1). The increase or decrease in GSI is correlated with the increase or decrease in the body wt. with respect to the increase or decrease in ovarian weight was observed in the present investigation. Reduction in ovarian weight with respect to the reduction in the body wt. was observed in rats by Arora *et al.* (1971) after *Embelia ribes* extract treatment; Adhikari *et al.* (1990), Choudhary *et al.* (1991) after *Piper betle* leaf stalk treatment; Pandey *et al.* (1992) after *Hibiscus rosa sinensis* flower extract exposure and Sreenivasulu *et al.* (1992) after CDRI 85/287 compound treatment and stated that reduction in wt. of ovaries was correlated with the antiestorgenic activity as also indicated by the absence of developing, primary, Graafian follicles and Corpora lutea in the ovaries of rats treated with *Carica papaya* seed extract in the present investigation.

Normal ovarian cycle showed ovarian histological architecture with all types of primary, developing and Graafian follicles with normal ovulation and corpora lutea formation and normal vascular stroma was observed in the present investigations as observed by Robert (1961); Copen (1964); Rough (1968); Craigmyle and Presely (1975); Bingle and Fransworth (1980); Sexena (1980); Browder (1984), while in treated groups ovaries showed dose and duration of treatment dependent gradual histopathological changes from STT to LTT of doses. The ovaries of *Carica papaya* seed extract treated rats showed somewhat lesser deformities. The ovaries showed deep red colour morphologically in comparison to pinkish colour of normal and control rats suggested the severe haemorrhage in ovarian tissue in the present investigation. Degenerative changes in ovarian tissue affected the fertility of corresponding LTT animal as observed by Radhakrishnan and Alam (1975) in female albino rats after *Embelia ribes* treatment.

Carica papaya seed extract treated albino rats showed somewhat mild to moderate damage in the ovaries of MTT to LTT albino rats. The severe degeneration resulted in follicular atresia as also reported by Arora *et al.* (1971) in rats after *Embelia ribes* and Kholkute and Udupa (1976) in female rats after *Hibiscus rosa sinensis* and by Singh and Singh (1991) in albino rats after treatment with *Abrus precatorius, Daucus carota* and *Embelia* herbal preperation respectively and reported gonadotoxic nature of treated agents.

There was gradual decrease in the number of developing Graafian follicles from STT to LTT and the decrease in number was fully dependent on the number of doses and duration of administration as observed by Modi (1991) in albino rats. The number of degenerative follicles increases in such a way that no follicle attain such maturity [due to degeneration of other follicular structure] which is essential for ovulation in normal developing ovaries and instead of ovulation it showed the degeneration and resorption of ovum along with zona pellucida, corna radiata and discus proligerous showed the dose dependent toxic effect as also observed by Singh & Singh (1991) in albino rats after *Abrus precatorius,Daucus carota and Embelia ribes* exposure and Pandey *et al.* (1992) in female rats after *Hibiscus rosa sinensis* extract exposure, and by Shukla (1995) in rats, mice, guinea pig and golden hamsters after *Pueraria tuberosa* butanolic extract treatment respectively.

Membrana granulosa, theca interna and theca externa of atretic Graafian follicles showed pycnosis as observed by Arora *et al.* (1971) with *Embelia ribes* Kholkute and Udupa (1976) with the treatment of *Hibiscus rosa sinensis* LINN. flower extract to female rats and by Singh and Singh (1991) in albino rats after *Abrus precatorius, Daucus carota* and *Embelin* exposure, respectively.

Between the inner and outer structure of follicle prominent basement membrane delimited the granulosa layer from the outer stromal tissue as observed in guinea pig by Suzuki and Yamaguchi (1980) while bred them on less calcium diet. This basement membrane became irregular, ruptured at places and loosened due to *Argimone maxicana* seed extract treatment (Bhide, 1999), while in the present investigation degenerative changes were well marked in severely affected atretic Graafian follicles.

In earlier treated stages the degenerataive changes were clearly marked in stromal tissue. Pycnosis and clumping of stromal tissue resulted into syncytium at places or giant nucleated cells within stroma of LTT albino rats as observed in LTT with *Argimone maxicana* treatment by Bhide (1999) in albino rats. The vacuolization and space formation in stromal tissue was the ultimate consequences due to degenerative action of antifertility agent . The rate of vacuolization is directly proportional to the rate of degeneration of ovarian tissue as observed by Kholkute and Udupa (1976) and Kholkute and Uza (1976) in female rats after *Hibiscus rosa sinensis* flower extract administration, Adhikari *et al.* (1990) and Pandey *et al.* (1992) in the ovaries of rats treated with *Piper betle* leaf stalk extract and by Gupta and Bhide (2000) in albino rats after LTT of OVRAL 'G'.

Stromal tissue also showed the presence of large number of blood spaces which were probably formed by the haemorrhage of blood vessels due to intoxication of antifertility agent. There was dose dependent significant increase in the vacuolization of stromal tissue as observed by Singh and Singh (1991) in albino rats after treatment with *Abrus precatorius, Daucus carota* and *Embelia ribes* seed extract exposure while usual compact cortex and loosely packed medulla was seen in normal, control and placentract treated dams as observed by Robert (1961), Copen (1964), Browder (1984), Craigmyle and Presley (1975), Ghandhi *et al.* (1991) in rats and hamsters after *Ruta graveolens* treatment.

The antifertility agents extracted from plants also act as antagonistic to estrogen as observed by Sreenivasulu et.al (1992) after CDRI 85/287 exposure and Dhar *et al.* (1993) in female rats after *Ischnochiton camptus* exposure respectively while *Carica papaya* seed extract treatment in long term treatment exhibited some antioestrogenic activity while inhibiting the normal oestrous cycle in the severely affected female rats as also observed by Adhikari *et al.* (1990) in female albino rats and Choudhary *et al.* (1991) in rats after *Piper betle* leaf stalk extract ingestion and have noticed the changes in ovarian steroidogenesis with altered oestrous cycle and degenerative changes in ovarian tissue respectively while Deans (1952) in albino rats and Hiremath *et al.* (1996) in female albino rats reported weak oestrogenic activity after *Stringa densiflora* extract administration .

Degeneration of Graafian follicle resulted into inhibition of ovulatory activity. Ovulation was totally arrested and was the main potent cause of total absence of corpora lutea in the ovaries in the present investigation as observed by Bhide (1999) in albino rats after *Argimone maxicana* treatment, Gupta and

Bhide (2000) in albino rats after OVRAL-G treatment respectively, while normal, control and placentract treated rats showed progressive development of all types of ovarian follicles as also reported by Sangha and Guraya (1989), Swariz and Gray (1989) and Bhide (1999) in normal ovaries of rats respectively.

In the present investigation necrotic degenerative changes were well marked in granulosa lutein cells as well as in theca lutein cells, resulted into multinucleated giant cells and giant nucleated cells which were formed by the fusion of most of the adjacent lutein cells or their nuclei respectively after treated with *Carica papaya* seed extract. Bhide (1999) also reported atretic corpora lutea in albino rats after treatment with *Argimone maxicana* seed extract and reported that due to main cause of antiovulatory activity such female rats did not conceive even though kept with fertile male resulted into zero per cent fertility. Necrotic changes in ovarian tissue and luteolytic effect with reduction or total loss of libido (sex desire) were well marked in albino rats as reported by Choudhary *et al.* (1991) after treated with *Alstonia scholaris*. Luteolytic activity was remarkable in LTT of *Carica papaya* seed extract treatment where total absence of corpora lutea was the prominent feature of ovarian architecture in the present investigation. Lerner (1979) in mammals also reported the contraceptive activities of plant extract as also observed by Kalla and Frick (1989) in human beings after gossypol treatment and Ghandhi *et al.* (1991) in female rats and hamster after treatment with benzene and chloroform extract of *Ruta graveolens* respectively.

Degeneration, disintegration and resorption of corpora lutea were clearly observed due to intoxication of *Carica papaya* seed extract, and due to total resorption of corpora lurtea in LTT albino rat blastocyst implantation was not possible and the effect was totally blastotoxic as reported by Bhide (1999) and Gupta and Bhide (2000) in albino rats after *Argimone maxicana* and OVRAL 'G' treatment respectively, while Johnsson *et al.* (1979) in mature female rats after ingestion with antagonists and Choudhary *et al.* (1991) in albino rat not only reported the luteolytic activity but also observed the abortificent and anti-implantation activity in rats after administration of leaf extract of *Alstonia scholaries, Cleistanthus collinus, Terminalia bellirica and Murraya peniculata* root extract exposure but Bhide (1999) reported the marked enhancement in the development of corpora lutea in the ovaries of albino rats after placentract treatment as reported in normal and control group in the present investigation.

So studies revealed that some plant extracts are contraceptive and act as gonadotoxic, blastotoxic or luteolytic agents so to control the population of rodents [as they are prolific breeders]. These plant extracts were used on *Rattus rattus* (Dixit 1977] with *Malva viscus conzatti* flower extract, male rats (Rao 1988) after *Solanum xanthocarpum* alcoholic seed extract and in albino rats [Mathur et.al 1990] with *Costus speciosus* rhizome extract while Oluwole and Bolarinwa (1995) studies the leucopenic properties of *Jatropa curacus* extract in rats and Keshri *et al.* (1995) reported the contraceptive activity by the efficacy of *Nigella sativa* seed extract in rats which was directly proportional to the antifertility activity of *Carica papaya* seed extract treatment in the present investigation.

So overall total arrest of development of various sized ovarian follicle resulted into increase number of atretic follicle in albino rat (Deans 1951) Antiovulatory, blastotoxic and contraceptive nature of antifertility agent resulted into 100% loss of fertility in severely affected animals and the normal ovarian architecture was restored after discontinued the effects for 30 days, so these effects were transient and were restored normal fertility in the corresponding affected female albino rats due to restore of normal estrous cycle.

Biochemical studies form an important part of investigation in animals especially when it is in relation to the effect of various drugs or antifertility agents. There were practically no such studies on the detection of +vely & -vely charged protein and lipid fractions in relation to antifertility agents by paper electrophoresis has been done so far. (Here P.F.=Protein fraction & L.F.= Lipid fractions).

In the present investigation ovaries of control female rats showed 6 +vely and 7 -vely charged P.F. and 6 +vely and

6 -vely charged L.F. The ovaries of *Carica papaya* seed extract treated female rats showed the presence of maximum 6 +vely and 7 -vely charged P.F. and 5 +vely and 4 -vely charged L.F. and minimum 2 +vely and -vely charged P.F. and 3 +vely and -vely charged L.F. In those samples had minimum P.F. (2 +vely and -vely charged) and L.F. (3+vely and -vely charged) about 1, 2, 4, 6, 7, 8 P.F. and 4, 5, 6, 7, 8 L.F. were totally absent (Electrophoretograms no. 1 & 2) and Table 2 &3.

It could be concluded that depletion in ovarian tissue P.F. & L.F.of treated female albino rats is correlated with the adverse effect of antifertility agents. While in some treated group first increase in protein or lipid fraction was observed which was more probably due to oedematous conditions of the affected organ and further decrease in protein and lipid fraction were due to necrosis in the affected organs as evidenced by the histopathological changes in the affected organs e.g. ovaries in the present investigation.

It can be concluded from the present investigation that the *Carica papaya* seed extract is proved to be an effective potent antifertility agent and should be used to control the fecundity of the pest animals like rodents.

CONCLUSIONS

When each group of 5 sexually mature female albino rats were ingested with 0.2 ml. *Carica papaya* seed extract (in triplicate)/100g. body wt./dose/rat upto 90 days daily, they showed gradual decrease not only in body weight but also in FFI as evidenced by Table No.1 Degenerative changes in ovarian tissue of treated female rats were more pronounced in LTT of *Carica papaya* seed extract. Presence of reduced number of developing follicles and large number of atretic follicles was the specific feature of ovarian architecture from MTT to LTT while gradual decrease in the number of corpora lutea was observed from STT to LTT. Stroma showed the active degeneration, disintegration and resorption indicated by the significant increase in vacuolization. Stroma also showed the presence of giant nucleated cells which were formed by the fusion of a number of neighbouring cells.

Ovarian tissue showed dose and duration of treatment dependent gradual decrease in number of +vely and -vely charged protein and lipid fractions from 20 days to 90 days of *Carica papaya* seed extract treatment as evidenced by electrophoretograms no. 1 & 2. It can be concluded by the present investigation that *Carica papaya* seed extracts be proved as an effective antifertility agent.

ACKNOWLEDGEMENTS

Authors are very much thankful to Prof. D.P. Gupta, Head of the Department of Zoology Dean of Life Sciences, Dr. H.S. Gour Vishwavidyalaya for providing the necessary laboratory facilities.

REFERENCES

Adhikary, P.; Banerjee, I; Choudhary, D.; Das, A.K.; Deb, C.C. and Mukherjee, S.R. (1990) Effect of oral administration of stalk of leaves of Piper betle on estrous cycle and its antifertility activity in rats. Ind. J. Physio. Appl. Sci. 44(3):116-123.

Arora, R.B.; Ghatak, N. and Gupta, S.P. (1971) *Antifertility activity of* Embelia ribes. J. Res. Indian. Med. 6:107-111.

Bahar, A. and Garg, D. (1998) Antifertility activity of emarginatosides B and C of fruits of Sapindus emurginatus. Med. From. Plant Sci. 20(2):362-363.

Bhide, M. (1999) Effect of some antagonist on the ovarian cycle of albino rats. Project report. Dr. H.S. Gour V.V. Sagar (M.P.).

Bhide, M. (2000) Dectection of protein fraction in blood serum of albino rats or other animals by paper electrophoreses. Work book developed during refresher course in Zoology. Published by Academic Staff College: Dr. H.S. Gour V.V. Sagar 3-6.

Bingle, A.S. and Fransworth, N.R. (1980) Botanical sources of fertility regulating agents.; Chemistry and Pharmacology. Proc. Horm. Biochem. Pharmacol. 1:149-225.

Browder, L.W. (1984) Developmental Biology, II edition Saunders. Philadelphia 220-703.

Choudhary, D.N.; Adhikary, P.; Banerji, J. Jana, S; Chatterjee A. and Mukherjee, S.R. (1991) *Pregnancy interceptive effect of* Piper betle. Fitoterapla. 62 (5): 397-402.

Choudhary, D.N.; Singh, J.H. and Singh. B.P. (1991) Effect of some medicinal plant on fertility of albino rats.Osooan. J. Pharmaco. 23(4): 253-257.

Copen,W.M. (1964) Bailey's textbook of histochemistry 15th edition the Williams and Wilkins Company, Calcutta. 483-560.

Craigmyle, M.B.L. and Presley (1975): Embryology II edition published by U.S.A. by Williams and Williams Company. Baltimor. 70.

Deans, W.H. (1951): Histochemical characteristics of atretic follicles in rat ovaries. Ant. Res. 111: 504.

Deans, W.H. (1952): Histopathological observations in ovary and oviduct of the albino rats during oestrous cycle. Amar. J. Anat. 91: 363.

Dhar; J.D. Setty, B.S.; Singh, K.; Jain, S. and Bhakkumi, D.S. (1993): *Contraceptive potential of* Ischnochiton camptus *Biological Memoire. 19(2): 95-103.*

Dixit, V.P. (1977): *Effect of Malva viscus conzatti flower extract on ovarian function to the house rat.* Rattus rattus. *Refescens and the gebril.* Merione hurricane Jerdon. Ind. J. Exp. Biol. 15: 506.

Edmond, J.F. (1950): The care and breeding of laboratory animals. John. Wiley and Sons. Inc. New York Company. & Hall Ltd. London.

Ghandi, M.; Lal, R.; Sankarnayana, A. and Sharma, P.L. (1991): *Post Coital Antifertility Activity of Ruta graveolens in female rats and hamsters. J. Ethinopharmacol.* 34(1): 49-60.

Gupta, A. and Bhide, M. (2000): Synergistic action of Levonorgestrel and Ethinylestradiol on the ovarian tissue of albino rats. 21st "Annu. Sess. of the Academy of Environmental Biology will be held at Narendra Dev University of Agricultural and Technology. Faizabad. (In Press).

Hiremath. A.O.; Ushara, S.P.; Swamy, H.K.S.; Badami, S.M.; Patil, S.B. and Londonkar, R.L. (1996): Postcoital antiferility activity of the plant *Stringa densiflora* on female albino rats. Intern. J. Pharm 34 (1): 48-52.

Johnsson, H.J.; Kell, J.E.; Gaddy, R.E.; Loadholi, L.B. Hennigar, G.R. and Walker, E.M. (1979): Prolonged ingestion of antagonist and effect on progesteron level and reproduction in the mature female rats. Arch. Environ. Contamin. Toxicol 3: 479-490.

Kalla, N.R. and Frick, J. (1989): Clinical studies with gossypol. Advances Ingyneco. and Obstetricres. Series. 2: 273-277.

Keshri, G.; Singh, M.M.; Laxmi, V. and Kamboj, V.P. (1995): *Postcoital contraceptive efficacy of* Nigella sativa *in rats. Ind. J. Physio. and Pharm.* 39(1): 59-63.

Kholkute, S.S. and Udupa, K.N. (1976) *Effect of Hibiscus rosa sinensis on pregnancy of rat. Plant. Medica.* 29: 321-329.

Kholkute, S.S. and Uza, K. (1976) *Effect of Hibiscus rosa sinensis flower extract on female rats. Ind. J. Exp. Biol.* 14:175.

Lerner, L.J. (1979) Contracepton, chemical control of fertility. Ed. Daniel. Ledmicer. Mercel Dekkerinc N.1: 161.

Mathur, A.; Choudhary, S. and Mathur, R. (1990) *Toxic effect of Costus speciosus extract in albino rats.* Environ. Exp. Toxicol, (Proc.). 290-300.

Modi., S. (1991) Histopathological studies on the effect of some pesticides on the reporudctive physiology of albino rats. Ph. D.Thesis DR. H.S. Gour V.V. Saugor, M.P. 470003 India.

Oluwole, F.S and Bolarinwa, A,F. (1995) *Possible leucopenic properties of Jatropha curacus extract. in rats.* Him. J. Env. Zool. 9:73-74.

Pandey, D.; Bisht, M. and Nabhachandra, Y. (1992) *The antiestrogenicity of Hibiscus rosa sinensis flowers estract on female rats.* Him. J. Envir. Zool. 6: 50-55.

Pearse, A.G.E. (1960) Histochemistry theoretical and applied. J. & A. Churchill, London.

Radhakrishnan,N. and Alam, M. (1975) Effect of Embelin on fertility of female albino rats. Ind. J. Exp. Biol.; B: 70-71.

Rao, M.V. (1988) *Effect of alcoholic extract of Solanum xanthocarpum seeds on adult ablino rats. Ind. J. Exp. Biol.* 26:95-98.

Robert, R. (1961) Laboratory manual of vertebrate embryology, Radiological Research Laboratory Columbia University, New York Allied Pacific.

Rough, R. (1968) The mouse, it's reproduction & Development. Bargess Minneapolies Minnesota. 182-198.

Sangha, G.K. and Guraya, S.S. (1989) DNA, RNA & Protein change in rat ovarian follicles. Proc. Ind. Nat. Sci. Acad. Part. B. Biol. Sci. 55(30): 155-158.

Saxena, R.C. (1980) Cytological and histochemical studies on the cortical zone of oocytes. Ph.D. Thesis. University of Saugor (M.P.) India.

Sharma, J.D. (1988) Forensic Science and Toxicology, Suvidha Law House Publisher, Bhopal.

Shukla, S. (1995) *Toxic potential of a contraceptive plant* Pueraria tuberosa Dc. Ind. J. Toxicol. 2 (182): 57.

Singh, K. and Singh, S.P. (1991) Effect of a herbal preperation on reproductive organs of female albino rats. Him. J. Env. Zool. 5(2): 82-86.

Singh, S.P. and Singh, S.P. (1992) *Anti-implantation effect of Carica papaya seed powder in female albino rats.* Ind. J. Physical and Natural Sci. 11(12): 22-24.

Sreenivasulu, S.; Singh, M.M. Dwivedi, A. Setty, B.S. and Kamboj, V.P. (1992) CDRI. 85/287,a novel antiestrogen and anti-implantation agent, Biological profile and interaction with the estrogen receptors in immature rat uterus. Contraception. 45(1): 81-92.

Suzuki, S. and Yamaguchi, H. (1980) Morphological changes of ovarian follicles of guinea pigs bred on a low calcium diet. Exp. Path. 18: 372-375.

Swariz, W.J. and Gray, M.M. (1889) Chlordecone induced follicular toxicity in mouse ovaries. Reprod. Toxicol. 3(3): 203-206.

3

Lac Insects and Their Host-Plants

K. Krishan Sharma and *K.K. Kumar*

Lac insects are well known for their commercially important products *viz.* resin, dye and wax. These are degenerate scale insects belonging to the family Kerriidae (Tachardiidae) which constitute a specialized and isolated group in the superfamily coccoidea of the order Homoptera. Lac insects are distributed in all zoogeographical regions except Palearctic. They, however, are concertated in tropical and subtropical regions (Varshney, 1976). Kerr (1782) was the first to study a species of lac insect in detail and named it *Coccus lacca* which is now known as *Kerria lacca* (Kerr). Lac insects of the world are represented by nine genera and 87 species (Sharma and Ramani, 1999), of which two genera and 19 species are reported from India. Lac insects recorded from India, their important host-plants and localities (Varshney, 1976, 1986, 1992) reported from are given in table 1.

Lac, the only resin of insect origin, is obtained from the secretion of specialised scale insects known as lac insects. Our country is previliged in being the world leader in production as well as export of lac. About 50% of the total world demand is fulfilled by India. Lac insects are exploited for their products of commerce *viz.*, resin, dye and wax in Jharkhand, Chhattisgarh, West bengal, Orissa, Assam and some other states of our country. Lac is a source of subsidiary income to more than three million tribal families engaged in lac cultivation. About 30% of their total agricultural income is contributed by lac in Jharkhand. Of the 87 species falling under

Table 1. Lac insects recorded from India, their important host-plants and localities reported from (Varshney, 1976, 1986, 1992).

Sr. No.	*Lac insect*	*Type host*	*Other important reported hosts*	*Type locality*	*Other places reported from*
1	**2**	**3**	**4**	**5**	**6**
A	***Kerria* spp.**				
1	*K. albizzae* (Green)	*Landolphia* sp. and *Albizzia stipulata* (= *A. chinensis*)	*Croton caudatum*, *Litchi chinensis*, *Pterospermum acerifolium*	Peradiniya (Sri Lanka)	Darjeeling (WB), Jharkhand
2	*K. brancheata* Varshney	*Schleichera oleosa*	*Albizzia lucida*	Ranchi (Jharkhand)	
3	*K. chamberlini* Varshney	*Ficus infectoria* (= *F. virens*)	*Flemingia macrophylla*	Jodhpur (Rajasthan)	
4a	*K. chinensis chinensis* (Mahd.) (= *Laccifer chinensis*, *L. Siamensis*, *L. longispina*)	not traceable for *chinensis*, *Schleichera* sp. for *siamensis*, *Cajanus indicus* (= *C. Cajan*) for *longispina*	–	Nowgong, Garo Hills (Meghalaya) Mt. Sutep near Chiengmai (Thailand) Haflong (Assam)	
4b	*K. c. kydia* (Misra)	*Kydia calycina*	–	Silcher (Assam)	–

(Table Contd...)

Table 1. (Contd...)

1	2	3	4	5	6
5	*K. communis* (Mahd.)	*Ficus mysorensis* (= *F. drupacea* var. *pubescens*)	Several others	Mysore (Karnataka)	Bangalore, Lingal, Hyderabad, Bombay, Goa, Madras and Travancore
6	*K. ebracheata* (Chamberlin)	Not traceable	*Ficus elastica*, *Ziziphus jujuba* (= *Z. mauritiana*)	Manbhum (Bihar)	Bangalore
7a	*K. fici fici* (Green)	*Ficus religiosa* and *F. beng(h)alensis*	*Samanea saman*, *Peltophorum inermis*, *Ficus carica*, *F. lacor*, *F. Lucescens*, *F. rumphii*, *Ziziphus mauritiana*, *Butea frondosa* (= *B. monosperma*)	Monghyr (Bihar)	Coimbtore (TN), 24 Parganas (WB), Rajasthan, Jammu & Kashmir, Delhi, Allahabad (UP)
7b	*K. f. jhansiensis* (Misra)	*Ziziphus jujuba* (= *Z. mauritiana*)	–	Jhansi (UP)	–
8	*K. indicola* (Kapur)	*Ziziphus jujuba* (= *Z. mauritiana*)	–	Jamunia ?	–
9a	*K. lacca lacca* (Kerr)	*Ficus religiosa* and *F. indica* (=*F. beng(h)alensis*), *Z. mauritiana*, *Butea monosperma*	*Several others*	Uncertain	All India Occurrence

(Table Contd...)

Table 1. (Contd...)

1	2	3	4	5	6
9b	*K. l. ambigua* (Misra)	Jheolia ?	Many others	Guna near Jhansi (UP)	–
9c	*K. l. mysorensis* (Mahd.)	*Shorea talura* (=*S. roxburghii*)	–	Mysore (Karnataka)	Bangalore
10	*K. nagoliensis* (Mahd.)	*Schleichera trijuga* (= *S. oleosa*)	*Acacia fernesiana*	Raipur and Sohagpur (Chhattisgarh)	–
11	*K. nepalensis* Varshney	*Litchi cinensis* (*Nephelium litchi*)	–	Valmikinagar (Bhainsalotan) Champaran (Bihar)	–
12	*K. pusana* (Misra)	*Ziziphus jujuba* (= *Z. mauritiana*) *Butea frondosa* (= *B. monosperma*)	–	Pusa (Darbhanga, Bihar)	Eastern India
13	*K. Sharda* (Mishra & Sushil)	*S. oleosa*	–	Mayurbhanj (Orissa)	–
14	*K. rangoonensis* (Chamberlin)	*Quisqualis* sp.	*Thea chinensis*	Rangoon (Myanmar)	NE India

(Table Contd...)

Table 1. (Contd...)

1	2	3	4	5	6
B	**Paratachardina spp.**				
1	*P. lobata* (Green)	*Flueggea leucopyrus* & *Flacourtia ramontchi*	Several others	Peradeniya (Sri Lanka)	Coimbtore, Madanpally (TN), Bangalore (Karnataka), Cuddappa (AP), Calcutta (WB)
2	*P. mithila* Varshney	*Photonia notoniana* var. *macrophylla*	–	Shillong (Meghalaya)	–
3	*P. silvestrii* (Mahd.)	*Ixora* parviflora	*Santalum album*, *Guazoma tomentosa*, *Ficus benjamina* and others	Mysore (Karnataka)	Bangalore (Karnataka),
4	*P. ternata* (Chamberlin)	*Acacia sundra* (*A. chundra*)	–	Travancore (Kerala)	–
5	*P. theae* (Green)	*Thea chinensis* & *Cinchona calisaya*	*Ficus retusa*, *F. Religiosa*, *Gardenia florida*, *Mallotis philippensis*, *Psidium guajava*, *mangifera indica*, *Machillia* sp., *Michellia longifolia*	Darjeeling (WB) & Assam	Calcutta, Kalimpong Hills (WB), Rhenok (Sikkim)

nine genera recorded from the world, 19 species belonging to two genera, namely, *Kerria* and *Paratachardina,* are found in India. Lac insects have been reported to thrive on more than 400 plant species generally growing in the forests which have varied economic, medicinal and social significance. Products obtained from lac insects find multifarious applications in our daily life. Lac cultivation and processing and product development units provide livelihood to millions of people inhabiting the lac growing region. Integration of lac-culture with agriculture particularly in non-lac growing regions of the country will not only enhance the profitability of the farmers but also help in conserving our fast depleting forest reserves, thereby checking the environmental degradation and reclaiming waste and barren land. An attempt has been made in this chapter to throw light on lac-culture with respect to its utility in sustained economic returns, employment generation, lac based industries and potential of lac-culture in environmental management.

POTENTIAL OF LAC AND ITS BY-PRODUCTS

Lac has been in use in India and abroad since vedic period. Besides its mention in Atherveda, Mahabharata (laksha-Griha), a mention has been made in Periplus (Greek literature) written about 80 AD that lac dye was exported to Aduli on the African coast of the Red Sea; also in the Aine-Akbari (1590 AD) that during the region of Akbar the Great, coloured varnishes used to be prepared with this resin and pigments for decorating the screens of public buildings (Agarwal *et al.*, 1998).

The importance of lac has assumed special significance in the present age. It forms renewable source, it is non-toxic and biodegradable. The various modern-day applications of lac and its by-products are recorded below:

Composition and Applications of Lac and its by-products (Agarwal *et al.*, 1998).

A. *Lac-Resin: (An ester complex of long chain hydroxy fatty acids and sesquiterpenic acids)*

1. **Varnish and printing ink industry**: Furniture polish (French polish), Floor polish, Sealers, Pattern

p[aints for wood, Flexible and fast-drying agent for printing ink, Heat and water-proof varnish (Melfolac), Shellac etch primer, Water-thinned shellac red-oxide primer

2. **Electrical Industry**: Air-drying and baking-type insulating varnish, coating of isolators etc., Micanites (sheets, tubes and moulded insulators), Cement for sockets of electrical lamps, Protective coating for PCB's, Shellac bond powder, Anti-tracking insulating varnish, Coating of spark plugs
3. **Adhesive Industry**: Gasket cement, Optical cement, Sealing wax, Grinding wheel, Hot-melt adhesive, Polymer adhesives, Adhesives for sichips and solar cells
4. **Food Industry**: Coating of fruits, Coating of chocolates, lozenges, coffee beans, coating of aluminium foils for making pizzas, non-toxic ink of marking food stuff, internal can coating
5. **Leather Industry**: Adhering coating of leather with metallic and plastic foils, Top-dressing material
6. **Pharmaceutical Industry**: Coating of tablets, Microencapsulation of vitamins, Coating of medicines for slow-release/delayed action
7. **Cosmetic Industry**: Hair spray, lacquer, Eye shadows, Microencapsulated perfumes, Lipstick, nail polish, mascara etc.
8. **Miscellaneous**: Proofing and stiffening in hat making, Jewellery, Polishing stone, Sharpening stone, Impact and water-resistant coal block as domestic fuel, High efficiency igniter-cum-fuel, Lac-coated fertilizer

B. *Lac dye: (anthraquinone derivative)*

Dyeing of wool, and silk, Soft drink formulations, Pill coating, Sausage finishing, Confectionery and chocolate coating, Finished food coating

C. ***Lac Wax: (Mixture of higher alcohols, esters, acids and hydrocarbons)***

Shoe, automobile and floor polishes, Bottle sealer, Tailor's chalk, Crayons (esp. for writting on glass), Electrical potting compounds, Lipstick, Fruit coating.

It is interesting that out of hundreds of documented Unani/Ayurvedic formulations, some prescribed for anti-inflammatory and anti-obese action etc. contain lac as an ingradient (Ramani and Siddiqui, 2000). Tribals also use lac insect in various forms for curing different diseases (Sharma, 1993) as detailed in the following table.

POTENTIAL OF LAC CULTIVATION AND PROCESSING IN EMPLOYMENT GENERATION

Our country is the largest producer of lac in the world accounting for about 50% of the total world lac production. At present, production of raw lac in India is approximately 16,000 metric tonnes per year. The major lac producing states are Jharkhand (57% of the country's lac production), Chhattisgarh (23%), West Bengal (12%) and minor lac growing states like Orissa, Gujarat, Maharashtra and Assam. Over two million tribals inhabiting these states are engaged in lac cultivation. Lac culture is of considerable importance in tribal economy. It is one of the major source of subsidiary income to them. About 30% of their total agricultural income comes from lac cultivation.

India is dependant on export market as most of the lac (70-80%) produced in the country is exported to various countries. Export of the lac is mainly in the form of semi-refined (Seedlac) or refined (Shellac) form. Export of value added products obtained from lac and its by-products such as bleached lac, aleuritic acid, isoambrettolide, lac dye and lac wax etc. have shown a healthy increase in the recent past. Our country earns about 80 crores of foreign exchange by exporting lac and its products. About 200 lac processing and product developement industries are there in the country engaging about 2500 labourers, thus generating about one million mandays in a year.

Although two genera and 19 species of lac insects have been recorded from India, only *Kerria lacca* and *K. nagoliensis* occupy the status of commercial lac insects, contributing about 90% of the national production (Mishra *et al.*, 2000c). Species belonging to *Paratachardina* genus produce a hard, horny

	Name of formulation	Form	No. of ingradients	Percentage of lac	Action
A. Medicinal uses	Dawa-ul-luk	Semi-solid	9	2.4	Anti-inflammatory
	Majoon Dabeed-ul-ward	Semi-solid	21	1.0	Diuretic, Anti-inflammatory, Haematogenic
	Majoon Mohazzil	Semi-solid	4	8.3	Anti-obese
	Sufoof-e-Mohazzil	Powder	6	11.1	Anti-obese
	Long Looks	Powder (capsule)	11	3.5	Height gain

	Name of insect	*Part used*	*Mode of administration*	*Disease*	*Tribe (district) state*
B. Ethno-medico-entomology	Lac insect	Entire	Boiled in mustard oil, mixed in sheep fat and rubbed extensively	Filariasis	Chik-Baraik (Ranchi, Kharia (Singhbhum), Ho (Hazaribagh) in Bihar and Konda Kapus (East Godawari), Konda Redis and Koya (West Godawari) in Andhra Pradesh
		Extract	Boiled in mustard oil, rubbed externally twice or thrice on chest	Chest pain	Munda (Ranchi, Singhbhum) in Jharkhand and Sugalis (Warangal) in Andhra Pradesh
		Entire	Boiled in cow fat and rubbed on ribs or/and muscles once daily for 5-7 days	Rib pain and internal injuries	Chero (Palamau), Chik-Baraik (Ranchi) in Jharkhand and Valmiki (Vishakhapatnam) in Andhra Pradesh
		Entire	Boiled in sheep milk and taken orally for 3-5 days	cramps, muscles and muscular pain	Kurmali, Majhi and mahli (Ranchi) in Jharkhand and Yenadis (Krishna) in Andhra Pradesh

substance which is unsoluble in alcahol. These are uni-voltine and generally treated as parasites of economically important plants such as tea and sandal. However, recently they have been found to be very potential bio-control agents for weeds (Campbell *et al.*, 1994) and need to be nurtured as such. Lac insects under *Kerria* genus are generally bi-voltine except *K. lacca mysorensis* found on *jalari (Shorea talura)* and *K. sharda* on *kusum (Schleichera oleosa)* (Mishra and Sushil, 2000) which are tri-voltine. Indian lac insect, *Kerria lacca* (Kerr), the most important and widely exploited insect for lac cultivation can further be distinguished into two strains or infra sub-species forms, the *rangeeni* and *kusmi* on the basis of difference in life cycle, host preference and quality of lac produced. *Rangeeni* strain is characterised by unequal duration of bi-voltine life cycle and non preference of *kusum Schleichera oleosa* as a host whereas, *kusmi* strain by more or less equidurational life cycle preferring *kusum* as a host.

Strain	*Crop*	*Duration*
Rangeeni	*Katki* - rainy season	Oct. - Nov. to Jun. - Jul.
	Baishakhi - summer season	Jun. - Jul. to Oct. - Nov.
Kusmi	*Aghani* - rainy season	Jun. - Jul. to Jan. - Feb.
	Jethwi - summer season	Jan. - Feb. to Jun. - Jul.

Lac insects thrive only on certain types of plant-hosts called lac-hosts. More than 400 plants have been recorded as lac-hosts so far (Sharma and Ramani, unpublished). Of these, *Palas (Butea monosperma)*, *Kusum Schleichera oleosa)* and *ber (Ziziphus mauritiana)* are the major hosts of national importance, besides there are some 20-30 minor hosts of regional importance. Lac insects settle closely on tender twigs, suck the sap of the plant and grow to form encrustation.

Lac cultivation is simple, does not need any large investment and requires only part-time attention. Sustained production and steady returns can be achieved by improved methods of cultivation. Thus, lac cultivation can be an attractive avocation (Choudhary *et al.*, 1999; Mishra *et al.*, 1999, 2000a, 2000b and Kumar *et al.*, 1996) as illustrated in table 2.

Important lac growing regions (States) of the country and various lac-hosts exploited for lac cultivation

Area	*State*	*Important lac-hosts*
Major lac growing region	-Jharkhand	• *Butea monosperma, Croton oblongifolius, Ficus spp., Protium serratum, Schleichera oleosa, Ziziphus mauritiana*
	-Chhattisgarh	• *Acacia catechu, Z. xylopyra, Butea monosperma, Schleichera oleosa, Ziziphus mauritiana*
	-West Bengal	• *Butea monosperma, Ficus spp., Samania (=Albizia) saman, Ziziphus mauritiana*
Minor lac growing region	-Orissa	• *Schleichera oleosa, Ziziphus mauritiana*
	-Maharashtra	• *Acacia catechu, A. nilotica, Butea monosperma, Ziziphus mauritiana*
	-Gujarat	• *Acacia catechu, Butea monosperma, Ziziphus mauritiana*
	-Uttar Pradesh	• *Acacia nilotica, Z. xylopyra, Ziziphus mauritiana*
	-Assam	• *Albizia lucida, Cajanus cajan, Grewia spp., Ficus spp., Flemingia macropyylla, Kydia calycina, Leea crispa*
Negligible contribution now	-Rajasthan	• *Acacia spp., Ficus spp., Ziziphus mauritiana*
	-Punjab	• *nilotica, Ficus spp., Ziziphus mauritiana*
	-Andhra Pradesh	• *Butea monosperma, Samania (=Albizia) saman*
	-Karnataka	• *Acacia spp., Ficus spp., Shorea talura,*
	-Tamil Nadu	• *A. nilotica, Butea monosperma, Shorea talura, Ziziphus mauritiana*

LAC PRODUCTION POTENTIAL, MEDICINAL AND OTHER UTILITIES OF IMPORTANT LAC-HOSTS

Most of the lac-hosts are multi-purpose trees. Besides being used for lac culture, many of these are source of fodder, fruits, fuel, timber etc. Many parts of various hosts are sources of medicine in ayurvedic formulations and are extensively used by tribals for the cure of different ailments. Lac production potential, medicinal and other utilities of these have been widely studied (Agarwal, 1997; Ambasta *et al.*, 1986; Anonymous, 1948-76; Chopra *et al.*, 1956 & 1969; Glover, 1937; Ramanchandra *et al.*, 1986; Roonwal *et al.*, 1958; Saha *et al.*, 1998, Srivastava and Kumar, 1982; Warrier *et al.*, 1994-96). Nomenclature of some of these lac-hosts has been revised in accordance with Bennet, 1987.

Acacia auriculaeformis A. Cunn. (Mimosaceae)

Habit and Habitat: Medium sized tree attaining a height of 6-12 mt, gregarious in nature; rarely distributed in forests, found on hilly slopes

Lac Production Potential: used for rainy season crops and suitable for intensive lac cultivation on plantation basis, used in alternation with *kusum* or *palas*

Medicinal and other utilities: Bark contains tanin with high tan/nontan ratio. It is used as a substitute for washing hair, Bark extract employed as a protective coating for boats and fishing nets; Seeds contain a fatty oil. A useful sand binder

Acacia catechu Willd. (Mimosaceae)

Habit and Habitat: Moderate sized deciduous tree; widely distributed in the forest areas, particularly in drier regions.

Lac Production Potential: Both the strains of lac insect thrive well on this species during rainy season yielding 2-4 kg sticklac per tree. Comes of use when alternate hosts are available for growing summer crop of lac.

Medicinal and other utilities: Bark: Astringent, given as tonic, useful in melancholia, conjunctivitis and haemoptysis; Gum: Antimicrobial, useful in diarrhoea, dysentery and

Table 2. Economics (per annum) of lac production on important lac-hosts alone or in combination.

Lac host	*Trees/ha*	*Expenditure (Rs)*	*Income (Rs)*	*Profit (Rs)*	*Mandays generated*
• *Palas (Butea monosperma)* for Sticklac production	734	35325	55650	20325	110
• *Palas (Butea monosperma)* for Broodlac production	734	24911	48250	23339	146
• *Kusum (Schleichera oleosa)*	70	31944	97380	65436	200
• *Ber (Ziziphus mauritiana)*	494	48200	92090	43890	150
• *Akashmani (Acacia auriculaeformis) + Kusum (Schleichera oleosa)*	640 + 16	14014	44800	30786	90
• *Akashmani (Acacia auriculaeformis) + Palas (Butea monosperma)*	640 + 160	9834	24500	14666	90
• *Galwang (Albizzia lucida) + Bhalia (Flemingia macrophylla)* with intercropping of Turmeric *(Curcuma domestica)* alone or in combination with Colocacia *(Colocacia pachyrrhizus)*/Sweet-potato *(Ipomoea batatas)*	1200 + 3600	58210	91260	33050	

diabetes; The *katha:* Acrid, bitter, thermogenic, digestive, appetiser, aphrodisiac, vulnerary, anthelmintic, depurative and tonic. It is useful in vitiated conditions of *kapha* and *vata,* laryngopathy, flatulence, anorexia, ulcers, wounds, helminthiasis, leucoderma, leprosy, skin diseases, urorrhea, colporrhagia, erysipelas and odontopathy; Another product called *kheersal,* met with in some older trees is used medicinally for sore throat and cough.

Source of catechu, obtained from natured of 20-30 years old trees. *Catechu* is marketed in the form of Katha, used in pan preparation and as cutch used for printing and dyeing purposes. *Wood* used for posts, rice-pestles, oil and sugarcane crushers, bows spear handles, ploughs, boats, furniture, combs, stocks and felloes of wheels, carts, bedpost-legs, sword handle and tool handles, tent pegs and rolling pins; excellent for spokes and hubs of wheels. Gum, yielded by the tree, called Kharigum is a seed substitute of gum Arabic.

A. Nilotica *(Linn.) Delile (syn.* A. arabica *Willd.) (Mimosaceae)*

Habit and Habitat: Trees of moderate size with spreading crown and feathery foliage; Thinly distributed on waste and undulated land in deep forests and top of hills. It prefers riverain alluvium subject to inundations and black cotton soil.

Lac Production Potential: *Rangeeni* strain thrives well on this species. Is a host of regional importance especially in Gujarat, Rajasthan and Punjab

Medicinal and other utilities: *Bark:* Astringent, demulcent, cooling, styptic, emolient, vulnerary, constipating, depurative, aphrodisiac, diuretic, expectorant, alexeteric, emetic and nutritive; its powder used as tooth powder, extract in asthma and bronchitis; antimicrobial and used in hair wash; decoction of bark used as a gargle. It is useful in vitiated conditions of *kapha* and *pitta*, haemorrhages, wounds, ulcers, helminthiasis, ascites, chronic dysentery, diarrhoea, leprosy, leucoderma, skin diseases, burning sensation, strangury, cough, bronchitis, leucorrhoea, haemorrhoids, proctoptosis, seminal weakness, pharyngodynia, utervesical disorders, oral ulcers and odontopathy; *Gum:* Sweet, astringent, cooling, emollient, expectorant, constipating, liver tonic, aphrodisiac, haemostatic, antipyretic and tonic. It is useful in vitiated conditions of *vata* and *pitta*, cough, asthma, diarrhoea, dysentery, seminal weakness, haemorrhages, leprosy, pharyngodynia, pneumonosis, haemorrhoids, urinogenital discharges, burns, colic, intermittent fevers, general debility and diabetes mellitus; *Leaves:* Antiseptic; *Flower:* Used by tribal women for deranged periods; *Pods:* Decoction used in urino-genital diseases and in impotency.

***Albizzia lucida* benth. (Mimosaceae)**

Habit and Habitat: A medium size deciduous tree, have high coppicing capacity and may be trained in to bushes; sub-Himalayan tracts and North Eastern region of India.

Lac Production Potential: Used for summer season *kusmi* lac crop, suitable for intensive lac cultivation on mixed plantation basis with *bhalia* (*Flemingia macrophylla*).

Medicinal and other utilities: *Wood* used for posts, rafts and scantlings.

Butea monosperma (Lam.) Taub. ***(Syn. B. frondosa*** Koenig ex Roxb., ***Erythrina monosperma)*** (Papillionaceae/Fabaceae)

Habit and Habitat: A medium sized deciduous tree, with a small somewhat crooked trunk, 10-15 ft in height and 5-6 ft in girth. Distributed through the plains of india from sea-level to the hills upto 4000 ft. Absent in most arid regions, typical of grassland; frequently gregarious, as in the northern India.

Lac Production Potential: Lac-host of all India importance. Preferred by *rangeeni* strain of lac insect Productivity ranges from 1-1.5 kg of sticklac per tree per year and on plantation basis 4-5 q/ha/yr.

Medicinal and other utilities: Seeds: Anthelmintic, purgative, ophthalmic, depurative, tonic, sedative and when pounded with lemon juice they act as powerful rubefacient and for Dhobie's itch; useful in arthritis, flatulence, constipation and diabetes; *Bark:* Astringent, emollient, appetiser, digestive, tonic, anthelmintic, used for piles, tumors, cold, cough, fever and menstrual disorders. It is useful in vitiated conditions of *pitta* and *kapha*, anorexia, dyspepsia, diarrhoea, dysentery, haemorrhoids, bone fractures, gonorrhoea, hepatopathy, ulcers, hydrocele and diabetes; Gum: Astringent, in diarrhoea, dysentery, haemoptysis, haematemesis, diabetes, leprosy, skin diseases, ulcer, pharyngodynia, general debility, hyperacidity, dyspepsia, fever and haemorrhoids, used in tonics, poultices and septic throat; Leaves: Astringent, anti-inflammatory, anodyne, flatulence, worm infestations, tonic, leaf juice given with curds and yellow zedoary in heat eruptions of children; useful in treatment of pimples, colitis, tumours and piles; Flowers: Astringent, cooling, diuretic, depurative, aphrodisial and antiseptic, good for skin troubles, very effective for feet sores. They are useful in vitiated conditions of *pitta* and *kapha*, diarrhoea, haemorrhoids, menorrhagia, strangury, fever, leprosy, skin diseases, swellings, hyperdipsia, haemoptysis, arthritis, burning sensation, bone fractures and are very efficacious in birth control; Shoots: Decoction considered emollient and sedative; used in piles. The ash of tender branches is useful in abdominal disorders such as flatulence, colic etc.

Leaves much used throughout the country for making plattes, cups etc., dried leaves used as *beedi* wrappers, young *shoots* yield a fibre used for ropes. Root bark yields a coarse fibre used for cordage and caulking boats. Trees yield a gum called butea gum or *Bengal kino*. Flowers yield a brilliant and fugtive yellow colouring matter. Seeds yield a fatty oil (18%). Wood used mainly for well curbs and water scoops; also employed on a cheap board wood and for structural work. Wood pulp is suitable for newsprint manufacture.

Cajanus cajan (Linn.) Millsp. (Syn. _C. indicus_ Spreng.) (Papillionaceae/Fabaceae)

Habit and Habitat: An annual or perrennial shrub, 4-10 ft high. Cultivated nearly throughout India as a pulse crop. Luxurient vegetative growth under humid conditions.

Lac Production Potential: It is an important lac-host in North East India. Performs best under humid environment.

Medicinal and other utilities: Seeds and leaves: Made into a paste, warmed and applied over the mammae to promote lactation; Leaves: Astringent, sweet, diuretic, laxative, cooling, anti-inflammatory and anodyne, and are useful in oral ulcers, odontalgia, gingivitis, strangury and inflammations; Seeds: Astringent, acrid, sweet, cooling, anthelmintic, resolvent, pectoral, constipating, alexeteric, febrifuge and expectorant. They are useful in vitiated conditions of *pitta*, intestinal worms, oral ulcers, tumours, bronchitis, cough, vomiting, haemorrhoids, fever and cardiac diseases.

Extensively eaten as a dal, rich in protein, contains globulins, cajanin and concajanin. Widely cultivated as a pulse crop. Leaves used in Madagascar (Malagasy) for rearing silkworms. Green leaves and tops used as fodder, also as green manure. Green pods used as a vegetable. Husk a useful fodder.

***Croton oblongifolius* Roxb. (Euphorbiaceae)**

Habit and Habitat: A small or medium-sized deciduous tree, often gregarious; leaves turning brilliant red before falling. Found in sub-Himalayan tracks from Uttar Pradesh eastwards.

Lac Production Potential: Takes both the strains of lac insects particularly during rainy season.

Medicinal and other utilities: Bark, root, fruits and seeds: Purgative in snake bite; Bark and root: Alterative; Bark: used in extenal applications for sprains, useful in liver diseases; Seeds and their oil are considered purgative also used for stupefying fish and for insecticidal purposes. Fatty oil from the seeds occasionally employed as rubefacient;

Wood used for turnery and for cement banels, rubber boxes and inside fittings of opium chests. Bark yields a cordage fibre; leaves chopped for fodder.

***Ficus lucescens* Blume (Syn. *F. Infectoria* Roxb., *F. Lacor* Buch.-Ham.) (Moraceae)**

Habit and Habitat: A large spreading, deciduous, quick growing usually epiphytic tree without aerial roots. Found in Outer Himalayas, Bengal, Orissa and western peninsula; usually in dry, deciduous forests.

Lac Production Potential: lac is grown on this species in Chhattisgarh, Andhra Pradesh and Bengal. Used as summer brood preserver for rangeeni strain of lac insect.

Medicinal and other utilities: Decoction of bark used as gargle and vulnerary. In chinese medicine, bark diaphoretic.

Succulent leaves are used as fodder, fruits edible, wood and bark suitable for paper pulp

Ficus beng(h)alensis **Linn. (Moraceae)**

Habit and Habitat: Large sized tree, nearly evergreen with spreading branches, Epiphytic in nature also. Thinly distributed, occurring in forest, personal land holdings and as avenue tree.

Lac Production Potential: Used as summer brood preserver for rangeeni strain of lac insect.

Medicinal and other utilities: All parts of the plant: Astringent, acrid, sweet, refrigerant, anodyne, vulnerary, depurative, anti-inflammatory, ophthalmic, styptic, antiarthritic, diaphoretic, antidiarrhoeal, antiemetic and tonic; Latex: Useful in neuralgia, rheumatism, lumbago, bruises, nasitis, ulorrhagia, ulitis, odontopathy, haemorrhoids, gonorrhoea, inflammations, cracks of the sole and skin diseases; administered with sugar for cough in children; Infusion of bark: Tonic, astringent, used in dysentery, diarrhoea, diabetes; Seeds: Cooling, tonic and laxative; Leaves: Tonic and cooling, applied as poultice for abscesses; since the leaves emit antimicrobial activity, eating on the plates of these leaves checks mouth sores and even cure them. Good for ulcers, leprosy, allergic conditions of the skin, burning sensation and abscesses. The buds are useful in diarrhoea and dysentery; extract of leaves in cholera; Root fibres: In gonorrhoea; Aerial roots: Young tips useful in obstinate vomiting and leucorrhoea and are said to be used in osteomalacia of the limbs; extract for promoting hair growth; Bark: Useful in burning sensation, haemoptysis, haemorrhages, diarrhoea, dysentery, diabetes, enuresis, ulcers, skin diseases, gonorrhoea, leucorrhoea and hyperdipsia; Fruits: Refrigerant and tonic and are useful in vitiated condition of *pitta*.

Fruits edible in times of scarcity. Wood suitable for paper pulp; also used for furniture, well-curbs, yokes, cart-shafts. Fibre from ark and aerial roots made into coarse ropes.

Ficus racemosa **Linn. (Syn. F. Glomerata Roxb.) (Moraceae)**

Habit and Habitat: Large sized tree, nearly evergreen with spreading branches, epiphytic in nature also. Thinly distributed, occurring in forest, personal land holdings and as avenue tree.

Lac Production Potential: Used as summer brood preserver for rangeeni strain of lac insect.

Medicinal and other utilities: Leaves: Antimicrobial; Bark: Astringent, given to cattle when suffering from rinder-pest, antidiabetic, refrigerant and useful as a wash for wounds, highly efficacious in threatened abortions and also recommended in uropathy; Root: In dysentery; Sap of root: In diabetes; Leaves: Powdered leaves mixed with honey is given in vitiated conditions of *pitta* . A decoction of the leaves is a good wash for wounds and ulcers; Tender fruits: Used in vitiated conditions of *pitta*, diarrhoea, dyspepsia and haemorrhages; Ripe fruits: Astringent, stomachic, refrigerant and carminative and are useful in menorrhagia and haemoptysis; Latex: is aphrodisiac, and is administered in haemorrhoids, diarrhoea and piles.

Ficus religiosa **Linn.**

Habit and Habitat: Large sized tree, nearly evergreen with spreading branches. Epiphytic when young. Moderately distributed, occurring in forest, personal land holdings and as avenue tree. Holds a religious status throughout the country.

Lac Production Potential: Brood preserver tree for summer crop of rangeeni strain. Owing to religious status systematic cultivation of lac is not practiced. However, collection of lac from natural infestation is reported.

Medicinal and other utilities: Bark: Astringent, sweet, used in gonorrhoea, infusion used in ulcers and skin troubles, cooling and aphrodisiac and an aqueous extract of it has an antibacterial activity against *Staphylococcus aureus* and

Escherichia coli. It is used in the treatment of gonorrhoea, diarrhoea, dysentery, haemorrhoids and gastrohelcosis. A paste of the powdered bark is a good absorbant for inflammatory swellings and good for burns; Fruit: Laxative and digestive; the dried fruit pulverised and taken in water cures; Seeds: Cooling, alterative and laxative; in menses for regulation; Leaves and young shoots: Purgative and are also recommended for wounds and skin diseases; Infusion of bark: Given internally in scabies; Latex: Good for neuralgia, inflammations and haemorrhages

Fruits and tender buds eaten in times of scarcity. The leaves are lopped for fodder. Hardened latex used for filling up cavities in hollow ornaments, also source of bird lime. Wood used for packing-cases and occasionally for felloes of wheels, spoons and bowls; also suitable for match boxes. Bark contains tannin. It yields a fibre, formerly used for paper. The tree is a good climate purifier releasing good quantity of oxygen during both day and night.

***Ficus semicordata* Buch.-Ham (Syn. *F. cunia* Ham. ex Roxb., *F. conglomerata* Roxb.) (Moraceae)**

Habit and Habitat: A moderate sized usually evergreen tree, not epiphytic. Distributed throughout the greater parts of India.

Lac Production Potential: Lac is produced on this species in Assam, Bengal and Punjab. Used as summer brood preserver for rangeeni strain of lac insect.

Medicinal and other utilities: Fruit: Given in aphthous complaints; Fruit and bark: Made into a bath and used as a cure for leprosy; Juice of roots: Given in bladder complaints and boiled in milk, in visceral complaints.

***Flemingia macrophylla* (Willd.) O. Ktze ex prain (Syn. *Moghania macrophylla* Kuntze, *F. congesta* Roxb.) (Papillionaceae/Fabaceae)**

Habit and Habitat: Bushy shrub attaining a height of 1-3 mt, gregarious in nature. Distributed in forests and hilly slopes.

Lac Production Potential: Used for rainy season lac crops. Suitable for intensive lac cultivation on plantation basis

Medicinal and other utilities: Plant: Useful in spleen complaints, small pox, cholera, dysentery and blindness; Roots: Used as external application to ulcers and swellings, mainly of the neck; Pods: "Warrus" obtained from dried pods employed as a cosmetic, anthelmintic and for cough and colds.

***Grewia glabara* Blume (Syn. *G. Didgma* or *laevigata* auct. Non Vohl., *G. Disperma* Rottl. ex Spreng.) (Tiliaceae)**

Habit and Habitat: A small to medium sized tree, found in outer Himalayas from the Uttar Pradesh eastwards, Assam and South India. Occurring usually in shady places.

Lac Production Potential: Important lac-host in North East India. In Assam lac is grown commercially on this species on a fairly large scale.

Medicinal and other utilities:

Wood used for turnery and for cement barrels, rubber boxes and inside fittings of opium chests. Bark yields a cordage fibre; leaves lopped for fodder.

***Kydia calycina* Roxb. (Malvaceae)**

Habit and Habitat: A small to moderate sized tree with grey bark exfoliating in irregular flakes or long strips. Found in mixed deciduous forests, not in arid region.

Lac Production Potential: Important lac-host in North East India

Medicinal and other utilities: Leaves: Pounded and made into a paste applied in rheumatism and lumbago; also chewed for promotion of salivary secretions.

Wood used for planks, agricultural implements, match boxes, splints and light packing cases. Also suitable for veneers and plywood, brush backs, turnery, bobbins and shuttles, shoe heels and cheap grade pencils. Wood yields pulp for news print in admixture with bamboo pulp. Bark yields fibre used for coarse ropes. Leaves lopped for fodder.

Leea asiatica **(Linn.) Ridstale (Syn. *L. crispa* Linn.) (Vitaceae)**

Habit and Habitat: A stiff erect shrub, 4-8 ft. High, Stem jointed and swollen above, the joints furnished with cusped wings.

Lac Production Potential: Important lac-host in North East India.

Medicinal and other utilities: Tubers: Used as a remedy for guinea worms; Leaves: Antiseptic, paste applied to wounds.

Berries eaten.

Leea compactiflora **Kurz. (Syn. *L. robusta* Roxb., *L. diffusa* M. Laws) (Vitaceae).**

Habit and Habitat: A shrub 6-10 ft high. Found in North eastern states.

Lac Production Potential: Important lac-host in North East India

Medicinal and other utilities: Root: Extract applied externally as anodyne, internally to cattle for diarrhoea. Pounded and applied externally to allay pain.

Ougenia oojienensis (Rxb.) Hoechr. (Syn. O. dalbergiodes Benth.) (Papillionaceae/Fabaceae)

Habit and Habitat: A moderate sized deciduous tree. Found in sub-Himalayan tracts on slopes upto 5000 ft. All over North India, not in Bihar and Orissa.

Lac Production Potential: Lac grown in Chhattisgarh and Maharashtra. In Jharkhand it is a good host for preserving summer brood of *rangeeni* strain to maturity.

Medicinal and other utilities: Bark: Astringent, acrid, cooling, stimulant, anti-inflammatory, constipating, urinary astringent, anthelmintic, soporific, depurative, styptic, febrifuge and rejuvenating. It is useful in vitiated conditions of *pitta*, burning sensation, inflammations, diarrhoea, dysentery, urorrhagia, diabetes, verminosis, leprosy, leucoderma, haemorrhages, fevers, ulcers, gonorrhoea, anaemia and general debility; when incised furnishes a kino-like exudation which is used in dysentery and diarrhoea; decoction given when the urine is highly coloured; Gum: Used in diarrhoea and dysentery.

Wood used for carts and carriages, especially for shafts, axels and hubs, for agricultural implements, sugarcane crushers, tool handles, bed legs, tent poles and pegs and for cars and other articles where strength and toughness are required. Also used for posts, rafters, struts and door and window frames; although heavy used for furniture and cabinets. Shuttles, spindles, bobbins and pickerarm, discdowels and cooperage. Yields a bast fibre used for cordage. Bark used as febrifuge and fish poison; contains tannin (7%). Tree lopped for fodder.

Protium serratum Engl. (Syn. *Bursera serrata* Wall. Ex Colebr.) (Burseraceae)

Habit and Habitat: A moderate to large sized tree, evergreen or nearly so. Used for growing *kusmi* lac in Jharkhand, also in Assam.

Lac Production Potential: Tree coppices well and is a good alternate most for kusmi strain of lac insect, especially for growing aghani crop. Quality of lac produced is comparable to that produced on *S. oleosa*.

Medicinal and other utilities: Fruits edible.

***Samanea (= Albizia) saman* Merrill (Syn. *Pithecolobium (Pithecellobium) saman* Benth., *Enterolobium saman* Prain) (Mimosaceae)**

Habit and Habitat: A large deciduous tree with a short stout bole, often attaining a height of upto 20 mts and a girth of 8 ft and over. Distributed in sub-Himalayan tracts from Sutlej to Nepal, Bengal, Chhotanagpur, Madhya Pradesh and the Peninsula in general. Common on well drained boulder deposits, often occurring in quantity along sides of ravines or along high banks, marking edges of terraces, on sandstone or boulderbeds.

Lac Production Potential: Preferred by rangeeni strain of lac insect. Productivity ranges from 10-15 kg per tree and in plantation condition 5-7 q/ha/yr.

Medicinal & other utilities: Leaves and pods: used as fodder; tree yields a gum; wood suitable for furniture but seldom used.

***Schleichera oleosa* (Lour.) Oken. *(Syn. S. trijuga* Willd. & Klein) (Sapindaceae)**

Habit and Habitat: A medium sized to large deciduous or nearly evergreen tree. Upto 30 m in height. Found in sub-Himalayan tracts from Kashmir to West bengal and in Central and Penninsular India upto an altitude of 900 m.

Lac Production Potential: A very good host for kusmi strain of K. Lacca. Average productivity of a tree ranges from 20-50 kg of best quality sticklac and in plantation condition 5-7q/ha/yr.

Medicinal and other utilities: Bark: Astringent, rubbed up with oil used as a cure for itch; pain in the back and loins, inflammation and ulcers, antipyretic and is useful in pruritus, adenitis, ripening boils, nostalgia, arthralgia, malaria, inflammations and ulcers; Powdered seeds: Applied to ulcers of animals and for removing maggots; Oil of the seeds: Bitter, sour, sweet, stomachic, anthelmintic, purgative, trichogenous and tonic. It is useful in pruritus, burns, acne, dermatopathy, scald, trichopathy, ulcers, cephalalgia and vitiated conditions of *vata*. Used for the cure of itch and acne, efficient and stimulating agent for the scalp, both cleansing it and promoting growth of hair; Oil as a cure for skin troubles.

Wood used for oil and sugar mills, rice pounders, pestles, axles and hubs, fellows and stocks of cart wheels, shafts, mortars, violin bows and agrricultural implements and tool handles. Treated timber available for construction and cabinet work and wagon building; also used for road paving, block flooring in mills and godwns and as pit props, side-props in shafts and galleries in mines. It is a good fuel and makes excellent charcoal. Seeds yield a fatty oil called Macassar oil which is used for dressing; also used for culinary and lighting purposes, as a lubricant for machinery. It is used for massage in shemalism. For edible purposes the oil is properly refined to remove cyanogenetic glucosides. A common adultrant of mustard oil. When in suitable proportions (2-15%) with oils consisting longely of oleic (and linoleic) acids it considerably improves the lathering properties of laundry soaps, imparts a better feel, provides immunity from rancidity. Raw fruits pickled, ripe ones eaten as such. Young shoots eaten, also lopped for fodder. Bark contain 4% tannin. Flowers yield a dye.

***Zizyphus mauritiana* Lam. (Syn. *Z. Jujuba* Lam. Non Mill.) (Rhamnaceae)**

Habit and Habitat: Small to moderate sized deciduous tree, with short bole, spreading rounded crown and drooping branches armed with stipular spines. Spread throughout the greater part of India, thrives in comparatively dry region. Often gregarious in tree or bush form; on sandy or shingly alluvium and on arable land, and on a variety of soils, open waste lands and on abandoned cultivation in dry localities; often on sandy or gravelly silluvium of dry riverbeds, frequently associated with *Acacia catechu*.

Lac Production Potential: Preferred by both the strains of lac insect; productivity ranges from 2-7 kg of sticklac per tree and in plantation condition 4-5 quintals/ha/yr.

Medicinal and other utilities: Leaves: Bitter, cooling, astringent, anthelmintic, diaphoretic and antipyretic and are useful in stomatitis, ulorrhea, wounds, sypluilitic ulcers, asthma, typhoid, fever, diarrhoea and obesity. To form a plaster in strangury, eaten with catechu as an astringent; Fruits of wild trees: Cooling, anodyne and tonic, used in the preparation of *Joshanda,* a medicine used in chest troubles, kernals sedative, used as a soporific and to stop vomiting. Also employed as an antidote to aconite-poisoning and for abdominal pains in pregnancy, considered to purify blood and aid digestion; purgative, mucilaginous, pectoral, styptic, aphrodisiac, invigorating, depurative, appetiser and tonic. They are useful in vitiated conditions of pitta, burning sensation, hyperdipsia, consumption, vomiting, constipation, flatulence dyspepsia, nausea, leprosy, skin diseases, pruritus, wounds and ulcers, haemorrhages and general debility; Seeds are acrid, sweet, astringent, soporific and tonic. They are useful in encephalopathy, ophthalmopathy, dipsia, cough, asthma, wounds, vitiated conditions of *pitta*, burning sensation, diarrhoea, vomiting, leucorrhoea, general debility and insomnia and in diarrhoea; Root: Bitter, cooling, anodyne and tonic and are useful in vitiated condition of *pitta*, fever, wounds and ulcers and cephalalgia. Used in decoction in fever and as a powder applied to old wounds and ulcers; Bark: Astringent, constipating and tonic and is useful in dysentery, diarrhoea, gingivitis and boils. The plant is mentioned in old literature as contraceptive.

Fruits eaten fresh, dried like raisins, candied, stewed or smoked. Tree also yields fodder and is grown for wind breaks. Large sized fruits, which have just begun to turn yellow are chosen for canlying. Badari is mentioned in old literature in the list of contraceptives; seeds yield a fatty oil. Leaves eaten with catechu as an astrigent; considered diaphoretic. Wood used for wells, axe and hoe-handles, gum stocks, sandals, yokes, toys, parts of wheels and turnery.

CONCLUSIONS

Lac-culture in India is not just a industry but a tradition. The rich production potential within the country and a steady demand for the commodity from abroad has promoted commercialisation of this activity. Lac today is an indispensible and growing part of tribal culture and could be a medium for transformation of rural economy. The industry has rich natural resources in host-plants and manpower. As lac-culture requires only biological components, there are no environmental hazards, rather it helps in conservation of forests. Promotion of lac-culture, especially in minor lac growing regions of the country would not only help in reviving the rural economy but also in development of the ecological environment.

REFERENCES

Agarwal, S.C., Kumar, P. and Goswami, D.N. (1998) *Safari of ILRI through seven decades*. 84 pp. Indian Lac Research Institute, Ranchi.

Agarwal, V.S. (1997) *Drug Plants of India* Vols 1&2. Kalyani Publishers, Ludhiana.

Ambasta, S.P., Ramachandran, K., Kashyapa, K., and Chand, R. (1986) *The Useful Plants of India*. Publications and Information Directorate, New Delhi. p. 918.

Anonymous. (1948-76) The Wealth of India, Raw Materials. Vol. I-XI.

Bennet, S.S.R. 1987. *Name Changes in Flowering Plants of India and Adjoining Regions*. Triseas Publications, Dehradun. p. 772.

Campbell, M.H., Holtcamp, R.H., McCormick, L.H., Wykes, P.J., Donaldson, J.F., Gullan, P.J. and Gillespie, P.S. (1994) Biological control of native shrubs, *Cassinia* spp. using native scale insects, *Austrotachardia* and *Paratachardina* sp. (Hemiptera: Kerriidae) in New South Wales. *Plant Protection Quarterly*, 9: 64-68.

Chaudhary, S.G., Naqvi, A.H., Sen, A.K., Mishra, Y.D. and Kumar, P. (1999) Package of practices for lac cultivation on *Akashmani (Acacia auriculaeformis)*. *Tech. Bull. No.* 2 Indian Lac Research Institute, Ranchi. p. 8.

Chopra, R.N., Chopra, I.C. and Varma B.S. (1969) *Supplement to Glossary of Indian Medicinal Plants*. Publications and Information Directorate, New Delhi. p. 120.

Chopra, R.N., Nayar, S.L. and Chopra, I.C. (1956) *Glossary of Indian Medicinal Plants*. Council of Scientific and Industrial Research, New Delhi. p. 330.

Glover, P.M. (1937) *Lac Cultivation in India*. Indian Lac Research Institute, Ranchi. p. 148.

Kerr, J. (1782) Natural history of lac insect which produces gumm lacca. *Phil. R. Soc. London*, 71 [1781] (2):374-382.

Kumar, P., Singh, B.P. and Srivastava, S.C. (1996). *Package of practices for kusmi lac cultivation on mixed plantation of bhalia and galwang*. Indian Lac Research Institute, Ranchi. p. 8.

Mishra, Y.D., Bhattacharya, A., Chaudhary, S.G. and Kumar, P. (1999) Advanced broodlac production technology on *kusum*. *Tech. Bull. No.* 1. Indian Lac Research Institute, Ranchi. p. 8.

Mishra, Y.D., Chaudhary, S.G., Bhattacharya. A., Bhagat, M.L. and Kumar, P. (2000a) Advanced technologies for lac production on *palas* in hot areas. *Tech. Bull. No.* 3. Indian Lac Research Institute, Ranchi. p. 8.

Mishra, Y.D., Chaudhary, S.G., Kumar, P., and Sushil, S.N. (2000b) Package of practices for *rangeeni* lac production on *ber*. *Tech. Bull. No*. 4. Indian Lac Research Institute, Ranchi. p. 4.

Mishra, Y.D., Kumar, P. and Sushil, S.N. (2000c) *Lac production in India - a perspective*. p. 22. Indian Lac Research Institute, Ranchi.

Mishra, Y.D. and Sushil, S.N. (2000) A new trivoltine species of genus *Kerria* Targioni-Tozzeti (Homoptera: Tachardiidae) thriving on *Schleichera oleosa* (lour.) Oken from eastern India. *Oriental Insects*, 34: 215-220.

Ramani, R. and Siddiqui, S.A. (2000) Medicinal uses of lac. *ILRI Newsletter*, 4(4):4.

Ramachandra, K., Kashyapa, K. and Chand, R. (Eds.) (1986) *Useful Plants of India*. Publications and Information Directorate, CSIR, New Delhi. p. 918.

Roonwal, M.L., Raizada, M.B., Chatterji, R.N. and Singh, Balwant. (1958) *Descriptive account of the host-plants of the lac insect* Laccifer lacca *(Kerr), and allied plants in the Indian region Parts 1 & 2*. Indian Lac Cess Committee, Ranchi, India p. 140.

Saha, S.K., Goswami, D.N., Srivastava, B.C., Prasad, N. and Majee, R.N. (1998) Industrial potential of lac and some natural products of Chhotanagpur. In *Proceedings of the Symposium Bioresources of Chotanagpur and Their Industrial Significance*. R. Ramani and S.C. Agarwal (eds.). Indian Lac Research Institute, Ranchi. pp. 92-96.

Sharma, V.P. (1993) Ethnomedicoentomological studies on the aborigins of Bihar and Andhra Pradesh, India. *Ann. Entomol*., 11(1): 55-58.

Sharma, K.K. and Ramani, R. (1999) An update on synoptic catalogue of lac-inscts (Homoptera: tachardiidae). *J. Bombay Nat. Hist. Soc*., 96(3):438-443.

Sharma, K.K. and Ramani, R. *Updated list and notes on the hosts of lac insects*. Under preparation.

Srivastava, B.C. and Kumar, P. (1982) Medicinal potentials of some Indian lac host plants. *The Eastern Pharmacist*. Sep. 1982: 47-51.

Varshney, R.K. (1976) Taxonomic studies of lac insects of India. *Oriental Insects*, Supplement No. 5. pp. 1-97.

Varshney, R.K. (1986). A report on the valid name of host plants of Indian lac insects (Homoptera: Tachardiidae) listed by Roonwal and Singh (1958). *Indian J. Forestry* 9(2): 116-122.

Varshney, R.K. (1992) A checklist of the scale insects and mealy bugs of South Asia Part I. *Records of the Zoological Survey of India*, Occasional paper No. 139. pp. 1-152.

Warrier, P.K., Nambiar, V.P.K. and Ramankutty, C. (Eds.) (1994-96) *Indian Medicinal Plants* Vols. 1-5. Orient Longman Ltd., Madras.

4

Pollination Potentiality of Honeybees in Crop Production

D. Rajagopal and *N. Nagaraja*

INTRODUCTION

Angiosperms are credited with the most beautiful gift of nature, the flowers which have greater significance in the ecosystem and human society. In plants, pollination is an important event as a prerequisite of sexual reproduction. The pollen grains mature inside the anther and the anther ruptures and the ripped pollen grains are discharged through dehiscence. The female sexual organ is the pistil, with an ovary at the base and a style arises from it and terminates as stigma. The receptive stigma has a sticky surface on which pollen grains adhere and the male nuclei subsequently move through a pollen tube to the ovary where the fertilization takes place. The process of transfer of pollen grains from the anther of a flower on to a receptive stigma of the same or another flower, belonging either to the same or a different plant of the similar kind is called 'Pollination'. If the pollen automatically gets deposited on the stigma of the same flower, it is called automotive "Self pollination" or "Autogamy". However, other plant needs a pollinating agent to transfer pollen from the anther to the stigma of the same or another plant of the same species is called "Cross-pollination" or "Allogamy". In grasses and trees, light and dry pollen grains are produced and are generally wind pollinated. However, few species of plants are pollinated by birds, animals, water gravity, vibrations etc.

Several countries all over the world are making desperate efforts to achieve self-sufficiency in food production. This was done by geographical expansion of the area under cultivation and better management of natural resources. These include use of better quality seeds, bringing possible wastelands under cultivation, use of manures and fertilizers and expansion of irrigation area. However in past, food production has come to a point of stagnation in some countries and it is no longer possible to enhance the productivity levels in certain cultivated crops without insect pollinators. Therefore, steps should be taken for full utilization of natural resources. One such resource is an increase in the yield of different cultivated crops through pollination by honeybees.

The vital role of honeybees in the pollination of a large number of agricultural and horticultural crops is often underestimated. The main role of honeybees is pollination and hive products like honey and bee wax are of secondary value. The income by agriculture by the use of honeybees in crop pollination is at least 20 times many greater than the value of honey and wax production. Many plants do not yield seeds or fruits without cross-pollination by honeybees and other wild insects. Cross pollination of various agricultural, horticultural and forest crops by honeybees is one of the most effective and cheapest methods of increasing the yield. Other agronomic practices like manuring, fertilizers, and irrigation are quite cost effective. The self-sterile and self-fertile varieties, which require cross-pollination, would also produce more and better quality seeds and fruits, which are pollinated by honeybees and other insects. Legumes are capable of fixing nitrogen, thereby enriching the soil. Without legumes, soil in many areas would be quickly depleted and become unproductive. Generally, inadequately pollinated plants have an uneven fruit or seed set. On the other hand, adequate pollination assures a good quality and quantity of crops.

The National Commission on Agriculture drew up a highly ambitious plan to develop beekeeping across the country. According to the commission every village with cultivated crops should have on an average of 10 bee hives. The underlying principle evidently was that pollination should be sited proximate to the target crops, practically depending with the need to move beehives to the cropped areas. The salient features of the strategy spelt out by the commission involved increasing the number of bee hives, of which there were only 0.5 million at that time and would be 6 million in 24 years (1976-2000); establishing a network of 125 queen bee multiplication stations serving 5,000 villages; ensuring to raise 2,000 queens in a year to meet the target of 0.25 million hives every year; establishing 3,000 honey houses for testing, processing, packing and storing the huge quantities of honey estimated at 60,000 tones per year and marketing the honey products through 30,000 markets by the year 2000. In other words, it was a comprehensive and complete master plan for substantially raising nation-wide agricultural productivity.

HISTORY OF POLLINATION

Though honeybees are known since ancient times, their utility in pollination is more or less a recent knowledge. There is evidence from 1500 BC, which was interpreted as some are widely staminate (male) in date palm flowers over the pistillate (female) one. The idea of the occurrence of sex in plants was given by Theophastus in the fourth century B.C. Aristotle (384-322 BC) has given the idea of a relationship between bees and flowers; but it was described later the detailed structure and the role of insects in pollination. During 682 AD, an Arabic writer quoted that the date palm (*Phoenix dactylifera*) had to be artificially fertilized. It was also known that Chinese and Japanese gardeners developed asters, and many other flowers through cross-pollination. In 1682, Grew, a botanist, stated that pollen must reach the stigma before seeds develop and it was realized only in 1760s that bees and other insects could transfer the pollen to the stigma. Sprengel (1793) put forward a theory stating that the plant anatomy and physiology is related to the structure and behaviour of the insects, which pollinate the flowers. In 1799, Thomas Andrew Knight showed the value of cross-pollination and its association with hybrid vigour. He postulated that nature adopted the pollen to adhere to insect bodies. During the end of nineteenth century reproduction in plants and the mechanism of pollination was discovered by Muller (1882-83). Darwin (1887) proved and popularized the importance of pollination in the maintenance of vigour, weight and height of plants and used both hand and insect pollination techniques and helped greatly in understanding the theories of plant perception and maintenance of vigour through cross-pollination.

The practical use of honeybees for the cross-pollination of cultivated crops and increasing their yield with the practice of moving honeybee colonies to the crops during bloom stage. Waite (1895) found that the causes of crop failure in an orchard of Barlett pear trees in Virginia, USA was due to self sterility and reported that cross-pollination was necessary for a good fruit set, because pear pollen was too heavy to be transferred by wind and recommended only honeybees for pear pollination. Benton (1896) also suggested five to six colonies of *Apis mellifera* per 100 trees for the pollination of self-sterile commercial varieties of apple.

The classical studies on the foraging behaviour of a bee from different flowers were made by Casteel (1912) and Hodges (1952). Dolgova (1928) assessed the daily cycle of richness of nectar supply in local herbaceous plants by observing the number of honeybee visits at different times of the day under standard conditions. Although Free (1970) considered relative effectiveness of pollinators on the basis of abundance, foraging rate and foraging methods, other factors such as body size, loose pollen on body and number of visits required for maximum pollination are equally important to determine the efficiency of pollination (Adlerz, 1966; Badan and Anderson, 1962; Drake, 1948). Free (1960) observed the nectar gathering honeybee on the anthers of apple, where it pushed its tongue and the front part of the body towards the nectaries, thus touched the stigma and stamens for pollination. Bohart and Nye (1960) observed the similar behaviour of bees on carrot bloom.

Kapil *et al.* (1971) observed that nectar collectors viz., *Apis dorsata* and *Apis florea* on rapeseed and mustard had a tendency to approach the flowers below the petals, through the gap between them and thus avoided contact with genital column of the flower. Mishra and Kaushik (1992) have attempted to compare the pollination efficiency of bee species on the basis of actual field observations on their abundance, the foraging behaviour siliqua and seed set in *Brassica compestris* which were allowed by the visit of *Apis* sp. using all these parameters. Dhaliwal and Sharma (1973) determined the foraging trend of *Apis cerana* on different days after being in new location. Mishra *et al.* (1976) observed the total colonies required per unit area from the appropriate distance at which fruit/seed set is significant. The behaviour of bees on cotton crop is such that they seldom touch the stigmas while leaving the flowers so it favours cross-pollination (Tanda and Goyal, 1979).

POLLINATING AGENTS

The plants either wild or cultivated, transportation of pollen from one flower to another is essential. This is done by wind, gravity, vibrations, rain, birds, small mammals and insects. The wind carries small and light pollen grains from plant species like pines, maize and graminaceous plants, etc. There are about one quarter million species of flowering plants on earth with amazing complex relationships to bees and other less important pollinators including beetles, moths, butterflies, birds and bats (Dowden, 1964). Cross-pollination provides greater genetic variability in the offspring than that by self-pollination. It gives the plants a greater opportunity to produce new varieties to adapt new environments and occupy different ecological areas. The value of insect pollination upon which man can exert much influence is not limited to cultivated crops. The most drastic effect of the absence of pollinating insects would be in the uncultivated area whereas the soil enriching plants would die out and insect pollination is still extremely important in grasslands, temperate forests and deserts. Many species of insects visit the flowers of commercial crops to seek nectar or pollen or both (Mc Gregor, 1976). However, relatively, few are consistently good pollinators. The most efficient pollinators carry plenty of pollen grains on their bodies, brush on the stigmas of flowers, visit several flowers of the same species and bees whose bodies are covered with branched hairs, are the most important. Various hairy flies also carry more pollen as honeybees but they do not forage as consistently as bees. Both pollen and nectar provides protein and carbohydrates to bees and the pollination of a crop is sometimes dependent on one or few local insect species. For example, thrips are the main pollinators of oil palm (*Elaeis guineensis*) in Malaysia. Many solitary bees show preferences for particular plant species and are well adapted to pollinate them. In Jamaica, a small solitary bee, *Exomolopcis pulchella* appears to be the sole pollinator of hot sweet pepper and egg plant (Free, 1975). A related bee, *Exomolopsis* spp. pollinates solanaceous vegetable crops including sweet pepper (Anais and Torregrossa, 1978). In Jamaica, skipper butterflies are the sole pollinators of papaya. They appear in considerable numbers at dusk, forage on both hermophrodite and male flowers. Flies are of considerable importance in pollinating Umbelliferae and some of the Cruciferae. They are considered to be highly important in pollinating mangoes and are connected with the pollination of apples, avocados, berries, cherries, pears and numerous other crops. Lepidoptera mainly visit actinomorphic flowers, collected only nectar and adults satisfy their own needs.

Honeybees are the most efficient pollinators of several cultivated and wild plants because of their following characteristics: (i) body parts are specially modified to pick up many pollen grains, (ii) flower fidelity and constancy, (iii) potential for long foraging hours, (iv) maintainability of high populations when and where needed, (v) adaptability to different climates and niches. As a result of cross-pollination by bees, reproductive and adaptive heterosis or hybrid effects occur in plant progeny. Such hybrid effects bring the following qualitative and quantitative changes in the economic and biological characters of plants viz., stimulates germination of pollen on stigma of flowers and improves selectivity in fertilization; increases viability of seeds, embryos and plants; more nutritious and aromatic fruits are formed; increases the vegetative mass and stimulates faster growth of plants; increases number and size of seeds and yield of crops; enhances resistance to diseases and other adverse environmental conditions; increases nectar production in the nectaries of plants; increases fruit set and reduces fruit drop and increases oil content in oil seed crops.

The effectiveness of pollinating agents depends on: (a) Frequency and kind of stigmatic contact: This can be a major factor in pollination of large flowers or flowers with nectaries on other parts. Some flowers need stigmas to be contacted for a bee to transfer enough pollen to form a full sized fruit. A few non *Apis* pollinators such as *Xylocopa* and wasps make a hole or a slit at the base of the flowers of red clover, field beans and blue berries. On small and open flowers such as Umbelliferae, Asteraceae and many of the less specialized flower families, the stigmatic contact is achieved early by most visitors. (b) Amount and availability of pollen: On large and relatively unspecialized flowers, bumble bees, honeybees and solitary bees have average 20-90,000 (may 0.5 million) pollen grains on their body, compared to only a few on thrips. On kiwifruit(*Actinidia chinensis*), the pollen carrying ability of Hylaens, with 100-500 pollen grains each, was similar to that of smaller house flies and syrphids.

POLLINATION REQUIREMENTS OF CROPS

Bees visit flowers to collect nectar and pollen. They are attracted to flowers and recognize them by their colour, shape and odour (Gould and Gould, 1988). They are able to distinguish only four types of colours such as yellow, blue-green, blue and ultraviolet. However, when flowers have more than one colour, bees readily change among them and may ignore colour as a distinguishing feature. Though bees are able to learn the general shape of flowers and plants, their visual tendency is small. However, bees have highly developed sense of smell. Once bees have become conditioned to a flower species they will even continue to visit the same flowers. Honeybees communicate the direction, distance, odour and source of food to other bees by various kinds of dances (Frisch, 1971). This communication system of honeybees allows them to get efficiently distributed on nectar and pollen plants within their flights range. Bees have highly developed sense of smell and can readily be trained to associate forage with a particular scent or mixture of scents (Ribbands, 1955).

A more comprehensive understanding of the pollination requirements of individual crops is needed, including the number of bee visits to individual flowers, the number and placement of pollen grains required for optimum effect on yield, stigma receptivity, pollen tube growth, effects of weather and a complex of factors relating to the dynamics of bee activity on crops. More irrigation of cucumbers stopped bee visits and reduced successful pollen tube growth. If pollination is critical, overhead irrigation should be done at night or otherwise time to avoid conflict with pollination. Placement of honeybee colonies in or near the crop can be an important aspect of efficient pollination and requirements vary for different crops. Reformation on the foraging areas of individual bees is useful to indicate best variety interplanting arrangements in fruit orchards, row arrangement and spacing for crop hybridization by bees and isolation distance required to prevent contamination of bee pollinated crops. Individual honeybees are generally known to restrict foraging to a smaller area although transfer of pollen from one stigma to another may occur. Also, the foraging area of individual honeybees increase greatly with succeeding visits to the same crop and bees are capable of transporting pollen on their bodies from one foraging trip to the other.

Nectar and its Collection

The nectaries may occur in many parts of the flower including receptacles, petals, sepals, bases of filaments and pistil. Nectar secretion is influenced by the maturation of the stigma, stamens and also often by the age of the flower. The nectar secretion is greater during summer than other periods, reflecting the fact that the nectar sugars are products of photosynthesis, which inturn are influenced by sunlight (Shuel, 1955). Soil moisture, atmospheric pressure, size of nectaries and position of the flower on the plant may also influence the amount of nectar secretion. Nectar contains mainly sugars, but small amounts of other substances contribute to its aroma and the characteristics of the honey. These include organic acids, volatile oils, polysaccharides, proteins, enzymes and alkaloids. The three main sugars of nectar are sucrose, fructose and glucose. The mouth parts of the bee come together to form a tube through which nectar or other sweet liquid is sucked. In the anterior part of the bees abdomen, the alimentary canal is enlarged into a crop or honey stomach in which the nectar is temporarily stored. On returns to its hive the forager passes the nectar through bees and stores in the comb. There may be great differences in the average sugar concentrations of nectar of different species. Although different species and varieties may have nectar with different sugar concentrations, even within a single flower, the attractiveness of a species may differ at different times of the day and at different stages of flowering.

The composition of nectars from different plants is similar in that the main constituents are sugars and water. The total concentration of sugars varies widely from species to species. The sugar concentration of nectar varies widely in a variety of plant species. For example, it is only 6 per cent in Bombax sp., whereas 79 per cent in *Grevillea robusta.* There seem to be three main patterns of nectar sugar composition (Percival, 1965) with sucrose dominant, with fructose + glucose dominant and balanced i.e. with the three sugars present in similar amounts. In addition to these three and other sugars, nectar contains small amounts of chemical compounds that include aromatic substances, minerals, organic acids and amino acids. Certain amino acids present in honey, especially proline are introduced by the bees during their elaboration of the honey and are thus not plant specific.

Nectar secretion follows a daily cycle, usually with a peak at some time during daylight for plants pollinated by day fliers or during night for plants pollinated by night flying moths or bats. Shuel (1975) suggests that the association between honey yield and temperature, often reported in temperate climates is more correctly based on an association between nectar production and sunshine. He suggested that there may be temperatures at which the enzymes involved in nectar secretion are activated. Minerals in the soil, or added in fertilizers, can have a considerable effect on nectar secretion. The sugar concentration of nectar may change after secretion has occurred. It can be increased by evaporation of water from nectar at low atmospheric humidity and decrease by absorption of water at high humidity, or by dilution if the rain falls in to the flower.

Honey is the unfermented, sweet substance produced by honeybees from the nectar of blossoms or from secretions of or on living parts of plants, which they collect, transform and combine with specific substance and store in honey combs.

Pollen and its Collection

Pollen contains a wide variety of nutrients that support growth and development in living organisms. Many of the nutrients especially certain amino acids are essential for the development of honeybees from the egg to the adult stage and pollen is the only source of bees. The larvae obtain pollen in brood food produced by young worker nurse bees. Foraging bees collect the pollen and in the process transfer some grains to the stigma of the same or other flowers, thus resulted in pollination. A pollen forager packs the pollen into a compact load on each hind leg and flies back to the hive with it. Colonies do not normally store pollen greatly in excess of their requirements, and pollen-producing plants are important throughout the bees' active season.

The pollen is mainly composed of protein, fat, carbohydrates and various inorganic substances. The pollen analysis gives 16-30 per cent protein, 1-7 per cent starch, 0-15 per cent sugars, 3-10 per cent fat and 1-9 per cent ash. The pollen forms an important part of the larval food of solitary and social bees whose hairy bodies are well adapted for pollen collection. Bumblebees and honeybees possess special modifications for packing pollen, which is transported back to their colonies as pollen pellets on their hind legs. The selection of pollen by bees did not seem to be influenced by their age, colour, moisture or protein content. Though intensity of odour is known to be responsible for pollen attractiveness, this odour could be identified and synthesised for increasing the attractiveness of crops and also making pollen substitutes, to feed colonies during dearth period.

Irrespective of weather conditions, foraging bees pack pollen into their pollen baskets and their bodies are often covered with several thousand grains of pollen. Even though nectar gatherers have relatively more pollen on their heads, the actual amount of pollen or nectar and pollen gatherers was similar. Hence, pollen gatherers are as likely as nectar gatherers to pollinate the crops.

The foragers which leave their hives carry several thousands of pollen grains on their bodies, some of which are viable and it is possible to cross-pollinate on visit of different crop varieties. However, due to pollen transfer within the hive, where forager keeps constant visits to a cultivar over several trips thus cross-pollination occurs. This could explain how honeybees may cross-pollinate fruit trees and hybrid seed crops.

Flower Preference of Diffcrent Bee Species

The relationship between bee and flower species is behavioural and physiological rather than morphological. But, some monotrophic and oligotrophic bees exhibit special morphological adaptations for gathering nectar and pollen from the flowers. Such bee-flower relationships are found in the bee families like Panugidae, Andrenidae, Melittidae, Anthophoridae and Megachilidae. Both bumble bees and honeybees prefer segmented flowers with a disrupted outline and bilaterally symmetrical (Free, 1970).

When a flower and foraging insect are not well adapted to each other, insect visits may occur without accompanying pollination. Honeybees are sometimes too small to touch the anthers during nectar collection and sometimes their tongues are too short to reach the nectaries of flowers with long corolla and relatively few would enter flowers to collect the pollen.

Although bee species are polytrophic, most individuals keep to single flower during a single trip. No bees constantly collect different pollen at different times of the day and many bees return with pollen and others with nectar only. Whenever possible, bees prefer to forage near their hives and this preference is enhanced by the tendency of successful foragers to recruit more bees when working near rather than distant food sources. Bees tend to forage near their hives during unfavorable weather conditions.

Floral Attractiveness and Floral Competition

Honeybees locate flowers by sight and odour. They respond to visual nectar guides and colour of petals. After locating flowers, bees respond to floral odours including olfactory nectar guides, as well as to pollen odours (Hopkins *et al.*, 1969). Bees are attracted to flowers that are available in large numbers that have readily accessible pollen or nectar (Kevan and Baker, 1983). The quantity of nectar, the concentration and type of constituent sugars in floral nectars determine attractiveness to bees. Although pollen and nectar are present at different times of day in different plant species bees are able to locate and collect these foods (Frisch, 1967). The proportion of pollen to nectar that collected is largely determined by the needs of the colonies.

FORAGING ACTIVITY, BEE FLORA AND HONEYBEE POLLINATING CROPS

Bees on a site, searching for forage without fixed patterns of flight and flower visitation, can often be more effective pollinators than bees whose foraging patterns are already fixed. There is evidence that foragers can recruit others to a specific area of an apparently uniform crop.

Foraging Activity

The bees communicate information of the distance, direction and quality of food by round or waggle dances, the former for short distances from the hive and the latter for long distances. The tempo of the dance depends upon the amount of food at the source. The tempo of the dance decreases with the increase in distance.

Generally, only a limited number of flowers open at one time on most plants. Even though bees do not keep to herbaceous plant until they have collected all the food from it, visit only a small proportion of the available flowers. This behaviour further facilitates cross-pollination and bees are especially important for the pollination of self sterile species. For limited periods, bees tend to forage in a comparatively small area, although they move frequently from one plant to another. The bumble bees also tend to restrict their foraging to small parts of a large crop. It has been supposed that bumble bees have larger foraging areas and move from plant to plant more frequently than honeybees though their foraging areas are similar (Free, 1968).

Pollen and Nectar Foragers

The number of foragers is correlated with atmospheric temperature, soil temperature, relative humidity, light intensity and nectar sugar concentration. The population dynamics were directly related to the temperature and inversely to the relative humidity. The number of pollen foragers shows maximum variation during different times of the day and on different days of the year compared to that of nectar foragers and of non-foragers (Reddy and Bai, 1979). Generally the diurnal peak of pollen collection is followed by that of nectar collection. *Apis cerana* carries pollen load ranging from 2 to 190 mg in weight (Rahman and Rahman, 1993). They carry 27 to 35 per cent of their body weight as pollen load. The ratio of the load to body weight is higher in *A. cerana* than in *A. mellifera*. In Jammu and Kashmir, a forager carried 5 to 88 per cent of its body weight as nectar load, amounting to a mean of 46.7 mg nectar load of mean volume of 31.2 ml (Singh, 1971). Foragers of colonies with brood are stimulated to collect pollen. The proportion of bees collecting pollen directly correlates with the level of egg laying by the queen (Cale, 1968) and the amount of brood present (Free, 1967) thus, providing colonies with extra brood frames for pollen collection. The odour of brood contact with bees that are tending brood increases pollen collection. Pollen collection increases with removal of pollen from colonies. When pollen traps are attached to hive entrance, pollen loads are removed from returning bees. Providing pollen or pollen supplements to colonies to encourage brood rearing results in decreased pollen collection by the bees. Besides their value as honey producers, nectar-collecting bees are also important pollinators of crops such as sunflower, pear, peach, sweet cherry, etc. Honeybees make their collecting trips more frequently in the morning than in the later part of the day. Young bees returning with their first loads are said to show great excitement, while the older bees move about leisurely. After a suitable cell has been selected, which may take some time the bee grasps one edge with its forelegs, while the apical end of the abdomen rests on the outer side of the cell. The hind legs hang free within the cell, the pollen masses about even with its margin. The pollen masses are then forcibly thrust out of the baskets by the middle legs. The bee then usually departs, leaving to another worker the packing of the pollen in the cell. The pellets are broken up and pressed down ward and sugar with perhaps some liquid is added to preserve them.

Foraging speed (time spent per flower) and foraging rate (number of visitor per min.) depends upon the foraging behaviour of insects and floral structure of the crop concerned, particularly the corolla depth (Gilbert, 1980). More is the nectar available, more will be the time spent per flower but both the parameters might be affected by the length of proboscis (Inouye, 1980). The number of flowers visited by an insect species also depends upon the type of floral reward (working for nectar and/or pollen), bloom and the density of flowers on a particular cultivar. Foraging speed varies among the insect species. *A. cerana* nectar foragers visited 17.5 and 25.0 flowers per minute in the morning and after noon whereas *A. mellifera* visited 25.8 and 33.6 flowers per minute during the same hours (Gupta *et al.*, 1984).

Bees generally tend to forage in a comparatively small area, although they move fairly frequently from one plant to another. Foraging areas are particularly restricted in size during calm, sunny weather when abundant nectar and pollen are present but when forage is sparse, either because few flowers are present or the flowers are yielding little nectar or pollen, foraging areas of the bees are larger and the bees spend less time per flower and are more easily distributed. Observations of various workers have established that even when many flower species are growing together, the honeybee is usually forage constantly.

Foraging Efficiency of Bee Colonies

Bees generally prefer to forage near their hives and this preference is enhanced by the tendency of successful foragers to recruit more bees when working near rather than distant food sources. Depending upon the crop, most of the bees have been found to forage within 100 meters of their hives. The number of foraging bees on a crop diminishes with increase in the distance from the hive. The effective pollination of a crop depends upon the distribution of honeybee colonies in it. It is important to find the maximum size of groups that can be distributed equally throughout the crop concerned so that the overlap between adjacent groups is sufficient to prevent the number of foragers decreasing mid-way between them. Hives should be placed at such a site so that bees can fly out against the prevailing winds and back to hive in the direction of the wind. Honeybees forage a considerable distance from their hives. During food collection seasons, colonies located around 3 km radius in which flowering crops occurred produced more honey, but the gain in weight of colonies more than 3 km from the area decreases with increase in distance. Distance had a greater effect on the crop and decreased with distances greater than 0.8 km from nectar source; however, the gain in weight of colonies situated in the irrigated area during the poor season was less than that of best season (Eckert, 1933). Even when colonies are placed at short distance from crops, the proportion of foragers that visit the crops can be greatly diminished. Probably the tendency to forage near the hive is greater for pollen gatherers than for nectar gatherers, because a bee usually collects a pollen load more quickly than a nectar gatherer and spends a greater time during each trip.

Foraging Range of Bee Colonies

The flight range of bees varies between 100 to 1000 m, depending upon the type of bee as also the local vegetation and environmental factors. The foraging range of the plains type was estimated to be 600 to 1600 m and that of the hills type 900 to 3700 m (Shah and Shah, 1980). When colonies are moved into crops the foraging range of the bees tends first to the vicinity of their hives and later gradually expands. Perhaps the bees from the original colonies forage close to their hives, though foraging was better in the field. Other results indicated that bees from colonies located some distance away from the field were more inclined to select the best area of the crop.

The arrangement of colonies in crops is important in ensuring uniform distribution of foraging bees. The foraging area of a colony depends on many factors i.e., availability of the amount of nectar and pollen per unit area, weather conditions and geographical features of the area. Ideally, colonies should be distributed equally as convenient to both growers and beekeepers with equal distribution of the foraging bees. During peak flowering, the number of bees per plant decreases greatly as the distance from the colonies increases.

The honeybee population needed to meet the pollination requirement of the crop. However, at depends mainly upon the density of flowers, their attractiveness, competing insects and crops, type of crop—whether self or cross-pollinated, the age of the crop and location of the crop. Younger fruit trees will have fewer flowers to be pollinated and need only fewer colonies of bees. The number of colonies required for optimum pollination of crops also depends upon their pollination requirements. Orchards having more number of self fertile cultivars require less number of honeybees colonies per hectare as compared to self fertile cultivars. When it is known that a crop is difficult to be pollinated because the species is relatively unattractive to honeybee then more than 2.5 colonies per hectare are some times recommended. In contrast, when a crop is attractive but its flowers are sparse or can be self-pollinated, less than 2.5 colonies per hectare are suggested.

Strength of bees in the hive will also determine the colony requirements of the particular crop. Sheeslay and Poduska (1970) compared the size of honeybee colonies in almond orchards. Those with 6-7 frames of bees collected 1.5 times and those contained with 8 or more frames of bees almost 3 times as much pollen as colonies with 4-5 frames. So 4 to 5 frames were taken as standard. Generally colonies for orchard pollination should have minimum of 4 to 6 brood combs.

Bee Flora

Bees and flowering plants are interdependent forming a mutually compatible system (Kevan and Baker, 1983). The flora from which bees collect good amount of nectar as well as pollen are often called nectar flora or bee flora, because this type of flora is most suitable for beekeeping. Pollen sources are essential for full explanation of honey flow because large field force is required to collect long honey stores.

The qualities of good bee flora are long flowering period, high density of flowers per unit area of the plants, nectar with high sugar concentration and easy accessibility of the nectaries to the honeybees and easy in collection of nectar.

Beekeeping maps and floral calendars are of special value for the development of beekeeping in any region. In view of the availability of bee flora, the flora of different states of India are presented (Table.1).

Honeybee Pollinating Crops

Bees and flowering plants are interdependent forming a mutually compatible systems and cross-pollination of flowers of entomophilous crops by insects is the most effective and cheapest method of increasing yield. Table 2 gives information on bee pollinated crop plants.

Fruits and Nuts

Tropical and subtropical fruits are very diverse in Asia. Even though the pollination requirements of many fruits are partly available, the oranges are greatly benefitted by honeybee pollination. Apples generally require interplanting of varieties along with cross pollination by insects. Peaches and plums are mostly self-incompatible and require cross pollination to produce fruits and offer good nectar and pollen to bees. Apricots are available in pollinator requirements depending on the variety or cultivar. Cherries cannot self pollinate, but some are self compatible. Thus, inter-plantings of varieties and the presence of bees for pollination are advised. Mango generally requires insects for pollination, papaya has a complex breeding system of male hermaphrodite and female flowers in which the best fruit results from cross pollination and bees are recommended as pollinators. In cashew, the cross pollination is important apparently accomplished by small insects. The almond is dependent on insects, primarily bees for fruit set.

Litchi (*Litchi chinensis*) is one of the most popular orchard crop with delicious fruits and is native of South-East Asia and is commonly grown as orchard crop. Honeybees are predominant among the floral visitors of litchi. Bee pollination resulted in an average yield of 64.45 fruits per branch that weighed 20.55 g, while the flowers that had no bee visits yielded

Table 1. Important bee flora from different states of India

Sl. No.	Scientific name	Common name	States from where commonly grown
1	**2**	**3**	**4**
1	*Acacia* sp.	Acacia	Punjab, Karnataka
2	*Aegiceras corniculatum* (L.) Blanco	Mangrove	West Bengal
3	*Allium cepa* L.	Onion	Uttar Pradesh (Jeolikote)
4	*Anacardium occidentale* L.	Kaju	Kerala, Karnataka, Orissa
5	*Antigonon leptopus* Hook & Arn.	Coral creeper	Bihar
6	*Azadirachta indica* A. Juss.	Neem	Uttar Pradesh
7	*Brassica campestris* L.var. *sarson*	Sarson	Uttar Pradesh, Bihar, West Bengal, Assam
8	*Brassica juncea* L. Cosson raya	Raya	Punjab, Haryana
9	*Brassica oleracea* L. var. *botrytis*	Cauliflower	Himachal Pradesh
10	*Cajanus cajan* (L.) Millsp.	Arhar	Bihar, Uttar Pradesh, Haryana
11	*Cajanus sinensis* (L.) Osb.	Santra/Orange	Punjab
12	*Cocos nucifera* L.	Coconut	Karnataka, Orissa, Andhra Pradesh

(Table Contd...)

Table 1. (Contd...)

1	2	3	4
13	*Coffea arabica* L.	Coffee	Karnataka
14	*Coriandrum sativum* L.	Dhania	Maharashtra, Punjab
15	*Dalbergia sissoo* DC.	Shisham	Bihar, Uttar Pradesh, Himachal Pradesh, Punjab, Haryana
16	*Eucalyptus* spp.	Safeda	Haryana, Punjab, Uttar Pradesh, Bihar, Karnataka, Himachal Pradesh
17	*Eugenia* spp.	Bhedas, Gudda Panneralu	Karnataka
18	*Gossypium* spp.	Cotton	Punjab, Haryana, Uttar Pradesh
19	*Guizotia abyssinica* Cass.	Niger	Maharashtra, Andhra Pradesh, Karnataka
20	*Helianthus annuus* L.	Sunflower	Maharashtra
21	*Hevea brasiliensis* Muell. Arg.	Rubber	Karnataka, Kerala
22	*Litchi chinensis* Sonner.	Litchi	Punjab, Uttar Pradesh, Bihar, Himachal Pradesh
23	*Madhuca longifolia* (Koenig.) J.F. Macbr.	Mahua	Bihar, Uttar Pradesh, Karnataka
24	*Malus domestica* Borkh.	Apple	Himachal Pradesh, Uttar Pradesh, Jammu & Kashmir

(Table Contd...)

Table 1. (Contd...)

1	2	3	4
25	*Mangifera indica* L.	Mango	Karnataka, Kerala, Himachal Pradesh
26	*Medicago sativa* L.	Lucern	Maharashtra, Punjab
27	*Melilotus* sp.	Clover/Melilot	Punjab
28	*Moringa pterigosperma* Gaertn.	Drum stick tree	Bihar
29	*Polygonum* sp.	Polygonum	Punjab
30	*Pongamia pinnata* (L.) Pierre	Karanj	Bihar
31	*Prunus amygdalus* Batsch	Almond	Himachal Pradesh
32	*Prunus armeniaca* L.	Apricot	Himachal Pradesh
33	*Prunus domestica* L.	Plum	Himachal Pradesh, Tamil Nadu
34	*Psidium guajava* L.	Guava	Maharashtra, Himachal Pradesh
35	*Pyrus communis* L.	Pear	Himachal Pradesh, Tamil Nadu
36	*Pyrus persica*	Peach	Himachal Pradesh, Punjab

(Table Contd...)

Table 1. (Contd...)

1	2	3	4
37	*Schefflera wallichiana* Harms.	Doddabettu/ Pongabettu	Karnataka
38	*Sesamum indicum* L.	Gingelly/Til	Bihar
39	*Syzygium aromaticum* (L.) Merrill & Perry	Jamun	Karnataka, Kerala
40	*Tamarindus indica* L.	Tamarind/Imli	Karnataka, Kerala
41	*Terminalia arjuna* (Roxb.) Wight & Arn.	Arjun tree	Karnataka, Kerala, Uttar Pradesh, Maharashtra, Assam, Punjab
42	*Terminalia bellerica* Behda/Thari (Gaertn.) Roxb.	Behda/Thari	Karnataka, Kerala
43	*Terminalia tomentosa* Wight & Arn.	Mathi	Karnataka
44	*Trifolium alexandrinum* L.	Berseem	Bihar
45	*Ziziphus mauritiana* Lam.	Ber	Maharashtra, Haryana, Orissa, Uttar Pradesh

Table 2. Important crops benefited by honeybee pollination

Sl. No.	*Scientific name*	*Common name*	*Family*	*Use*
1	**2**	**3**	**4**	**5**
1	*Abelmoschus esculentus*	Okra	Malvaceae	Vegetable
2	*Allium cepa*	Onion	Liliaceae	Vegetable
3	*Allium ampeloprasum*	Leek	Liliaceae	Vegetable
4	*Allium sativum*	Garlic	Liliaceae	Spice
5	*Amomum aromaticum*	Bengal cardamom	Zingiberaceae	Vegetable
6	*Amomum compactum*	Large cardamom	Zingiberaceae	Spice
7	*Anacardium occidentale*	Cashewnut	Anacardiaceae	Fruit
8	*Arachis hypogaea*	Peanut	Leguminosae	Oil
9	*Areca catechu*	Betalnut palm	Araceae	
10	*Beeta vulgaris*	Beet	Chenopodiaceae	Vegetable
11	*Brassica campestris*	Field mustard	Brassicaceae	Oil
12	*Brassica juncea*	Indian mustard	Brassicaceae	Oil
13	*Brassica rapa*	Turnip	Brassicaceae	Oil
14	*Brassica nigra*	Black mustard	Brassicaceae	Spice
15	*Brassica oleracea var. botrytis*	Cauliflower	Brassicaceae	Vegetable

(Table Contd...)

Table 2. (Contd...)

1	2	3	4	5
16	*Cajanus cajan*	Pigeon pea	Leguminosae	Pulse
17	*Capsicum annum*	Red pepper	Solanaceae	Spice
18	*Carica papaya*	Papaya	Caricaceae	Fruit
19	*Carthamus tinctorius*	Safflower	Asteraceae	Oil
20	*Cichorium intybus*	Chicory	Asteraceae	Beverage
21	*Citrullus vulgaris*	Water melon	Cucurbitaceae	Fruit
22	*Citrus aurantifolia*	Lime	Rutaceae	Fruit
23	*Citrus amara*	Bitter orange	Rutaceae	Fruit
24	*Citrus aurantium*	Sour orange	Rutaceae	Fruit
25	*Citrus limetta*	Sweet lemon	Rutaceae	Fruit
26	*Citrus limon*	Lemon	Rutaceae	Fruit
27	*Citrus sinensis*	Sweet orange	Rutaceae	Fruit
28	*Cocos nucifera*	Coconut	Aracaceae	Oil/beverage
29	*Coffea arabica*	Arabian coffee	Rubiaceae	Beverage
30	*Coriandrum sativum*	Coriander	Apiaceae	Pulse
31	*Cucumis anguria*	Gherkin	Cucurbitaceae	Vegetable

(Table Contd...)

Table 2. (Contd...)

1	2	3	4	5
32	*Cucumis sativus*	Cucumber	Cucurbitaceae	Vegetable
33	*Cuminium cyminum*	Cumin	Apiaceae	Spice
34	*Cucumis melo*	Melon	Cucurbitaceae	Fruit
35	*Cucurbita pepo*	Pumpkin	Cucurbitaceae	Vegetable
36	*Daucus carota*	Carrot	Apiaceae	Vegetable
37	*Dolichos uniflorous*	Horsegram	Leguminosae	Cereal
38	*Elettaria cardamomum*	Cardamom	Zingiberaceae	Spice
39	*Gossypium arboreum*	Cotton	Malvaceae	Fibre
40	*Guizotia abyssinica*	Niger	Asteraceae	Oil
41	*Helianthus annuus*	Sunflower	Asteraceae	Pulse
42	*Ipomea batatus*	Sweet potato	Convoluvlaceae	Vegetable
43	*Lavandula augustifolia*	Common lavender	Lamiaceae	Essence
44	*Lavandula latifolia*	Broad leaved lavender	Lamiaceae	Essence
45	*Linum usitatissimum*	Flax or Linseed	Linaceae	Fibre
46	*Lycopersicon esculentum*	Tomato	Solanaceae	Vegetable
47	*Mangifera indica*	Mango	Anacardiaceae	Fruit

(Table Contd...)

Table 2. (Contd...)

1	2	3	4	5
48	*Malus domestica*	Apple	Rosaceae	Fruit
49	*Medicago falcata*	Lucern	Leguminosae	Forage
50	*Medicago lupulina*	Black medick	Leguminosae	Forage
51	*Medicago sativa*	Alfalfa	Leguminosae	Forage
52	*Melilotus indica*	Indian sweet clover	Leguminosae	Forage
53	*Mentha spicata*	Spearmint	Laminaceae	Essence
54	*Metroxylon sago*	Sago palm	Palmaceae	Condiment
55	*Momordica charantia*	Bitter gourd	Cucurbitaceae	Vegetable
56	*Moringa oleifera*	Drumstick	Moringaceae	Vegetable
57	*Litchi chinensis*	Litchi	Sapindaceae	Fruit
58	*Nephelium longana*	Longan	Sapindaceae	Fruit
59	*Ocimum basilicum*	Basil	Laminaceae	Essence
60	*Ocimum sanctum*	Holy basil	Laminaceae	Medicinal species
61	*Olea emopaea*	Olive	Oleaceae	Oil
62	*Punica granatum*	Pomegranate	Punicaceae	Fruit
63	*Phaseolus aureus* Roxb.	Green gram	Fabaceae	Cereal
64	*Phaseolus radiatus*	Black gram	Fabaceae	Cereal

(Table Contd...)

Table 2. (Contd...)

1	2	3	4	5
65	*Pisum sativum*	Field pea	Leguminosae	Vegetable
66	*Psidium guajava*	Guava	Myrtaceae	Fruit
67	*Prunus armerica*	Apricot	Rosaceae	Fruit
68	*Prunus amygdalus* Bartsch	Almond	Rosaceae	Fruit
69	*Prunus persica*	Peach	Rosaceae	Fruit
70	*Raphanus sativus*	Radish	Cruciferae	Vegetable
71	*Rubus fruticosus*	Black berry	Rosaceae	Forage
72	*Sesamum indicum*	Sesame	Pedaliaceae	Oil
73	*Solanum melongena*	Brinjal	Solanaceae	Vegetable
74	*Solamum tuberosum*	Potato	Solanaceae	Vegetable
75	*Tamarindus indica*	Tamarind	Leguminosae	Multiple
76	*Tectona grandis*	Teak	Verbanaceae	Spice
77	*Trifolium alexandrium*	Berseem	Leguminosae	Forage
78	*Trichosanthes argnina*	Snake gourd	Cucurbitaceae	Vegetable
79	*Trifolium fragiferum*	Strawberry clover	Fabaceae	Forage
80	*Trifolium hybridum*	Hybrid clover	Fabaceae	Forage
81	*Trifolium protense*	Red clover	Fabaceae	Forage
82	*Vitis rotudifolia*	Muscadine grape	Vitaceae	Fruit
83	*Vitis vinifera*	Grape	Vitaceae	Fruit

26.49 fruits weighing 13.98 g per branch (Table 3). Fruit yield per hectare was 70 tonnes in bee pollinated plots compared to 36.8 tonnes in self pollinated plots (Mahanta and Rahman, 1997).

Table 3. Effect of bee pollination on the fruit set of *Litchi chinensis*

Treatment	*Initial number of flowers*	*Number of fruits set*	*Per cent fruit set*	*Average fruit weight (g)*	*Average number of fruits per branch*
Bee Pollination	97.43	3.2	3.28	20.55	64.43
Open Pollination	85.56	2.6	3.03	19.75	56.75
Self Pollination	78.82	1.5	1.90	13.98	26.49
CD (P=0.05)			1.22		

Kiwifruit (*Actinidia chinensis*), also known as Chinese gooseberry, is functionally delicious and flowers are visited by pollinators only for pollen as there is little nectar and so are not particularly attractive to bees. It has been suggested that bees can be stimulated to collect pollen by feeding colonies with sugar syrup (Goodwin, 1997).

Peach is self-compatible; however, cross-pollination increases yield and improves fruit quality. Cross-pollination is brought about by insect pollinators that visit flowers for pollen and nectar. Among pollinating insects, honeybees in particular, are attracted to these crops and are reported to be of great benefit (Choi and Lee, 1988; Free, 1993; Mann and Singh, 1983).

The particular requirements of the different kinds of citrus vary from almost complete self sterility in some cultivars to complete self fertility in others. While in the self sterile cultivars pollen must be transferred from those of another compatible varieties for maximum fruit production and in self fertile cultivars the plant is benefited if pollen is moved from flower to flower within the cultivars or within the species.

Research results revealed from different countries showed that the higher yield, better size and shape of strawberry fruits depend upon the adequate pollination of its flowers by different insects (Crane, 1990). Among different species of pollinating insects, domesticated species of honeybees such as *A. cerana* and *A. mellifera* are reported to be the principal pollinators of strawberry flower (Singh, 1979).

Varieties of plum are self incompatible which require cross pollination to produce better fruits. Peach varieties are self compatible; however, cross-pollination has been reported to increase yield and improve fruit quality (Free, 1993). Cross-pollination of peach and plum is brought about by insect pollinators that visit their flowers for pollen and nectar. Among pollinating insects, honeybees in particular are attracted to these crops and are reported to be of great benefit (Mann and Singh, 1983).

Vegetables

Insect cross-pollination is important in obtaining seed production of vegetables and their flowers are often excellent forage for bees. Tomatoes are not automatically self pollinating but movement of the plant causes pollen from the anthers on to the stigma. The cucurbitaceous vegetables such as pumpkins, squash, cucumber and gherkins, balsam pear, muskmelon, watermelon and gourds require pollinators for fruit set. Honeybees are the principal pollinators of the periodic and semi-seasonal crops. Seed set and seed weight were significantly higher in bee pollinated than in bagged flowers in cauliflower (Kumar *et al.,* 1988). The seeds obtained from selfing were higher in weight (Table 4). Therefore, required bee pollination be provided during blooming for the qualitative and quantitative seed production. Among pulses, cowpeas are highly attractive to bees. Pollination is best effected by larger bees rather than by smaller bees. Carrot is known to be a cross-pollinated crop and its flowers are protandrous. Since production through self pollination is limited and wind pollination is not of much importance, pollination by insects particularly the honeybees assumes greater significance. The reports show that the seed set was more in cages with bees than in cages without bees (Table 5). Similarly, the seed index was found to be higher in plots with bees than in cages without bees and was highest in open plots. The quality of seed assessed

Table 4. Effect of bee pollination on pods, seeds and seed weight in different cultivars of cauliflower

Cultivars	Pod formation (%) a				Number of seeds/pod b				Seed weight (mg)/100 seeds			
	OP	NC	MB	Average	OP	NC	MB	Average	OP	NC	MB	Average
Snow ball-16	*80.43* (63.97)	58.92 (50.86)	73.95 (60.68)	71.10 (58.50)	8.43 (2.86)	4.73 (2.15)	3.23 (1.77)	5.46 (2.26)	257.08 (16.03)	213.55 (14.61)	106.43 (10.31)	192.35 (13.65)
Sel- 12	88.48 (70.99)	79.34 (63.30)	78.73 (63.23)	82.18 (65.84)	11.51 (3.39)	7.16 (2.65)	6.28 (2.50)	8.32 (2.85)	368.63 (19.18)	279.18 (16.70)	347.13 (18.63)	331.65 (18.17)
Sel- 1	80.59 (64.20)	80.03) (63.82)	83.02 (66.11)	81.21 (64.71)	8.40 (2.90)	7.19 (2.63)	5.52 (2.33)	7.04 (2.62)	185.00 (13.39)	228.15 (15.10)	177.30 (13.31)	196.82 (14.00)
K - 1	97.16 (81.72)	89.22 (73.47)	87.33 (60.98)	91.24 (72.06)	10.18 (3.18)	10.78 (3.28)	7.97 (2.80)	9.64 (3.09)	157.58 (12.55)	151.85 (12.31)	112.50 (10.61)	140.64 (11.82)
Sel - 25	82.91 (66.20)	87.23 (70.25)	74.55 (60.65)	81.56 (65.73)	10.19 (3.17)	7.15 (2.63)	7.89 (2.79)	8.41 (2.86)	256.08 (16.07)	239.53 (15.47)	219.15 (14.80)	238.25 (15.45)
Average	85.91 (69.42)	78.95 (64.34)	79.52 (62.35)		9.74 (3.10)	7.40 (2.67)	6.18 (2.44)		244.87 (15.48)	222.45 (14.84)	192.50 (13.53)	
CD 0.05 Cultivars				5.52				0.41				0.40
Modes of pollination				NS				0.29				0.34
Interaction (Modes of pollination X cultivars)				NS				NS				0.73

Figures in parentheses: a= angular transformation; b= and c= √ n transformations
OP = Open pollination NC = Nylon net caged; MB = Muslin bagged

Table 5. Effect of bee pollination on the seed production of *Dacus carota* L.

Particulars	*Caged plots*		*Open*	*Cage without bees**	*Critical difference (P=0.05)*
	Apis mellifera	*Apis cerana indica*			
Weight (g) of seed Per 2 sq. m area	735 (64.42)*	715 (59.95)	755 (68.90)	447	224.1
Seed set: Seeds matured/ 100 florets	89.4 (61.66)	87.8 (58.77)	92.1 (66.37)	55.3	28.7
Seed index: Weight/1000 seed (g)	2.335 (3.23)	2.328 (2.91)	2.340 (3.44)	2.262	N.S.
Germination (%)	92.6 (9.32)	90.3 (6.61)	95.2 (12.39)	84.7	N.S.

* Figures in parentheses indicate the percentage increase over the characteristic in plots caged without bees.

in terms of per cent germination in cages with bees was found to be better (Goyal *et al.*, 1989).

The flowers of cauliflower and radish produce good amount of nectar and pollen and are quite attractive to pollinating insects particularly honeybees. Many cultivated varieties are partially or completely self sterile and require cross-pollination to produce fruit and seed (Verma and Pratap, 1994). The honeybee (*Apis cerana indica*) pollinated flowers of onion (*Allium cepa*), the number of seeds per flower, seed set and seed germination were significantly higher than in net caged flowers (Kumar *et al.*, 1989).

Beverage, Condiment and Spices

Coffee production can considerably be enhanced by augmented pollination of honeybees and Coffee flowers are also a good source of nectar. Honey bee pollination is indispensable for cardamom plant. Greater cardamom (*Ammomum subulatum* Roxb.) is an important spice crop cultivated in the sub-Himalayan region. The dependence of greater cardamom on insects for its pollination is proved by the fact that the per cent fruit set, green and dry fruit weight and number of seeds per capsule were more in open pollinated inflorescence (Table 6), which had access to all insect pollinators than those in bagged inflorescence, which had no access to insects (Verma, 1987). Coriander is grown extensively in India and seems to benefit from cross-pollination, but they offer good bee forage and benefit from out-crossing.

Oil Seed Crops

The rape seed, mustards and related *Brassica* are grown extensively for oil seed production in Asia. *Brassica campestris* and *B. juncea* are dependent on insects for pollination and are excellent source of pollen and nectar for bees. Turnip (*Brassica rapa*) is a self incompatible crop and needs pollen carriers for cross-pollination of its flowers (Sihag, 1985). However, all are not equally important. Some make a better contribution to pollination because of their suitable attributes. These include their abundance, foraging behaviour, maximum number of pollen grains carried on their body, foraging rate and activity duration. Sunflower is being grown more and more in Asia.

Table 6. Effect of bee pollination on yield of greater cardamom at Jeolikote, Uttar Pradesh

Year	Number of honeybee colonies	Mean per cent fruit set		Total fruit weight (kg)		Number of seeds per capsule	
		Open pollination	*Bagged infloresence*	*Green*	*Dry*	*Open pollination*	*Bagged inflorescence*
1980	107	80.67	6.12	110.250	27.500	3.5	1.6
1981	110	81.34	5.44	112.000	28.200	3.6	1.3
1982	115	82.34	7.20	120.500	30.250	3.8	1.4
1983	110	81.00	5.54	115.250	28.800	3.5	1.4
1984	25	42.00	5.67	59.750	14.200	2.1	1.5
1985	20	36.70	6.00	53.500	13.500	1.9	1.4
1986	82	65.25	6.72	108.300	26.300	3.4	1.3

A field experiment conducted during summer 1995 on honeybee pollination at Bangalore using CMS 234A and 234B (female parent and maintainer) parents of sunflower showed that the filled seed weight, grain filling percentage and oil content were considerably increased in the plots pollinated by the honeybees (Table 7). The seed weight was maximum in honeybee pollinated plot followed by hand and bee pollinated, pollination by other pollinators and exclusively hand pollinated plots in case of female parent (Rajagopal *et al.*, 1999; 2000).

Table 7. Yield parameters of CMS 234A (Female parent) sunflower

Treatment	*Filled seed weight (g)*	*Unfilled seed weight (g)*	*No. of filled seed*	*No. of unfilled seed*	*Filling (%) (%)*	*Oil content from filled seeds*
Check	5.16	1.89	53.17	330.70	13.85	39.1
Hand pollination only	12.98	0.64	360.14	46.44	88.57	38.4
Pollination by small insects	13.59	0.65	330.40	48.36	87.23	45.2
Open Pollination	25.40	1.02	501.13	38.00	92.95	53.1
Hand and open pollination	24.73	0.73	534.73	73.70	87.88	46.0

Sunflower is a good source of nectar and pollen for bees and yields more oil when cross-pollinated. Even though new self compatible varieties are available, they benefit from cross-pollination and oil yields are comparatively greater. Self incompatible varieties require insect pollination and bees are the primary pollinating agents. Niger is a good bee plant and seed yield greater when bee hives are placed in niger fields. The pollination studies conducted on sesamum and niger show the significance of honeybee pollination in increasing the yield though a number of other pollinators visited the flowers did not have much impact on the pollination. Therefore, there should be sufficient number of bee colonies in the vicinity of the crops during the flowering period (Panda *et al.*, 1988). Flax is grown for oil and fibre and benefits from outcrossing by bees. The coconut is grown over a large area in South India, which provides food, oil and fuel. Evidence is accumulating slowly that coconut production would be improved by augmentation of honeybees for pollination.

Forage Crops

Alfalfa or lucerne (*Medicago* sativa) is grown in drier parts of tropical Asia, where it requires insects for pollination. Honeybees and leaf cutter bees are the most effective ones. Berseem has a wide variety of pollinators including honeybees on which it is dependent for seed set. Similarly subabul is also required honeybees for seed production.

Fibre Plants

The most important fibre giant is cotton. However, it is not thought to depend on insects for cross-pollination. In fact, cross-pollination has been considered detrimental, because the resultant seeds and progeny are not necessarily bred exclusively from the desired cultivars.

MANAGEMENT OF BEES FOR POLLINATION

Foraging Strength of Colonies

It is important to have colonies at full strength for crop pollination. A single store colony should have a blancket of bees covering the tops of every frame. The colony size is usually assessed as the number of combs occupied by bees and brood. A hive used for pollination should contain at least seven Langstroth frames and 60 per cent of brood in all stages of development with minimum 25 per cent of brood should be in the egg or larval stage. The colony should also consist of young prolific queen with at least ten Langstroth frames covered worker bees, with enough honey and pollen stores.

Concentration of Honeybee Colonies

The number of colonies needed varies on local conditions including abundance of honeybees and other pollinators already present in the area, size of crop and the area of competing crops of the same or different species. However, the honeybee colonies needed for pollination depends on the density of flowers, their attractiveness and competing insects or crops.

Bees flew long distance to collect pollen and nectar, but they usually collect their floral rewards close to their hive. Thus, locating colonies close to target crops appears to be important in maintaining bees on crops. Colonies placed near crops collect more pollen and nectar (Free, 1963). Foragers spend less time for collecting a load of pollen than a load of nectar. The number of bees per unit area of target crop and amount of pollen collection decreases as the distance from a hive increases (Alpatov, 1984). Yields of target crops are also higher when colonies are close to them. However, some bees tend to forage relatively close to the colony, while others ignore rewarding local floral sources. During unfavourable weather conditions, bees forage less and tend to forage close to their colonies (Jay and Jay, 1984). Although recommendations are available about the number of colonies required to pollinate in given areas of crops, opinions differ about when and where to sight the colonies. The colonies are normally placed equidistant from each other and maintain even distribution of bees on target crops.

The number of average bee colonies needed for effective pollination ranges from 3 to 9 per hectare for *A. cerana* and 2 to 5 colonies for *A. mellifera*. The number however depends upon floral density competing insects and peculiarities of floral biology of individual crops. Taking even the minimum value of 3 colonies *A. cerana* and 2 colonies of *A. mellifera* per hectare, the total number of colonies needed for crops production in India will be 150 million of *A. cerana* or 100 million of *A. mellifera*. The number of colonies of *A. cerana* and *A. mellifera* presently in India are 9,56,745 and 1,00,000 respectively (Goyal and Gupta, 1994). Thus there is need to increase the number of bee colonies to meet pollination requirements of crops.

The number of honeybee colonies required for pollination and the expected increase in production of important crops or benefited by bee pollination in India is presented (Table 8).

Migratory Beekeeping

Several vegetation regions of the country exhibit short or long gaps in flowering. Accordingly, although bee forage is

Table 8. Estimated number of honeybee colonies required for pollination and the expected increase in production of the important crops in India

Crop	*Scientific name (Family)*	*Population requirement*	*Expected increase in yield due to cross-population*	*Number of colonies required per hectare*	*Number of colonies needed for pollination of total area*	*Estimated number of colonies needed for pollination of total area*	*Present production (tonnes)*	*Estimated increase in production by optimizing bee pollination (tonnes)*	*Present rate per tonne (Rs)*	*Expected increase in national income (Rs)*
1	**2**	**3**	**4**	**5**	**6**	**7**	**8**	**9**	**10**	**11**
Almond	*Prunus amygdalus* (Rosaceae)	Highly cross pollinated	20	5-8	17759	115433	1925	385	80000	30800
Apple	*Malus sylvestris* (Rosaceae)	Highly cross pollinated	20	2-3	178305	410101	861404	172280	5000	861400
Citrus	*Citrus* spp. (Rutaceae)	Often cross pollinated	20	2-3	382820	957050	2250053	450011	10000	4500110
Coconut	*Cocos nucifera* (Palmaceae)	Transfer of pollen essential	5	2-3	1510000	3775000	1428402	7142	30200	2156884
Grape	*Vitis* spp. (Vitaceae)	Often cross pollinated	10	2-3	15389	38473	251035	50207	10000	502070

(Table Contd...)

Table 8. (Contd...)

1	2	3	4	5	6	7	8	9	10	11
Guava	*Psidium guajava* (Myrtaceae)	Cross pollinated	10	2-3	176853	442133	1112582	111258	2000	222516
Mango	*Mangifera indica* (Anacardiaceae)	Bee visits helpful in pollination	3	2-3	1232913	3082283	10250396	310512	10000	3105120
Papaya	*Carca papaya* (Caricaceae)	Pollen transfer essential	10	2-3	32000	80000	270000	27000	3000	81000
Mustard rape	*Brassica* spp. (Cruciferae)	Mostly cross pollinated	20	3-5	5720000	22880000	5150000	1030000	11400	11742000
Sesamum	*Sesamum indicum* (Pedakuaceae)	Often cross pollinated	20	2-3	2433000	6082500	667000	1334002	13530	1804902
Sunflower	*Helianthus annuus* (Compositae)	Cross pollinated	20	2-4	2096900	6290700	1184800	236960	6700	1587632
Cotton	*Gossypium* spp. (Malvaceae)	Often cross pollinated	10	2-6	7400000	29600000	1653200 (Lint)	165320	20400	3372528
Total						73753673		2694475		29966962

available for a major part of the year, there are often one or more floral dearths of short or long duration. Migratory beekeeping can be practiced to overcome these deficiencies in bee forage availability. In Punjab, the stationary beekeeping colonies fail to produce any honey (Goyal, 1978), while in Himachal Pradhesh this leads to colony losses during long dearths. In the plains these losses were over 93 per cent, in the mid-hills almost 67 per cent and in the upper hills with a long period of forage availability. The losses amount to less than 22 per cent. Inter migration between the plains, lower and upper hills helps cut down these losses (Shårma, 1958). Migration management is important in commercial beekeeping, where colonies are taken from a low productive area to high productive area. Migration to the rubber plantation area along the lower elevation of the Western Ghats in Karnataka, Kerala and Tamil Nadu is a standard management for honey production.

Moving colonies for pollination services include the use of mechanical hive loaders together with various hive parts. Colonies are moved at night and if the day time movement is necessary, the colonies should not be exposed to hot dry conditions and if necessary the colonies should be sprayed with water.

A more effective way to cope with the competing plant problem is by colony movement. If colonies are moved more than two miles from the old location, worker bees need to reorient themselves. If the movement into the area is delayed until 15 to 20 per cent of the blossoms are open and colonies are interspersed among the crop, pollination will be more complete than if colonies are moved in advance of blossom time. Temperature at blossom time is also critical, especially for early blooming crops such as apples, pears and almonds. Normally, a worker confines foraging for nectar and pollen to a plant species at a time.

Conditioning of Colonies

Floral fidelity or constancy is commonly referred to a bees restriction to one species of flowers on one trip or several trips. It is a distinct behavioural phenomenon from visiting a single or some closely related species of plants. Honeybees are polytrophic but when an individual honeybee finds a source of nectar or pollen, they continue to gather from one source until they find out or recruited to another source. This behaviour is advantageous to the plant as it facilitates cross-pollination. Whenever bees tend to visit the same plant species with the result that the proportion of the foragers of a colony is often related to the properties that have visited previously. Hence it is important that colonies should not be taken to a crop until it is flowering sufficiently (Free, 1959).

Increasing Attractiveness of Crops

Attempts were made to direct the bees to certain crops by feeding the colonies with sugar syrup containing the scent of the target crop. In this technique the flowers are immersed in sugar syrup for few hours and later strain the flowers and feed the same to the colonies. It is reported that honeybees can be encouraged to visit a crop by spraying sugar syrup and commercially available Bee-Q. It is possible to applying the attractive odours present in pollen to a crop which would increase the population of foraging bees. Another possible way to increase pollination is to increase the attractiveness of the crop itself to bees (Shuel, 1989). Long-term selective breeding of certain species might produce strains whose nectar or pollen is more accessible to honeybees. It might also be possible to select strains that secrete more nectar probably of genetic variability in nectar secretion due to effect of environment on the sugar concentration of nectar. The most effective method of improving a crop is to select plants directly for attractiveness to honeybees. The dominant factors affecting bee visits appear to be nectar abundance and sugar concentration and chemical attraction of pollen. Some plant breeders of bee pollinated crops have recently become much more aware of the need to keep bee pollinated crops attractive to bees to avoid seed production problems later. Aroma may be used by bees as a label to locate and mark flowers which yield a reward of nectar or pollen.

Attractants

Sugar syrups, in combination with floral scents, are used to induce bees to visit crops. Scout bees of fed colonies search for the target crop and communicate this information to potential foragers and bees. However, feeding scented sugar syrups outside the hives is more effective in directing bees to crops as measured by number of bees performing 'recruitment dances' (Fisch, 1967) and increases in number of bees visiting target crops. Distilling essential oils from blossoms and feeding sugar syrup to colonies increase the pollen collection and crop yields (Heinrich, 1981).

It has been suggested that odours from Nasonov gland located on the dorsal side of the abdomen of worker bees might attract bees to crops. The citral and geraniol of nasonov gland mask the repellent odours and attract bees to crops and increase crop yields (Praagh and Ohe, 1983). Pheromones and chemical attractants may be useful in inducing bees to pollinate crops. Bees prefer to visit feeding dishes that have been previously visited by bees, possibly because a chemical from their tarsi (foot-print pheromone) is deposited at the site. Anethole (1-methoxy-4(1-propenyl) benzene which is not known as pheromone, also attracts bees to crops. The use of bee pheromones to maintain bees on crops and controlled release system for applying pheromones on plants is now well developed.

Selection and Breeding of Honeybees

Selection and breeding of bees have advanced through the development of instrumental insemination techniques for queen bees. Bees that have a limited flight range might be used on crops requiring pollination. It may be possible to select specific strains of bees with normal wings that tend to forage close to their hives. Pollen load analysis shows that bees from individual colonies have preferences over various pollen yielding crops. These studies suggest that the selection of bees that prefer certain nectars, particularly where nectar collectors are known to be important pollinators of crops should also be undertaken for breeding purposes.

Plant breeders have been actively modifying crops for many purposes, generally associated with improving yield and quality or adaptations to mechanised agriculture. Apiculturists should be more intimately associated with breeding programmes involving bee pollinated crops to maintain attractiveness and usefulness to bees and to elucidate ways in which pollination affects breeding techniques and selection of genetic material. Breeding crops which are more attractive to bees has enormous possibilities. For instance, soybean breeders are potential in developing hybrid soybeans with honeybees providing cross-pollination of inbred lines.

Hybrid crops of particular interest in pollination include carrot seed production which relies an insect crossing of male fertile to male sterile lines. Sunflower is a source of nectar and honeybee visits to improve seed yield. Present interest in hybrid sunflower may lead to greatly expanded use of their plants for oil production and increased demand for bee pollination. Safflower cultivars vary in their degree of self fertility and the degree to which they are self-pollinated. Gynoecium lines of musk melon have been developed and hybrid seeds are available for commercial trials.

Special Bee Management

Certain amount of cross pollination in variety of crops is necessary to prepare bee colonies for pollination.

(i) The area of the crop to be pollinated is so large that many colonies are needed during the flowering period.

(ii) In the newly reclaimed land and where wild bees are not available.

(iii) The crop flowers for only a short period which comes up early in the active season.

(iv) Bee population in normal colonies are small and are able to fly only during occasional short warm periods in the day.

(v) When the crop is grown in a greenhouse.

(vi) Pesticide applications would kill the wild bees.

Colonies managed for pollination should have a large population of foraging bees during the flowering season and colonies can be stimulated to collect pollen by feeding them with sugar syrup. Pollen collection by the bees quickly increases if the young brood is added to a colony (Free, 1967).

Frames of brood are usually inserted from other colonies that are not being used for pollination. The foragers themselves deposit their pollen loads into store in cells immediately and direct contact between foragers and brood seems to stimulate pollen collection. The beekeeper may also create a pollen dearth in colonies by removing combs containing stored pollen, or by trapping the pollen before it enters the hive.

The hives used for pollination in green houses, 4-8 frames of bees per colony are often enough. The beekeeper must check regularly that the colonies have enough food. It is often better if each hive has two flight entrances, one into the greenhouse and another outside. So that the bees can get access to other plants in flower. Bees from a hive just taken into a green house are likely to become disoriented, flying against the glass on plastic, but young bees that fly only inside the green house become accustomed to it.

MANAGEMENT OF OTHER INSECTS FOR POLLINATION

Insects other than bees are often essential pollinators of several plant species. However, they lack sufficient body hairs and necessary behaviour patterns and few of them transfer pollen from the anthers to the stigmas of the flowers. Furthermore, unlike the bees which forage consistently to obtain sufficient food for their young, most other insects forage to satisfy their own immediate needs on a variety of foods other than flowers. Probably, many of the most important, supplementary pollinators are various species of Diptera, including those belonging to the genera *Eristales, Syrphus, Platycheirus, Rhiugia, Calliphora, Lucilia, Sarcophaga, Bibio, Dilophus* and *Bombylius*.

It is evident that intensive cultivation of the land has destroyed many natural food sources and nesting sites of wild pollinating insects. Regular removal of hedge rows around field crops has destroyed flowering plants that help to provide food for beneficial insects. The use of selective herbicides results in a virtual absence of flowering weeds within cereal crops. Efforts should be made to favour planting trees and shrubs, from which pollinating insects are known to obtain large amounts of nectar and pollen, on road side.

Bees that are commercially useful for crop pollination must fulfil certain requirements: (a) they must be social bees or solitary bees that nest gregariously and can be mass-reared, so that large numbers can be made available at flowering time; (b) they must be amenable to mass rearing which produces a maximum adult population during the flowering; (c) they must be sufficiently resistant to diseases and parasites, under the conditions of mass rearing; (d) they must rear brood successfully in artificial nests; and (e) they must forage on and pollinate the commercial crop. Artificial rearing has important advantages over reliance on wild bee populations. Since the bees are on the crop during flowering, it is easier to protect them from insecticide applications whereever possible. It is also economically advantageous to maintain and increase wild populations of the bees and plants that provide forage for the whole of the bees life cycle and by using only pest control methods that safeguard the bees.

Managing Bumble Bees

Bumble bees (*Bombus*) are the social bees native to the north temperate zones, which have been reared as alternative crop pollinators to honeybees. Many bumble bees pollinate a wide variety of crops. Long tongued species such as *Bombus terrestris* and *Bombus hestorum* are especially useful for pollinating flowers whose corolla is so long that honeybees cannot reach the nectar in it. Wild nests may be searched and transferred to boxes for rearing them (Donovan and Wier, 1978).

The lucerne *(Medicago sativa)* is attracted by *Megachile rotundata*. It is sometimes difficult to get *Apis mellifera* to forage and pollinate lucerne, but two other species of bees, which can be mass reared are highly effective pollinators of this crop, *Megachile rotundata* and *Nomia melander*. *M. rotundata* is a highly gregarious solitary bee from the old world known in North America as the lucerne leaf cutter bee normally nesting in plants. The pupae are wintered in dry cool storage (1-4^0 C) in spring. About three weeks before flowering, they are incubated at about 29^0 C and 50-70 per cent. Relative humidity conditions are regulated so that the bees start to emerge when flowering starts. After emergence, the adult females mate and then make their nests in the stems, lining each cell with pieces of lucerne leaf. The bees forage for pollen and nectar on the lucerne flowers and pollinations occurs later.

The alkali bee (*Nomia melander*) is a highly gregarious solitary bee native to North America, nesting in fine soils, pollinate lucerne most efficiently and high seed yields can be produced. Many of the megachilid bees are suitable for mass rearing, because of their natural nesting sites either in soil, wood or stem. In the Anthophoridae, the species of *Ceratina* are small, brightly coloured metallic bees of which some species have been used to pollinate lucerne. *Peponapis* and *Xenoglossa* are new world bees that forage on flowers of cucurbits and are especially well adapted for pollination.

Aids of Pollination

Pollen inserts

A pollen insert also known as a pollen dispenser is designed to apply viable pollen to worker bees as they leave the hive. Its major use is to provide cross-pollination in fruit orchard which lacks adequate pollinators. The accelerating trend towards dense planting of dwarf fruit trees provides a situation where mechanical harvesting from solid block plantings would be of great interest. Pollen dispensers provide only possible means of pollinating solid block plantings of apples and other crops which require cross-pollination and dispenser perfection would be a key factor in drastically changing fruit production methods. There should be no problem in developing dependable dispensers and techniques which will apply viable pollen to outgoing bees.

Pollen dispensers (hive inserts) apply pollen to bees which can cross-pollinate crops when few pollinizer varieties are available. Thus dispensers may increase pollination efficiency of bees without maintaining them on the target crop. It has been suggested that dispensers may stimulate foraging activity and induce bees to forage for the type of pollen in the dispenser (Lotter, 1960). However, cage trials and behavioural studies of pollen in dispensers on bees, on stigma and on floral surfaces show that pollen dispensers are more valuable as pollination aids, for maintaining bees on crops. Dispensable pollination units (DPU) are small combless colonies hived in inexpensive containers that are tracked on target crops and then destroyed after flowering is over (Free,1979). Originally DPUs consist of wooden packages covered with wire screen and wrapped up in tar paper. DPU colonies with laying queens collect more pollen and nectar than queenless units and show greater foraging activity (Kauffeld *et al.*, 1970). Little pollen but much nectar is collected by the bees of DPUs . Therefore, it may be important to use DPUs in crops that are pollinated mostly by nectar collecting bees. Further, the percentage of flowers of a crop that is open may differ greatly during the different stages of its flowering period and the number of colonies needed will vary accordingly.

Attempts were made to saturate all the bee crops in an area by providing large number of colonies. Moreover it has become apparent recently that the receptivity of flowers of many plant species diminishes soon and pollination on the day a flower opens is likely to be the most successful and sufficient pollinators should be present to ensure pollination. Attempts were also made to relate the number of foraging honeybees often given number of flowers needing pollination, with seed or fruit set (Degrandi-Hoffman *et al.*, 1988). To do this accurately, it is necessary to know the rate at which the flowers are visited, the number of visits, needed per flower, the pollinating efficiency per visit the number of pollen grains deposited on the sigmas per visit, the effective pollination period of the flower including the time of pollen availability and stigma receptivity, the movement of honeybees on flowers and the number of foraging hours per day and the number of flowers available per unit area at different environmental conditions. The number of bee visits depends largely on quality and quantity of nectar and pollen availability. So, the estimates of the number of honeybee colonies needed must take into account. In addition, the number of other insect pollinators present, the size of the colony population, the proportion of nectar and pollen gatherers; the percentage of honeybee foragers that visit the target crop; and the percentage of those visiting the target crop species are also to be considered.

POLLINATION HAZARDS AND PREVENTION

Honeybees have a preference for specific flowers. They do not distinguish between weeds and cultivated crops. It is believed that sugar concentration and balance among glucose, fructose and sucrose influence foragers to prefer one species over the other. The nectar of pears is low in sugar (5-10 %) while apple bloom almost during the same time, produces nectar which contains about 40 per cent sugar. Foraging workers have been known to abandon pears for other plants. Sugar contents of sweet and red clover is not dramatically different but foragers prefer sweet clover possibly because of the more sugar content.

Pesticides Toxicity in Honeybees

Pesticide contaminated natural ecosystem is known to cause several long term ecological effects such as reduction in pollinator density, increased inbreeding, increased homozygosity, reduced fecundity and population size, reduced genetic variation and alteration in breeding systems of the plants. The pesticide induced changes in natural ecosystems are thus ecologically catastrophic, economically disastrous and evolutionarily unpredictable. In an agricultural ecosystem, decline in crop and honey production is the most frequently observed effects of pesticides.

Rapidly changing agricultural practices and a substantial decline in bee flora have caused an increasing dependence of honeybees on cultivated crops. A large number of honeybees are subjected to mortality affecting directly honey production and indirectly seed production which constitute the major products of pollinators' activity (Naim, 1982). The long-term qualitative effects of insecticides on bee pollination such as slow poisoning of brood through ingestion of contaminated pollen and nectar and exposure of foraging bees to insecticides through translocation in nectar are of crucial importance in maintaining plant-pollinator dynamics.

Bee poisoning is the accidental killing of bees through the use of insecticides in pest control. Pesticides used for plant protection are generally hazardous to honeybees. Mostly bee poisoning occurs when insecticides are applied to the crop during blooming. Drifting of toxic sprays or dust into the adjoining crops in bloom is another hazard. Bee comes in contact with insecticide residues on plants and get poisoned either by taking the contaminated dew, pollen or nectar.

Systemic insecticides are extremely toxic to honeybees. These insecticides provide long range control of certain insect and mite species because they are translocated to all parts of the treated plants including the nectar and pollen, and remain there upto several weeks. It has been shown that systemic insecticides applied as sprays on thc plants or as granules on the soil can penetrate floral nectar in sufficient quantity to kill honeybees (Jaycox, 1964). Insecticide poisoning of honeybees collecting pollen and nectar from Coconut (Beetsma, 1975), *Brassica* (Hameed *et al.,* 1973) and Rubber (Deodasan, 1972) has been investigated.

Pyrethroids elicited a sub-lethal toxic response whereby the bees minimized their contact exposure to the toxicants by returning to the colony for a period of recovery. The repellent effect on bees is analogous to the knock down effect because after exposure the bees return to the colony and are relatively inactive for sometime. Thus the pyrethroids may be used to control insect pests in situation where exposure to honeybees is inevitable. They are low hazard only when used properly i.e. day time application and the use of oil based adjuvants may increase the hazard enormously. If properly used, pyrethroids can greatly reduce the conflict between beekeepers and the users of insecticides

Preventive Measures for Bee Safety

Bee losses due to indiscriminate use of pesticides is a serious threat to the beekeeping industry. The following measures can be adopted for bee safety.

- Identifying insecticides which are safe or have a low toxicity to bees.
- Identifying pesticides with a short residual action.
- Insecticides toxic to bees should not be applied when the crops are in blooming stage because bees and other pollinators visit most of the crops during their blooming.

- Beekeeper can either remove the colonies or confine the bees for short periods when pesticides are applied on the crops.
- Pest management strategy should be integrated with the pollinating behaviour of honeybees.
- Imparting training, education and awareness to farmers and beekeepers.

(a) *Habitat destruction*

Habitat destruction affects pollinators by destruction of food sources, nesting or oviposition sites and resting or mating sites. The destruction of food sources is best illustrated by examples of the removal of vegetation, which provides pollinator food when the crops are not in bloom from agricultural areas. Very often the vegetation which is removed is regarded as unwanted as weeds or competition for the crop plants, yet is invaluable to pollinator and other beneficial insects.

(b) *Natural enemies of pollinators*

Mite attack of honeybees has evoked major concerns as tracheal mites, *Varroa and Tropilaelaps* have spread at alarming rates. It has been suggested that many amateur and small scale beekeepers will abandon their activities because of the additional complexities of bee management associated with monitoring for mite attack and controlling them once detected. *Apis cerana* in India is affected by Thai sacbrood virus disease. Therefore, management of pests, predators and diseases is required in all beekeeping areas (Sihag, 1991).

(c) *Diversification of pollinators*

Although it must be conceded that honeybees are the most valuable pollinators in agriculture, they are not the full source for crop pollination. The examples of greater efficiencies of honeybees for pome fruit and blue berry pollination, leafcutter bees for alfalfa pollination, bumble bees for tomatoes and other solanaceous crops in green houses and carpenter bees for passion fruit. The lack of pollination brought about by honeybees on oil palm, various annonaceous fruit crops, red clover and other crops are even more extreme as sound examples of the need to consider alternative pollinators for many crops.

(d) Habitat management for wild pollinators

The natural pollinators of wild and agricultural plants include a wide variety of organisms besides bees. There is a wide array of bees and non-bee pollinators capable of pollinating agricultural crops. Over a hundred different insect species can be observed on flowers of some fruit trees though not all contribute significantly to their pollination. Maintaining such a diverse insect fauna increases the chances of sufficient pollination without the need for additional pollinator population. In order to have sufficient wild pollinators, their habitats must be preserved and maintained. That means adult and larval stages of the pollinators need to find their food. Some insects require certain soil conditions to survive during one of their life stages. For migratory species like some humming birds or Asian and African honeybees, the habitats needed at each extreme of the migratory range should be preserved in order to assume that sufficient numbers return during the next migratory season.

REFERENCES

Adlerz, W.C., 1966. Honeybee visit numbers and water melon pollination. *J. Econ. Ent.*,59: 28-30.

Alpatov, V.V., 1984. Bee races and red clover pollination. *Bee Wld.* 29: 61-63.

Anais, G. and Torregrossa, J.P. 1978. Possible use of *Exomalopsis* species in pollination of solanaceous vegetable crops in Guadaloupe (French Antilles). *Proc. IV. Intl. Symp. Pollination*, pp. 321-329.

Badan, K.L. and Anderson, S.R., 1962. Effect of pollen and nectar collecting honeybees on seed yield of birds foot trefoil, *Lotus corniculatus* L. *Crop Sci.*, 2: 148-149.

Beetsma, J. 1975. Beekeeping in Surinam. Maandschur. *Bijent.* 77: 187-188.

Benton, F. 1896. The honeybee, a manual of Instruction in Apiculture. *US Dept. Agr., Div. Ent. Bul.*, 1: p. 118.

Bohart, G.E. and Nye, W.P., 1960. Insect pollinators of carrots in Utah, Bull. *Utah Agri. Exptl. Stat.*, 41: 1-16.

Cale, G.H. 1968. Pollen gathering relationship to honey collection and egg laying in honeybees. *Am. Bee. J.*, 108: 8-9.

Casteel, D.B., 1912. The behaviour of the honeybee in pollen collecting. *U.S. Bull. Ent.* No. 21, p. 36.

Choi, S.Y. and Lee, M.L., 1988. Studies on the diurnal flight activities of honeybees on peach flowers. *Korean J. Apic.,* 3: 81-89.

Crane, E., 1990. Bees and beekeeping. *Heinmann. Oxford*, pp. 260-269.

Darwin, C. 1887. The different forms of flowers on plants of the same species. *John Murray*; London.

DeGrandi-Hoffman, G., Terry, I. and Huber, R.T. 1988. Incorporating fruitset estimates with thrips management to create a decision support system for apples. *Hort Sci*., 23: 571-574.

Deodasan, A. 1972. Rubber plantation and beekeeping. *Indian Bee J*. 34:(1-2) 38-39

Dhaliwal, H.S. and Sharma, P.L., 1973. The foraging range of the Indian honeybee on two crops. *J. Apic. Res.,* 12: 131-134.

Dolgova, L.P., 1928. Effect of some factors upon the visiting of honey plants. *Opait pas*, 176-179, 248-253 (In Russian).

Donovan, B.J. and Wier, S.S. 1978. Development of hives for field population increase and studies on the life cycles of the four species of introduced bumble bees in New Zealand. *N.Z.J. Agric. Res*. 21: 733-756.

Dowden, A.O. 1964. The secret life of flowers. *The Odyssey Press*, New York, p. 45.

Drake, C.J., 1948. Influence of insect on alfalfa seed production. *J. Econ. Ent.,* 41: 742-750.

Eckert, J.E. 1933. The flight range of honey bee. *J. Agric. Res.* 47: 257-285.

Free, J.B. 1959. The effect of moving colonies of honeybees to new sites on their subsequent foraging behaviour. *J. Agric. Sci.,* Camb., 53: 1-9.

Free, J.B., 1960. The behaviour of honeybees visiting the flowers fruit trees. *J. Anim. Ecol.,* 29: 385-395.

Free, J.B. 1963. The flower constancy of honeybees. *J. Anim. Ecol.*, 32: 119-131.

Free, J.B. 1967. Factors determining the collection of pollen by honeybee foraging. *Anim. Behav*., 15(1): 134-144.

Free, J.B. 1968. The pollination of strawberries by honeybees. *J. Hort. Sci*., 43: 107-111.

Free, J.B. 1970. The effect of flower shapes and nectar guides on the behaviour of foraging honeybees. *Behaviour*. 37: 259-285.

Free, J.B. 1975. Observations on the pollination of papaya (*Carica papaya* L.) in Jamaica. *Trop. Agriculture*, 52: 275-279.

Free, J.B. 1979. Managing honeybee colonies to enhance the pollen gathering stimulus from brood pheromones. *Appl. Anim. Ethol.*, 5: 173-178.

Free, J.B., 1993. Insect pollination of crops (Second edition) *Academic Press*, London, UK.

Frisch, K. Von. 1967. The dance language and orientation of bees. *Oxford Univcrsity*, London.

Frisch, K. Von. 1971. Bees, their vision, chemical senses and language. Ithaca *Cornell Univ. Press.* p.1 57.

Gilbert, F.S., 1980. Flower visiting by hover flies (Syrphidae). *Ann. Rev. Ecol and Syst.,* 6: 139-170

Goodwin, R.M., 1997. Feeding sugar syrup to honeybee colonies to improve pollination: A review. *Bee Wld,* 78: 56-62.

Gould, J.L. and Gould, C.G. 1988. The honeybee. *New York Scientific American* Library.

Goyal N.P., 1978. Performance of *Apis mellifera* and *Apis indica* as observed in Punjab plains. *Indian Bee J.,* 40: 3-5.

Goyal, N.P., Singh, M. and Kandoria, J.L., 1989. Role of insect pollination in seed production of carrot, *Daucus carota* Linn. *Indian Bee J.*, 51(3): 89-93.

Goyal, N.P. and Gupta, J.K., 1994. Development of beekeeping with *Apis mellifera* L. in India. Present status and future scope. *Indian Bee J.,* 56: 127-131.

Gupta, J.K., Mishra, R.C. and Kumar, J., 1984. Plentrumthus as forager for *Apis cerana indica* F. and *Apis mellifera* L. *Apidologie,* 15(1): 72-82.

Hameed, S.F.; Adlakha, R.L. and Giamzo, S.P. 1973. Relative toxicity of some insecticides to workers of *Apis mellifera. Madras Agric. J.* 60(7): 552-556

Heinrich, B. 1981. The mechanisms and energetics of honeybee swarm temperature regulation. *J. Exp. Biol.*, 91: 25-55.

Hodges, D., 1952. The pollen loads of the honeybee. International Bee Research Association, London, p. 120.

Hopkins, C.Y., Jevans, A.W. and Boch, R. 1969. Occurrence of octa-deca-trans-2. Cis-9, cis-12-trienonoic acid in pollen attractive to honeybee. *Can. J. Biochem*. 47: 433-436.

Inouye, 1980. The effect of proboscis and corolla tube lengths on patterns and rates of flower visitation by bumble bees. *Oecologia*, 45: 197-201.

Jay, D. and Jay, C. 1984. Observations of honeybees on Chinese gooseberries (Kiwi fruit) in New Zealand. *Bee Wld.*, 65: 155-166.

Jaycox, E.R., 1964. Effect of honeybees on nectar from systemic insecticide treated plants. *J. Econ. Entomol.*, 57: 31-35.

Kapil, R.P., Grewal, G.S., Kumar, S. and Atwal., A.S., 1971. Insect pollinators of rape seed and mustard, *Indian J. Ent.,* 33: 61-66.

Kauffeld, N.M.; Levin, M.D.; Roberts, W.C. and Moeller, F.E. 1970. Disposable pollination units, a reviewed concept of crop pollinaton. *Am.Bee J.* 110: 88-89.

Kumar, J., Gupta, J.K., Mishra, R.C. and Dogra, G.S., 1988. Pollination studies in some cultivars of cauliflower (*Brassica oleraceae* var. *botrysis* L.). *Indian Bee J.*, 50(4): 93-95.

Kumar, J., Mishra, R.C. and Gupta, J.K., 1989. Effect of honeybee pollination on onion (*Allium cepa* L.) seed production. *Indian Bee J.,* 51(1): 3-5.

Kendall, D.A. and Solomon, M.E. 1973. Quantities of pollen on the bodies of insects visiting apple blossom. *J. Appl. Ecol.*, 10: 627-634.

Kevan, P.G and Baker, H.G. 1983. Insects as flower visitors and pollinators. *Ann. Rev. Entmol.* 28: 407-453.

Lotter, J.V. 1960. Recent developments in pollination techniques of deciduous fruit trees. *Decid Fruit Grow*. 10: 182-311.

Mahanta, M. and Rahman, A., 1997. Studies on pollination of Litchi (*Litchi chinensis* Sonn.) by honeybees (*Apis cerana* F.). *Indian Bee J.,* 59(2): 74-75.

Mann, G.S. and Singh, G., 1983. Activity and abundance of pollinators of plum in Ludhiana (Punjab). *Am. Bee J.*, 123: 595.

McGregor, S.E. 1976. Insect pollination of cultivated pollination crop plants. Agriculture Handbook. 496. USDA Agricultural Research Service Washington. DC.

Mishra. R.C., Dogra, G.S. and Gupta, P.R., 1976. Some observations on insect pollination of apple. *Indian Bee J.,* 38(1-4): 20-22.

Mishra, R.C. and Kaushik H.D., 1992. Effect of cross-pollination on yield and oil content of *Brassica* spp. and *Eruca sativa* with pollination efficiency of honeybees. *Ann. Ent.,* 10(2): 32-37.

Naim, M. 1982. Integration of bee pollination with pest management. *Indian Bee J.*, 44(1): 10-11.

Panda, P., Sontakke, B.K. and Sarangi, P.K., 1988. Preliminary studies on the effect of honeybee (*Apis cerana indica* Fabr.) pollination on yield of sesamum and niger. *Indian Bee J.*, 50(3): 63-64.

Pankeratova, E.P., 1958. Data on the biology of flowering and pollination of carrots. *Dokl. Tsksha.* 36: 118-123.

Percival, M.S., 1965. Floral Biology, *Oxford; Pergamon Press*, p. 243.

Praagh, J.P. Van. And Ohe, W. Vonder., 1983. The role of scents in pollination by the honeybee. *Acta Hortic.*, 65-67

Rajagopal, D., Veeresh, G.K., Chikkadevaiah, Nagaraja, N. and Kencharaddi, R.N., 1999. Potentiality of honeybees in hybrid seed production of sunflower (*Helianthus annuus* L.). *Indian J. Agri. Sci.*, 69(1): 40-43.

Rajagopal, D., Kencharaddi, R.N. and Nagaraja, N., 2000. Utilization of honeybees in sunflower hybrid seed production. In: *Asian Bees and Beekeeping, Progress of Research and Development* (Eds. Mastuka, M. Verma, L.R., Wongsiri, S., Shrestha, K.K. and Pratap, U.) Oxford and IBH Publishing Co. Pvt. Ltd., New Delhi, p. 186.

Rahman, S. and Rahman, A., 1993. Comparative pollen gathering activity of *Apis cerna indica* F. and *A. mellifera* L. under Jorhat conditions of Assam (India). *Indian BeeJ.*, 55(3-4): 42-46.

Reddy, C.C. and Bai, A.R.K., 1979. Foraging and brood rearing activity of *Apis cerna indica* during 1975-77. *J. Apic. Res.*, 18: 7-10.

Ribbands, C.R., 1955. The scent perception of the honeybee. *Proc. Royal . Ent. Soc.*, (B): 143: 367-369.

Savoor, R.R. 1998. Pollinator management: An ecofriendly green revolution eludes India. *Curr. Sci.,*. 74(2): 21-125.

Shah, F.A. and Shah, T.A., 1980. Flight range of *Apis cerana* from Kashmir. *Indian Bee J.*, 42: 48.

Sharma, P.L., 1958. Sugar concentration of nectar of some Punjab honey plants. *Indian Bee J.*, 20(7): 81-91

Sheeslay, B. and Poduska, B., 1970. Relative values of honeybee colonies for almond pollination. *Gleanings Bee Cult.*, 98: 486-491.

Shuel, R.W. 1955. Nectar secretion. *Ame. Bee. J.*, 95: 229-234.

Shuel, R.W., 1975. The production of nectar. The hive and honeybee (eds. Dadant and Sons), pp. 264-282

Shuel, R.W. 1989. Improving honey production through plant breeding. *Bee Wld*. 70: 36-45.

Sihag, R.C., 1985. Floral biology, melittophily and pollination ecology of cultivated cruciferous crops. In: *Recent advances in pollen research* (ed. Varghese, T.M., Allied Publishers, New Delhi, India), pp. 241-268.

Sihag, R.C., 1991. Ecology of European honeybee, *Apis mellifera* L. in semi-arid subtropical climates. II. Seasonal incidence of diseases, pests, predators and enemies. *Korean J. Apic.*, 6(1): 16-26.

Singh, G., 1971. Sucrose solution imbibing capacity of *Apis indica* F. (Hill strain) in relation to various sucrose solution concentrations and body weight. *Indian Bee J.*, 33: 1-54.

Singh, Y., 1979. Pollination activity on strawberry at Jeolikote. (Dist. Nainital, India). *Indian Bee J.*, 41: 17-19.

Tanda, A.S. and Goyal, N.P., 1979. Insect pollination in Asiatic Cotton. *J. Apic. Res.*, 18: 64-72.

Verma, S.K., 1987. Preliminary studies on the effect of honeybees on the yield of greater cardamom. *Indian Bee J.*, 49: 25-26.

Verma, L.R. and Pratap, U., 1994. Foraging behaviour of *Apis cerana* on cauliflower and cabbages and its impact on seed production. *J. Apic Res.*, 33: 231-236.

Wait, M.B., 1895. The pollination of pear flowers. *Bull. Div. Veg. Physiol. Path. U.S. Dept. Agric.*, 5.

5

Livestock Resources of Indian Himalayas: Present Status, Constraints and Future Thrusts

S.K. Verma and *K.S. Dadhwal*

The Himalayas are supposed to be the highest and youngest and most unstable mountain ecosystem in the world. These are spread across eight Asian countries including India, China, Nepal, Afghanistan, Pakistan, Bangladesh, Bhutan and Myanmar. In India, the Himalayas cover an area of 53.7 million ha, which is 16.4 per cent of the total geographical area of the country. The Indian Himalayas cover 14 states (fully or partly) and 93 districts in two distinct geographical flanks of the western and eastern region. The western Himalayan ranges extend from Jammu and Kashmir, Himachal Pradesh, Shiwaliks of Punjab and Haryana and Uttaranchal up to the western border of Nepal. The eastern Himalayas cover northeastern hill states of Arunachal Pradesh, Manipur, Meghalaya, Mizoram, Nagaland, Sikkim, Tripura and parts of Assam and West Bengal.

About 15 per cent area of the Indian Himalayas is permanently covered with snow, which provide perennial flow to the tune of 1204 km^3. annually to the vast Indo Gangetic plains (Chandra, 1994). The western Himalaya is relatively dry and cold, while the eastern Himalayan region is extremely wet. Average annual rainfall in the region varies from <150 to 4000 mm, representing the lowest and the highest rainfall zones of the country. The mean annual temperature of the region ranges from 8°C to 22°C. Wide range of variation in altitude, latitude and longitude generate vast diversity in microclimate, habitats, soils, vegetation and livestock in the region. (Samra *et al.*, 1999) (Table 1)

ANIMAL SCENARIO IN INDIAN HIMALAYAS

Animal-agriculture has played a major role in growth and development of Himalayan region. Food crop production is a major land based activity in the Himalayan region. Traditional cultivation practices coupled with harsh climatic conditions produce low crop yields. Productivity of different field crops shows wide variation in the region.

Although the region has plenty of water resources but more than 75 per cent of its gross cropped area is rainfed. Typographical limitations, small and fragmented land holdings as well as lack of awareness are important reasons for small area under irrigation. ***In-situ*** moisture conservation and rainwater harvesting and recycling techniques are not popular in the farming community. This has led to high runoff during rainy season and depletion of soil fertility through increasing soil erosion. Therefore, the productivity of food crops is very low as compared to the national average.

Due to the low productivity of crops, hill farmers are not able to fulfill their basic needs; therefore, they are more dependent upon livestock for subsistence. So in hill farming system, livestock acquire special importance. The majority of hill farmers operate mixed livestock-crop farming. The number of livestock per hectare of cultivated land as well as per thousand human populations is much more in hill districts as compared to plains. Livestock supplement the income of the farmers up to 26 per cent in mid-hills and are considered as a capital asset (Verma, 2001[b]) (Fig 1). But the production potential of livestock is very low as compared to national production due to various problems. Livestock is considered as an incidental proposition but not as an animal husbandry enterprise though livestock sector plays a unique role in the Himalayan economy. Cattle and buffalo have been a source of food. Sheep and goat provide fibre and meat. Horses and ponies are a mean of transport.

Table 1. Ecosystem diversity in the Indian Himalayas

Attribute	*North-eastern Himalayas*	*Western Himalayas*
Geographical area (million ha)	20.22	33.49
Climate	Extremely wet	Dry and cold
Latitude	21°5'-29°3' N	29°31'-36°58' N
Longitude	85°5'-97°5'E	73°26'-83°30'E
Elevation (m, amsl)	200-8598	400-8611
Annual Rainfall (mm)	2800-12000	350-3000
Mean Temperature (°C)	8-22	8-22
Physiography	Gentle slopes	Steep slopes
Soils	Deep to shallow and acidic to neutral in reaction	Shallow and neutral to acidic in reaction
Vegetation	Tropical forest	Cold loving and draught resistance
Farming System	Shifting cultivation and settled	Settled and migratory
Livestock	Pig, Mithun, Yak, Cattle and poultry	Cattle, Sheep, Goat, Buffalo and Yak
Dominant land ownership	Community and private	Government and private
Ethnicity (dominant religions)	Christian, Muslim and Buddhist	Hindu, Muslim and Buddhist
Major tribes	Bhutia, Lepchas, Chakma, Naga and Mizo with a large number of sub tribes	Bhutia, Gaddis, Gujjar, Rajees, Tharus, Buxas and Jaunsarees
Property inheritance system	Matriarchal	Patriarchal
Marriage system	Polygamy in some tribes	Polyandry in some tribes
Minerals	Petroleum and natural gas, coal, silmanite, limestone and glass	Gypsum, limestone, dolomite sand and pyrite

(*Source:* Samra *et al.*, 1999)

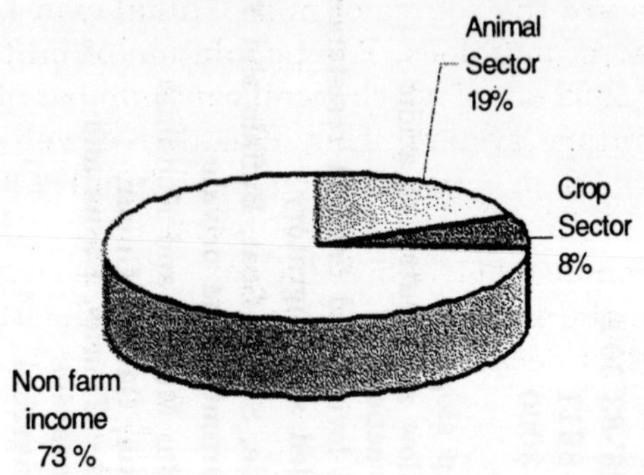

Fig 1. Contribution of livestock sector in farm economy in mid hills of Uttaranchal

LIVESTOCK WEALTH OF HIMALAYAN REGION

Himalayan region has a large variety of livestock *viz.* cattle, buffaloes, sheep, goats, horses, ponies, pigs, mules, donkeys, yaks and mithuns. In addition poultry and fishes are also another species throughout the Himalayan region. Each species has its own importance.

Cattle

Himalayan region has a large number of cattle populations. The population of cattle was 10.57 million in 1992. The size of hill cattle reduces with the increase in altitude. These are of small stature having thick neck, under developed udder, bright eyes and attentive ears. Average weight of the cattle is 225 kg. Due to its low weight, it can walk on the steep slopes easily. Both bullocks and cows contribute to the hill farmer and in that way assets to the Himalayan economy. The composition of the cattle population has changed significantly due to one or another reason. The proportion of cows and heifers is increasing since 1972, while the number of bullocks has declined drastically. Among females, the population of adult female has increased. There has been a significant growth in the number of crossbred cows, while the proportion of native cows has declined.

Buffaloes

Buffaloes are the source of milk. Himalayan region has a large number of buffaloes. The population of buffaloes was 2.71 million in 1992 out of which significant number of buffaloes is under migratory system. They can thrive well up to an elevation of 3500 m amsl and on poor quality feeds and concentrates. Most of the buffaloes found in the Himalayan region are of non-descriptive breed. Considerable changes have taken place in the population of buffaloes in the Himalayan region during last decade.

Sheep

Sheep is the source of mutton and wool. Sheep population in Himalayan region was 4.47 million in 1992. In the upper Himalayan region sheep is a vital source of income and employment generation for marginal and landless families.

Goat

Goat is a major source of meat. Himalayan region has a sizable stock of goats. The population of goats in this region was 5.06 million in 1992.

Horses and Ponies

Himalayan region has a reasonable stock of horses and ponies. Their number aggregated 2.07 lakhs in 1992. Horses and ponies are means of transport in this region.

Pigs

Pigs are mainly found in the northeastern Himalayan region. Most of the farmers in the western Himalayas normally do not rear pig because of social *taboos*. The population of pigs was 1.83 million in 1992.

Yaks

Yak is a source of milk and transport in the temperate and cold regions of the northeastern Himalayas. The population of yak in the region was 0.58 lakh in 1992.

Mithuns

Mithun is found in the northeastern Himalayan region. It is a source of meat and transport. It is also a symbol of social status among the tribes in Arunachal Pradesh. The population of mithuns was 1.54 lakh in 1992.

CONSTRAINTS IN LIVESTOCK PRODUCTION SUB-SECTOR IN INDIAN HIMALAYAS

High Population of Low Yielding Animals

Indian Himalayan region has a very huge population of low yielding animals as compared to the rest of the country. Majority of the livestock includes local breeds of cattle, buffaloes, goats and sheep. Milk, wool and mutton productivity of these animals is low. All the animals except buffaloes are dependent on grazing. The improved breeds constitute only about 10 per cent of the total livestock population in the region. The people prefer tiny low-productive animals for easy maintenance. Inbreeding in these animals has resulted in low productivity due to the absence of new germplasm.

High Mortality

There are number of diseases such as foot and mouth disease, hemorrhagic septicemia, mastitis, brucellosis, tuberculosis, black quarter etc. that affect livestock production and sometimes cause enormous economic losses due to the death of animals. The livestock infant mortality is high due to lack of proper management and health care facilities. Incidences of disease epidemics cause high death rate in the livestock herd. Undernourishment and inadequate housing also increase mortality in the animals.

Malnutrition

Deforestation and over-grazing have induced fodder scarcity in terms of quantity as well as quality in the region. Due to the increasing demand for foodgrains for ever-increasing human population, non-cultivable and forest land is being converted into cultivable land reducing the availability of fuelwood and fodder. Poor economic conditions of farmers prevent them for feeding the required quantity of concentrates to the animals. Cultivation of fodder is almost absent in the region. In the middle hills of Uttaranchal, only about 75 per cent of the livestock's total feed requirements are met. Good quality green fodder is generally available only from July to

November (the wet season). Thus the major problem is the lack of sufficient quantity of dry matter and green fodder during the long dry season from December to June. Thus the energy and protein requirements are met only up to 76 and 53 per cent, respectively. This forces the farmers to resort to selective feeding of productive livestock leaving the growing and currently non-productive livestock to feed for themselves. Consequently, the livestock are under-nourished and have poor health, low fertility and productivity. This results in unusually high age at 1st calving and increased calving interval making livestock rearing an uneconomical enterprise.

Low Productivity

Milk is the main product from the buffaloes and cattle. Low milk yield of the local cow (200-250 litres/lactation) and buffaloes (500-550 litres/lactation) is only due to poor feeding and management practices. (Verma, 2001[a])

Inaccessibility

In the entire Himalayan region transportation facilities are less developed as compared to the rest of the country. Difficult circumstances of the hill region hamper the staff working for the development of different infrastructure. Monitoring of different schemes is adversely affected due to the lack of transport system.

Lack of Marketing Infrastructure

Marketing of the livestock and livestock products is the most unexplored area. Marketing infrastructure is under-developed for disposal of different livestock product. There is no functioning cooperative in more than 85 per cent villages for the sale of fresh milk. The purchase and sale of livestock have been conventionally done through middlemen.

Low animal-agriculture productivity, industrial backwardness, poverty, lack of social services, inadequate infrastructure, lack of income-earning opportunities and general backwardness of the economy has led to significant lack of employment opportunities in all the sector of economy and have turned out to be the major cause of migration of rural youth to the cities/down plains. As the land for crop production is limited and it cannot be increased further, therefore, there is an ample scope for the promotion of animal husbandry in Indian Himalayan region, which will contribute a lot in rural economy.

STRATEGIES FOR DEVELOPMENT OF LIVESTOCK SECTOR IN INDIAN HIMALAYAS

Improvement in the Feed Resource Base

Feed shortage is a major problem for livestock farmers throughout the Indian Himalayas. The shortage of animal feed is acute during the dry period and winter and livestock are generally underfed to the extent of 33 per cent. The fact that animals are undernourished has resulted in late maturity, high calf and adult mortality, poor lifetime performance and infertility in cattle and buffaloes. Shortage of animal feed can be overcome by increasing the production from fodder trees and grasses in community controlled areas and by enhancing the productivity of fodder grasses and fodder trees from private farms. Uncultivable community and private land, marginal land, terrace risers should also be used for planting of fodder species (trees, grasses and shrubs). Besides this promotion of efficient use of agricultural residues including ammonia (urea) treatment of crop residues and dry forest grasses as well as their supplementation with urea-molasses mineral blocks is advocated. Surplus crop residues produced in lower Himalayan region should be used for creation of fodder banks.

Improvement in Animal Productivity

Poor feeding practices, local non-descriptive breeds and their sub-optimal nutrition are the primary reasons behind poor livestock productivity. Scientific animal breeding programme can enhance productivity of the animals. Bulls of high production potential of local breeds can be used for improving the livestock productivity. Besides this, balanced feeding and proper livestock management practices can also enhance the productivity of animals.

Improvement in Livestock Support Services

There is a significant shortage of livestock support services. One of the reasons for farmers' reluctance to adopt improved animal husbandry practices is the insecurity of their investment. Animals, particularly improved breeds, are more prone to diseases and cannot be sustained without proper health care services. The number of skilled livestock technicians is also very small and extension and live agencies are under-staffed. The ratio of veterinarian to livestock is such that each veterinarian has to serve about 12000 livestock. To ensure support services, human resource development has to be done. Para-vets can also be deputed to serve hill farmers in an effective manner. Arrangements should also be made to start mobile veterinary clinics.

Promotion of small-livestock for Marginal and Disadvantaged Farmers

Small livestock such as rabbit, goats and poultry have been an important source of immediate cash income for every household in the hills. There is a considerable scope for developing small livestock to help and improve the income levels of resource-poor farmers because it requires less capital and land investment. There is a great scope for backyard poultry keeping in view the geographical situation and participation of women in animal-agricultural production system. Micro-livestock should be included in policy agenda. Government and non-government organizations should introduce such schemes that provide credit support for disadvantaged groups. Small enterprises based on goat and poultry raising for easy cash flow should be encouraged. Training should be provided to women for small-scale poultry raising and marketing live animals and products. Establishment of a regular and reliable marketing network for disposal of marketable surplus at remunerative prices is also needed.

Improvement of Soil Fertility through Better Management of Livestock and Forage Crops

Livestock contributes in upholding of soil fertility. However, there is still a lot of scope for contributing to the sustainable management of soil through management of livestock. In a semi-stall fed system, about 44.1 per cent of manure produced during the dry time is lost while animals are grazing in the forest or on community fallow land (Anonymous,

1980). Even in an entirely stall-fed system, potential fertilizer is lost due to poor housing structure in efficient FYM preparation, and improper application methods therefore, efficiency of the nutrient recycling of the crop-forest-livestock system needs to be improved. For efficient nutrient recycling, knowledge and technologies should be disseminated among farmers.

FUTURE THRUSTS

- Augmenting feed resources of the region both from within and from surplus areas.
- Improved utilization of existing feed resources.
- Breed improvement efforts to reduce age at 1st calving and calving interval to enhance milk, wool and meat productivity.
- Creating marketing infrastructure for livestock products.
- Training of unemployed educated youths in para-veterinary services and marketing of livestock products.

CONCLUSIONS

It can be concluded that Indian Himalayan economy is basically dependent upon livestock sector and there is an ample scope for livestock development, which includes improvement in feed resource base and livestock support services through which animal productivity can be enhanced. Promotion of micro livestock is another potential option for socio-economic upliftment of hill farmers.

REFERENCES

Anonymous, 1980. Progress report of "Ford Foundation Funded Project" on "Integrated natural and human resource planning and management in the hills of UP". GBPUAT, Pantanagar.

Chandra, C. 1994. Water resources of India. In: Tiwari, D.N. (Ed.) Himalayan ecosystem. International Book Publishers, Dehradun. pp. 212-145.

Samra, J.S.; Dhyani, B.L. and Sharma, A.R. 1999. Problem and prospects of natural resource management in Indian

Himalayas—A base paper. Hill and Mountain Agro-ecosystem Directorate, NATP, CSWCRTI, Dehradun. p. 149.

Verma, M.L. 2001[a]. Reforms in the concept of rural development in the field of animal husbandry. Workshop on reforms in the concept of rural development. Pantnagar, GBPUAT.

Verma, S.K. 2001[b]. Evaluation of livestock production sub-system under mixed farming in Garhwal region of Uttaranchal. Thesis, Ph.D., GBPUAT, Pantnagar.

6

State of Coral Reefs in India

T.K. Ghosh

INTRODUCTION

A coral reef is 'a ridge or mound of limestone, the upper surface of which lies, or lay at the time of its formation near the level of the sea, and is predominantly composed of calcium carbonate secreted by organisms, of which the most important are corals' (Vaughan 1911). Reef building is an interaction between biological and geological processes, the former represented by primary (corals) and secondary (other calcareous organisms including coralline algae) reef framework growth, and the latter by sedimentation and early cementation. The coral reef ecosystem is one of the most complex, rich and productive environments. It is unique among the marine association and it plays an important role in sustainable biological diversity, global biogeochemical cycle and the economy of many countries. The massive reef structure is built over thousands of years by tiny coral polyps aided by minute dinoflagellate algae that live in their tissues.

Coral reefs have been in existence in this earth for over 400 million years. The annual growth rate of reefs, especially the massive forms, is very low. Guppy (1989) observed that the massive porites had a gain of 12.9 to 19.0 mm in height per year, and a 24 to 40-fold increase in weight in 15 years. The branching forms grow faster and may grow by 4-5 cm in height and diameter per year, depending upon the species and prevailing environmental conditions. It has also been estimated that it takes about 1000 years to accumulate 3-6m of limestone.

Even some fast growing reefs can accumulate only 8-15m in the same period (Barnes and Hughes 1982). Many species can experience a total cessation of growth either temporarily or permanently after attaining specific size (Mayer 1918). Besides the slow growth, there is an overall tendency of coral reefs to die out. Every post-monsoon period witnesses a destruction of corals of the Palk Bay & Gulf of Mannar (Pillai 1967), and in the Gulf of Kachchh (Chhaya 1997).

Coral reefs are a boon to many countries. If conserved properly, they increase shore stability by offering protection to the coast against natural disasters like cyclones and storms. They even provide economic stability by playing an enhancing role in fisheries and providing opportunities for development of tourism (Mukesh 1997). Coral reef species also provide great contributions to the world of science and medicine. Kainic acid is produced by certain reef organisms and can be used as a diagnostic chemical to investigate Huntington's chorea, a nervous system disease. Other chemicals produced by reef species are used in cancer and AIDS research. Corals produce a natural sunscreen, and their porous limestone skeletons are promising for bone grafts in humans.

The coral reefs occur around the world in shallow seas of less than 100m depth and clear tropical waters. They lie mainly between the tropics of cancer (23°30' N lat) and capricorn (23°30' S lat) where the temperature never falls below 20°C and the salinity is not lower than 30 ppt. The optimal growth occurs at 23-25°C within a salinity range of 32-42 ppt. Approximately, about 600,000 Km^2 or 0.17per cent of the ocean surface is occupied by coral reefs and about 54 per cent of the total coral reefs are found in Mediterranean and Indian Ocean regions. Of the remaining, Pacific, Atlantic, Caribbean, Red Sea and Persian Gulf reefs account for 25, 6, 9, 4 and 2 per cent respectively (Smith 1978, UNEP 1988).

Coral reefs are disappearing at an alarming rate throughout the tropics. In southeast Asia where approximately 30 per cent of the world's coral reefs are found, about 60 per cent of these reefs are already destroyed or on the verge of destruction (Wilkinson *et al.*, 1993). The remaining reefs are also threatened and it can be predicted that most of the reefs in the region will be eradicated within the next 40 years (Lundin and Linden 1993). The growth rate of the coral reefs is affected by many factors, such as contaminated water, pollution in the vicinity, phytoplankton blooms, turbidity, oil spillage, sedimentation, growth of algae, release of effluents, pollution caused due to cement factories etc. (Chhaya 1997, Mukesh 1997). All these reduce the amount of light available to corals thereby reducing reef growth.

As corals are highly specific in their requirement of temperature, salinity and water clarity for their growth, any environmental changes consequent to improper exploitation will seriously impede the recovery of the reefs. Many world reefs are already in this state and Indian reefs are no exception (Wafar 1986). The increased turbidity of coastal waters and the consequent damage to the coral reefs and sea grass beds is ranked as possibly the greatest threat to the coastal waters (UNEP 1982, ISRO 1993).

PRODUCTIVITY OF CORAL REEF

Coral reefs are highly productive marine ecosystems, with annual gross production rates in the range of 2000-5000 gCm^{-2} (Mann 1982). Such a high rate of productivity is due to the efficient retention and recycling of nutrients within the reef system. A summary of gross production and respiration rates (community metabolism) of several atolls and fringing reefs may be found in Qasim *et al.* (1972), Sournia (1977) and Lewis (1977). The diverse primary producers in the coral reefs are briefly summarized below.

Zooxanthellae: These are dinoflagellates, named as *Gymnodinium microadriaticum*. In coral polyps, they occur in enormous numbers, varying from 0.9 to 2.48 X 10^6 cells cm^{-2}, with an average of 1.45 X 10^6 cells cm^{-2} (Drew 1972). The chlorophyll content of zooxanthellae varies from 5 to 15µg 10^{-6} cells (Titlyanor 1981) and their production rate is around 0.9 gC m^{-2} d^{-1} (Scott and Jitts 1977), which is about 3 times higher than phytoplankton production on unit area.

Boring filamentous algae: These symbiotic algae, belonging predominantly to the genus *Ostreobium*, live buried below the coral tissue and skeletal matter. Due to low light energy level reaching to these algae (Halldal 1968), production by these algae is less than 10 per cent of the total productivity of the coral (Kanwisher and Wainwright 1967).

Calcareous algae: The calcareous algae of the genera *Lithothamnion, Porolithon* and other corallinaceae are important secondary frame builders in a coral reef. The production rates of the coralline algae vary between 0.66 gC m^{-2} d^{-1} and 5.7 gC m^{-2} d^{-1} (Littler 1974, Marsh 1970).

Benthic macroalgae: The sea grasses form extensive beds in shallow lagoons, and the net production of such beds is in the order of 400-4000 gC m^{-2} yr^{-1} (Lewis 1977). The importance of sea grasses in reef productivity becomes evident when the production of different components on a reef is compared. In Kavaratti atoll, the gross production of the sea grass bed was estimated as 12.0 gC m^{-2} d^{-1} (Qasim and Bhattathiri 1971) while that of the reef was 6.15 gC m^{-2} d^{-1}, and of the sea weeds, 3.84 gC m^{-2} d^{-1} (Qasim *et al.* 1972).

Benthic microalgae: These algae have increasingly come to be known as important primary producers on a reef. Sournia (1976) found that the photosynthetic pigments and productivity of lagoon sands can be 1000 and 50 times more respectively, than those of phytoplankton per unit area. Production of benthic microalgae (sand flora and periphyton) on several pacific and Indian Ocean reefs can be as high as 2-6 gC m^{-2} d^{-1}, a figure comparable to gross production in many reefs (Wafar 1986).

Phytoplankton: Phytoplankton production in coral reefs does not assume any importance, but represents only the production rates of the surrounding waters, being often only a few tens of mgC m^{-2} d^{-1}. On unit area basis, phytoplankton production is only about 1 per cent of the total benthic production (Sournia 1977).

High primary productivity of coral reefs and the low nutrient concentrations in surrounding waters would suggest that the reefs and corals have means of conserving and efficiently recycling the nutrients within the ecosystem. This was investigated by Wafar *et al.* (1985) in Kavaratti atoll using stable isotopes of nitrogen ($^{15}NO_3$, $^{15}NH_4$ and ^{15}N-urea). Their results demonstrated that, even at the maximum rate of uptake, the zooxanthellae could obtain all their N requirements directly from coral excretion alone, and that nearly 80 per cent of the N synthesized by zooxanthellae might be translocated back to the corals within 4 h. Such kind of recycling of nitrogen, as for phosphorus (Pilson and Betzer 1973), explained the higher productivity of coral reefs in the midst of oligotrophic waters.

Detritus production is an important pathway in the transfer of energy in reef ecosystems. The major source of detritus is the mucus produced by the corals. This mucus entraps other particles and is subsequently colonized by heterotrophic microorganisms. Together they form what is commonly known as the 'organic aggregates' of reef waters. These organic aggregates are important food sources for many reef invertebrates, particularly the zooplankton (Gottfried and Roman 1983). Reef fish and shrimps can also ingest coral mucus (Johannes 1967). The trophic significance of organic aggregates extends beyond the reef too (Johannes 1967, Qasim and Sankaranarayanan 1970), which otherwise would not be possible, with sparse phytoplankton. Qasim and Sankaranarayanan (1970) measured the production of particulate organic carbon in organic aggregates over the Kavaratti atoll; where production was 95 mgC m^{-2} h^{-1} and was about 20 per cent of gross production by the reef.

RESOURCES FROM CORAL REEFS

Food Fishes: Fish production from coral reefs can be quite high, upto 20 tonnes km^{-2} yr^{-1} (Marten and Polovina 1981) and heavily fished coralline shelves are able to produce sustained harvests of 4-6 tonnes km^{-2} yr^{-1} (Munro 1984). The estimated potential fish yield from all world reefs varies from 6 million (Smith 1978) to 9 million tonnes yr^{-1} (Munro 1984), or 10-15 tonnes km^{-2} yr^{-1}, equivalent to 9-12 per cent of all oceanic fish caught. These yields are comparable with intensively fished shelf areas in the world. Indian coral reefs have great potential for fish yield. It is estimated that Indian reefs, together with their shelves, lagoons and submerged banks, cover over 18,000 km^2 and have a potential fish yield of 0.2 million tonnes or about 10 per cent of the annual marine fish production (Qasim 1997). While Lakshadweep with reef area of 4200km^2 can yield 40,000-60,000 tonnes yr^{-1}, Andaman islands with 11,000 km^2 area of coral reefs can yield 110,000 - 170,000 tonnes yr^{-1}. Similarly, 30,000-40,000 and 1,000-15,000 tonnes yr^{-1} of fish can be harvested from the reef areas of Nicobar islands (2700 km^2) and Gulf of Mannar & Palk Bay (100 km^2) respectively (Wafar 1986).

Aquarium Fishes: Commercial exploitation of aquarium fishes from coral reefs has gained importance only recently. However, the potential for revenue from this industry is great, as shown by the Philippines, where the revenue from aquarium fish trade had increased from US$120,000 in 1970 to US$2.5 million in 1979 (Albaladejo and Corpuz 1981). Asia is the largest producer with a contribution of about 68 per cent aquarium fishes, followed by South America (25 percent) and others (less than 20 percent). In Asia, countries like Singapore, Philippines, Hong Kong, Japan, Thailand and Sri Lanka are main producers of ornamental fishes (Devraj 2000). Though Indian reefs sustain an abundant variety of aquarium fishes, no organized effort has yet been made to exploit these resources. Over 300 species of ornamental marine fishes are known in India, of which majority are from coral reef areas.

Corals: Corals are mainly used as building blocks, in the construction of roads and as lime in cement industries. In the Gulf of Kachchh, coral sands are dredged for use in cement industry. Estimates of the quantity dredged vary from 0.6 (Patel 1985) to 1 million tonnes yr^{-1} (Rashid 1988).

Molluscan resources: Molluscs, a major group of invertebrates on coral reefs, are exploited largely for food and shell resources. While molluscan fishery in many reef areas is only a sustenance fishery, especially for the meat, in terms of shell resources these are of commercial importance. Commercial molluscan fisheries in Indian reefs exist for pearl oysters and chanks. The chank fishery for the sacred chank *Xancus pyrum* is a regular one.

VALUE OF CORAL REEFS

Fisheries, Aquarium, and Curio Trades: A single reef can harbor many thousands of organisms and species, many of which can be harvested sustainably. Many are edible and regularly harvested for consumption. Potential harvests of reef fish are estimated at 9-million (M) t yr^{-1}, an eighth of the current world fish harvest (Munro 1984). An average maximum sustainable yields (MSY) of 15 Mt km^{-2} yr^{-1} of all reef organisms is suggested by Munro and Williams (1985), although total MSYs calculated for reef fish range from 0.5-3.6 Mt km^{-2} yr^{-1}. Reef fish, mollusks and corals are also extensively used in the aquarium and curio trades, for which individual specimens may realize high prices. The world import value for the expanding marine aquarium trade was estimated to be US$ 24-40 million yr^{-1} in 1985 (Wood 1985).

Pharmaceutical and other Industrial Uses: Coral reef organisms have excellent potential for pharmaceutical and industrial applications (Patel and Bhaskaran 1978, Jacob 1973, Krishnamurthy 1969, Hornell 1916). Corals, gorgonians and sponges contain many biologically active compounds of considerable potential value. Some skeletons are used in bone grafts (Jeyabhaskaran *et al.* 1996).

Construction: Coral is used extensively as building blocks, aggregate and for production of lime. However, because usually the more compact, slow growing corals are used, and because mining is destructive, it is generally considered a non-sustainable use. Its immediate market value may be high, but this fails to account for the considerable loss of other reef benefits.

Tourism: There is little doubt that tourism yields the greatest direct financial benefit of all reef uses. Many small island nations, such as the Maldives, depend heavily on reef-based tourism for economic development. Revenues are directly generated from reef-related tourism, from SCUBA (Self-Contained Underwater Breathing Apparatus) diving and Marine Park entrance fees. In addition, indirect tourism revenues, such as accommodation, food and travel costs are also attributable to reefs where their expenditure is directly related to reef activities.

Using the above assumptions, Mattson and DeFoor (1985) estimated that in 1984-85, coral reefs in two marine parks in

Florida generated an income of US\$ 47.6 millions. These parks, John Pennecamp Coral Reef State Park and Key Largo National Marine Sanctuary attract one million visitors each year. By dividing revenues generated by the area of reef used, the authors determined coral reef values of US\$ 15.75$m^{-2}yr^{-1}$ from direct revenues, and US\$ 85 $m^{-2}yr^{-1}$ from gross revenues.

Education: Financial benefits arise through education programme expenditures. Such expenditures are, however, likely to be accounted for when valuing tourism and research revenues. As for research, possibly the greatest value arises in the form of social benefit. This includes the first hand experience and knowledge gained, together with the increased sense of environmental awareness acquired. To determine the extent of this benefit, a survey could be carried out, assessing people's opinions before and after visiting reef sites.

Social values: Local communities living nearby and utilizing coral reefs gain additional benefit similar to that of tourism consumer surplus. This 'social value' is referred to in the literature under a variety of guises all of which are very much interrelated. It includes cultural, heritage, spiritual and aesthetic benefits. Since the different attributes making up this social value for coral reefs are inextricably linked, there seems little point in attempting to value them individually. No quantifications seem to exist for this social value of reefs.

Biological support: Coral reefs interact in a variety of ways with other ecosystems having indirect economic implications. They support fisheries both offshore and in nearby sea grass beds, lagoons and mangroves. Pelagic juvenile stages produced by many reef organisms drift across to other ecosystems due to currents and either act as a food source for commercial fish, or settle and mature until harvested by fishermen. Mature individuals of some species also migrate daily between reefs and other ecosystems. Colonies of seabirds feed heavily on reef fish, and turtles feed and breed on reefs and reef islands.

Physical protection: One of the more significant but perhaps least obvious benefits of coral reefs is the physical protection they afford to coastlines. Coral reef organisms actively produce calcium carbonate skeletons, which form an effective regenerating barrier that dissipates wave energy. This protection provides measurable benefits. In regions prone to severe storms and strong prevailing currents, reefs can significantly reduce coastal erosion. Many beaches, stretches of coastal land and indeed whole islands owe their existence to coral reefs.

IMPACT OF GLOBAL WARMING ON CORAL REEFS

It has been estimated that coral reefs act as a sink for 111 Mt of carbon per year, the equivalent of 2 per cent of present output of anthropogenic CO_2. In the event of global warming and predicted increases in reef production, this figure may even rise to 4 per cent within 100 years (Kinsey and Hopley 1991).

Tropical coral reefs are under severe threat from rising ocean temperatures, both directly and indirectly through, the impact of global warming on other species that live in or feed on the reefs. Oceanologists regard some ten per cent of coral reefs as lost and believe another 30 to 60 per cent could be reduced to dead mounds of under sea chalk within the next two decades (Anon. 2000). Temperature anomalies have surged in the Pacific since 1970, as shown by the recurrent El Nino phenomenon.

Coral bleaching is a major symptom of reef decline. The coral polyps expel their symbiotic algae and take on a bleached appearance, with the gleaming white chalk skeleton showing through. Since bleaching has become widespread, 32 countries in the Pacific and Indian Oceans, in the Red Sea, the gulf, the Mediterranean and the Caribbean reported the phenomenon after the 1997-98 El Nino winter. In some parts of the Indian Ocean, such as around the Maldives, coral mortality has reached almost 90 per cent of the total and all coral species are affected. Many corals currently exist in temperatures at the upper limit of their tolerance. Bleaching occurs when water temperatures are as little as one-Celsius degree above the average temperature of the warmest month. If the temperature rises still further, or if the warm period lasts for longer than two months, the coral is irreparably damaged.

Oceanologists believe that if the atmospheric temperature is increased by two-Celsius degrees in the next 50 years, as predicted by climatologists in 1995, coral bleaching will accelerate. Coral reefs have existed in the seas of the tropics for 200 million years, and they attained their current level of species diversification around 500 million years ago. During this period they survived ice ages and considerable variations in sunlight. During the last ice age, for example, the sea level fell by 100 meters, and all existing coral reefs were destroyed. When the sea level rose again 10,000 years ago, the reefs began to spread out again from the coastlines and from under sea volcanoes. They have been growing continually over the past 6,000 years. It is difficult to comment what the consequences would be if all the current living reefs in the tropics were to die.

DISTRIBUTION OF CORAL REEFS IN INDIA

The coral reefs of the mainland coast of India are located chiefly along the SE coast in Gulf of Mannar and Palk Bay, and in the Gulf of Kachchh in the Gujarat state (Fig. 1). Some patch growths also occur along the west coast between Mumbai and Malvan. The coral reefs along the mainland are mostly fringing or patch reefs built on shallow waters around near shore islands. The Andaman and Nicobar islands in the Bay of Bengal form the major continental island reefs of India. The Lakshadweep is the only atoll formation of Indian waters. An estimate of the reef flat areas of India (DOD and SAC 1997) by remote sensing showed the extent of reef flat in Gujarat coast as 148.4 km^2, that of Tamil Nadu coast as 64.9 km^2, Lakshadweep 136.5 km^2 and that of Andaman and Nicobar islands as 495.7 km^2. In addition to that, coral knolls and lagoon reefs also form roughly 50 km^2 of reef formations. The coral reef and their resources are fast dwindling in Indian waters. Except, perhaps, for some islands in the Andaman and Nicobar Islands, no pristine reefs exist today in this country. Overexploitation, and unwanted interference and lack of effective management in the past may be the major reasons for this sad state of affairs. In Lakshadweep, the reefs form a major life supporting resource to the locals as important as the coconuts. The coral fauna of the Gulf of Kachchh in Gujarat

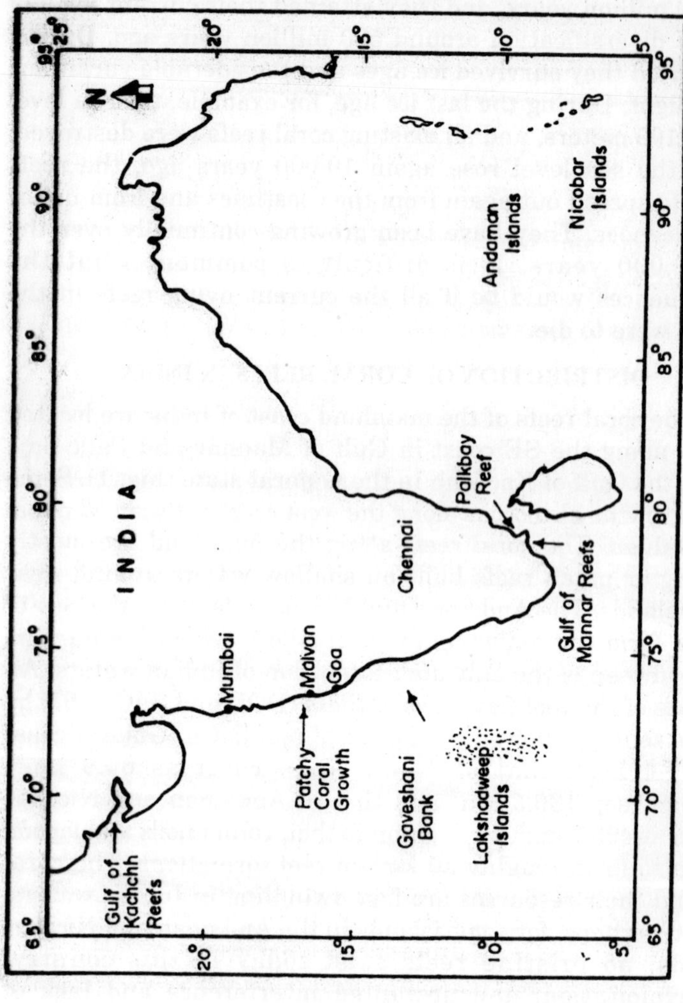

Fig. 1: Location of occurrence of reef-building corals and reef-formations in Indian waters

is comparatively less diverse compared to other parts of India. Remose corals such as *Acropora* sp, *Pocillopora* sp, *Stylophora* sp and *Seriatopora* sp are not found at present though semi-fossilized specimens of *Acropora* sp are found on some beaches. In this context, the present state of the coral reefs primarily at four major areas in India, *viz.* the Gulf of Kachchh, Lakshadweep, Gulf of Mannar and Palk Bay, and Andaman and Nicobar islands, has been discussed.

Gulf of Kachchh

The Gulf of Kachchh is located in the northwest of Indian coast. The intertidal zone of the Gulf including 42 islands along the coast of Jamnagar in the State of Gujarat, between 22°15' to 23°40' N lat. and 69°00' to 73°35' E long. was declared India's first 'marine protected area' through a series of notifications between 1980 and 1982. The first notification dated 12-08-80 made a sanctuary of approximately 221 km^2 and it was later extended by a second notification dated 20-07-1982 to include 237 km^2 more. As a result of voluntary effort and interest in the conservation of the sanctuary, a decision was taken in January 1981 to upgrade a portion of the sanctuary, and the adjacent areas to the extent of 163 km^2 by a notification dated 20-07-82 to the status of a national park. The Marine National Park and Sanctuary (MNPS) in the Gulf is the first of its kind in India.The International Union for Conservation of Nature & Natural Resources (IUCN) has declared this park as the threatened protected area and has put in red data book.

The Gulf of Kachchh is a unique ecosystem in the Indian sub-continent by virtue of its deeply indented coastline (Rashid 1985). The tidal range of the Gulf varies between four and seven meters, with fast moving currents of 2.4 m a second that divert the sediment outflow of the Indus river away from the Gulf into the high seas. But despite this natural protective barrier, the Gulf water is very turbid with the visibility of less than a meter. The poor condition of the Gulf waters is attributed to terrigenous input arising from anthropogenic disturbances. The Gulf in the Arabian Sea separates Saurashtra peninsular from Rann of Kachchh. There are a number of islands in the Jamnagar coast, surrounded by reefs. The important islands are Pirotan and Karumbhar (Chavan 1983, Pillai *et al.* 1979). The velocity of current in this region is about 2.4 m per second and the coral reefs are acting as dynamic barrier (Nair 1984). Pillai (1996) has recorded 37 species under 24 genera, indicating that the Gulf of Kachchh is not very rich in diversity in corals. Among these, 33 species under 20 genera are hermatypes, and 4 species under 4 genera are ahermatypes. However, Patel (1985) had recorded 44 species of scleractinian corals and 12 species of soft corals from the Gulf. The coral reef area, which was 116 km^2 during 1982, has been reduced to 53 km^2 during 1985. The main reason for the reduction of area is deposition of silt on coral reefs, converting it to mud flats (Bahuguna *et al.* 1989). Gulf of Kachchh reefs mostly consist of dead coral boulders and pebbles. There are 42 islands in the Gulf, of which a few are big, some small, while others are tiny. Some are submerged and not visible during high tide, while many others are well above the high tide level (Pillai *et al.* 1979). Most of these islands or group of islands have live and dead corals and coral reefs. Dead and semi-fossilized forms of the *Acropora* coral are commonly found (Table 1) on many reefs in the Gulf. Some of the reef areas on Pagar, Borea, Pirotan, and Goose bets support good colonies of live corals. Corals still occur in 33 out of 42 islands (bets) and on some reefs in coasts. Vast areas of dead reefs with live patches at fringe of islands are an indication that the loss of coral cover is of recent origin. Majority of coral reef area in the Gulf is under mud and sand cover. Earlier, Wafar (1977) estimated dead and live coral reefs of about 130.3 km^2.

During 1998-2000, a total of 12,7,28 and 28 species of stony corals, soft corals, benthic marine algae and benthic fauna respectively were recorded during underwater survey (5-25 m depth) near Pirotan, Karumbhar, Boria, Beyt Shankhodar, Mungra, Mundra and Mandvi islands (Deshmukhe *et al.* 2000). The study concluded that abundance of live corals decreased from the inner to the outer reefs from 75 to 80 per cent at Pirotan and Karumbhar, and around 50 per cent at Boria and Beyt Shankhodar.

Central West Coast

In view of low salinity (20-25 ppt), very high turbidity, silting and strong wave action during monsoon months (June-August) along the central west coast, the occurrence of reef-

Table 1. Scleractinian corals recorded from the seas around India

Species	*Lakshadweep*	*Gulf of Kachchh & Palk Bay*	*Gulf of Mannar*	*Andamans & Nicobars*
1	2	3	4	5
Family: Thamnasteriidae				
Psammocora contigua	+	-	+	+
Psammocora digitata	+	+	-	-
Psammocora haimeana	+	-	-	-
Psammocora profundacella	+	-	-	+
Family: Pocilloporidae				
Stylophora pistillata	+	-	-	+
Seriatopora crassa	-	-	-	+
Seriatopora hystrix	-	-	-	+
Seriatopora stellata	-	-	-	+
Pocillopora brevicornis	-	-	-	+
Pocillopora damicornis	+	-	+	+
Pocillopora ankeli	-	-	-	+
Pocillopora ligulata	+	-	-	-
Pocillopora meandrina var. nobilis	-	-	-	+
Pocillopora verrucosa	+	-	+	+
Pocillopora eydouxi	+	-	+	+
Pocillopora danae	-	-	+	-
Pocillopora acuta	-	-	-	-
Madracis sp.	-	-	+	+
Family: Acroporidae				
Acropora intermedia	+	-	+	-
Acropora formosa	+	-	+	+
Acropora valenciennesi	-	-	+	-
Acropora abrotanoides	+	-	-	-
Acropora gravida	-	-	-	+
Acropora efflorescens	+	-	-	+
Acropora conigera	+	-	-	+
Acropora obscura	-	-	+	-
Acropora teres	+	-	-	-

(Table Contd...)

1	2	3	4	5
Acropora nasuta	+	-	-	-
Acropora secale	-	-	-	+
Acropora corymbosa	+	-	+	-
Acropora hyacinthus	+	-	+	+
Acropora indica	+	-	+	-
Acropora millepora	-	-	+	+
Acropora pinguis	-	-	-	+
Acropora cervicornis	-	-	+	-
Acropora plantaginea	-	-	+	-
Acropora exigua	-	-	+	-
Acropora spicifera	-	-	+	-
Acropora surculysa	-	-	+	-
Acropora multiformis	-	-	+	-
Acropora cyclonica	-	-	+	-
Acropora polymortha	-	-	+	-
Acropora multicaulis	-	-	+	-
Acropora thurstoni	-	-	+	-
Acropora pharaonis	-	-	+	-
Acropora haimei	-	-	+	-
Acropora valida	-	-	+	-
Acropora digititers	-	-	+	-
Acropora erythraea	-	-	+	-
Acropora brevicollis	-	-	+	-
Acropora palifera	+	-	-	+
Acropora nobillis	-	-	+	+
Acropora humilis	+	+	+	+
Acropora diversa	-	-	+	+
Acropora hebes	+	-	-	-
Acropora variabilis	-	-	+	+
Acropora squarrosa	+	+	+	-
Acropora hemprichi	+	-	-	-
Acropora forskali	+	-	-	-
Acropora rambleri	+	-	-	+
Acropora granulosa	+	-	-	-
Acropora dumosa	-	-	-	+

(Table Contd...)

1	2	3	4	5
Acropora echinata	+	-	-	-
Acropora multi-acuta	-	-	-	+
Astreopora myriophthalma	+	-	+	-
Astreopora listeri	-	-	-	+
Montipora subtilis	-	-	+	-
Montipora granulosa	-	-	+	-
Montipora explanata	-	+	+	-
Montipora exserta	-	-	+	-
Montipora digitata	-	-	+	+
Montipora divaricata	-	-	+	+
Montipora cocosensis	-	-	-	+
Montipora turgescens	-	+	+	+
Montipora manauliensis	-	-	+	-
Montipora monasteriata	-	+	+	-
Montipora verosa	-	+	+	-
Montipora spumosa	-	-	+	-
Montipora tuberculosa	+	-	+	-
Montipora jonesi	-	-	+	-
Montipora edwardsi	-	-	+	-
Montipora verrucosa	-	-	+	-
Montipora peltiformis	-	-	-	+
Montipora elschneri	-	-	+	-
Montipora spongiesa	-	-	+	-
Montipora informis	-	-	+	-
Montipora verrilli	-	-	+	-
Montipora hispida	-	+	+	-
Montipora foliosa	-	+	+	+
Montipora composita	-	-	+	+
Family: Agariciidae				
Pavona explanulata	-	-	-	+
Pavona xarifae	-	-	-	+
Pavona varians	+	-	+	+
Pavona decussata	-	-	+	+

(Table Contd...)

1	2	3	4	5
Pavona praetorta	-	-	+	-
Pavona clavus	-	-	-	+
Pavona maldivensis	+	-	-	-
Pavona duerdeni	+	-	+	+
Pavona divaricata	-	-	+	-
Pavona sp	-	+	-	-
Pachyseris rugosa	-	+	+	+
Pachyseris speciosa	-	-	-	+
Leptoseris papyracea	-	-	-	+
Leptoseris fragilis	-	-	-	+
Gardineroseris planulata	+	-	-	+
Coeloseris mayeri	-	-	-	+
Family: Siderastreidae				
Siderastrea savignyana	-	+	+	-
Siderastrea lilacea	-	-	+	-
Pseudosiderastrea tayami	-	+	+	+
Coscinarea monile	-	+	+	-
Family: Fungiidae				
Cycloseris cyclolites	-	-	+	-
Cycloseris sinensis	-	-	-	+
Cycloseris distorta	-	-	-	+
Cycloseris hexagonalis	-	-	-	+
Cycloseris costulata	-	-	-	+
Cycloseris sp	+	-	-	-
Fungia scutaria	+	-	-	+
Fungia paumotensis	-	-	-	+
Fungia somervilli	+	-	-	+
Fungia echinata	-	-	-	+
Fungia repanda	-	-	-	+
Fungia danai	+	-	-	+
Fungia horrida	-	-	-	+
Fungia fungites	+	-	-	+
Fungiacyathus symmetrica	-	-	-	+

(Table Contd...)

1	2	3	4	5
Herpolitha limax	-	-	-	+
Herpetoglossa sp	-	-	-	+
Polyphyllia talpina	-	-	-	+
Podabacia crustacea	+	+	-	-
Family: Protidae				
Goniopora stokesi	+	-	+	+
Goniopora tenuidens	-	-	-	+
Goniopora nigra	-	+	+	-
Goniopora minor	+	+	-	-
Goniopora planulata	-	+	+	+
Goniopora djibontiensis	-	-	+	-
Goniopora duotaciata	-	-	+	-
Porites solida	+	-	+	+
Porites lobata	-	-	-	+
Porites minicoiensis	+	-	-	-
Porites sonaliensis	-	-	+	-
Porites lutea	+	+	+	+
Porites lichen	+	+	+	-
Porites exserta	-	-	+	-
Porites andrewsi	+	-	-	-
Porites eridani	-	-	-	+
Porites compressa	-	+	+	-
Porites mannarensis	-	-	+	-
Alveopora daedalea	-	-	-	+
Family: Favidae				
Plesiastrea versipora	+	+	+	+
Favia stelligera	+	+	+	+
Favia pallida	+	-	+	+
Favia speciosa	+	+	+	+
Favia favus	+	+	+	+
Favia rotumana	-	-	-	+
Favia valenciennesi	-	-	+	+
Favia valenciannegii	-	-	+	-

(Table Contd...)

1	2	3	4	5
Favites abdita	+	-	+	+
Favites halicora	+	-	+	+
Favites complanta	+	+	+	+
Favites flexuosa	-	-	-	+
Favites pentagona	+	-	+	-
Favites virens	-	-	+	-
Favites melicerum	+	+	+	-
Goniastrea retiformis	+	-	+	+
Goniastrea incrustans	-	-	+	-
Goniastrea pectinata	+	+	+	+
Platygyra daedalea	+	-	+	+
Platygyra sinensis	+	+	+	+
Platygyra lamellina	-	-	+	-
Leptoria phrygia	+	+	+	+
Oulophyllia crispa	-	-	-	+
Hydnophora microconos	+	-	+	+
Hydnophora grandis	-	-	+	-
Hydnophora exesa	-	+	+	+
Hydnophora laxa	-	-	-	+
Diploastrea heliopoa	+	-	-	+
Oulastrea crispata	-	-	-	+
Leptastrea bottae	+	-	-	-
Leptastrea purpurea	+	+	+	+
Leptastrea tarnsversa	+	-	+	+
Cyphastrea microphthalma	-	-	+	+
Cyphastrea serailla	-	+	+	-
Cyphastrea chalcidicum	-	-	+	-
Echinopora lamellosa	+	-	+	+
Echinopora horrida	-	-	-	+
Echinopora gemmacea	-	-	+	-
Family: Trachyphyllidae				
Trachyphyllia geoffroyi	-	-	-	+

(Table Contd...)

1	2	3	4	5
Family: Rhizangiidae				
Culicia rubeola	-	-	+	+
Cladangia exusta	+	-	-	-
Family: Oculinidae				
Galaxea fascicularis	+	-	+	+
Galaxea clavus	-	-	+	+
Family: Merulinidae				
Merulina ampliata	+	-	+	+
Scacophyllia cylindrical	-	-	-	+
Family: Mussidae				
Lobophyllia corymbosa	+	+	-	+
Acanthastrea simplex	-	+	-	-
Acanthastrea echinata	+	-	-	-
Symphalia nobilis	+	-	+	+
Symphalia radians	+	+	+	+
Symphalia recta	-	-	+	-
Family: Pectinudae				
Mycedium elephantotus	-	+	+	+
Mycedium tubifex	-	-	+	-
Pectinia lactuca	-	-	-	+
Family: Caryophylliidae				
Caryophyllia clavus	+	-	-	+
Caryophyllia arcuata	+	-	-	+
Caryophyllia scanthocyathus	-	-	-	+
Deltocyathus andamanensis	-	-	-	+
Paracyathus indicus	-	-	-	+
Paracyathus profundus	-	-	+	-
Paracyathus andersoni	-	-	+	-
Paracyathus stokesi	-	+	-	-
Polycyathus verrilli	-	+	+	+
Polycyathus andamanensis	-	+	+	+
Heterocyathus aequicostatus	-	-	+	+
Stephanocyathus nobilis	+	-	-	-

(Table Contd...)

1	2	3	4	5
Euphyllia glabrescens	+	-	-	+
Plerogyra sinuosa	-	-	-	+
Physogyra lichtensteini	-	-	-	+
Family: Flabellidae				
Flabellum pavonium	+	-	-	-
Placatrochus laevis	-	-	-	+
Family: Dendrophylliidae				
Balanophyllia imperialis	-	-	-	+
Balanophyllia scabra	-	-	-	+
Balanophyllia affinis	-	-	+	-
Endopsammia philippinensis	-	-	+	-
Heteropsammia michelini	-	-	+	+
Heteropsammia cochlea	-	-	+	-
Tubastrea aurea	-	+	+	+
Dendrophyllia coarctata	-	-	+	-
Dendrophyllia arbuscula	-	-	-	+
Dendrophyllia minuscula	-	+	-	+
Dendrophyllia micranthus	-	-	-	+
Dendrophyllia indica	-	-	+	-
Dendrophyllia robusa	-	-	+	-
Enallopsammia amphelioides	-	-	-	+
Enallopsammia marenzelleri	-	-	-	+
Turbinaria crater	-	+	+	+
Turbinaria undata	-	-	+	-
Turbinaria peltata	-	+	+	+
Turbinaria mesenterina	+	-	-	-
Turbinaria veluta	-	-	-	+

+: Recorded, -: Not recorded

Source: Pillai 1983, Upreti and Shanmugaraj 1998, Wafar 1986, Deshmukhe *et al.* 2000

building corals in this region is of ecological significance. Certain varieties of corals thrive here and judging from the size of the colonies, these are believed to be growing since many years. The tolerance of reef corals is further evidenced by the occurrence of *Pseudosiderastrea* and *Porites* in Mumbai waters which, besides experiencing monsoon effects, are also subjected to heavy oil, sewage and industrial pollution (Wafar 1986). Qasim and Wafar (1979) recorded the occurrence of coral patches in the intertidal regions of Ratnagiri, Malvan and Redi. Malvan is considered to be a richer region having about 208 species of macroscopic invertebrates. Bakus (1994) recorded fossil reefs near Vijayadurga and Vengurla. The recorded corals were from the genera *Coscinarea, Turbinaria, Fabites, Porites, Pseudosiderastrea, Cyphastrea, Goniastrea* and subgenus *Synaraea*. Of these, the genus *Pseudosiderastrea* was a new record from Indian waters. *Porites* was the dominant form at all the three sites, followed by *Goniastrea* and *Pseudosiderastrea. Fabites* was recorded only at Malvan with very few colonies.

A submerged bank with living corals, named as 'Gaveshani Bank', at 13° 24' N and 73° 24' E, about 100km offshore to Mangalore and about 38 m depth, was discovered in 1977 (Nair and Qasim 1978). The corals recorded from this bank were *Euphyllia fimbriata, Leptoseris papyracea, Stylocoeniella armata, Porites lichen* and *Cyphastrea seriala.* Two genera, *Porites* and *Cyphastrea* were common between Gaveshani bank and the intertidal areas of the central west coast.

Lakshadweep Archipelago

The Lakshadweep archipelago is biologically significant in view of its isolation from major continental coastline as well as for its rich and varied marine life. It consists of 12 atolls, 3 reefs and 5 submerged banks with a total of 27 islands (Anonymous 1984). The early studies on reefs of this area are those of Gardiner (1903, 1906) from Minicoy, which include detailed descriptions of the reefs and some reef corals. Alcock (1902) visited several islands of this area at the end of 19th century and described many atolls. Excluding Minicoy, which is the largest atoll in the Lakshadweep archipelago, all other atolls are located within the geographical limits of 10°-12°N and 71°40'-74°E. Almost all the atolls have a NE-SW orientation, with the low-lying island on the east, a broad well developed reef on the west, with a lagoon in between, connected to the open ocean by one or more channels. The width of the lagoon varies from 1 to 4 km, and most of the lagoons have also coral knolls arising from the lagoon floor.

Till date there are no comprehensive studies of coral fauna of Lakshadweep. Present knowledge is primarily based only on Minicoy (8°18' N and73° E) at the south and Kiltan at the north and a total 28 genera comprising over 80 species of stony corals were recorded (Table 1). Of these, 36 species were new records from Minicoy. Subsequently, Rotrigues (1997) surveyed coral reefs surrounding Agatti island during 1993-1995, and recorded 41 species of corals dominated by *Pocillopora damicornis, Porites lutea* and *Acropora humilis.* The maximum percentage cover of corals was recorded in the lower reef flat and the minimum in the upper reef flat. Future investigations all around Lakshadweep will bring forth many more unrecorded species and genera. The octocorals are represented by *Heliopora coerulea* that forms a very conspicuous element in the inner lagoon reefs throughout Lakshadweep. The gorgonids are not found in shallow waters. *Millipora* represents the hydrozoans with three known species. The genus *Acropora* (Plate 1) is the richest, as is the case with all the Indian Ocean reefs, and forms about 25 per cent of the

Plate 1: The corals of the genus *Acropora* exposed to sunlight during low tide near Agatti Island, Lakshadweep

total species known from Lakshadweep. A notable feature of the coral fauna of Lakshadweep is the absence of foliaceous forms such as *Montipora foliosa* and *Echinopora lamellosa*. *Montipora*, a very common genus in Indian Ocean reefs, is rare and only one encrusting species is recorded. The massive species are mainly those of *Porites lutea* and *P.solida*. Ramose *Porites* is essentially a component of the lagoons in Kiltan, Minicoy and Kavaratti atolls.

Southeastern Indian Reefs

***Palk Bay reef*:** There is only one reef in the Palk Bay lying in east-west direction along the mainland at Mandapam and the Rameswaram island. The reef extends between lat.9°17'N and long.79° 8' to 79°17' E. The coral reef in the Palk Bay starts from Munakad as a wall like formation (1-2m broad) and east upto Tonithurai to a distance of about 5.5 km, where the width of the reef attains more than 300m east to Pamban pass. The reef again starts near Thangachimadam and ends near Rameswaram (Mahadevan and Nair 1969, Pillai 1969). Systematic investigations of the hermatypic corals have been carried out in this reef (Pillai 1971,1972; Wafar 1974). Pillai (1969) recorded 63 species of corals, belonging to 22 genera, from this reef. The dominant genera were *Favia, Acropora, Leptastrea, Porites, Pocillopora* and *Goniastrea*. The genera *Acropora* and *Montipora*, of the family Acroporidae, were the most diverse with 25 species; while *Porites* was the most important genus in terms of abundance and reef building.

***Gulf of Mannar reefs*:** Gulf of Mannar is the first Marine Biosphere Reserve in south and southeast Asia. The Biosphere Reserve which lies between India and Srilanka covering an area of about 10,500 km^2, was set up on Feb.18, 1989 jointly by Government of India and state of Tamil Nadu.The reef formation in the Gulf of Mannar is of fringing type developed around a chain of 21 islands from Tuticorin 8°48'N; 78°9' E to Shingle island 9°14' N; 79°14' E, lying at an average distance of about 8 km from the mainland (Table 2). Additionally, a 2 km shore reef exists in the Kilakkarai (Reddiah *et al.* 1974). Investigations revealed that eastern sides of all these reefs support active coral growth, while the western and northern shores are quarried for coral blocks. The steep

Table 2. List of Islands in the Gulf of Mannar

Sl.No.	*Name*	*Area (ha)*	*Circumference (m)*	*Distance from main land*
1	Van Tivu	16.00	2,015	6 km from Tuticorin
2	Koswari	19.50	2,160	7 km from Tuticorin
3	Kariya Challi	16046	1,610	15 km from Tuticorin
4	Villangu Challi	0.95	614	15 km from Tuticorin
5	Uppu Thani	29.94	2,292	8 km from Vembar
6	Pulvini Challi	6012	1,372	18 km from Vembar
7	Nalla Thanni	110.00	4,700	2 km from main land
8	Anaipar	11.00	1,605	9 km from Kilakarai
9	Vali Munai	6.72	1,170	9 km from Kilakarai
10	Appa Tivu	28.63	4,840	8 km from Kilakarai
11	Poovarasan Patti	0.25	–	Lying midway between Appa &Vali Munai islands
12	Talairi	75.15	8,338	10 km from Kilakarai
13	Valai	10.15	1,889	10 km from Kilakarai
14	Mulli	10.20	1,712	9 km from Kilakarai
15	Hare (Musal)	129.04	11,520	7 km from Mandapam camp
16	Manoli	25.90	2,958	6 km from Mandapam camp
17	Manoli Putti	2.34	940	5 km from Mandapam camp
18	Poomarichan	16.58	2,500	3 km from Mandapam camp
19	Pullivasal	29.95	5,520	3 km from Mandapam camp
20	Krusadai	65.80	5,193	7 km from Pamban & 500m from Kundugal point
21	Shingle	12.69	1,736	4 km from Pamban

southern side slopes have free circulation with open oceanic water; however, northern side of the island has poor circulation of water. Pillai (1971,1972) has carried out detailed systematic studies in select islands, *viz.* Shingle, Krusadai, Pullivasal, Pulli, Manauli, New and Hare, over a period of several years. He recorded 88 species, and together with those of the Palk Bay reef, the total number of species known for Mandapam area was 117, distributed among 32 genera, of which 110 species belonging to 25 genera were hermatypic. *Acropora* was recorded as the most diverse genus with 25 species, followed by *Montipora* (20 species) and *Porites* (12 species). Subsequent survey by Zoological Survey of India during 1993-1997 recorded only 21 species (from 13 genera) of corals in the Gulf of Mannar (ZSI 1998). However, Tamil Nadu Forest Department compiled the biological wealth of the Reserve, and recorded 137 species of corals in the Gulf of Mannar (Upreti and Shanmugaraj 1998).

Andaman and Nicobar Islands

The Andaman and Nicobar islands are located between lat.6° and 14°N, and long.91° and 94°E. Coral reefs are stretched over an area of 11,000km² in the Andamans and 2,700km² in Nicobar islands (Krishnakumar 1997). The Andaman sea is relatively extensive basin with a maximum depth of 4360m and an uneven bottom topography. The depth in the Bay of Bengal side of the islands is 70-80m, while in Andaman seas, it varies from 50-75m. The mean salinity varies between 29.7 and 32.8 ppt (Rangarajan and Marichamy 1972).

Andaman and Nicobar islands comprise 136 species of scleractinian coral species (Table 1) belonging to 59 genera. Species of the genera *Alveopora, Coeloseris, Seriatopora, Plerogyra, Physogyra* and *Oulastrea* were common in these areas. Presently only some sites in the Andaman and Nicobar islands remain in pristine condition and the live corals show patchy growth in the reef area (Devraj 1997). The growths of corals are better in the western side than the eastern side of the island (Mukesh 1997). According to Vineeta Hoon (1997) the Andaman reefs are in a better condition than the reefs in Lakshadweep and Gulf of Mannar. WWF of India monitors the National Marine Park at Wandoor with an extent of 281.5km² enclosing 14 small islands.

There are several reports towards damage and death of corals due to siltation and effluent discharges from the shore areas of south and middle Andamans (Dorairaj *et al.* 1987, Wood 1989, James *et al.* 1990). In some areas, like Red Skin Island, corals in the reef slope are badly damaged mostly due to boat anchorage. Tourist activities like trampling, overturning the coral blocks, snorkeling and SCUBA diving have damaged the corals in Jolly Boys island (Dorairaj and Soundarajan 1997). The mud deposits have been found in the reef area near Port Blair, Flat Bay, Reef island etc. (Wood 1989). Corals, particularly branching types, are prone to breakage, especially in the areas open to tourists or fisher folk and others. The "Crown of Thorns" starfish *Acanthaster planci* feeds on coral polyps and often a plague of them is observed in many parts of the Indo-Pacific reefs causing mortality to vast areas of reef corals and their subsequent degradation. The species was previously recorded from several localities of Andaman and Nicobar islands but aggregation of the starfish, which is extensively damaging the reef corals, was never noticed prior to eighties. Large numbers of *A.planci* have been recorded in the National Marine Park at Wandoor by Forest Department and Fisheries department. Crown of Thorns are more common on *Acropora* sp. thickets than *Porites* sp. community. If conditions are favorable, hundreds of thousands of them can settle on the reef and cause total bleaching of the reef (James *et al.*1990).

REEF ASSOCIATED BIODIVERSITY

As the coral reefs are the sites for high biological productivity, the reef's biodiversity is quite amazing. It houses various species of benthic algae, seaweeds, sea grasses, lobsters, sponges, echinoderms, fin fishes, crabs, gastropods and cephalopods. The states of biodiversity at various sites are delineated hereunder.

Gulf of Kachchh

This Gulf area, with its necklace of islands, had much richer marine and land fauna about a century ago. Earlier, Baroda State had invited Mr. James Hornel from Ceylon to survey and study Okhamandal coast. He revealed in 1905 that he had never found before such a rich marine biota in so confined a place. Further survey of the marine fauna in the Gulf was conducted in early fifties by Gideon *et al.* (1956). Certain coral reef areas have been brought under the Marine National Park and Sanctuary for preservation. Important varieties of flora and fauna found in the MNPS are as follows:

Phylum Porifera: About 70 species of sponges are reported in the region. They occur in a fantastic array of colors. Most of the sponges are found in shallow water, feeding on microscopic plants, animals and detritus.

Phylum Coelenterata: Jellyfishes, sea anemones and corals primarily represent this group. Animals of this group are most abundant and widely distributed. Ecologically, most important animals in this group are corals. More than 37 species of corals are recorded in this area. Phylum coelenterata includes three classes. The class Hydrozoa includes hydroids; 17 species of which have been recorded by Mr. James Hornel in the Gulf of Kachchh near Khamandal. This class also includes the genus *Physalia* that is found near Okha and Mithapur. Many species of the classes Scyphozoa and Anthozoa have also been reported from the Gulf of Kachchh.

Phylum Echiuroidea: This group is represented in the MNPS mainly by species of Bonellia. Two important members of Bonellia are: (1) *Ikedella misakiensis* - found probably at only 3 places in the world and (2) *Ikedessma pirotanensis* - only found in the Gulf of Kachchh.

Phylum Annelida: This large group of worms has many forms found burrowed in sand or under dead and live corals e.g.: nereis, polychaetes etc.

Phylum Crustacea: Barnacles, lobsters and crabs mainly represent this group of animals. Among these, barnacles and crabs are most abundant. About 30 different species of crabs are found in this area.

Phylum Mollusca: This group is mainly represented in the MNPS by snails, mussels, oysters, shells, octopus, sea horse, chiton, sepia etc. The oysters collected from Gulf of Kachchh had been increased steadily from 1781 in 1979-80 to

1275, 2773, 3713, 3727 and 3042 (partial) in 1980-81, 1981-82, 1982-83, 1983-84 and 1984-85 respectively (Vohra 1999).

Phylum Echinodermata: These spiny skinned marine animals are mostly bottom dwelling and many species of this group are found in the MNPS, e.g.: star fishes, brittle star, sea urchin, feather start, *Paberia* sp. etc.

Phylum Chordata: This phylum included many important classes like pisces, reptiles, aves and mammals of which many important species like cartilaginous fishes, bony fishes, prawns, green sea turtle, sea snakes, olive ridley turtle, dolphin, whale etc. have been recorded in the MNPS. The Park and Sanctuary area supports approximately 200 species of fishes and 27 species of prawns. Eagle ray fish, stiny ray fish, electric ray fish, shark etc. of cartilaginous type, and butterfly fish, sole fish, porcupine fish, puffer fish etc. of bony type, are abundant in the Gulf of Kachchh. Naik *et al.* (1994) recorded 94 species of water birds and 76 species of terrestrial birds in the Gulf.

Marine Algae: The Gulf of Kachchh is rich in algal diversity (120 species), dominated by the species of *Sargassum, Ulva, Enteromorpha* and *Cladophora* (Bahuguna *et al.* 1992). A large number of species of green algae, brown algae, red algae, etc. has been recorded in the Park area. Air temperature shows an inverse relationship with the diversity and total biomass. At most of the places the net primary productivity is maximum for the brown algae species.

Lakshadweep

Around 603 species of fish (both reef dwelling and oceanic) have been recorded from Lakshadweep, however, comprehensive biodiversity studies are lacking. Varieties of ornamental fishes, such as species of *Abudelduf, Amphirprion, Apogon, Coris, Balistes, Platax* and several other beautiful coral fishes are common in Lakshadweep (George *et al.* 1986). Bhaskar (1984) recorded four species of turtles, *viz. Eretmochelys imbricata, Lepidochelys olivacea, Chelonia mydas* and *Dermochelys coriacea*. Thomas (1979) reported 41 species of sponges from Minicoy, including typical coral and shell-boring species, *viz. Spirastrella cuspidifera, S. inconstance* and *Cliona* sp.

In view of shallowness, the lagoons support rich growth of macrophytes. Totally, 69 varieties of marine algae belonging to 52 genera were recorded from the Lakshadweep islands (Jagtap 1987). The lagoon areas of the islands have seagrass beds dominated by *Thalassia hemprichii* (James *et al.* 1987). The thraustochytrid fungus (*Corallochytrium limacisporum*) has been found in great numbers in the waters of coral reef lagoons of Lakshadweep.

Gulf of Mannar and Palk Bay

The Gulf of Mannar Marine Biosphere Reserve is endowed with two important ecosystems, coral reef and sea grass. In these ecosystems different flora and fauna with varying ecological habitats are represented in great numbers. TN Forest Department (Upreti and Shanmugaraj 1997) recorded 3600 species in the marine reserve. Further studies carried out by ZSI (ZSI 1998) concluded presence of 1060 species in the reserve.

Besides 108 species of sponges belonging to Demospongia, there are numerous species of coral-boring and shell-boring sponges in the Gulf. Fourteen, 119,56,6 and 12 species of gorgonids, polychaetes, crabs, lobsters and penaeids respectively have been recorded in the Gulf (Devraj 1998). Echinoderms are represented by Crinoidea (9 species), Ophiuroidea (19 species), Asteroidea (26 species), Echinoidea (19 species) and Holothuroidea (22 species). The most common echinoderms in this area are *Holothuria arta, H.scabra, H. edulis* and *Protereaster lincki*. Amongst the mollusks, 6 species of pearl oysters dominated by *Pinctada fucata*, and 5 species of chanks, commonly being *Xanchus pyrum* var. *obtuse*, are most important. The branching corals *Acropora corymbosa* and *Pocillopora damicornis* have large number of molluscan associates as compared to other corals (Ramesh 1996).

There are beautifully colored fishes inhabiting the reef corals. The fishes belong to the genera *Lytijanus, Chaetodontopis, Chaetodon, Holocenturs, Pomacanthoides, Linophora, Thallasoma, Abudefduf* and *Zanclus*. Two species of eels, *Gymnothorax undulatur* and *G. punctatur*, are densely populated in the reef flat (Reddiah 1970). While the dolphins

Tursiops aduncus and *Delphinus delphis* are commonly sighted in this area, the dugongs are often attracted in the pasture grounds near the islands. A total of 61 species of marine macro algae belonging to three major groups, *viz.* green algae (14 genera and 28 species), brown algae (8 genera and 13 species), and red algae (17 genera and 20 species) has been recorded. The dominant genera of seaweeds are *Gracilaria, Gelidiella, Hypnea, Sarconema, Hydrodathrus, Caulerpa, Sargassum* and *Tubinaria.*

Andaman and Nicobar Islands

In the islands, there are 120 species of algae and 70 species of sponges. Among the reef inhabiting invertebrates, the major groups are gastropods, bivalves and echinoderms comprising 35, 25 and 25 per cent respectively (Dorairaj and Soundararajan 1997). The *Turbo* sp. are always seen underneath the rocks, coral stones or buried in the mud. Besides *Turbo* sp., the shells of *Trochus niloticus, Cassis cornuta, Xancus sp., Lambis lambis, L.chiragra, Murex* sp., *Nautilius.*sp., *Conus* sp. and *Cypraea* sp. are also commonly available (Appukuttan 1977). The fish assemblage on the reef region was moderate on Twin islands, followed by Red Skin, Tamugil, Boat, and Jolly Boys islands. Altogether, there are 571 species of fishes recorded in the reef region. An analysis of fish assemblage indicated 32, 19,18,12, 4 and 3 per cent angel fish, snappers and fusilers, surgeon fish, coral fish, spine foot and wrasses respectively in this region. Out of the 55 fish species recorded in the park area, many of them are of highly priced ornamental fishes and tuna live bait fishes (Dorairaj and Soundararajan 1997). The islands have the nesting beaches of various turtles.

DEGRADATION

Coral reefs are disappearing at an alarming rate throughout the tropics. In Southeast Asia where approximately 30 per cent of world's coral reefs are found, about 60 per cent of the reefs are already destroyed or on the verge of destruction (Wilkinson *et al.* 1993). Humans, pollution, sedimentation and overexploitation directly or indirectly induced the major causes for the destruction of the reefs in this region. It is evident from literature and satellite data that vast areas of coral reefs in the Indian waters are dead. Even in areas normally referred to as coral reefs or patches, living coral covers rarely exceed 20-30 percent. The principal causes are delineated hereunder.

Prolonged Exposure: Prolonged exposure of corals is detrimental to their survival and very few forms can survive in the high tidal amplitude.

Siltation: Siltation is another cause for degradation of reefs. It is also responsible for selective destruction of small polyped forms that cannot cope up with the deposition of silt. Sediments can, at times, smother them, clogging up the mouths and tentacles and hindering respiration and feeding.

The incidence of siltation is present everywhere in varying degrees, increasing from the inner to the outer reefs. In Wandoor Marine Park area in South Andamans, siltation caused mass mortality to ramose corals. Sea erosion was rampant and a vast area in May 1989 was found covered by silt along the shore, killing corals. The reef of Chideathope is also mostly dead. Quarrying of sand from the near shore area was believed to be the reason for sea erosion and subsequent siltation (Pillai 1996).

Mining: Mining operations also resulted in suspension of enormous quantities of sand and soil particles, which covered corals and resulted in their degradation. Govt. of Gujarat leased out some area under MNPS for industrial activities before declaration of the sanctuary. As a result, large areas of Narara and Pirotan coral reefs were ruthlessly exploited. Removal of calcareous sands from different coral ridges and inshore areas played an important role in rapidly rising sedimentation leading thereby to decrease and loss of precious marine life from Gulf of Kachchh. Similarly, introduction of large mechanized fishing vessels and added demand for cargo transport has resulted in the dredging of lagoons in Minicoy, Kavaratti, Amini and Kiltan islands. The lagoon corals in these areas are almost dead. The coral reefs in Palk Bay and Gulf of Mannar were quarried for industrial purposes from early sixties in between Mandapam and Tuticorin. Pillai (1973) estimated the exploitation of corals to the tune of 250m^3 per day in this area. It is believed that some of the islands near Tuticorin are almost vanished due to quarrying (Pillai 1996).

Industrial Effluent: Release of pollutants from a number of industries, oil pipeline terminal, oil refineries etc. on the Kachchh coast adds to the degradation process of corals. The effluents from timber factories around Port Blair and in Middle Andaman are reported to have caused damages to coral growth (Arif *et al.* 1987).

Salt Works: The salt production method involves taking up saline seawater at 35 ppt in small ponds, which is then allowed to evaporate. The concentration of brine water goes on increasing as it moves from reservoirs to different ponds till crystallization occurs. It is from the crystallizing pond where salt is obtained (Plate 2) and the liquid concentrate (bittern) at about 250 ppt is released into the sea. Bittern is a highly concentrated form of seawater, rich in marine chemicals like potassium, bromium, sodium sulphate and magnesium salts. This concentrated water, when discharged in shallow seawater by the salt manufacturers, causes pollution and the delicate reefs are exposed to drastic rise in salinity. It is estimated that 944 litters of bittern is produced in every tonne of salt production.

Plate 2: Crystallizing pond showing formation of salt in the Gulf of Kachchh at Jamnagar

Growth of Algae: Growth of algae on reefs and the algae blades spreading over corals, trap the TSS and TDS released in the effluents of the industries. The pollutants interfere light penetration and prevent photosynthesis, thus hampering the primary productivity in the coral reef areas. This affects the entire ecological pyramid of the reefs.

Destructive Fishing Practices: Although fishing should not be allowed in the Park and Sanctuary area as per Wildlife (Protection) Act, 1972 and its amendment in 1991, but this practice could not be fully stopped in the area. Fishermen fix their nets on coral reefs in intertidal zones. Often they turn coral stones to fix the nets. Movements of people on and turning of coral stones damage coral along with other marine organisms. Often fishermen use bombs made of potassium nitrate to cause large explosions, which kills all the fishes in the surrounding area and reduce nearby corals to lifeless rubble. Cyanide, one of the most toxic poisons known, is also being used to harvest live fish. Fishermen stun fish by squirting cyanide into the reef areas where these fish seek refuge. They then rip apart the reefs with crowbars to capture disoriented fish in the coral where they hide. In addition, cyanide kills the coral polyps, the symbiotic algae and other small organisms necessary for healthy reefs.

PROTECTION OF CORAL REEF ECOSYSTEMS

Coral reef ecosystems are a myriad of interwoven and interdependent habitats and associated organisms. Despite ecological significance and beauty of reefs, they are threatened around the globe by human activities, especially over fishing and pollutions. The most critical and pervasive threat to sensitive coral reef ecosystems worldwide is the combined impact of living marine resource extraction (fishing, collecting and harvesting) and water quality degradation. Intact coral reef systems thrive in clear, clean water.

Marine "no-take" reserves, areas closed to fishing, are a proven tool for the protection of coral reef systems. Their benefits to both humans and reef communities are intuitive and well documented. These are:

Protect Diversity of Life: In a reef ecosystem each species plays an important role. Parrot fish, conch, and grazers keep reefs free of overgrowth of algae and decaying matter. Interactions in a reef have achieved a complex balance by removing too many of one kind of species. Reserves restore nature's balance by leaving some areas intact without any species being removed intentionally.

Restoring fish population: Within a reserve, fish population increases, and individual fish has a chance to grow larger. Reserves also protect the spawning stock of the fishery. A certain minimum number of adults are needed to ensure that the population will thrive. Reserves ensure survival of healthy adult populations that allows the production of numerous and robust offspring, thus maintaining the species for future generations.

Replenishing other areas: Because of the increased numbers of fish in the reserve, some fish may migrate out of the closed area, and be available for capture. More important, the larvae of the fish in the reserve are taken by ocean currents and dispersed through the region, thereby increasing the density of fish beyond the reserve area. Accordingly, the reserves help replenishing fisheries outside the protected areas.

Expand knowledge and understanding of reef systems: By using reserves as control areas, one can conduct research and acquire knowledge about the interaction between species. Reserves can also help to understand the impacts of humans on ecosystems, thus making management of fisheries and habitat protection more effective.

Enhance enjoyment and tourism: Observing jewel-colored fish swim among the corals, watching large grouper wander through the reefs, and enjoying the natural beauty of the marine world are important parts of the experience for snorkels and divers, and thus critical to regions where tourism is an important part of the economy.

CONSERVATION OF CORAL REEFS IN INDIA

There are a few laws in India to protect coral reef areas, such as the Environment Protection Act, 1986 and the Coastal Regulation Zone (CRZ) Act, 1991, issued under the broad Environment Protection Act (EPA) as well as the Wild life

(Protection) Act (WPA), 1972. The other laws that would have a bearing on coral reef area conservation are: Indian Forest Act, 1927, the Forest Conservation Act, 1980 and the Indian Fisheries Act, 1897. Various State Fisheries Acts and most recently, the State Coastal Management Authority would be relevant for conservation and management of coral reefs.

Marine protected areas are not well managed in India. There are five marine protected areas in India, viz. Gulf of Mannar Biosphere Reserve (GoMBR: 10,500km^2), Gulf of Kachchh Marine National Park (GoKMNP: 400km^2), Mahatma Gandhi Marine National Park (MGMNP: 282km^2) also known as Wandoor Marine National Park in Andamans, Great Nicobar Biosphere Reserve (GNBR: 885km^2), and Rani Jhansi Marine National Park (RJMNP) in Ritchie's Archipelago in Andamans. These MPAs are still vulnerable to impacts from the crown-of-thorns star fish and bleaching which are beyond local management. There is a need for studies on reef resource use, particularly harvesting of reef fish and economic links between resource exploitation and coastal communities. A national policy to bring all coral reefs under one central authority was necessary, and accordingly Coral Reef Monitoring Action Plans (CRMAPs) have been prepared and launched by the Indian Coral Reef Monitoring Network (ICRMN) for all reef areas except the Gulf of Kachchh.

In Indian coral reef ecosystem, there is no detailed biodiversity assessment data, which are the basic requirements for comparing the changes caused due to anthropogenic activities and natural calamities. Data on the potential harvest rates, carrying capacity, habitat degradation and species interaction of the coral reef ecosystems need to be collected. Zoning plan on coral reef ecosystems for conservation and management purposes, and legal status for each category should be implemented.

REFERENCES

Albaladejo V.D. and Corpuz V.I. (1981). A market study of the aquarium fish industry of the Philippines: an assessment of the growth and the mechanics of the trade; Proc. Fourth Int. Coral Reef Symp. 1: 75-81.

Alcock A. (1902). A naturalist in Indian seas; or four years with the Royal Indian Marine Survey Ship "Investigator". John Murray, London, p. 328.

Anonymous (2000). Global warming, threat to coral reefs, The Hitavada, April 18, 2000.

Anonymous (1984). The coral paradise, Lakshadweep Secretariat, p. 51.

Appukuttan K.K. (1977). Trochus and Turbo fishery in Andamans. Seafood Export J. 9(12): 21-25.

Arif M.M., Warwadekar, Y.M. and Ravindran, E.K. (1987). Endangered coral reefs of Bay islands and their ornamental fishes. Proc. Symp. Management of Coastal Ecosystems and Oceanic Resources. Cent. Agric. Res. Inst., Port Blair.

Bahuguna, A., Ghosh, A., Nayak, S., Patel, A. and Agarwal, J.P. (1992). Ecological status of the coral reefs of the Gulf of Kachchh and Lakshadweep. Proc. Coral Reef Distribution of Gulf of Kachchh, Workshop on remote sensing for sustainable development, SAC, Ahmedabad.

Bahuguna, A., Nayak, S., and Agarwal, J.P. (1989). Monitoring of ecological changes in the Marine National Park using satellite data IRS-UP/SAC/MCE/SN/24/8, SAC (ISRO), Ahmedabad.

Bakus P. (1994). Coral reef ecosystem, Oxford & IBH Publishing Co. Pvt. Ltd. New Delhi, 1994, p. 232.

Barnes R.S.K. and Hughes, R.N. (1982). Coral reefs, p. 154-193 in: An introduction to marine ecology. Blackwell Scientific Publication, London.

Bhaskar S. (1984). The distribution and status of sea turtles in India, pp. 23-25, Proc. Workshop on sea turtle conserv., Madras.

Chavan S.A. (1983). Marine Park in the Gulf of Kachchh. Bombay Nat. Hist. Soc. Centenary Seminar: Conservation in developing countries, Problems and Prospects. Bombay, p. 7.

Chhaya N.D. (1997). Minding our marine wealth, an appraisal of Gujarat's coastal resources. Centre for Environment Education, Ahmedabad.

Deshmukhe G., Ramamoorthy, K. and Sen Gupta, R. (2000). On the coral reefs of the Gulf of Kachchh. Curr. Sci. 79: 160-162.

Devraj K.V. (2000). Ornamental fish culture—a potential money-spinner. Souvenir, The Fifth Indian Fisheries Forum, Asian Fisheries Society and Assoc. of Aquaculturists, CIFA, Bhubaneswara, pp. 24-29.

Devraj K.V. (1997). A brief on the contribution of the Central Marine Fisheries Research Institute to research and knowledge of coral reefs of India. Proc.Reg. Workshop on Conser. & Sust. Management of the Coral Reef Ecosystem. M.S. Swaminathan Res. Foundation, Chennai, pp. 21-25.

Devaraj M. (1998). Conservation and sustainable management of the marine living resources of the Gulf of Mannar Marine Biosphere Reserve. Key note address, Workshop on 'Coastal biodiversity of the Gulf of Mannar', Feb., 1998, Chennai.

DOD and SAC (1997). Coral reef maps of India. Department of Ocean Development and Space Application Center, Ahmedabad, India.

Dorairaj K. and Soundararajan, R. (1997). Status of coral reefs of Mahatma Gandhi Marine National Park, Wandoor, Andamans. Proc.Reg. Workshop on Conser. & Sust. Management of the Coral Reef Ecosystem. M.S. Swaminathan Res. Foundation, Chennai, pp. 55-63.

Dorairaj K., Soundararajan, R. and Singh, N.T. (1997). Corals of Andaman and Nicobar islands: A status report. Central Agricultural Institute, p. 29.

Drew E.A. (1972). The biology & physiology of alga invertebrate symbioses. II. The density of symbiotic algal cells in a number of hermatypic hard corals and alcyonarians from various depths; J. Exp. Mar. Biol. Ecol. 9: 71-75.

Gardiner J.S. (1903). Introduction, p. 1-11 *in:* J.S.Gardiner *ed.* The fauna and geography of the Maldive and Laccadive Archipelagoes, Cambridge University Press, Cambridge

Gardiner J.S. (1906). Notes on the distribution of the land and marine animals with a list of the coral reefs, p. 1046-1057 *in:* J.S.Gardiner *ed.* The fauna and geography of the Maldive and Laccadive Archipelagoes, Cambridge University Press, Cambridge.

George K.C., Thomas, P.A., Appukuttam, K.K. and Gopakumar (1986). Ancillary living marine resources of Lakshadweep. Mar. Fish. Inform. Serv.T and E Ser. India, No.68: 46-50.

Gideon P.W., Menon, P.K.B. and Rao, S.R.V. (1956). On the marine fauna of the Gulf of Kutch: a preliminary survey. J. Bombay Nat. Hist. Soc. 54(3): 690-706.

Gottfried M. and Roman, M.R. (1983). Ingestion and incorporation of coral - mucus detritus by reef zooplankton; Mar. Biol. 72: 211-218.

Guppy H.B. (1989). The Cocos Keeling Islands. Scottish Geographic Magazine, 5 (b): 569-588.

Halldal P. (1968). Photosynthetic capacity and photosynthetic action spectra of endozoic algae of the massive coral, Favia: Biol. Bull. 134: 411-424.

Hornell J. (1916). The utilization of corals and shells for lime burning in the Madras presidency. Madras Fish. Dept. Bull. 8:105-125.

ISRO (1993). Coral reefs of the Gulf of Kachchh, Gujarat. Space Application Centre, Ahmedabad, Sci. Note SAC/RSA/RSAG/DOD - COS/SN/04/93.

Jacob P.G. (1973). Corals-raw materials for industries. Seafood Export J. 5: 35-38.

Jagtap T.J. (1987). Distribution of algae, seagrass and coral communities from Lakshadweep islands, Eastern Arabian Sea. Indian J.Mar. Sci., 16:256-260.

James D.B., Pillai, C.S.G., and Gopakumar, G. (1990). A case study of infestation of Acanthaster planci in Andaman waters. CMFRI Marine Fisheries Information Service, 72:1-11.

James P.S.B.R., Pillai, P.P. and Jauprakash (1987). Impression of a recent visit to Lakshadweep from the fisheries and marine biological perspectives. CMFRI Marine Fisheries Information Service, 106:1-3.

Jeyabaskaran R., Lyla, P.S. and Ajmal Khan, S. (1996). Coral: The excellent bone graft material. Seafood Export J. 27(5): 27-28.

Johannes R. E. (1967). Ecology of organic aggregates in the vicinity of a coral reef; Limnol. Oceanogr. 12: 189-195.

Kanwiser J. W. and Wainwright S. A. (1967). Oxygen balance in some reef corals; Biol. Bull. 133: 378-390.

Kinsey D. W. & Hopley D. (1991). The significance of coral reefs as global carbon sinks - response to Greenhouse. Palaeogeogr. Palaeoclimatol. Palaeoecol. 89, 363-377.

Krishnakumar N. (1997). The coral reef ecosystem of the Andaman and Nicobar islands, problems and prospects and the World Wide Fund for Nature-India initiatives for its conservation. pp. 29-47, *in* Vineeta Hoon *ed*. Proc. Regional Workshop on the Conservation and Sustainable Management of Coral Reefs, MSSRF and BOBP, FAO/UN.

Krishnamurthy N. (1969). Industrial uses of corals in India. Proc. Symp. Corals and Coral reefs., Mar. Biol. Assoc. Ind. Souvenir, pp. 18-24.

Lewis, J.B. (1977). Process of organic production on coral reefs, Biol. Rev. 52: 305-307.

Lundin C.G. and Linden O. (1993). Coastal ecosystem: attempts to manage threatened resources. Ambio, 22 (7): 468-473.

Littler M.M. (1974). The productivity of Hawaiian fringing reef crustose corallinereceae and an experimental evaluation of production methodology; Limnol. Oceanogr. 18: 946-952.

Mahadevan S. and Nair, N. (1969). Distribution of the coral reefs in the Gulf of Mannar and Palk Bay and their exploitation and utilization. Proc. First Intern. Symp. Corals and Coral reefs, India, pp. 181-190.

Mann K.H. (1982). Ecology of coastal waters: A systems approach, Stud., Ecol. 8, 160-182.

Marsh J.A. Jr. (1970). Primary productivity of reef building caleareosis red algae; Ecology 51: 255-263.

Marten G.G. and Polovina J. J. (1981). A comparative study of fish yields from various tropical ecosystems (ICLARM/CSIRO Workshop on the theory & management of Tropical Multispecies Stocks.Cronulla, Australia) January 1981.

Mattson J.S. and Defoor J.A. (1985). Natural resource damages: Restitution as a mechanism to slow destruction of Florida's natural resources. J. Land Use Environ. Law 1 (3), 295-319.

Mayer (1918). Ecology of Murray Island Coral Reef. A paper prepared by the Department of Marine Biology, Carnegic Institute, 213: 1-48.

Mukesh, M.V. (1997). Sedimentology of the coral reef ecosystem in and the barrier islands of Chidambaranar Dist., Tamil Nadu, Ph.D. Thesis submitted to M.S. University, Tirunelveli.

Munro J.L. (1984). Coral reef fisheries and world fish production; ICLARM Newslett. 7: 3-4.

Munro J.L. and Williams D. Mc. B. (1985). Assessment and management of coral reef fisheries: biological, environmental and socio-economic aspects. Proc. 5th Coral Reef Congr. Tahiti (4): pp. 545-578.

Naik, R.M., Murthy, M.S., Mansrui, A.P., Parvez, R., Mundkur, T., Krishnan, S., Faldu, P.J. and Krishna, T.S.U.R. (1991). Coastal marine ecosystem and Anthropogenic pressure in the Gulf of Kachchh, Dept. Bioscience, Saurashtra Univ.

Nair R.R. (1984). The Indus paradox, New Sci. 1397: 41-42.

Nair R.R. and Qasim S.Z. (1978). Occurrence of a bank with living corals off the South-West coast of India, Indian J. Mar. Sci. 7: 55-58.

Patel M.I. (1985). Conservation of patchy corals of Gulf of Kachchh. *In*. Symp. Endangered marine animals and marine parks, Marine Biol. Assoc. India, Cochin, Jan.12-16. p. 56.

Patel M. I. and Bhaskaran M. (1978). Possibilities of establishing coral-based industries in Gujarat. Seafood Export J. 10(4): 25-27.

Pillai C.S.G. (1967). Studies on corals. Ph. D. Thesis, Kerala Univ., India.

Pillai C.S.G. (1969). The distribution of corals on a reef at Mandapam (Palk Bay), S.India, J.Mar.Biol. Assoc.India, 11:62-72.

Pillai C.S.G. (1971). Distribution of shallow water stony corals at Minicoy atoll in the Indian Ocean. Atoll Res.Bull.,Wash., 141: 1-12.

Pillai C.S.G. (1972). Stony corals of the seas around India. Proc.Symp.Corals and Coral Reefs, Mar.Biol.Assoc. India, Kochi, pp. 191-196.

Pillai C.S.G. (1973). Coral resources of India with special reference to Palk Bay and Gulf of Mannar, pp. 700-705, *in* Proc Symp. Living Resources of the seas around India.

Pillai C.S.G. (1996). Coral reefs of India, their conservation and management. pp. 16-31, *In*. N.G. Menon and C.S.G. Pillai *ed*. Marine Bioderversity Conservation and Management. CMFRI, ICAR, Kochi.

Pillai C.S.G., Rajagopal, M.S. and Varghees, M.A. (1979). Preliminary report on the reconnaissance survey of the major coastal and marine ecosystems in Gulf of Kachchh. Marine Fish. Infor.Serv. T. and E. Ser., CMFRI, Cochin, 14: 16-20.

Pilson M. E. and Betzer S. B. (1973). Phosphorous flux across a coral reef; Ecology 54: 581-588.

Qasim S. Z., Bhattathiri P.M.A. and Reddy C. V. G. (1972). Primary production of an atoll in the Lakshadweep; Int. Revue Ges. Hydrobiol. 57: 207-226.

Qasim S. Z. (1997). Glimses of the Indian Ocean. University Press, New Delhi.

Qasim S. Z. and Bhattathiri P. M. A. (1971). Primary production of a seagrass bed on Karavatti atoll (Laccadives), Hydrobiologia 38: 29-38.

Qasim S. Z. and Sankaranarayanan V. N. (1970). Production of particulate organic matter by the reef on Kavaratti atoll (Laccadives), Limnol. Oceanogr. 15: 574-578.

Qasim S. Z. and Wafar M.V.M. (1979). Occurrence of living corals at several places along the west coast of India, Mahasagar Bull. Nat. Inst. Oceanogr., 12: 53-58.

Ramesh D.A. (1996). Association of coral colony (Montipora digitata) and the fish fry (Epinephelus sp.) on the western side of Pulli island (lagoon and reef crest of Gulf of Mannar). Indian Hydrobiol., 2:21-23.

Rangarajan K. and Marichamy, R. (1972). Seasonal changes in the temperature, salinity and plankton volume at Port Blair, Andamans. CMFRI Technical ser. 19(12): 60-69.

Rashid M.A. (1988). Gujarat's Gulf of Kutch—a marine paradise. pp. 467-474 *In* Silas, E.G. (*ed*) Proc. Symp. Endangered Marine Animals and Marine Parks. The Marine Biol. Assoc. India, Cochin.

Reddiah K. (1970). The Appa island and its fringing reef in the Gulf of Mannar. J. Marine Biol. Assoc.India, 12:57-63.

Reddiah K., Subba Rao, N.V., Cherian,P.T., Halder, K.R. and Roy, T. (1974). Two Indian Ocean coral knolls. Indian J. Marine Sci. 3:67-71.

Rotrigues C.L. (1997). An analysis of the carrying capacity of Lakshadweep coral reefs. pp. 71-83, in Vineeta Hoon *ed.* Proc. Regional Workshop on the Conservation and Sustainable Management of Coral Reefs.

Scott B.D. and Jitts H.R. (1977). Photosynthesis of phytoplankton and zooxanthellae on a coral reef. Marine Biol. 41:307-315.

Smith S. V. (1978). Coral reef area and the contributions of reefs to processes and resources of the world's oceans; Nature (London) 273: 225-226.

Sournia A. (1977). Analyse et bilan de la production primaire dansles reefs corallines; Ann. Inst. Oceanogr. Paris 53: 47-73.

Sournia A. (1976). Primary production of sands in the lagoon of an atoll and the role of foraminiferan symbionts. Marine Biol. 37:29-32.

Thomas P.A. (1979). Demospongiae of Minicoy island (Indian ocean) Part. 1. Orders Keratosida and Haploscleriade. J. Mar. Biol. Assoc. India, 21(1&2): 10-16.

Titlyanor E. A. (1981). Adaptation of reef - building corals to low light intensity; Proc. Fourth Int. Coral Reef Symp. 2: 39-43.

UNEP (1982). Pollution and the marine environment in the Indian Ocean: UNEP regional sea reports and studies. No. 13, p. 70.

UNEP (1988). Coral reef of the world, Vol. 2: Indian Ocean, Red Sea and Gulf, Cambridge, U.K.

Upreti A. and Shanmugaraj (1998). Gulf of Mannar Marine Biosphere Reserve, Tamil Nadu Forest Department, Ramnathapuram, India.

Vaughan T. W. (1911). Physical conditions under which Paleozoic coral reefs are formed. Geoi. Soc. Am. Bull. 22: 238-252.

Vineeta Hoon (1997). Coral reef of India: review of their extent, condition, research and management status. pp. 1-27, *In* Proc. Regional Workshop on the Conservation and Sustainable Management of Coral Reefs.

Vohra, N.A. (1999). Status of fishery resources of Gulf of Kachchh and impacts of environmental qualities on fish production, ICMAM Workshop, Sept. 23-24, 1999.

Wafar M. V. M. (1977). Phytoplankton production of two atolls of the Indian Ocean; Mahasagar - Bull. Nat. Inst. Oceanogr. 10: 117-121.

Wafar M. V. M. (1974). Nematocysts in four species of corals. Mahasagar-Bull. Nat. Inst. Oceanogr., 7: 119-123.

Wafar M. V. M., Devassy V. P., Slawyk G., Goes J., Jayakumar A. and Rajendran A. (1985). Nitrogen uptake by phytoplankton and zooxanthellae in a coral atoll. Proc. of Fifth Int. Coral Reef Cong. Tahiti, Vol. 6, pp. 29-37.

Wafar M. V. M. (1986). Corals and coral reefs of India. Proc. Indian Acad. Sci. (Animal Sci./Plant Sci.) Suppl. pp. 19-43.

Wilkinson C. R., Chaou L. M., Gomez E., Ridzwan A. R., Soekarno S. and Sudra S. (1993). Status of coral reefs in Southeast Asia: Threats and responses in: Global Aspects of Corals: Health, Hazard and History, University of Miami.

Wood E. M. (1985). Exploitation of coral reef fishes for the aquarium fish trade. Marine Conservation Society, Ross - on - Wye, U.K.

Wood E. (1989) Corals: Wandoor Marine National Park, sane awareness series, INTACH, A&N, p. 14.

ZSI (1998). Ecology and Biodiversity of the Gulf of Mannar Biosphere Reserve, Marine Biol. Stn., Chennai.

7

Seaweeds: Fascinating Marine Bioresources

Lalita N. Sangolkar and *T.K.Ghosh*

INTRODUCTION

While considering utilization of vast aquatic resources, it is usually thought in terms of seafood like prawns, lobsters, crabs, fishes, oysters etc. Normally humans are concerned about plants and vegetation on land, but forget or do not really care about the aquatic plants – macroscopic and microscopic. Many of these can be extremely useful to us, such as seaweeds. Seaweed is the name implied to cover the macroscopic plants of the sea except the flowering plants. However, the submerged marine angiosperms, the seed producing plants, are the sea grasses. Seaweeds fall under the division thallophyta of plant kingdom and belong to three important groups, namely, rhodophyceae (red algae), phaeophyceae (brown algae) and chlorophyceae (green algae). Most of them are red (5500 species), brown (2000 species) or green (1200 species). Generally the weeds are attached to rocks and also grow on other plants as epiphytes. Along the coastline of India, the weeds are abundant where rocky or coral formations occur. This sort of substratum is found in the states of Tamil Nadu and Gujarat, and in the vicinity of Mumbai, Ratnagiri, Goa, Karwar, Vizhinjam, Varkala, Vishakhapatnam, and in the Lakshadweep and Andaman-Nicobar Islands.

Apart from being important as primary producers in marine ecosystems, the weeds act as a breeding and nursery ground for various marine organisms. The weeds contain protein, carbohydrates, vitamins, minerals, trace elements, iodine, bromine, manitol and some bioactive substances. This makes them useful in many maritime countries as a source of food, for industrial applications and as a fertilizer. The major utilization of these plants as food is in Asia, where seaweed cultivation has become a major industry. Meals prepared from seaweeds can also be utilized as supplements to the rations of cattle, poultry and farm animals. The weeds rich in iodine has medicinal properties and are used for controlling goiter disease caused by enlargement of thyroid gland. From time immemorial seaweeds have been used as manure in the coastal areas. The weeds are the only source for production of phytochemicals such as agar and algin and are exploited commercially. Use of the weeds for fermentation and pyrolysis has not been carried out on an industrial scale at present, but is possible options for the 21st century. The weeds have the potential to be used as a source of long-and short-chain chemicals with medicinal and industrial uses.

Surveys of seaweed resources of Indian coastal waters were carried out by several workers (Umamaheswara Rao 1970, Kaliaperumal *et al.* 1987, Chennubhotla *et al.* 1990, Rama Rao *et al.* 1993) in different regions along the coast of the mainland and also Lakshwadeep and Andaman-Nicobar Islands to locate the seaweed growing areas and to assess the standing crop of seaweeds. All these surveys were restricted to near shore areas of the coast (Subbaramaiah 1984). Though no overall picture of the estimates regarding standing crop of these weeds of any maritime state or of India is available, numerous surveys provide indication about the general situation of seaweed resources in many maritime states.

Since 1980, many agar and algin manufacturing seaweed industries are coming up in India (Madurai, Tiruchirappalli, Ahmedabad, Vadodara, Hyderabad etc.), which resulted in increased demand of raw material of agar and algin yielding seaweeds, and hence agarophytes and alginophytes are overexploited. In view of extensive and unrestricted commercial harvest of seaweeds throughout the year, there is depletion in the natural stock of these algae. So it is necessary to conserve these seaweeds by adopting rational way of commercial exploitation.

This chapter deals with occurrence and distribution of seaweeds in Indian waters, their properties and potential for different uses, and their harvest for commercial exploitation. The authors have attempted to bring this fascinating subject within the grasp of the general readers. Besides this, the authors also intended to throw light in the research areas, *viz.* how these bioresources can further be utilized for various end uses and need for their conservation and propagation.

OCCURRENCE AND DISTRIBUTION

Seaweeds are confined to narrow littoral and sublittoral belts of the marine environment. As in land plants, they require certain environmental conditions for proper growth and establishment in different regions of the coastline. However, the topography, physical nature of the substratum, salinity, currents, tidal action and other factors of the marine environment vary in different parts of the coast line and as a result of these fluctuations, marked changes occur in the distribution and abundance of different kinds of seaweeds. All along the 5,689 km long coastline of India, the important seaweed growing regions are the vicinity of Gulf of Kachchh, Gulf of Mannar, Palk Bay, Mumbai, Karwar, Ratnagiri, Goa, Visakhapatnam and coastal lakes like Chilika (Orissa) and Pulicat (Tamil Nadu) (Umamaheswara Rao 1970, Shaw *et al.* 2000). A list of seaweed recorded CSMCRI & CMFRI (1991) Rama Rao *et al.,* (1993), Chennubhotla *et al.,* (1987), Umamaheswara Rao, (1970) in Indian coastal water is presented in Table 1. The pear shaped aquatic lagoon, the Chilika Lake, is largest in Asia having an average water spread area of 1054 Km^2 (64.5 km length x 5 to 18.5 km width). However, over the years the salinity of Chilika Lake is decreasing and hence infested with freshwater macrophyte weeds resulting in reduction of seaweed species composition and density, which are of great concern. The growth of the species is salinity dependant and hence measures to facilitate entry of seawater into the Chilika Lake causing increase of salinity of the lake water, should be taken up. So also discharges of fresh water need to be reduced.

Harvestable quantities of seaweeds grow in other areas of the east and west coasts and in the Lakshadweep and Andaman

Table 1. Seaweed recorded in Indian coastal waters

SN	Species of seaweed	SN	Species of seaweed
Chlorophyceae (Green algae)		29	*Dictyophaeria cavernosa*
1	Acitabularia calyculus	30	*Ectochaete leptochaete (endoderma)*
2	Anadyomene stellata	31	*Enteromorpha clathrata*
3	*Avarainvillia nigricus*	32	*Enteromorpha compressa*
4	*Boergensenia forbessi*	33	*Enteromorpha flexuosa*
5	*Boodlea composita*	34	*Enteromorpha gujaratensis*
6	*Bryopsis hypnoides*	35	*Enteromorpha intestinalis*
7	*Bryopsis plumose*	36	*Enteromorpha linza (bicomuta)*
8	*Caulerpa fastigiata*	37	*Enteromorpha ovata*
9	*Caulerpa fergusonii*	38	*Enteromorpha plumosa*
10	*Caulerpa freycinetii*	39	*Enteromorpha prolifera*
11	*Caulerpa microphysa*	40	*Enteromorpha tubulosa*
12	*Caulerpa peltata*	41	*Halicystic boergesenii*
13	*Caulerpa racemosa*	42	*Halimeda macroloba*
14	*Caulerpa scalpelliformis*	43	*Halimeda macroloba*
15	*Caulerpa sertularioides*	44	*Halimeda tuna*
16	*Caulerpa taxifolia*	45	*Neomeris annulata*
17	*Caulerpa veravalensis*	46	*Ostroebium remenkei*
18	*Caulerpa verticillata*	47	*Spongomorpha indica*
19	*Chaetomorpha antennina*	48	*Struvea anastamosans*
20	*Chaetomorpha linoides*	49	*Udotea indica*
21	*Chamaedoris auriculata*	50	*Ulva beytensis*
22	*Cladophora crystalline*	51	*Ulva fasciata*
23	*Cladophora expansa*	52	*Ulva lactuca*
24	*Cladophora fascicularis*	53	*Ulva reticulata*
25	*Cladophoropsis sundanesis (reundanses)*	54	*Ulva rigida*
26	*Cladophoropsis zollingeri*	55	*Valonia aegagrophila*
27	*Codium dwarkense*	56	*Valonia utricularis*
28	*Codium iyengari*		

(Table Contd...)

(Table 1. Contd...)

SN	Species of seaweed	SN	Species of seaweed
57	*Valoniopsis pachynema*	85	*Hormophysa triquetra*
58	*Vaucheria piloboloides*	86	*Hydroclathrus clathratus*
59	*Wisllella ordinata*	87	*Iyengaria stellata*
Phaeophyceae (Brown algae)		88	*Myriactula arabica*
60	*Colpomenia sinuosa*	89	*Nemacystus deepiens*
61	Cystoceria trinodis	90	*Padina crassia*
62	Dictyopteris acrostichoides	91	*Padina gymnospora*
		92	*Padina pavonica*
63	Dictyopteris australis	93	*Padina tetrastromatica*
64	Dictyopteris woodwardii	94	*Peyssonella obscura*
65	Dictyota atomaria	95	*Pocockiella veriegata*
66	Dictyota bartayresiana	96	*Sargassum carpophyllum*
67	*Dictyota carvicornis*	97	*Sargassum cinetum*
68	*Dictyota ceylanica*	98	*Sargassum ilicifolium*
69	*Dictyota ciliata*	99	*Sargassum jhonstonii*
70	*Dictyota delicatula*	100	*Sargassum merifieldii*
71	*Dictyota dichotoma*	101	*Sargassum plagiophyllum*
72	*Dictyota dichotoma*	102	*Sargassum prismaticum*
73	*Dictyota maxima*	103	*Sargassum sinereum*
74	*Dictyota rivericata*	104	*Sargassum sinereum*
75	*Dilophus fasciola*	105	*Sargassum swartzii*
76	*Ecocarpus arabicus*	106	*Sargassum tenerrimum*
77	*Ecocarpus battersii*	107	*Sargassum vulgre*
78	*Ecocarpus columellaris*	108	*Sargassum wightii*
79	*Ecocarpus elachistae formis*	109	*Spacelaria furcigera*
		110	*Spacelaria tribuloides*
80	*Ecocarpus irregularis*	111	*Spatoglossum asperum*
81	*Ecocarpus oralis*	112	*Turbinaria indica*
82	*Ecocarpus rhodochortonoides*	113	*Turbinaria omata*
83	*Giffordia mitchellae*	114	*Zoonaria crenata*
84	*Giffordia sandriana*	115	*Zoonaria variegata*

(Table Contd...)

(Table 1. Contd...)

SN	Species of seaweed	SN	Species of seaweed
Rhodophyceae (Red Algae)		146	*Champia parvula*
116	*Acanthophora anceps*	147	*Champia somalensis*
117	*Acanthophora delilei*	148	*Chondria armata*
118	*Acanthophora dendrodes*	149	*Chondria cornuta*
119	*Acanthophora fragilissima*	150	*Chondria dasyphylla*
120	*Acanthophora muscoides*	151	*Chondrococcus hornemanii*
121	*Acanthophora spicifera*	152	*Coelarthrum muelleri*
122	*Acrocheatium carnea*	153	*Coelarthrum opuntia*
123	*Acrocheatium crassipes*	154	*Corynomorpha prismatica*
124	*Acrocheatium dwarkense*	155	*Cryptonemia coriacea*
125	*Acrocheatium erectum*	156	*Dictyurus purpurescens*
126	*Actinotrichia fragilis*	157	*Erythrotrychin carnea*
127	*Agardhiella robusta*	158	*Eucheuma spinosum*
128	*Agardhiella tenera*	159	*Galaxaura oblongata*
129	*Alwothamnion byssoides*	160	*Gastrodonium iyengarii*
130	*Amphiroa anastromosans*	161	*Gastroclonium textorii*
131	*Amphiroa anceps*	162	*Gelidiella acerosa*
132	*Amphiroa fragilissima*	163	*Gelidiella myrioclada*
133	*Asparagopsis taxiformis*	164	*Gelidiella tenuissima*
134	*Athrocardia capensis*	165	*Gelidiopsis variavilis*
135	*Athrocardia rubens*	166	*Gracilaria corticata*
136	*Botryocladia leptopoda*	167	*Gracilaria dura*
137	*Caloglossa leprieurii*	168	*Gracilaria edulis*
138	*Catenella impudica*	169	*Gracilaria millartedtii*
139	*Caulacanthas ustulatus*	170	*Gracilaria pygmaea*
140	*Centroceros clavulatum*	171	*Gracilaria verrucosa*
141	*Ceramium miniatum*	172	*Gracilariopsis sjoestedtii*
142	*Ceramium truncatum*	173	*Gracilona cuculia*
143	*Champia compressa*	174	*Gracilona debilis*
144	*Champia globulifera*	175	*Gracilona fatiifera*
145	*Champia indica*	176	*Gracilona verrucosa*

(Table Contd...)

(Table 1. Contd...)

SN	*Species of seaweed*	*SN*	*Species of seaweed*
177	*Gratèloupia indiaca*	207	*Laurencia pedicularioides*
178	*Grateloupia filicina*	208	*Liagara ceranoides*
179	*Gyptonemia undulata*	209	*Lithothamnion fruticulosum*
180	*Haloplegma dupereyii*	210	*Martensia fragilis*
181	*Haloplegma flagelliformis*	211	*Menbranoptera murrayi*
182	*Haloplegma floresia*	212	*Meristotheca papulosa*
183	*Haloplegma musciformis*	213	*Myriogramme bombayensis*
184	*Haloplegma pannosa*	214	*Myriogramme okhensis*
185	*Haloplegma porphyroides*	215	*Neurymenia fraxinifolia*
186	*Haloplegma valentiae*	216	*Plastisiphonia miniata*
187	*Haloplegma venusta*	217	*Polysiphonia coacta*
188	*Halymenia dilatata*	218	*Polysiphonia tuticorinensis*
189	*Halymenia sp.*	219	*Psendoglophloea fasclucularis*
190	*Helminthocladia australis*	220	*Rhodymenia australis*
191	*Helminthocladia calvadosii*	221	*Sarcodia furcellatum*
192	*Heloplegma duperreyi*	222	*Sarcodia indica*
193	*Herposiphonia stuposa*	223	*Sarconema filiforme*
194	*Holymenia porphyroides*	224	*Sarconema furcetum*
195	*Holymenia venusta*	225	*Scinaia bengalica*
196	*Hypnea esperi*	226	*Sarconema surcellatum*
197	Hypnea musciformis	227	*Scinaia furcellata*
198	*Hypnea cervicornis*	228	*Scinaia indica*
199	*Hypolossum spathulatum*	229	*Scinaia leotei*
200	*Jania adhaerens*	230	*Sebdenia pilydactyla*
201	*Jania iyengarii*	231	*Solieria robusta*
202	*Jania rubens*	232	*Spyridia aculeata*
203	*Laurencia ceylanica*	233	*Spyridia filamentosa*
204	*Laurencia cruciata*	234	*Taenioma perpusillum*
205	*Laurencia obtuse*	235	*Wrangelia argus*
206	*Laurencia pamosa*		

and Nicobar islands also. The horizontal distribution of different plants varied in relation to the natural habitats. For instance, *Gelidiella acerosa* and *Sargassum* species have normally been found in association with the coral formations of the Tamil Nadu and Gujarat coasts (Plate 1). Many *Gracilaria* species (Plate 2) have been reported from the localities between

Plate 1: Association of the seaweed with corals in the Gulf of Mannar, Tamil Nadu.

Plate 2: Luxuriant growth of seaweed *Gracilaria sp.* in the Gulf of Mannar, Tamil Nadu.

Mandapam and Cape Comorin in Tamil Nadu. Some species of *Sargassum, Gracilaria, Ulva, Enteromorpha* and *Chaetomorpha* which thrive in the littoral habitats, have a continuous distribution all along the intertidal rocky surfaces of the coastline. It was believed that *Porphyra* species, which is popularly known as "Nori" and is commonly used in Japan as a food item, was not available in Indian seas. However, over a decade ago, researchers from Fisheries Research Centre, Gujarat Agriculture University, Okha, reported the presence of *Porphyra* sp. in the Gulf of Kachchh (Chhaya 1997).

The surveys conducted so far towards the estimation of seaweeds (during the survey, samples for estimation of biomass are generally collected by SCUBA diving) resources in different regions indicated the range of values pertaining to total standing crop and the extent of occurrence of alginophytes and agarophytes. An estimate of one million tonnes of seaweed comprising 740 species belonging to 220 genera has been reported in Indian Ocean (Michanek 1975, Kaliaperumal *et al.* 1987, Kaliaperumal 1994). The density of seaweed was reported to be less, 43.502 tonnes/km^2 in deeper region as compared to that of 231.459 tonnes/ km^2 in shallow waters upto 5 m depth (Rama Rao *et al.* 1993). However, the deep-water areas at Alantalai-Manapad (Southern Zone) and Vembar-Nallatanni Tivu (Northern Zone) region in Tamil Nadu which were surveyed during 1988 for seaweed resources (Rama Rao *et al.* 1996) indicated greater abundance of deep-water algae vegetation as compared to the intertidal survey (upto 4m depth). In the intertidal survey, the number of species recorded in Alantalai-Manapad and Vembar-Nallatanni Tivu was 17 and 10 respectively (Anon. 1978), whereas it was 28 and 27 in deep water of these two areas. The composition of seaweed flora was found to be entirely different except for the common occurrence of *Spatoglossum asperum* in Alantalai-Manapad and *Sargassum plagiophyllum* and *Hypnea musciformis* in Vembar-Nallatanni Tivu. The quantity of economic seaweeds assessed during this survey revealed the feasibility of commercial exploitation of *Spatoglossum asperum*, *Halymenia* sp., and *Gracilaria textorii* for phycocolloid industry (Rama Rao *et al.* 1996). The coast of Tamil Nadu from Mandapam to Kanyakumari supports luxuriant

Table 2. Economically important seaweeds and their uses

Uses	*Name of species*
Agar	*Gelidiella acerosa, Gracilaria lichenoides, G. crassa, G.edulis, G. verrucosa, G. corticata, G. folifera, Gelidium sp.*
Agaroids	*Gigartina acicularis, Hypnea musciformis, Spyridia* sp., *Sarconema filiforme, Acanthophora* sp, *Laurencia* sp, *Chondrus* sp, *Agardhiella tenera, Fucheuma* sp.
Algin	*Sargassum tenerrimum, S. wightii, S. swartzii, S. cinereum, S. johnstonii, Turbinaria conoides, T. ornate, Cystophyllum* sp., *Hormophysa* sp., *Dictyota* sp., *Padina* sp., *Cystoseira* sp.
Food	*Ulva* sp., *Enteromorpha* sp., *Chaetomorpha* sp., *Caulerpa* sp., *Codium* sp., *Dictyota* sp., *Padina* sp., *Colpomenia* sp., *Hydroclathrus* sp., *Rosenvingea* sp., *Chnoospora* sp., *Sargassu* sp., *Turbinaria* sp., *Porphyra* sp., *Halymenia* sp., *Grateloupia* sp., *Gracilaria* sp., *Hypnea* sp., *Rhodymenia* sp., *Centroceras* sp., *Acanthophora* sp., *Laurencia* sp.
Protein and amino acids	*Ulva fasciata, U. rigida, Enteromorpha* sp., *Cladophora morumentalis, Spathoglossum variabile, Porphyra* sp., *Asparagopsis taxiformis, Centroceras clavulatum, Acanthophora muscoides, Codium* sp., *Monostroma* sp., *Hydroclathrus* sp., *Laminaria* sp., *Undaria* sp., *Eucheuma* sp., *Laurenica* sp., *Caulerpa racemosa, Gracilaria corticata, Sargassum wightiis*
Iodine content *(Iodine above 200mg/100 gm)	*Sargassum* sp., *Myriogloea sciurus, Padina australis, Asparagopsis* sp., **Udotea indica*, **Gracilaria lichenoides*, **Sarocnema furcellatum*
Liquid fertilizer	*Ascophyllum nodosum, Fucus serratus, Laminaria* sp., *Ecklonia maxima*
Allopathic medicine	*Sargassum* sp., *Hypnea musciformis, Laminaria* sp., *Enteromopha* sp., *Condoria armata, Centroceras clavulatum, Asparagopsis taxiformis, Sarconema* sp.
Ascorbic acid	*Sargassum myriocystum*

Source: Umamaheswara Rao (1970), Ganesan and Kannan (1994), Dave and Chauhan (1993)

growth of seaweeds having the total standing crop of 22,000 tonnes (wet) (Anon 1978). The survey conducted from Tuticorin to Tiruchendur during December 1986 to March 1987 by Chennubholta *et al.* (1990) indicated the total biomass of 9,100 tonnes (wet). However, only an area of 125 km^2 was found to support the seaweed vegetation out of the total 650 km^2 surveyed area. During a survey covering an area of 417.5 sq. km. of the deep-water area from Kilakkarai (Appa Tivu) to Rameswaram Island (Dhanushkodi) from December 1990 to January 1991, a total biomass of 18,162.5 tonnes (wet) of seaweeds were estimated. Twenty species were found in estimable quantities, of which the abundant 8 species were: *Halimeda macroloba, Spatoglossum asperum, Zonaria crenata, Sargassum illicifolium, Amphiroa fragillssima, Hypnea musciformis, Borryocladia leptopoda* and *Lyngbya majuscula* with a biomass of 1325, 9775, 650, 1550, 1925, 300, 862.5 and 1012.5 tonnes (wet) respectively. The species of *Hypnea* and *Sargassum* could be exploited for the manufacture of phytochemcials (CSMCRI & CMFRI 1991).

Taxonomic studies on seaweeds of Visakhapatnam coast have been carried out earlier (Umamaheswara Rao and Kaliaperumal 1980, Krishnamurthy and Baluswami 1982 and Umamaheswara Rao 1991). During the course of ecological studies, a few interesting algae were collected from Visakhapatnam coast and nearby localities, *viz.* Gangavarram and Bheemunipatnam, which are situated 15 to 30 km away from Visakhapatnam. The seaweed species *Boodleopis indica* which is new to science and two red algae, *viz. Jania ungulata* F. *brevior* (Yendo) Yendo and *Dsya sinicola* v. *abyssicola* Dawson were reported for the first time from India. Fourteen others, *viz. Enteromorpha plumose, Ulva lactuca, Acetabularia parvula solms, Hectonema enhal, Myrionema vulgare, Erythrotrichia carnea, Ahnfeltiopsis pygmeaus, Crouania attenuta, Antithamnionella elegans var. elegans, Plenosporium borreri, Griffithsia heteromorpha, Ptilothamnion cladophorae, Cologlossa leprieurii* and *Laurencia gracilis,* are first records from Visakhapatnam coast (Umamaheswara Rao 1999)

The survey of marine algal resources along the Gujarat coast of India has revealed great diversity in this region. A total of 35 marine algal species were recorded during a survey of the sub tidal flora of Dwarka. Maximum number of species was found at 5 to 8 m depth. The Maximum biomass 2.088 kg/m^2 was recorded at 5 m depth comprising *Caulerpa, Dictyota* and *Spatoglossum* as dominant genera. The dominance of genera suggested that they are less vulnerable to the grazing pressure. Species belonging to Dictyotaceae produce some phenolic compounds that protect the thallus from grazing. In caulerpales, a substance called caulerpicin, a toxic substance acts as a repellent for grazing organisms (Den 1972, Dhargalkar and Deshmukh 1996).

The intertidal and subtidal habitats of Andaman and Nicobar islands also support diverse flora of seaweeds. *Caulerpa mexicana* f. *indica* is described as new form and six other species of algae, *viz. Anadyomene wrightii* Harvey in Gray, *Chroodactylon ornatum* (C. Ag.) Basson, *Peyssonnelia simulans* Weber Van Bosse, *Mastophora rosea* (C. Ag) Setchell, *Amphiroa foliacea* Lamourx f. erecta weber Van Bosse, *Ceraminu fastigiatum* (Roth) Harvey f. flaccida Peterson were recorded for the first time from this region (Umamaheswara Rao 2000). All the 21 islands of the Gulf of Mannar, Tamil Nadu also harbor rich flora of seaweeds with diverse species composition. The seaweeds exploited from these islands are used for the production of agar and alginate in the country. The total standing crop of seaweeds from these islands was estimated at 11,650 tonnes (wet wt) consisting of 1,472 tonnes agarophytes, 7,141 tonnes of alginophytes and 3,037 tonnes of other seaweeds (Anon. 1978). The total biomass of seaweeds estimated in the Gulf of Mannar islands constituted 53% of seaweed biomass of Tamil Nadu Coast.

USES OF SEAWEEDS

At present in India, seaweeds are used as raw material for commercial production of phytochemicals such as agar and algin. Several protein rich seaweeds such as species of *Ulva, Enteromorpha, Caulerpa, Codium* and *Monostroma* (green algae); *Hydroclathrus, Laminaria, Undaria* and *Sargassum* (brown algae); *Phaephyra, Gracilaria, Eucheuma, Laurenica* and *Acanthora* (red algae) (Table 3) are used for human consumption in the form of soup, salad and vegetable. Food

Table 3. Standing crop of seaweeds in maritime states of India
(After Chhaya 1997, Kaliaperumal *et al.* 1987)

Maritime State	*Standing crop of seaweeds (tonnes wet wt)*
Gujarat (Gulf of Kachchh)	
• Agarophytes	205
• Alginophytes	10000
Maharashtra	278
Goa	2000
Tamil Nadu*	
• Agarophytes	73754
• Alginophytes	131588
Andhra Pradesh	7493
Orissa (Chilika)	5
Lakshadweep	7524
Total	232847

Wet to dry ratio is 10:1
* Between Rameshwaram and Kilakari, and Dhanushkoti and Hare island

products such as jelly, jam, chocolate, pickle and wafer could also be prepared from certain weeds. The weeds are also utilized in different parts of the world as feed for poultry and cattle, fertilizer for many land crops etc. as delineated hereunder. Most types of seaweeds are excellent sources of iodine and also provide a wide range of minerals including calcium, copper, iron, magnesium, potassium and zinc containing varying amounts of B-vitamins and B carotene.

Seaweeds for Industrial Gums

Agar: This commercially important product is a colloidal carbohydrate present in the cell walls of certain red seaweed species and it is a mixture of two polysaccharides, agarose and agaropectin. Important and commonly occurring agarophytes in India are *Gelidiella acerosa, Gracilaria* *lichenoides*, *Gracilaria crassa*, *Gracilaria verrucosa*, *Gracilaria corticata Gracilaria folifera* and *Gelidium* species (Umamaheswara Rao 1970). Obviously, the world's agar production depends mostly on *Gracilaria* species (Sasikumar *et al.* 1999), which is an important source of agar in India, Chile, Taiwan, China and other countries. About 25,540 tonnes per year dry matter of *Gracilaria* is currently being harvested from different parts of the world, of which India's contribution is about 600 tonnes (McHugh 1991). The yield and physical properties of the agar extracted from these red seaweeds vary particularly in *Gracilaria* species (Umamaheswara Rao 1970).

In general, yield of agar is high when biomass is low. Yield and characteristics of agar often vary between species and within species. However, environmental factors, e.g. nutrients (N & P), salinity and temperature contribute towards variation in gel strength and yield of agar. While gelling and melting temperature of agar is reported to be 35 to 41^{0}C and 67 to 76^{0}C respectively, the gel strength is 94 g/cm^2, the maximum being during summer. While low water temperature may be requisite for high gel strength, low salinity accelerates agar yield (Sasikumar *et al.* 1999). Agar yield in *Gracilaria edulis* was reported to be higher under reduced levels of nitrate (0.05 mM), ammonium (0.05 mM), and phosphate (0.001 mM). However, at higher nutrient levels quality of agar was improved with reduced SO_4 content (Hemlatha and Rengasamy 1993).

The agar is used as:

- In food technology–in bakery, and for the preparation of confectionery products, for jelling and solidifying; in pudding, cream, jelly and jam preparations; in canning of fish and meat; in the brewery industry, to clean and filter beers, wines and other alcoholic preparations and make them transparent
- As a culture medium in the pharmaceutical industry

In the cosmetics industry–for stabilizing various emulsions, and for the preparation of skin lotions, creams etc., to give them smooth texture.

- Agar is also used in textile (sizing, finishing), paper and photographic industries.

Several methods have been described by different workers, for example, in the pilot plant trials conducted by Visweswara Rao *et al.* (1965), pulverized weed was soaked overnight in fresh water before wet grinding and extraction of the agar. The method suggested by Srinivasan and Santhanaraja (1967) is more or less similar to the one described by Kappanna and Visweswara Rao (1963), but the seaweed is pulverized into fine powder before extraction. To eliminate the cost of freezing, Desai (1967) used 90% industrial alcohol for flocculating the agar from the filtrate. Recently CIFT (1997) has developed a process of producing good quality agar with high gel strength by alkali treatment instead of the conventional acid treatment method. The process developed at the Central Marine Fisheries Research Institute and by other investigators for the manufacture of agar is given in **Annexure 1 A & B**.

Alginates: Algin or alginic acid is the main polysaccharide occurring in the cell walls of brown algae. Species of *Sargassum*, *Turbinaria*, *Cystoseira*, *Hormophysa*, *Dictyota* and *Padina* are some of the weeds reported from the Indian waters. Of these, *Sargassum* is the principal source for the production of algin in the country. However, species of *Padina* are important source of manitol, alginic acid and iodine (Ganesan *et al.* 2000).

In general, alginates are used (Umamaheswara Rao 1970, Chhaya 1997) as:

- In food technology–in bakery, and for the preparation of confectionery, jelly, soup, margarine, ice cream and other dairy products;
- For protection of color pigment, so that color can stick to surfaces properly and does not dry up prematurely;
- In pharmaceutical industries–for homogeneous mixtures and to maintain their balance;
- In the paper and cardboard industry–for bringing softness, creating ease in printing and varnishing purposes;
- In dentistry–for the preparation of dentures and strong fixation of teeth;

ANNEXURE 1A

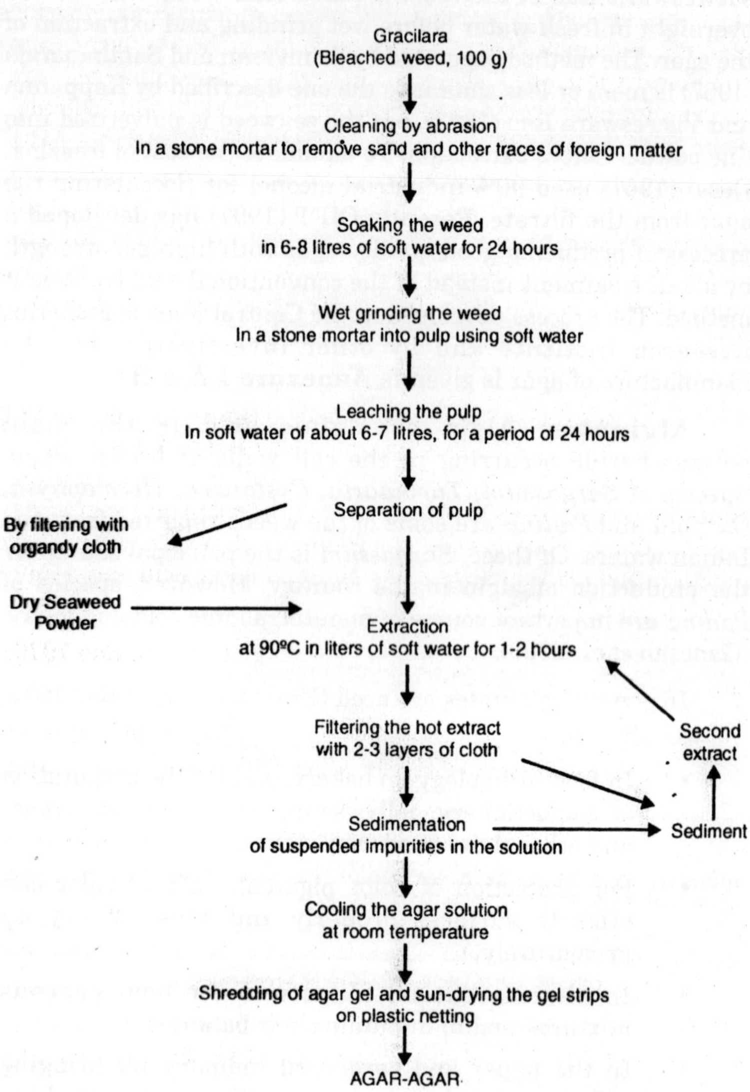

Cottage industry method for the Extraction of agar from *Gracilaria* species
(*After* Thivy 1960 and Umamaheswara Rao 1970)

ANNEXURE 1B

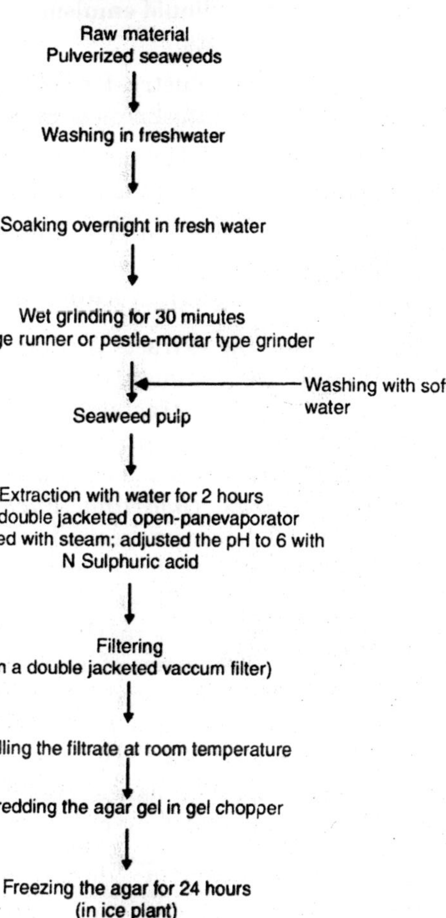

Method for agar manufacture on a commercial scale
(*After* Visweswara Rao *et al.* 1965 and Umamaheswara Rao 1970)

- In the cosmetic industry–for the preparation of creams, jellies, liquid emulsions, lotions, toothpastes, soaps, detergents;
- In the textile industry–for sizing, finishing, refining, etc. and adding thickness during the printing process;
- Alginates are also used in enameling, foundry, ceramics, rubber and leather industries.

The process developed at the Central Marine Fisheries Research Institute and by other investigators for the manufacture of algin is given in **Annexure 2A, B, C.**

Agaroids: Gel-like extracts produced from certain types of red seaweeds are commonly known as agaroids. The carrageenans, the water extractable cell wall polysaccharides which occur as a matrix material in numerous species of red seaweeds, are obtained from such seaweeds of the genera *Agardhiella, Chondrus, Gigartina, Fucheuma* and *Hypnea* (Kaliaperumal 1997). The carrageenans content of *Agardhiella tenera* from Okha, Gujarat, have been reported to be 35% of the algal dry weight (Parekh *et al.* 1993). Organic sulphate content is very much high in these compounds and the chemical nature and properties of agaroids are different from agar-agar. Pure solutions of agaroids are viscous and do not form gel when cooled as in the case of agar. However, various inorganic and organic solutes alter the properties and increase the gelling power of agaroids as observed in the case of *Hypnea* extractive. Carrageenan yielding plants seem to have not been reported from Indian waters, except for a rare and less abundant species of *Gigartina* (*G. acicularis*) occurring in the intertidal habitats. However, species of *Spyridia, Sarconema, Acanthophora, and Laurencia* growing along the Indian coast also give gel–like extracts (Umamaheswara Rao 1970).

In India, there are more than 25 actively functioning agar and algin industries situated in different places in the maritime states of Tamil Nadu, Kerala, Karnataka and Gujarat. The red algae *Gelidiella acerosa, Gracilaria edulis* and *G. folifera* are used for agar manufacture and brown algae *Sargassum* sp, *Turbinaria* sp. and *Cystoseira trinodis* for alginate. All these seaweeds are harvested since 1966 from the natural seaweed

ANNEXURE 2A

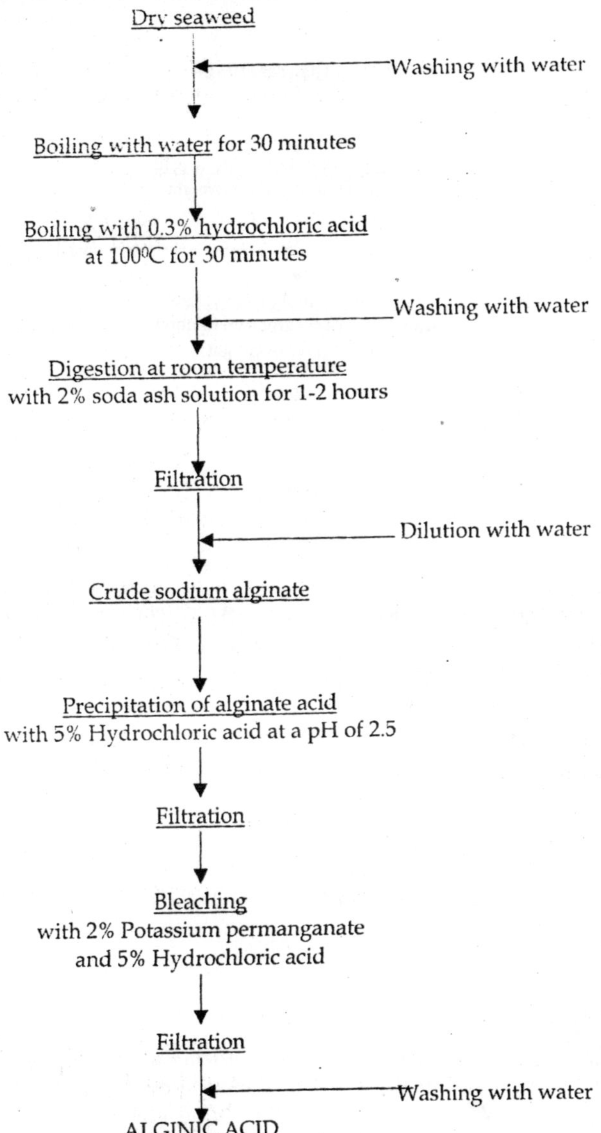

Process for extraction of alginic acid
(*After* Pillai 1957 and Umamaheswara Rao 1970)

ANNEXURE 2 B

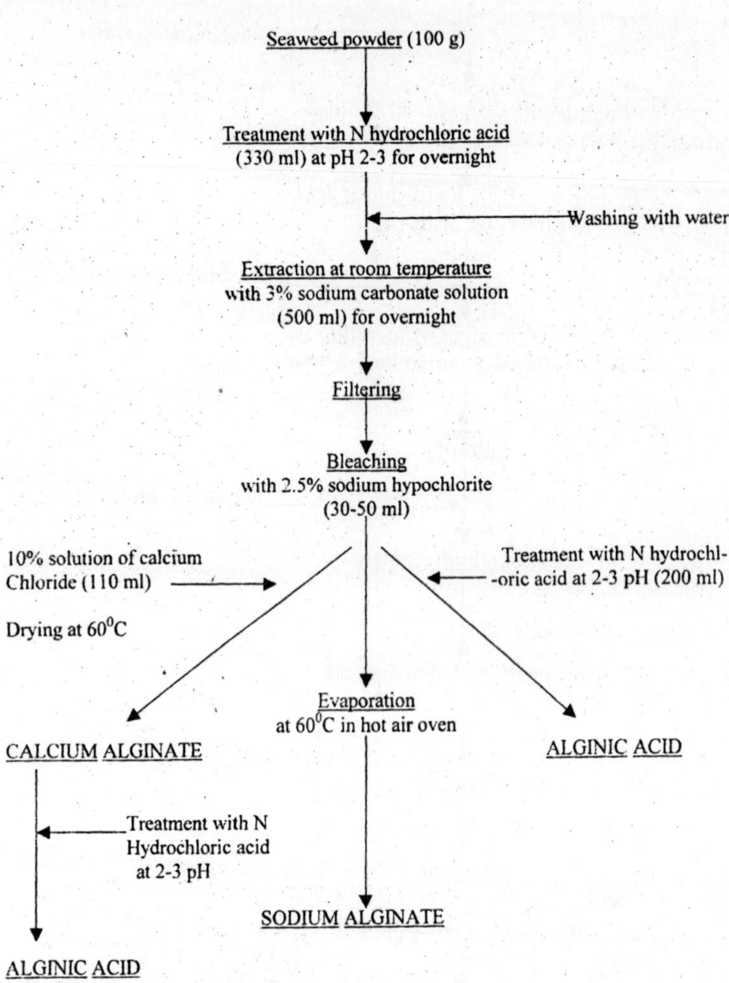

Methods for extraction of Sodium alginate Calcium alginate and Alginate acid
(*After* Visweswara Rao & & Mody 1964 and Umamaheswara Rao 1970)

ANNEXURE 2C

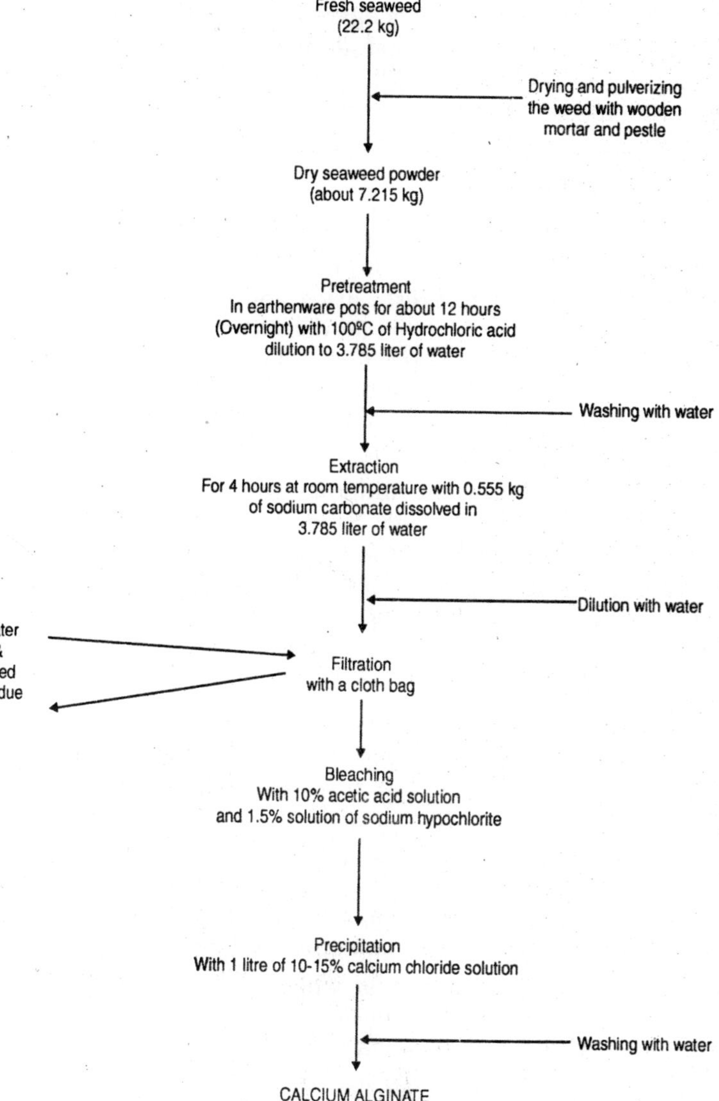

Flow sheet for the preparation of Calcium alginate on a Cottage-Industry basis
(*After* Sadasivan Pillai 1961 and Umamaheswara Rao 1970)

beds in different maritime states. Data collected by the Central Marine Fisheries Research Institute for 20 years from 1978 to 1997 show that the quantity of agarophytes landed ranged from 248 to 1289 tonnes (dry wt) and alginophytes from 651 to 5537 tonnes (dry wt) (Silas and Kalimuthu 1987, Kalimuthu *et al.* 1990, Kaliaperumal and Kalimuthu 1997). About 75% of these seaweeds are exploited from the Gulf of Mannar islands.

Seaweed as Human Food

Many seaweed species are used as food in India, Japan, Indonesia, Ceylon, China, Philippines and other countries of the Indo-Pacific region. Fresh, dried and processed seaweeds are utilized as human food and for decorating the dishes. There is growing need to assess the food value of seaweeds in view of their proven potential for use of food and fodder. The algal carbohydrates are not easily digestible and the food value of the seaweeds comes not only from the bulk of nutrients provided, but also from the minerals, trace elements, proteins and vitamins present in them. Several reports suggested that Vitamin C occurs richly in most seaweed. Some green and red seaweed such as *Ulva fasciata, Ulva rigida, Porphyra vietnamensis, Acanthophora muscoides* and *Centroceras clavulatum* are very rich in proteins and are extremely nutritious (Umamaheswara Rao 1970).

Seaweed as staple item of diet has been used in Japan and China for a very long time. There are large industries in Japan on edible seaweeds like *Porphyra* (Laver or Nori). Thin algal sheets are prepared by washing and drying *Porphyra* plants. The product thus obtained is an important food item. Nori is sold in sheets that may be toasted to give a green color and then flaked and added to sauces, soups and broths. Sometimes it is just soaked and eaten. Small, dry Nori sheets are used to wrap cold rice balls, which make a popular lunch-time snack for Japanese children. The food value of Nori lies in its high protein content (25-35% of dry weight), vitamins and mineral salts, especially iodine. Its vitamin C content is about 1.5 times that of oranges and 75% of the protein and carbohydrates are digestible by humans. While some 21 species are used in everyday cookery in Japan; in China, there are 8 to 11 commonly used species. The most important being

Porphyra sp. (Nori), *Laminaria* sp. (Kombu) and *Undaria* sp. (Wakame) in Japan and Kelp species *Laminaria japonica* (haidai) in China. Plants are dried after harvesting and either cut into strips or powdered. In Japan, kombu is used in the preparation of fish, meat dishes and also as a vegetable with rice. Powdered kombu is employed either in sauces and soups or is added to rice in the same way as curry. Some kinds are used in making an infusion similar to tea.

Vitamin C was analyzed by Sarojini and Sarma (1999) in samples of 24 species of macroalgae collected for one year from the intertidal habitats of Visakhapatnam coast. Chlorophyll-a and carbohydrate were also analyzed. In general, the algae belonging to Chlorophyceae and Phaeophyceae have higher annual mean contents of the vitamin (360 and 352 μg/g respectively) than Rhodophyceae (271 μg/g). The species, which are rich in vitamin C maxima was observed during summer for Chlorophyceae and Phaeophyceae, and during winter for Rhodophyceae. The vitamin C showed significant positive relationship with chlorophyll-a in Rhodophyceae and Chlorophyceae. It also showed a significant positive relationship with carbohydrate content of the latter. The distribution of vitamin C is associated with the morphogeneticaly important vegetative and reproductive phases indicating its key role in metabolic function (Sarojini and Sarma 1999).

Proteins, total carbohydrates and lipids were also analyzed in 23 marine algae collected along the coast of Visakhapatnam, for one year from May 1996 to April 1997 (Sarojini and Subbarangaiah 1999). Green and red seaweeds were found to be rich sources of proteins than the browns. The annual mean for proteins varied from 11.8 to 38.8% in Chlorophyceae, 19.4 to 29.0% in Phaeophyceae and 9.0 to 25.3% in Rhodophyceae. Higher concentrations were observed in *Bryopsis pennata* (38.8%), *Cladophora socialis* (33.5%), *Padina tetrastromatica* (29%), *Spongomorpha indica* (26.3%), *Sargassum vulgare* (27.7%) and *Cheatomorpha branchygonia* (25.2%). Higher content of proteins in Chlorophyceae has also been reported from other localities (Kaliaperumal *et al.* 1987, Chennubhotla *et al.* 1990, Ragothaman and Harinatha Babu 1992, Qari and Qasim 1993 and Venkataraman 1993). Rhodophyceae contained higher percentage of carbohydrates than Phaeophyceae and Chlorophyceae. The annual mean carbohydrates ranged from 20.2 to 30.8% in Chlorophyceae, 20.5 to 36.5% in Phaeophyceae and 7.2 to 67.5% in Rhodophyceae. The highest concentration was recorded in *Pterocladia heteroplatos* (75.0%) in November 1996. *Hypnea valentiae, Gelidium pusillum, Gracilaria corticata, Sargassum tenerrimum* were the other species that contained higher levels of carbohydrate. Majority of seaweeds showed carbohydrates maxima during February to May. This might be attributed to high salinity, temperature and nutrient availability due to up welling during summer, which influenced the photosynthetic activity. The highest lipid content was recorded in *Chaetomorpha antennina* (23.0%) in March 1997. *Caulerpa sertularioides, Enteromorpha compressa, Chaetomorpha brachygonia* and *Gracilaria corticata* were the other genera that contained higher lipids. Lipid maxima were observed during winter season and minima during rainy season. Higher content of lipids in green seaweeds might be due to their occurrence together with chlorophylls in chloroplast and this is in agreement with the earlier reports (Parekh and Chauhan 1987, Chennubhotia *et al.* 1990, Selvi *et al.* 1999). It may also be due to the presence of higher amount of fatty acids (Ramavat and Rao 1997).

Undaria pinnatifida (Laminariales) is widely used in Japan (where it is known as "wakame") and China ("qundai-cai") as food, which is a more important crop than *Laminaria* both in value and production. Hybrids with superior growth and nutritional characteristics have been developed in Japan. In 1976, about 20,000 tonnes (wet) were collected from wild sources and 127,000 tonnes (wet) were cultivated. The harvested algae are dried after washing in freshwater. After presoaking, the plant material is used as an additive to soups (wakame soup is served with virtually every meal in Japan); toasted (Yaki-wakame); used half resoaked, with boiled rice; and coated in sugar and tinned (Ito-wakame) (Guiry 2000). While a few of the above referred species are available in Indian coast, the species used as human food in India are *Ulva lactuca, Ulva fasilaciata, Ulva reticulata, Ulva rigida, Porphyra viethamensis* & *Gracilaria lichenoides* (Table 1). Since olden days *Gracilaria lichenoides* species is used for preparation of porridge or "Kanji" in the coastal districts of Tamil Nadu (Umamaheswara Rao 1970).

Seaweed as Poultry Feed

As seaweeds are cheap sources for minerals and trace elements, meals prepared from seaweeds can be utilized as supplements to the daily rations of the cattle, poultry and other farm animals. Seaweed meals can be prepared from species of *Ulva, Enteromorpha, Gracilaria, Hypnea, Sargassum, Padina, Dictyota* etc. by pulverizing the cleaned and washed weeds. In India, replacement of cereals by seaweed to some extent in cattle feed and poultry feed has been found to be feasible, practical and useful in many drought-prone areas of Gujarat, Saurashtra and Kachchh. In ten weeks, the food intake of one chick is 2.5 kg seaweed. However, the cereal content of this intake is 1.25 kg. If only 10 per cent of this is replaced by the weeds, i.e. 0.125 kg, it would result in saving of 125 kg of cereal in a unit of 1,000 birds over a ten-week period. On a large-scale, this would amount to a considerably large amount of cereals that could be diverted to human consumption (Chhaya 1997). Of course, it would be necessary to identify the type of seaweed to be used, ensure its hygienic utilization, and calculate the optimal amount to be harvested without disturbing its ecosystem.

Seaweeds as Medicine

Ayurveda has for long recognized the use of certain seaweeds in medicines for acidity and other digestive ailments, and treatment of blood impurities and fevers. Seaweed species of the genus *Sargassum, Hypnea, Lamineria*, and others are known to be used in allopathic medicines too. An amino acid called "laminine" is extracted from Laminaria, which is used in certain heart and blood pressure problems. It is also used in relaxing and conditioning the involuntary muscles of the alimentary canal. Problems of the central nervous system and respiratory process, certain types of tumors, etc. are treated with the help of specific types of seaweeds. Species of *Sargassum* are used for cooling and blood cleansing effect. *Hypnea musciformis* is employed as vermifuge or worm expelling agent while *Centroceras clavulatum* is useful as cathartic agent. Seaweeds, rich in iodine such as *Asparagopsis taxiformis* and *Sarconema* sp., can also be used for controlling goitre disease caused by the enlargement of thyroid gland (Chhaya 1997, Umamaheswara Rao 1970).

The late Dr. K. Rao of the Central Salt and Marine Chemicals Research Institute, Bhavnagar, and his team had isolated an antibiotic from *Enteromorpha* sp. Diago of Japan extracted "donionic acid" from *Condoria armata,* which is used in the treatment of roundworms. Some seaweeds with considerable amount of viscosity are used in gynecological applications. Seaweeds are also used in preparations of media cultures in the pharmaceutical industry.

Seaweed as Agriculture Fertilizer

Seaweeds have been used since long time as manure in coastal areas. As minerals and trace elements occur in water-soluble form, plants readily absorb these chemical constituents when seaweed manure is applied (Chennubhotla 1996). A wide range of beneficial effects have been reported from the use of liquid seaweed extracts, including enhancement of seed germination and plant growth (Blunden 1972, Blunden and Wildgoose 1977, Button and Noyes 1964, Bhosle *et al.* 1975, Rajeshwari *et al.* 1983, Mohan *et al.* 1994, Sekar *et al.* 1995 and Mostafa 1999), increased crop yields, resistance of plants to frost, increased uptake of inorganic constituents from the soil, more resistance to stress conditions, and reductions in storage losses of fruit. Wu *et al.* (1998) used the alkaline extract of *Ascophyllum nodosum* to suppress the root-knot nematode in *Arabidopsis thaliana.* Aqueous seaweed extract was employed extensively as fertilizer additive in foliar spray on agronomic and horticultural crop plants.

The analysis of liquid fertilizers from seaweeds has identified the presence of growth motivator's cytokines, gibberelic acid and auxin, in addition to nutrients and trace elements like nitrogen, phosphorus, potassium, zinc, iron, copper, manganese, calcium, magnesium etc. (Chhaya 1997). Booth (1996) observed that the value of seaweed as fertilizers was not only due to nitrogen, phosphorus and potash content but also because of the presence of trace elements and metabolites similar to plant growth regulators. Aqueous extract of *Sargassum wightii*, when applied as a foliar spray on *Zizyphus mauritiana* showed an increase in yield and quality of fruits (Rama Rao 1991). Blunden *et al.* (1997) showed that aqueous extract of *Ascophyllum nodosum* applied as foliar spray enhanced leaf chlorophyll levels in tomato, dwarf French bean, wheat, barley and maize. The effect of crude extract of green seaweeds *Cladophora dalmatica, Enteromorpha intestinali* and *Ulva lactuca*, and the red seaweeds *Corallina mediterranea, Jania rubens* and *Pterocladia pinnata*, the most abundant algae occurring along the cost of Alexandria, on seed germination, growth and some metabolites in faba beans (*Vicia faba* L.), were reported to increase protein content, soluble sugars and chlorophyll content in leaves (Mostafa and Alla Eldin 1999).

A number of organizations, *viz.* Central Salt and Marine Chemicals Research Institute (CSMCRI, Bhavnagar), Central Marine Fisheries Research Institute (CMFRI, Kochi), National Institute of Oceanography (NIO, Goa) and Gujarat Fisheries and Aquatic Sciences Research Institute (GFASRI, Okha) have been working on preparing liquid fertilizer from certain seaweeds. Experiments with this liquid fertilizer were carried out on many plant varieties; brinjals, tomatoes and cabbage; cash crops like groundnut, tobacco, banana; cereals like wheat, bajra; and on roses. Liquid fertilizer was applied at concentrations ranging from one to five percent, every 15 to 25 days on leaves or roots. Totally 3 to 4 sprays were applied and the results were encouraging. Over 10 tonnes of liquid fertilizers were manufactured and sold to farmers and fertilizer agents in different parts of Gujarat. Repeat orders and new enquires followed which are the indication of its acceptability among farmers and distributors of fertilizers (Chhaya 1997).

In other countries, liquid extracts of marine brown algae are marketed for use in agriculture and horticulture. Most of these extracts are prepared from dried *Ascophyllum nodosum* meal (e.g. "Maxicrop", manufactured in the United Kingdom), or from dried total drift (detached seaweeds), often referred to as "blackweed", but some utilize other species, such as *Fucus serratus* and *Laminaria* sp. (e.g. "SM3"; United Kingdom). One product, currently being marketed, is prepared from the stipes of *Ecklonia maxima* from South Africa ("Kelpak 66"). These are prepared from hot-water extracts of either the dried or wet seaweed, sometimes with the addition of sodium carbonate to aid extraction. "Maxicrop" is primarily used for gardens and glasshouse crops and is exported to a wide range of countries. It has been used on citrus fruits in Guyana, on citrus and grapes in Greece, on orchids in Belgium, on garden crops in Thailand, and on glasshouse crops in Iceland (Guiry 2000).

Detached seaweeds or "total drift" have been used for many years in several European countries for the making of "lazy beds". Soil or sand is layered with seaweed for vegetable production, particularly potatoes. Such organic material has proved very useful in very barren areas. More recent uses of seaweeds as fertilizers and soil stabilizers have been in the seeding of motorway embankments in Britain. Grass seed is mixed with a crude extract of brown algae to make a paste-like mixture, which is then sprayed onto disturbed areas. The paste keeps the seeds in place, retains moisture and binds the soil (Blunden 1999).

Seaweed as Bio-energy

With conventional and non-renewable resources of energy like oil, water and coal being rapidly depleted and posing their own sets of problems, it has become necessary to look for renewable and non-conventional sources for energy requirements. Energy can be obtained from seaweeds by a comparatively simple technology. The algae are collected, shredded and mixed with bacterial culture and cow dung in a biogas digester. The resultant outputs are fertilizer, light CH_4, and energy for cooking gas. There are many advantages in producing energy through this way for example, pure gas is obtained; the residue is used as fertilizer; it entails comparatively less labor cost with greater safety, and most importantly, this technique of producing energy is pollution-free.

Seaweed as Global Thermostat

The potential threat of global warming and the consequent climate change is currently a major international environmental issue. While nation's haggle and fight over technology transfer, nature may have its own answers to the problem. Seaweed may be one of them. Researchers in Europe and America who have been working on marine weeds believe that when these are exposed to high temperature, they produce a sulphur compound that gets converted to sulphur particles, around which clouds form. This can lower the earth's temperature.

The link between algae and climate involves dimethyl sulphide (DMS), the gas that gives sea air its bracing smell. It is created by enzymatic breakdown of the salt dimethyl sulphoniopropionate (DMSP). Marine algae produce DMSP to keep their osmotic balance of sea water, without which water would leave the cells of the algae, killing them. The process by which DMSP is released into the sea is not clearly understood as yet. Scientists feel that this occurs when zooplankton graze algae or when algae die. Algae transfer 20 to 30 million tonnes of sulphur from the oceans to the atmosphere every year. This indicates that seaweeds could serve as some kind of a global thermostat (Chhaya 1997).

HARVESTING OF SEAWEEDS

The major seaweeds harvested from different coasts in India are *Gelidiela* sp., *Gracilaria* sp., and *Turbinaria* sp. and *Sargassum* sp. The agar yielding *Gracilaria* sp. is reported to be collected throughout the year in Tamil Nadu. However, Gujarat does not really possess enough of them for commercial production purpose. The weed is sold fresh to dealers on the seashore itself, and they dry it on the beach before transporting to godowns. The weed is normally mixed with other types and is normally 50 per cent pure with about 25 per cent moisture. *Sargassum* is the major group of seaweeds in Tamil Nadu and Gujarat. Of these, *S. wightii* is the most important in Tamil Nadu while *S. tenerrimum* is in Gujarat. Algin industries insist on formalin-treated material because the stock has to be maintained and consumed all throughout the year. The material is harvested during the peak season, i.e. from November to January. Fresh plants are stocked in 2 per cent formalin for two hours and then dried in the sun. Formalin is supplied free by the concerned industry. Since there is no commercial algal culture worth the name, harvesting is dependent on natural collection from coastal areas and nearby islands in both Tamil Nadu and Gujarat. The weeds are collected by fishermen and fisherwomen manually and brought to shore on boats. In the process of collection not much care is taken and the weeds are uprooted from the substratum. The weeds should normally be cut about six to ten inches above the point of attachment to substratum leaving the roots and some filaments. This would enable the attached filament to grow again for further collection.

With the development of seaweed based industries the role of economically useful marine algae has assumed significant proportions in improving the economic conditions of the rural community in the coastal areas. The alginophytes available in harvestable quantities along the south-eastern coast of Tamil Nadu is a boon to the rural inhabitants on the shoreline villages to supplement their earnings. About 500 persons are daily engaged in the seasonal harvesting period of 5 to 6 months in the year for collection, and connected operations of the Alginophytes, while about another 500 persons are engaged in the harvesting and connected operations of Agarophytes in their respective seasons lasting 5 to 7 months. There is a marked betterment in living conditions of the rural community engaged in these bulk seaweeds collection (Patel *et al.* 1979).

Studies pertaining to various aspects such as seasonal changes in growth and fruiting behavior, spore output and germination effect of repeated harvesting on growth, yield and quality of phytochemicals of commercially important seaweeds revealed that the life spans of agar yielding and algin yielding seaweeds are six months and one year respectively. The young plants appear in April/ May and reach harvestable size during the period July/August and November-December/January. Based on these findings a harvest time-table was prepared (Kaliaperumal and Kalimuthu 1993 and Kaliaperumal 1993). A single harvest in a year during the period of October–December/January for species of *Sargassum*, *Turbinaria* and *Hypnea* and two harvests in a year during the period January-March and July-September for *Gelidella acerosa* and *Gracilaria* sp. are recommended.

The study of Ganesan *et al.* (2000) on biomass and sporulation suggested that a single harvest in a year during January and February may not disturb natural population of *Padina boergesenii* and *P. tetrastromatica* in Mandapam region of Tamil Nadu, India. The biomass ranging from 730 to 4200 g fresh wt with relative growth rate from 0.003 to 0.07 g fresh wt./day was reported for *Padina* species. Experimental studies carried out (Ganesan *et al.* 1999) with tetraspores of *Padina boergesenii* to understand the effect of ultraviolet radiation, light intensity, photoperiod, temperature, salinity and desiccation on the liberation of tetraspores indicated that UV-B (290 nm) radiation inhibits spore liberation. While UV – C (250 nm) enhances spore release, shorter period exposure of UV- A (360 nm) radiation increases spore liberation but larger period exposure inhibits the same.

Some species, like *Sargassum tenerrimum,* are still available in considerable quantities. Their harvesting season starts somewhere in November–December, after which they attain harvestable vegetative growth. These algae start disintegrating in the month of March-April, when they release spores in the ecosystem. These spores lie dormant for some period and then grow in favorable conditions. They attach themselves to the substrata with the help of root-like structures and grow. The length and size of *Sargassum* sp. is not as big as that of "kelp".

The experimental field cultivation of agar yielding seaweeds *Gelidiella acerosa* and *Gracilaria edulis* and carrageenan yielding seaweed *Hypnea musciformis* in the lagoon at Krusadai island by Central Salt and Marine Chemical Research Institute (Raju and Thomas 1971, Subbarmaiah *et al.* 1979) and the agaroids yielding seaweed *Acanthophora spicifera* in the lagoon at Hare islands by the Central Marine Fisheries Research Institute revealed that the lagoon environment of the Gulf of Mannar Islands is very ideal and potential sites for large scale cultivation of seaweeds. However, Effect of different environmental factors such as, salinity, seawater pH, nutrients (nitrogen and phosphate) and enriched media on the growth of *Gracilaria corticata* under controlled laboratory conditions indicated that maximum growth rates of this intertidal red alga

Gracilaria corticata were observed at 27μ EM-2 S-1 photon flux density, 30% salinity, 8.0 pH, 0.1 mM of $NaNO_3$, 0.2 μM of NH4Cl, 0.02 μM of disodium hydrogen phosphate and 0.03 mM of sodium β-glycerophosphate. Among the enriched media tested, peak growth was seen in Provasoli enriched seawater medium (Srinivasa Rao and Umamaheswara Rao 1998).

Since 1972, the Central Marine Fisheries Research Institute has been engaged in the cultivation of several economically important seaweeds and the method of cultivation of *Gracilaria edulis,* a fast growing species. These culture experiments have been done by introducing the fragments of the seed material in the twists of the coir ropes that are fabricated in the form of 5 x 2 m size nets and these coir nets were tied to wooden poles fixed in the coastal waters for further growth. The plants reach harvestable size after a period of 60 days. The yield from a coir net is approximately 30 kg fresh weight in the coastal waters of peninsular India, while the experiments conducted in Minicoy Lagoon (Lakshadweep) using long line coir ropes gave a maximum production value of 31-fold increase over the initial seed material introduced; thus indicating the Minicoy Lagoon to be more productive than main land water. The difference between coir net method and coir rope method is that the former gives a very good support to the growing plants and it also withstands for 3 harvests, while the latter does not give much support as it is at the mercy of the waves, tides and wind. But the long line rope culture method is suitable for bays and protected areas, which has given better results in Minicoy Lagoon than the net method. Further it brings down the operational cost also by this method as the charges for fabrication of the nets and the cost of poles for fixing the nets can be obviated by tying coral stone at either end (Chennubhotla 1996).

Use of growth promoting substances or hormones to improve the biomass production, in lagoons and brackish water areas is one of the various modern techniques in algal culture (Imelda and Chennubhotla 1999). The growth regulatory effects of gibberellic acid (GA) and 2,4-D (2, 4-dichlorophenoxy acetic acid) on a green alga *Ulva fasciata* and two species of red algae, *Gracilaria corticata* and *Hypnea valentiae* were studied. For a culture period of 30 days, the thallus segments showed varying growth rates on pretreatment with 6 to 20 mg/l of the above growth substances. These compounds stimulated growth and increased wet weight of the algae. At higher concentrations (16-20 mg/l) 2, 4-D was growth inhibitory, whereas GA showed significantly higher growth at 16-20 mg/l concentrations. It is concluded with the study that plant hormones like GA and 2, 4-D could be used for enhanced algal production in culture practices at different concentrations

CONCLUSIONS

It can be inferred that the culture and utilization of seaweeds can contribute to rural development and thus to, economy of the country as a whole. Along the coast of peninsular India, if seaweed farming is taken up in one hectare area, a return of Rs. 9,000/- is assured in a year for an investment of Rs. 36,000/- assuming the production rate to be 3-fold and that two harvests only can be taken in a year. While in Minicoy Lagoon the input requirement (infrastructure) is less and the production rate high (average of 15-fold) and hence the net profit works out to around Rs. 20,000/- for an investment of Rs. 25,000/- in a year with 3 harvests.

In order to conserve the natural stock of the commercially important agar and algin yielding seaweeds in Gulf of Mannar islands and also to get consistent crop every year, commercial exploitation of seaweeds should be made only during the peak growth period of the algae without affecting the island ecosystem. A trend in depletion of some of the Agarophyte resources has been noticed. Successful resource management and cultivation of selected seaweeds would solve the problem and generate further potentials for rural development. What is needed is a complete statewise and all-India picture, which will provide a clear understanding of standing crop and seasonal harvestable quantum. This is necessary because over 20 industries are exploiting algae for agar and an equal number for algin manufacture; and there will be more in future. There is thus a danger of too much harvesting and overexploitation. CMFRI, Kochi, CSMCRI, Bhavnagar, and various maritime states would have to come up with a collaborative report. It is understood that both CMFRI and CSMCRI are working in this direction. The principles of harvesting are clear. We have first to calculate what amount is exactly harvestable without disturbing the eco-balance. Nothing more should be harvested. Proper identification of the type of weed needed, meticulous collection methods during the right season, etc, is a must. A harvesting location should not be exploited year after year, but should be "rested" periodically for the species to stabilize in that area before being harvested again. Ignorance, carelessness or irresponsibility in such matters can lead to serious imbalances in the ecosystem. This can result in stress on the species and even its loss. Indiscriminate exploitation of weeds has actually created such a situation in the Gulf. Of the two hundred and more species of seaweeds mentioned, some have already become rare and others are on the verge of disappearing.

In recent years, many entrepreneurs are enthusiastic to start these industries, as a result of which there will be a great demand and competition for the raw material (seaweeds) will be felt, as a consequence of which the whole world may find it difficult to prepare certain life saving vaccines, since agar-agar is used as a culture medium for the bacteria and moulds. In order to meet this, a process to augment the supplies of these seaweeds by culture practices has to be developed. In view of the increasing demand for seaweeds, it has become imperative to culture them, as the natural stock is being heavily exploited throughout the world, and demand for algal products exceeds supply. Besides, increasing the total supply, seaweed culture is also attractive in that it has the capacity of providing high quality raw material. Use of growth promoting substances or hormones to improve the biomass production, in lagoons and brackish water areas is one of the various modern techniques in algal culture

In the context of this immense versatility of marine algae, universities and marine institutions need to take up algal studies on a priority basis. Seaweeds are the green gold of the ocean, but man cannot hanker only after the gold without sparing a thought towards keeping it green. Culture and utilization of seaweeds can contribute to rural development and thus the economy of the country as a whole. Special efforts should be made to explore the different parts of the Indian coast, including Lakshadweep, Andaman and Nicobar Islands and to estimate the harvestable quantities of agarophytes, alginophytes and other important seaweeds growing in these areas.

REFERENCES

Anon (1978). A report on survey of marine algal resources of Tamilnadu, 1971-1976, CSMCRI, Bhavnagar, pp. 1-137.

Blunden, G. (1999). Agriculture uses of seaweeds and seaweed extract. In Seaweed Resources in Europe: Uses & Potential – Guiry, M. D. & Blunden, G.; John Wiley & Sons, Chichester 65-81.

Bhosale, N.B., Untawla, A.G. and Dhargalkar, V.K. (1975). Effect of seaweed extract on the growth of Phaseolus vulgaris L. Indian J. Mar. Sci. 4: 208-210.

Blunden, G. (1972). The effect of aqueous seaweed extract as a fertilizer additive. In: Proc. 7th International Seaweed Symp. University of Tokyo Press, Tokyo pp. 584-589.

Blunden, G. and Wildgoose, P. B. (1977). The effects of aqueous seaweed extract and kinetin on potato yield. J. Sci. Food Agric. 28: 121-125.

Blunden, G., Jenkins, T. and Liu, Y-W (1997). Enhanced chlorophyll levels in plants treated with seaweed extract. J. Appl. Phycol. 8: 535-543.

Booth, E.; 1996. Some properties of seaweed manures. In: Proc. 5th International Seaweed Symp. Pergamon Press, London pp. 3459-357.

Booth, E. (1996). Some properties of seaweed manures. In: Proc. 5th International Seaweed Symp. Pergamon Press, London pp. 349-357.

Button, E.F. and Noyes, C.F. (1964). Effect of seaweed extract upon emergence and survival of seedlings of creeping red fescue. Agron. J. 56: 444-445.

Chhaya, N. (1997). "Seaweeds: Green Gold" in Minding our Marine wealth: An appraisal of Gujarat's coastal Resources", Environment & Development Series, Centre for Environmental Educations, Ahmedabad.

Chennubhotla, V.S.K. N. Kaliaperumal, S. Kalimuthu, J. R. Ramalingam, M. Selvaraj and M. Najmuddin (1987) Seaweed culture. CMFRI Bulletin, 41: 60-74.

Chennubhotla, V.S.K. Chennubhotla, V.S.K., N. Kaliaperumal, S. Kalimuthu, J. R. Ramalingam, M. K. Subbaramaiah, K. Rama Rao, and P.V. Subba Rao (1990) Seaweed resources of the Tuticorin Tiruchendur coast, Tamil Nadu India J. Mar. bio. Ass. India, 26 (1&2) 146-149.

Chennubhotla, V.S.K. (1996). Seaweeds and their importance. Bull. Cent. Mar. Fish. Res. Inst., 48: 108-109.

CIFT (Central Institute of Fisheries Technology) (1997). Technologies developed at CIFT, special bulletin 11, CIFT, Kochi.

CSMCRI & CMFRI (1991) A report on the Seaweed Resources of Tamil Nadu Coast, India III Sector. Valinokkam – Kalakkarai and Manapad – Kanyakumari 1989-90. ICAR. CSMCRI& CMFRI.

Den Hartog, C. (1972). In Marine ecology, Vol. 1 part 3 environmental factors, (Wiky Inte'saence, London, 1972, 1277)

Dhargalkar, V.K. and Deshmukh, G.V. (1996). Subtidal marine algae of the Dwarka Coast (Gujarat). Indian Journal of Marine Sciences, 25, 297-301.

Desai, B.N. (1967) Seaweed resources and extraction of alginate and agar. Proc. Semi. Sea Salt and Plant CSMCRI, Bhavanagar.

Dave, M.J. and Chauhan, V.D. (1993). Amino acid composition of Caulepra. Phykos 32 (1 & 2): 21-26.

Ganesan, M. and Kannan, L. (1994). Seasonal variation in the biochemical constituents of economic seaweeds of the Gulf of Mannar. Phykos 33 (1 & 2): 125-135.

Ganesan, M., Mairh, O.P., Eswaran, K. and Subba Rao, P. V. (1999). Effect of UV radiation and other environmental factors on the liberation of tetraspores from brown alga Padina boergesenii (Phaeophyta, Dictyotales). Indian Journal of Marine Sciences, 28, 50-54.

Ganesan, M., Mairh, O.P., Eswaran, K. and Subba Rao, P.V. (2000). Sseasonal variation in growth and spore production of marine brown algae Padina boergesenii (Dictyotales/Phaeophyta,) in Mandapam region, southeast coast of India. Indian Journal of Marine Sciences, 29, 253-257.

Guiry, M.D. (2000). Uses of Seaweeds. Seaweed site/ frame232965.html /Seaweed Home Page.

Hemalatha, R. and Rengasamy R. (1993). Studies on Gracilaria edulis (Gmelin) Silva. Phykos 32 (1 & 2): 65-75.

Imelda Joseph & Chennubhotla, V. S. K (1999). Gibberellic acid and 2, 4-D as growth regulators in laboratory culture of seaweeds India. Indian Journal of Marine Sciences, 28, 66-69.

Kaliaperumal, N. (1994). Seaweed resources of Tamil Nadu coast. Biology Education, 11: 281-293.

Kaliaperumal, N. and S. Kalimuthu (1997). Seaweed potential and its exploitation in India. Seaweed Res. Utilin., 19 (1 & 2): 33-40.

Kaliaperumal, N. (1993). Seaweed culture. In handbook on "Aqua farming- Seaweeds, Seaurchin and Seacucumber". MPEDA, Cochin pp. 9-22.

Kaliaperumal, N. and S. Kalimuthu (1993). Need for conservation of economically important seaweeds of Tamil Nadu and time-table for their commercial exploitation. Mar. Fish. Infor. Serv. T & E Ser., 119: 5-12.

Kaliaperumal, N., S. Kalimuthu and J.R. Ramalingam (1995). Economically important seaweeds. CMFRI Special Publ. 62: 1-36.

Kalimuthu, S., N. Kaliaperumal, J.R. Ramalingam and V.S.K. Chennubhotla (1990). Present status of seaweed exploitation and seaweed industry in India. Mar. Fish. Infor. Serv. T & E. Ser., 103: 7-8.

Kaliaperumal, N. (1997). Seaweed Industry in India Souvenir, CFCRI, Mandapam camp, pp. 55-56.

Kaliaperumal, N., Chennubhotla, V.S.K. and Kalimuthu, S. (1987). Seaweed resources of India. CMFRI Bulletin, 41: 51-54.

Krishnamurthy, V. and M. Baluswami (1982). Some species of Ectocarpaceae new to India. Seaweed Res. Utiln. 5: 105-112.

Michanek G. (1975) Seaweed resources of the ocean, FAO Fish. Tech. Rep. 138:1-126.

Mostafa, M.EL-Sheekh and Alla Eldin F.EL Saied (1999) Effect of seaweed extract on seed germination, seedling growth and some metabolic processes of Fabe beans (Vicia faba L.)Phykos 38(1&2): 55-64.

Mohan, V.R.; Venkataraman Kumar; Murugeswari, R. and Muthusamy, S. (1994). Effect of crude and commercial seaweed extract on seed germination and seedling growth in Cajanus cajan L. Phykos 33: 47-51.

Parekh R.G., Y.A. Doshi and V.D. Chaun (1993) Carageenan from Mrine red alga Agardhiella tenera (J.Ag) Schmitz. Phykos 32 (1&2) 95-98.

Parekh, R.G. and Chauhan, V. D. (1987). Lipid content of some Indian seaweed. Indian J. Mar. Sci. 16: 272-273.

Patel J.B., B.V.Gopal and Subbaramaiah K.(1979) Role of seaweed in the upliftment of rural economy in India. International symposium on marine algae of the Indian Ocean region 9-12 January 1979.

Qari and Qasim, R. (1993). Biochemical constituents of seaweeds from Karachi coast. Indian J. Mar. Sci. 22: 229-231.

Raikar, S.V., Iima, M. and Fujita, Y. (2001). Effect of temperature, salinity and light intensity on the growth of Gracilaria spp. (Gracilariales, Rhodophyta) from Japan, Malaysia and India. Indian Journal of Marine Sciences, 30, 98-104.

Raju, P.V. and Thomas, P. C. (1971). Experimental field cultivation of Gracilaria edulis (Gmelin) Silva. Bot. Mar., 14 (2) 71-75.

Rama Rao, K. Subba Rao, P.V., Mal, T.K., Subbaramaiah, K., Kaliaperumal, N., Kalimuthu, S., Muniyandi, K., Ramalingam, J.R., Krishnapillai, S. and Chennubhotla, V.S.K. (1996). Distribution of seaweeds off Alantalai-Manapad and Vembar-Nallatanni Tivu in Tamilnadu. Phykos 35 (1 & 2): 163-170.

Rama Rao, K. (1991). Effect of aqueous seaweed extract Zizyphus mauritiana Lam. J. Indian Bot. Soc. 71:19-21.

Ranghotaman, G. and Harinatha Babu, M.V. (1992). Studies on marine algae of Devka coast. Seaweed Res. Utiln., 15: 91-97.

Ramavat, B.K. and Rao, A.V. (1997). The total sterols and fatty acids composition of some marine algae of Saurashtra coast, west coast of India. Seaweed Rs. Utiln., 19: 75-79.

Rajeshwari, M.; Lakshamanan, K.K. and Chitra, A.S. (1983). Effect of seaweed on Tomato. Proc. Of National Semina on the production Technology of Tomato and Chillies, TNAU, Coimbatore, pp. 87-89.

Rama Rao, K. (1991). Effect of aqueous seaweed extract *Zizyphus mauritiana* Lam. J. Indian Bot. Soc. 71:19-21.

Rama Rao, K., Subba Rao, P.V., Thomas, P.C., Zaidi, S.H. and Subbaramaiah, K. & Kaliaperumal, N., Kalimuthu, S. Muniyandi, K., Ramalingam, J.R., Najmuddin, N. and Chennubhotla, V. S. K. (1993). Seaweed Resources off Tamil

Nadu Coast, Sector–IV Kilakkarai – Rameswaram Island (Dhanushkodi). Seaweed Res. Bulletin: 16 (1&2), 103-110.

Rama Rao, K., Subba Rao, P.V., Thomas, P.C., Zaidi, S.H. and Subbaramaiah, K. & Kaliaperumal, N., Kalimuthu, S. Muniyandi, K., Ramalingam, J.R., Najmuddin, N. and Chennubhotla, V.S.K. (1993). Seaweed Resources off Tamil Nadu Coast, Sector – IV Kilakkarai–Rameswaram Island (Dhanushkodi). Seaweed Res. Bulletin: 16 (1&2), 103-110.

Sarojini, Y. and Subbarangaiah G. (1999). Seasonal variation in biochemical composition of some macroalgae along Visakhapatnam, east coast of India. Phykos 38 (1 & 2): 71-79.

Sasikumar, C., Rao, V.N.R. and Rengasamy, R. (1999). The effect of environmental factors on the qualitative and quantitative characteristics of agar from the marine red alga Gracilaria verrucosa (Gracilariales, Rhodophyta). Indian Journal of Marine Sciences, 28, 270-273.

Sekar, R.; Thangaraju, N. and Rengasamy, R. (1995). Effect of seaweed liquid fertilizer from Ulva lactuca L. on Vigina unguiculata L. (WALP) Phykos 34: 49-53.

Selvi, M., Shakila, P. and Selvaraj, R. (1999). Studies on biochemical content of macroalgae from Cuddalore and Thirumullaivasal estuaries of Tamil Nadu. Seaweed Res. Utiln. 21: 99-103.

Silas, E.G. and S. Kalimuthu (1987). Commercial exploitation of seaweeds in India. CMFRI Bulletin, 41: 55-59.

Shaw, B.P., Raut, N.P., Barman, B.C., Choudhary, S.B. & Rao, K.H. (2000). Distribution of macrophytes vegetation in relation to salinity in the Chilka lake, lagoon, along east coast of India. Indian Journal of Marine Science. June 144-148.

Subbaramaiah, K., Rama Rao, K., Nair, M.R.P. (1979). "Marine algal resources of Lakshadweep". International Symposium on Marine Algae of the Indian Ocean Region, CSMCRI, Bhavnagar India.

Srinivasan, R. and Santhanaraja T. (1967) Studies on extraction and properties of agar-agar from the seaweed Gracilaria species in Madras state. Madras Jour. Fish., 3: 146-151.

Thivy, F. (1960). Seaweed utilization in India. Proc. Symp. Algology. ICAR, New Delhi. pp. 345-365.

Umamaheswara Rao, M. and N. Kaliaperumal (1980). Genus *Pterocladia* (Rhodophyceae) from India. Bull. Bot. Surv. India 22: 109-111.

Umamaheswara Rao, M. (1970), "The economic seaweed of India ICAR, Bulletin of the Central Marine Fisheries Research Institute, No. 20.

Umamaheswra M. Rao (1999) Additions to the algal flora of Visakhapatnam coast phykos 38 (1&2):113-126.

Umamaheswara Rao, M. (1991). On a new species of *Helminthocladia* from India. *Seaweed Res. Utiln.* 13: 169-174.

Umamaheswara Rao, M. (2000). Some marine algae from Andaman and Nicobar islands. Phykos 39 (1 & 2): 85-99.

Venkataraman, Kumar (1993). Biochemical constituents of marine algae from Tuticorin cost. Indian J. Mar. Sci. 22: 138-140.

Venkataraman, K. and Mohan, V.R. (1997). The effect of seaweed liquid fertilizer on Black Gram. Phykos 36 (1 & 2): 43-47.

Visweswara Rao and Mody A.C.(1964) Extraction of alginic acid and alginate from brown seaweeds Indian jr. tech. 3:261.

Visweswara Rao, A., Patel K.N. and Shah H.N.(1965) Manufacture of agar-agar from red seaweeds. Res. Ind., 10: 131-133.

Vasuki, S., Ganesan, M., and Subba Rao (1999). Seasonal growth and reproduction of marine red alga *Asparagopsis delilei* (Rhodophyta/ Bonnemaisoniales) from the Mandapam region, Southeast Coast of India. Indian Journal of Marine Sciences, 28, 60-65.

Vasuki, S., Ganesan, M., and Subba Rao (2000). Effect of plant growth regulators on the growth of the marine alga *Padina boergesenii* Allender & Kraft. Phykos 39, (1&2), 107-111.

Wu Y., Jenkins, T., Blunden, G, Menede, N. and Hankins, S.D. (1998) Suppression of fecundity of the root-knot nematode, Meloidogyne javanica in mixotrophic culture of *Arabodopsis thaliana* treated with an alkaline extract of *Ascophylum nodosum*. J. Appl. phycol. 10:91-94.

8

Potentials of Molluscs

Amita Saxena

Mollusca is the second largest phylum of invertebrates and has over 30,000 living and 35,000 fossil species. Their shells are hard and tiny, and as a result, they preserve well and make the best fossils among all the animals in the world. The phylum is extremely ancient and has left a very complete fossil record, which extends as far back as 550,000,000 years ago. Among marine biodiversity of molluscs 3370 species out of 5070 species are known in coastal wetland. They are divided on the basis of foot, mantle, shell, muscles, radula nervous and respiratory system. Some of these animals are of great use to mankind, they form an essential food item of modern man, their shells are used for making buttons and various ornamental objects. The pearl industry alone is a multimillion rupee business, and molluscs hold the unique distinction of being the only animals to produce such 'natural jewellery'. Molluscs have tremendous use for religious purposes and other gastropods for decoration and games.

Some molluscs have a symmetrical body having undergone torsion and detorsion and coiling but shell remained univalved shell may be conical or spiral, pointed, smooth or spiny, the vast majority of fresh water/marine gastropods have a spiral or discoidal coiled shell. However in five American genera (the limpets), the shell is in the form of a very low cone. The range in the maximum dimensions of the shell (length or width) of the mature animal is about 2 to 70 mm.

The muscular portion of the animal which projects from the shell is called the foot. For the most part it is inconspicuously colored, usually being grayish, brownish, or blackish, and often flecked or motted with yellowish or whitish. The ventral surface of the foot is flat and there is a more or less prominent head at the anterior end. The head bears two tentacles which range in shape from short, blunt, or near the base of the tentacles. The mouth is on the ventral surface of the head at the extreme anterior end and in contact with the substrate in most species, but in the viviparidae, pleurocoridae, amnicolidae, and valvatidae it is at the end of a nuclear proboscis, or rostrum. In the Ampullariidae and Neritidae the head itself, is widened into two flat lateral lobes, constituting the velum a structure which persists from the larval stage.

The foot consist mainly of muscle tissue, but it contains the anterior portions of the digestive tract and reproductive system as well as most of the nervous system, most.

The internal organs from the visceral hump which lies within the shell where its shape is a counter part of the coils. Forming a lining for the shell and enclosing the viscera is a thin layer of specialized tissue, the mantle. The shell grows by the addition of the shell material secreted by the edge of the mantle (collar) at the orifice of the shell. Although a few snails, such as ralvata have a delicate external gill, most fresh water gastropods have either internal gills which are specialized folds of the mantle for aquatic respiration, or an internal air filled "lung" consisting of a special portion of the mantle cavity surrounded by highly vascularised mantle. The respiratory opening (pneumostone) to either the internal gills or lung is usually situated on the right side at the edge of the shell. In those aquatic snails with a lung the edge of the mantle is often elongated into a respiratory siphon through which air is obtained at the surface film. The exposed portion of the body of a snail is securely fastened to the shell internally by a large columellar muscle, and by the contraction of this muscle the foot may be doubled up and withdrawn from view. Those snails with gills have a discoidal chitinous (rarely calcareous) operculum attached dorsally on the posterior part of the foot, and when the animal is withdrawn into the shell the operculm fits the opening and effectively closes it.

In some families external reproductive structures may be seen. In the amni colidae and valvatidae, for example more or less laterally in the region of the neck is the verge, a projecting male copulatery organ of variable structure and size. In the vivigaridae the right tentacles of the male is much larger than the left and serves as a penis sheath. There are some diagrams of gastropods which showed for beautiful torsion/distorsion thus used in ornamentation.

The wentle traps (*Epitonium* spp are superficially resemble to *Turritella* from which the shells are distinguished by closely set ribs or varies ornamenting all whorls.

Cowries (Cypraeidae) are amongst for beauty and variety of colouring of their splendidly polished shells render them conspicuous and valuable. Adults forms are different from others. Young form is elongated with a prominent conical spire, and a long wide mouth aperture bounded by a thin, sharp outer lip. A delicate periostracur covers the surface. In matured ones mantle flap from each side expands and becomes reflected over the back of shell, edges meeting a little to one side. The mantle and foot are even more vividly coloured than the shell and few objects are more beautiful than large cowry crawling in a coral reef pool the mantle bright with scarlet and yellow and beset with gracefully branched filaments. Cowries live on rocky ground, particularly in and about coal reef. They are often found hiding under boulders at low tide and appear limited on shallow water money cowry. (*Cypraea moneta*) abundant on reef, Eyed cowry (*C. Ocellata*), large Tiger cowry (*C. tigris*) covered with large, bordered spots, Black cowry (*C. mauriliana*) Mole cowry (*C.* talpa); serpent head cowry (*C.* caput-serpents), Arabian cowry (*C. arabica*).

Here certain diagrams of molluscan shell are given to show the beauty and diversity of shell torsion and distorsion.

The freshwater mussel is a familiar representative of the Phylum Mollusca. The family Unionidae is widely distributed all over the world and includes nearly all the large freshwater mussels or clams. The family consists of several genera and nearly 1,000 species of which a good number are represented in India. The commonest species in England is the Swan Mussel

Some Beautiful Molluscs

1. *Clanculus depictus* (A.Ad.)
2. *Lambis lambis* (L.)
3. *Cymatium piliare* (L.)
4. *Xancus pyrum* (L.)
5. *Cassis rufa* L.
6. *Cypraea tigris* L.
7. *Babylonia spirata* L.
8. *Cardita bicolor* Lam.
9. *Conus textile* L.
10. *Chlamys senatoria* (Gmelin.)

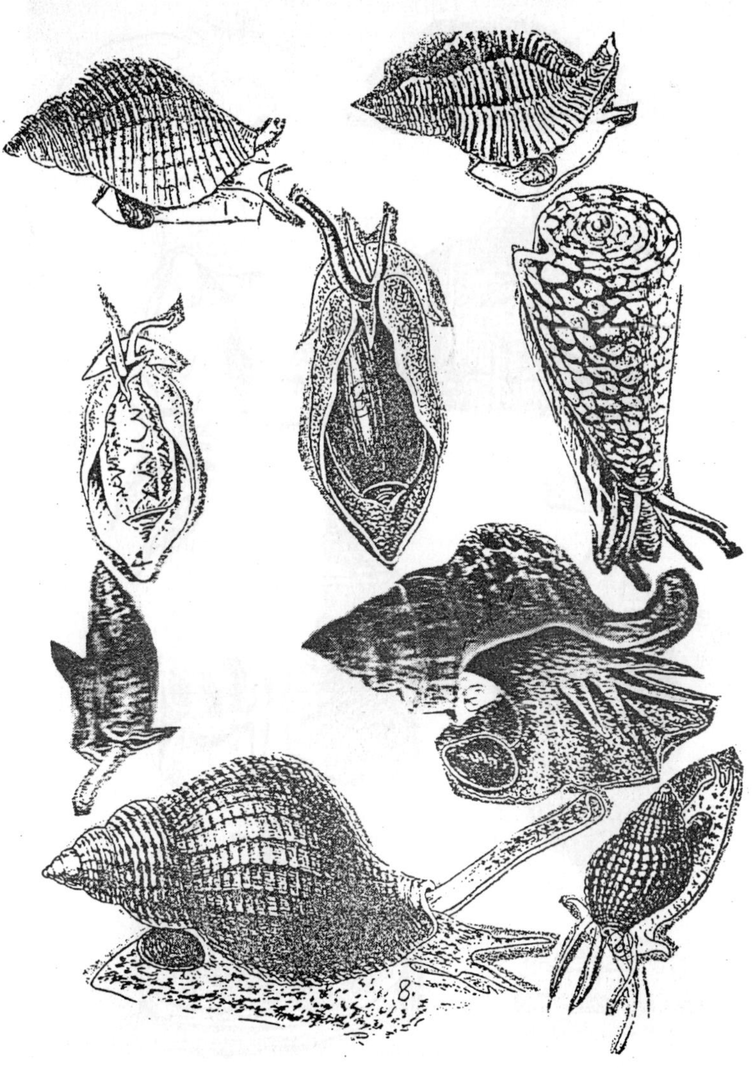

Marine Prosobranchi

1. Atlanlic dogwinkle *Nucella lapilus*, 2. *Ceratostoma* ***crinaceum*** 3. *Conus mormoreus* 4. *Oliva flamulata* 5. Black olive snail ***Oliva*** *maura* 6. Myrex dye *Truncularioptsis trunculus* 7. *Mitra cornicula* 8. Waved whelk *Buccinum undatus* 9. *Hinia reticulate*

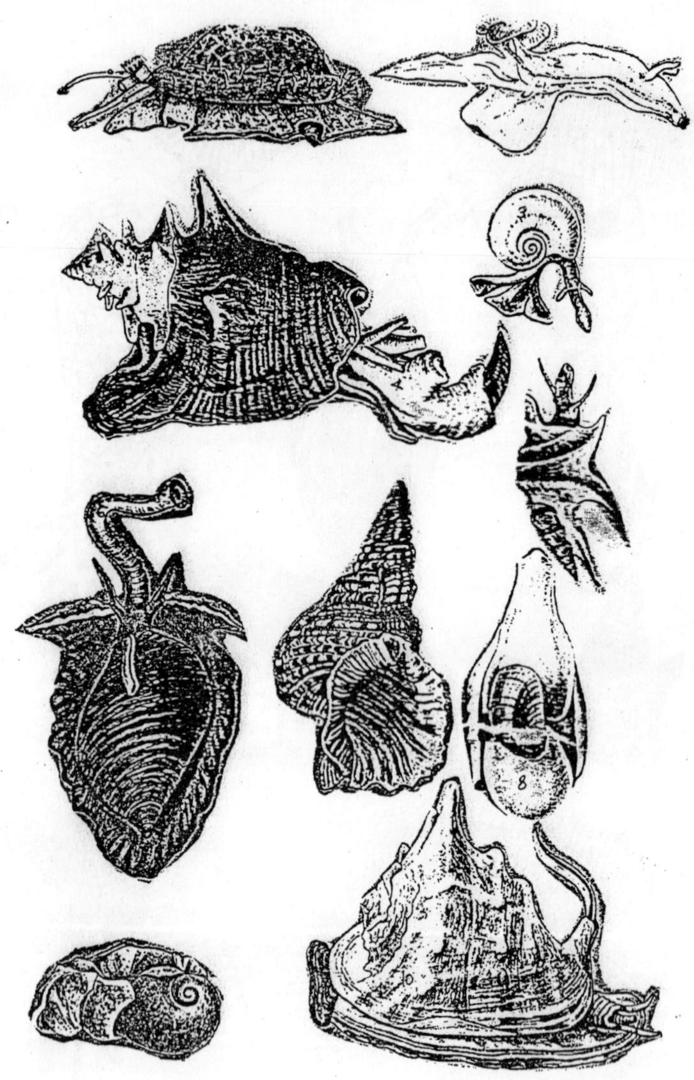

Marine Taenioglossa

Tiger cowries *(Cypraea tigris)* 2. *Carinaria mediterranea (pelagic)* 3. *Atlanta per (magic)* 4. Pink conch *Stombus gigas* 5. Pelican's foot *Aporrhais pespelecani* 6. Giant tun *Tonna galea* 7. Trumpet shell *Charionia tritonis 8. Sinum leachi 9. Naticarius uscorum* 10. Large helmet *Cassis cornuta.*

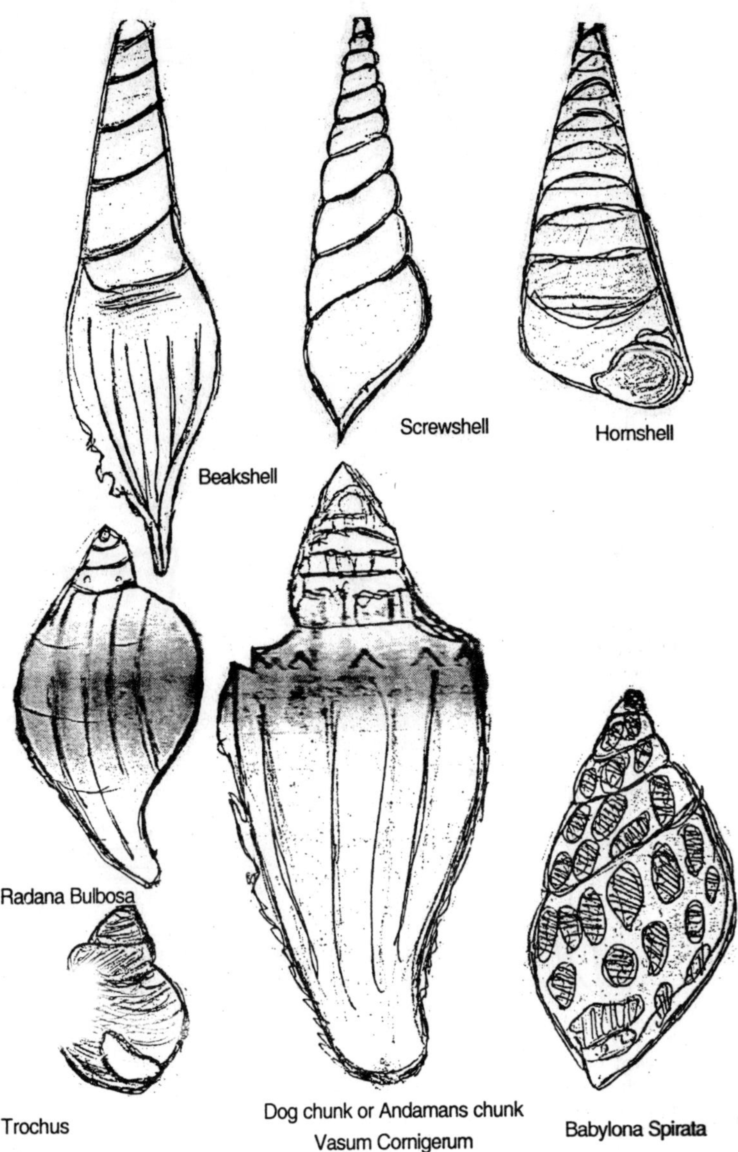

Few Gastropod Shell's

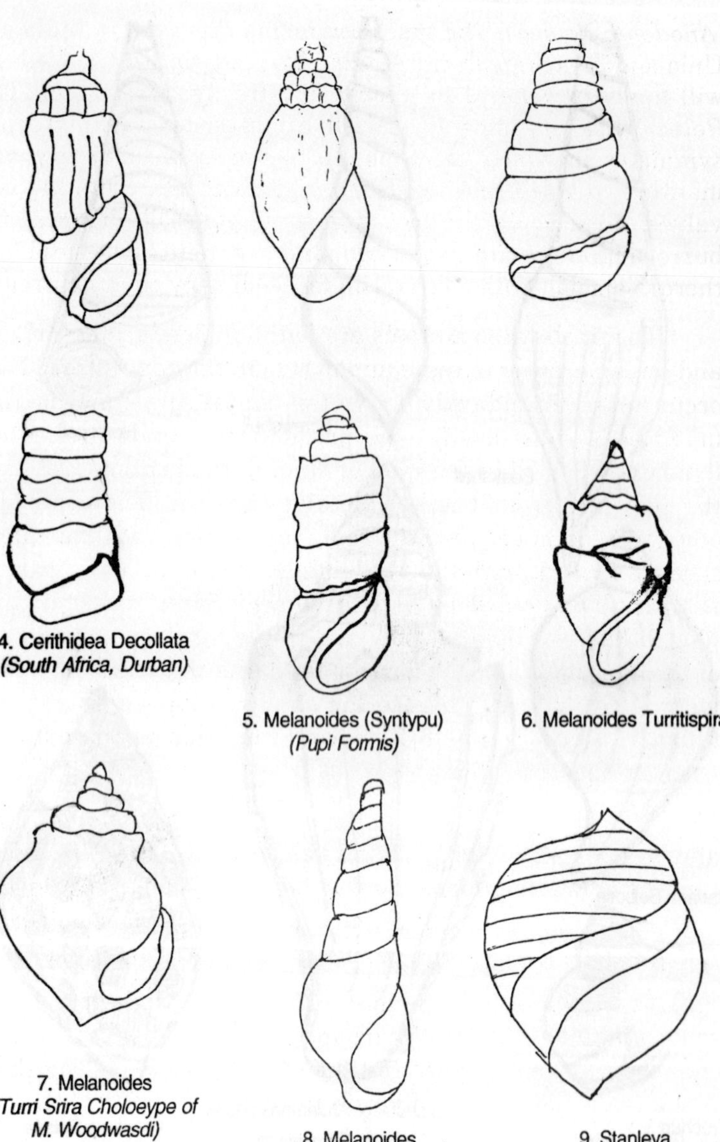

4. Cerithidea Decollata
(South Africa, Durban)

5. Melanoides (Syntypu)
(Pupi Formis)

6. Melanoides Turritispira

7. Melanoides
(Turri Srira Choloeype of M. Woodwasdi)

8. Melanoides
(***

9. Stanleya

(*Anodonta cygnea*). The types commonly dissected in India are Unio and *Lamellidens marginalis*. The description that follows will apply in general to almost any freshwater mussel. The Pelecypoda are molluscs with an internal and external symmetry in which the cephalic region is rudimentary, the mantle is divided into two lobes, right and left, secreting two valves of the shell, the foot is ventral, generally adapted for burrowing, there are two lateral and symmetrical ctendia and there is an elaborate mechanism for feeding by ciliary currents.

The freshwater mussels are found in ponds, lakes, rivers and streams, some in quite and others in flowing waters. They occur more abundantly in water containing lime as this material is necessary for the production of their shell. They live nearly buried in the mud or sand at the bottom, with only the posterior tips of their shell valves exposed. They may also occur wedged in between the rocks and stones, with the values slightly spread, and the two siphons exposed. They burrow slightly slowly crawl by extending their large muscular and plough-like foot between the two valves, or through the gape of the shell, and leave a furrow to marks the path they have followed. They may migrate the shallow places by night and retire to deeper places by day, and may change the habitats with the seasons.

The food consists of microscopic plants (diatoms, etc.), and animals (Protozoa, flukes, etc.) brought in through a current of water caused by the beating of cilia. The water current also brings in oxygen and carries away faeces, carbon dioxide and excretory and genital products.

The siphons open or close in response to light, touch and other stimuli. A slight disturbance causes the withdrawal of the foot and the siphons, and the closing of gape of the shell. Although a clam is usually safe in its tightly colored shell, it is nevertheless preyed upon by the starfish which has the ability to open the shell and devour the soft body parts.

There are many genera and hundreds of species of there molluscs of which several are found in India. The large lamellibranchs those are available here are found abundantly in the eastern districts of Uttar Pradesh and have now been named *Lamellidens marginalis.* These mussels live partially buried.

The shell of the lamellibranch consists of two valves, which are united along a straight hinge-line by a tough elastic substance that forms the hinge ligament. The ligament passes transversely from one valve to another. Its elasticity is responsible to open and close the shell. Along the hinge line each valve has teeth like projections that enable the shells to fit with each other. There are the hingeteeth. The hinge is dorsal and the gape is ventral. In Anodonta the connection between the two valves is afforded only by a ligament, but in *Unio* each is produced with strong projections and ridges, the hingeteeth, separated by grooves and sockets, and so arranged that the teeth of one valve fit into the sockets of the other. The same is the case with *Lamellidens.*

The valves are externally marked by a series of concentric lines, lines of growth. They lie parallel to the free edges of the shell. Alternate periods of slow and rapid growth can be marked out by their size. The umbo represents the oldest portion of the values. At this place the shell is the thickest. This is the first part of the shell to be recreated along the edges of which subsequent layers are added. The inner surface of the shell also shows similar markings. On a closer examination it is revealed that the shell consists of three layers:

(a) The periostracum is the outermost layer formed of a chitin-like chemical compound conchiolin. It is dark blue when wet and brown when dry. This thin and translucent layer is partly responsible for some variations in the colour of the shell.

(b) The middle layer lying beneath the periostracum is the primatic layer. It is compared of numerous calcareous prisms embedded in conchiolin.

(c) The last layer of the shell. The nacreous layer or nacre. It is also called mother of pearl and is composed of alternate 10 years of calcium carbonate and conchiolin.

The first two layers are secreted only by the edges of the mantle and hence show concentric markings of discontinuous growth. The inner pearly layer is laid down by the whole surface of the mantle and has smooth lustrous surface. This layer may also secrete a pearl as a protection against foreign body, such as a parasite. When a parasite, say a larval stage of a fluke, enters the mantle it becomes enclosed in a sac formed by the growth scar of the mantle eithelium, which later on secretes thin concentric layers of pearly substance around the body. Except for the concentric lines of growth the shell is smooth and there are no ridges or tubercles which are frequently found in the shells of other bivalve molluscs. The inner surface of the shell presents characteristic markings and structures:

(i) The hinge-teenth that interlock the valves tightly are more clear on the inner side.

(ii) The scars of the adductor muscles are other prominent structures. The scar of the posterior adductor muscle is larger and posterior. Posterior to the scar of the anterior adductor is the scar of the anterior retractor muscle of the foot. Ventral to this is the of the protractor muscle. Parallel to the margin between the adductor scars making the attachment of the retractor muscle of the mantle is a fine line, the pallial line indicating the attachment of the mantle.

The mantle consists of two thin lobes attached to one another and to the vinceral mass dorsally. The mantle secretes the two valves of the shell that lie over the two lobes. Normally the mantle is creamy white in colour but preservation spoils it. Certain parts of the mantle edges are so thickened and rolled as to form a dorsal exhalant aperture and a ventral inhalant aperture. The space inclosed between the two lobes of the mantle is known as the mantle or pallial cavity. Dorsally the pallial cavity is bound by the visceral mass.

The shells are enclosed agape by the elasticity of the ligament and closed by the contraction of large muscles. There muscles include:

The vascular system in BIVALVES is like GASTROPODS including considerable haemocoelic spaces around the viscera and in the foot.

The heart consists of a ventricle which comes to lie around the rectum during development and into which two auricles discharge. Blood is discharged both anteriorly and postcriorly from the Ventricle but the vessels (anterior and posterior aortae) cannot be homologous with the two aorotae of gastropods. The anterior passes forward and supplies blood to the head, the foot and the visceral mass while the posterior goes backwards and takes blood to the mantle and spihons. Blood is returned from the anterior part of the mantle, the viscera and the foot into a vessel which takes it to the kidney, thence to the gills and back to the heart.

The fresh water clams diocious, i.e., the sexes are separate, but there is no sexual dimorphism. Reproduction involves a parasitic larval stage, called the glochidium in fresh water clams and the veliger in marine forms.

The gonads either testes or ovaries, are a pair of large simple visceral branded structure, lying among the intestinal coils in the visural mass just above the foot. During breeding season, the gonads become greatly enlarged and conspicuous, when it is difficult to distinguish their paired nature. When mature, the testes are whitish and the ovaries reddish and the two sexes can be distinguised though not very distinctly. The epithelium lining the tubules of the gonads gives rise to the spermatozon in male and ova in female. The mature ovum is large, rounded, filled with a finely granular cytoplasm rich in yolk, and containing a vesicular nucleus with vacuole. The gonad of each side has a short duct, the vas deepens in male or the ovidul in female. It leaves the gonad from in upper side and open into the supra-bronchial chamber of the inner gill lamina by a genital aperture, just in front of the renal aperture of the ureter. There are no accessory reproductive organs. Males are more responsive than females to stimulation by rise in temperature and almost always initiates spawning. Presence of eggs stimulate other males.

In male sperms are shed through the ginital aperture into the supra-branchial chambers and released from the body with the outgoing water through the exhalent siphon. Some of these sperms will be taken into inhalent siphon of a female purely by chance. The eggs, shed through the genital apertures into supra-branchial chamber, do not leave the body but are carried through the ostia into water tubes of the outer gill laminae, where they are held by nucleus and fertilized by sperm from another clam, drawn in by water currents, thus forming the zygotes.

Development of the fertilized eggs or zygotes proceeds within the outer gill laminae, which greatly enlarge due to accumulation of a large number of embryos and serve as the breeding pouches or marsupia. The females are easily recognized at this stage by the swollen appearance of their outer grill plates. In anodanta, the eggs are usually fertilized in August and the embryos, which develop from them, remains in the gills of the mother throughout winter, escaping in spring as tiny larvae.

Anodanta will apply almost equally well to it. The later stages in the development, show a complication, due to a temporary parasitism of the larva, "Glochidium".

In larva escapes through the exhalant siphon of female mussel and finds a suitable fish host for further development. The larvae attach to gills or fins of fish for a period of 60-70 days drawing nourishment from host. At the end of parasitic life the juvenile mussel detaches from host and falls to bottom for further growth.

For rearing young mussels ponds should be cleared of weeds and excess bottom mud. To maintain hygiene and enrich status of pond, it should be treated with lime 200-250 kg/ha/yr cowdung (10,000 kg/ha/yr) urea (100 kg/ha/yr) and single superphosphate (100 kg/ha/yr) in instalments. Such doses will help to maintain good algal density in pond. In water having current or ponds having inlet and outlet facility and receiving canal water, metabolites from mussels are washed out. Regulate all water quality parameters specially food, pH and oxygen etc. and prevent any kind of pollution. The pH of soil and water should both be in range of neutral to slightly alkaline, pH 7.0-7.5 (soil under wet condition) and water 7.0-8.0, dissolved oxygen 5-8 mg/l, total alkalinity should be medium 70-150 mg/l. Calcium is a major ingredient of pearl and oyster shels, so its presence in adequate quality (10-20 mg/l) in water is considered to be essential. Cultured pond should be free from aquatic weeds. The optimum depth for fish-cum-peal culture varies between 2-3 meters.

In shallow pond the water may be very hot during summer thereby causing stress on fishes and pearl oysters while deeper ponds have difficulties in handling and nursing of oysters.

After 60-70 days the larvae get dislodged from fish.

In artificial pearl production two methods are used. In first a graft tissue is injected into another mussel. In second method a hard ball like material called bead along with flesh is pushed into its body. The ideal temperature of water should be 15°C-30°C. As the mussels become weak during breeding season this period may also be avoided for pearl culture. The chief ingredient for pearl culture is a healthy above two years old mussel. If a foreign object like an insect or a grain of sand enters the body of mussel some sort of secretion occurs nacre, this deposits layer upon layer leading to the formation of a pearl.

For artificial production of pearl healthy mussel immersed in a pot of water for 24 hours before introduction. Then washed afresh. Choose bigger one which will be donor. A thin and long portion from the side of shell (mantle epithelial layer) is cut out. After cleaning put this portion in glass/slide and cut into small pieces. Add 2-4 drops of water soluble Eosin. Now cover the material for freshness. The other mussel are kept in tub full of water in such a manner that shell valves placed upwardly. This makes valves of the shell to open easily. Sprinkle the methanol (menthol) to easy opening of valves. Then, as a mussel starts to openers its valves a piece of wood is inserted with the help of valve open and mussel is finally placed on mussel holder stand. The pieces of graft tissue may be introduced if mussel is a big sized one. After this operation the mussel should be removed from stand and put into water. The introjected mussels are washed thoroughly and kept in a tub of water for over night. Next day they are again washed before being dropped into pond.

According to second method a small hand round object made from conch shell to inserted along with the piece of graft tissue. At present small beads from conch shell are used. Alongwith the graft tissue this type of beads may be introduced in the mantle epithelial layer of mussel. Dorsal side of the graft should touch the bead. The graft tissue makes a sac within a short period. After that nacre starts formation over the bead to form pearl, this method lakes 18-24 months but in second method only one year is sufficient (Anjana Sen Gutpa 1994).

According to Sen Gupta (1994) there are three basic steps in freshwater mussel.

1. Preoperative management of mussels
2. Implantation management.
3. Post operative management of mussel.

1. Pre-operative Management of mussels

(i) Selection of mussels

Healthy Donor Age 2-4 years, size less than 14 cm but not less than 9 cm.

Healthy mother age-3-5 years, size 9.5-12 cm.

(ii) Identification of healthy mussel.

(a) Mussel when subjected to shock is able to close quickly.

(b) Shell wall ejecting water in a straight line to considerable distance.

(c) Adductor muscles must be strong.

(d) Strong foot free from any injury.

(e) Fully developed and devoid of injury in pallium and gills.

(f) No empty sound come out from mussel by gentle stroke or finger.

(iii) Collection and transportation

(i) Collected in early spring (Feb.-March) at 20°C and transported in 20°C-25°C.

(ii) Mussels are safely transported in semi-dry condition in cane/bamboo/plastic containers polythene with hydrilla.

2. Implantation Management

(i) **Season**. Early spring is the best time. Feb-April in congenial temperature.

(ii) **Cut piece implantation** should be with sharp knife and tweezers, mantle epithelium of pallium region (posterior to middle) is cut and carefully cleaned by a sponge and stretched and trimmed to 2 mm^2 or 3 mm^2. Cutpieces are kept soaked in reosporun or physiological saline.

There should be short duration (instant) in cutting the pieces and implantation.

(iii) Care should be taken to maintain proper space between implantation of cut pieces.

(iv) Opening of valves should not be above 1.5 cm.

(v) The region of implantation is washed with water and 0.5-1 cm. wound is made in conformity with diameter of nucleus. After mercurochrome soaked nucleus is securedly placed, the cut piece is placed in mouth of wound, a drop of Betnesol-N may be given in wound.

3. Post operative management

(i) Rearing Tanks: Water quality and management of aquatic weeds and predators are very important.

(ii) Rearing method: Rearing by suspension.

(iii) Rearing time and management are important for pearl.

(iv) Harvesting time: December-January best till early April.

(iv) Washing: Wash pearl with 3-5% of hydrogen peroxide or diluted Hcl.

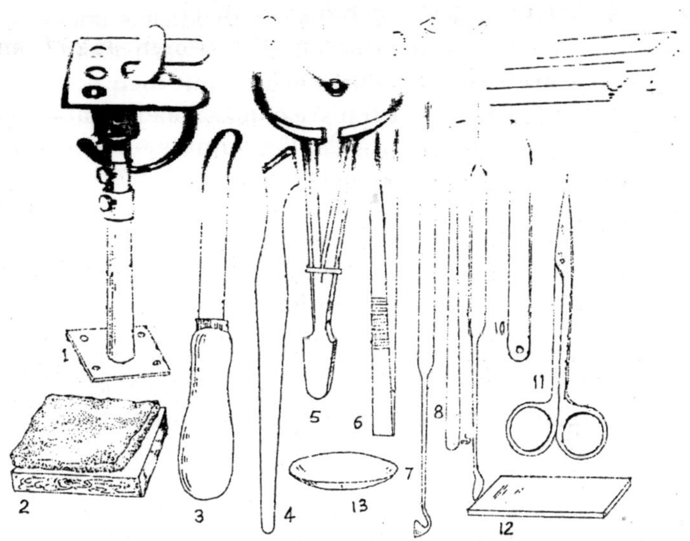

1. Mussel holder stand, 2. Wooden piece for tissue cleaning, 3. Knife for mussel cutting, 4. Tissue cutter, 5. Valve opener, 6. Forcep, 7. Bead holder, 8. Bent needle, 9. Bent needle, 10. Mussel opener, 11. Scissors, 12. Glass slide, 13. Watch glass, 14. Wooden peg.

The large family of Pteridae is of great importance of Indian Ocean (*Pinctada* spp) wingshells (Pterudsproper), Hammer oyster (*Malleus*) and *Vulsella*. Indian pearl oyster is (*Pinctada vulgaris*) smallest, it seldom exceeds 3.5 inch in height and compares unfavourably in size with huge gold lip *P. maxima*) from mergie and Australian waters. The shell is straight along the hinge line and is produced at each end into a short ear a modification of winglike projections. Ventrally shell is deep and rounded with a series of delicate finger-like at the ends of radial bands made up of older fingers now outgrown, worn and disused. These pearl oysters reach their limit of size in about 3½ years. If they survive longer the shells become worn, the fingers disappear and overall length and depth actually decrease. The thickness increases until death. This is the best criterion to determine the age. In life history and habits they are akin to mussels. A strong cable of byssal threads attaches them to rocks, stones and shells, they possess the power of casting of their capable at will, they are cable of crawling short distances and this power to shift their foothold sometimes enables them to avoid entombment through an overwash of sand. When young, they are restless and inclined to shift their position 7-8 times within few hours each place of attachment marked by a tiny tuft of golden byssal threads. The formation of one of these threads takes only a few seconds; the thread is formed from a glue-like secretion poured into a fine groove on underside of small tongue-shaped foot. On contact with water it coagulates into a strong slightly elastic substance, golden yellow in young oysters dark bronze green in adults.

Sexes are separate but probably alternate in same individual in successive seasons, fertilization takes place in water development rapid, a trochophose stage first then a veliger. At the end of about a week or even earlier if conditions are favourable larva settles upon the bottom and makes its first attachment. This condition is called 'spat', a satisfactory 'spatfall' is dependent upon a favourable run of currents at time of oysters spawn hence irregularly of formation of fishable beds of pearl oysters in Indian waters.

Indian pearl oyster is of comparatively rare occurrence; sand grains causation is much more common and is to be reckoned important; a more prolific source, however, is that due to irritation caused by fragments of nacre dislogged from the inner lining of the valves by muscular strain, which pass into the oyster soft tissues. These fragments of nacre may wander far; in their course they become invested with a containing sac of nacre-producing cells immigrant from mantle which lay down successive coats of nacre upon the nucleus thus forming a pearl, other pearls mainly of seed pearl kind are formed around nuclei consisting of microscopical concretion which arise within muscles at the place of insertion on shell. These may also wander into oyster's tissues and may even be found in 'nests' of as many as a dozen or more within the visceral mass. Thus pearl formation may be induced by several kinds of irritation.

The solitary handsome stoutly built shell larger black lipped pearl oyster *Pinctada margaritefera* with a shade of green on very dark ground colour decorated with several radial lines of white spots passing outward from umbo. Ears disappear in adult and outline becomes suborbicular. Nacre of this species is thick enough to be of value in button making but its dark smoky colour greatly impairs its market price, it is common in Persian Gulf.

Water depth, bottom characteristics, protection from wave action, tidal flow, height, tubidity, water quality, chemical parameters, predation, fouling, pollution and accessibility are the main factors for site selection. From intertidial region to areas extending upto 5 m depth can be considered for adopting suitable culture method salinity required 22-35 ppt and temperature 21-31°C. Special nursery ponds in inter-tidal region (can be constructed. Oyster spat out of hatchery are too small, to be grown in field without protection. They should be enclosed in velom screen bags of suitable mesh size and suspended from racks. A string can hold six shell valves containing 80-100 spats and 3-4 strings are enclosed in a bag. Clean the bags with regular intervals. After 40-50 days they are transferred to farm.

They are broadly grouped as bottom (on bottom) culture and off bottom culture practices. The off bottom culture methods are advantageous over the bottom culture in various ways:

1. Relatively rapid growth and good meat yield.
2. Facilities three dimensional utilization of culture area.
3. The biological functions of oyster such as filtration, feeding etc. are carried out independent of tidal flow, and
4. It minimises the silting and predatory problem. Other methods are:
   (i) Raft and string (ren) method.
   (ii) Rack and tray method.
   (iii) Stake method
   (iv) Raft method.
   (v) Long line method.

For farm maintenance one has to control crab, fishes, starfishes, polychaete and gastropods. There are mainly *Scylla*

Serrata Pagurus, Thais rudolphie. They kill young oyster gastropod, *Cymatium cingulatum* cause 15% mortality, *Gracilaria* spp alga affects the water flow. *Balanus amphitrite* is a fouler that attached to wooden structure, trays and oysters and compete for food. Farm management includes timely cleaning of oysters and rearing trays, racks, string, rafts. All the devices should be cleaned and poles and other spoiling materials are to be replaced. Harvesting is done manually. Good meat yield can be obtained during March-April and August-September.

The history of Broome Pearling has been one full of high's and low's and after World War II, it was at a particularly low point. Luckily, a demand for cultured pearls developed, no doubt as the economy picked up, and this revival brought about the development of the first pearl culture farm in Australia in 1956.

It was established at *Kuri Bay,* 420 km North of Broome, by a joint venture between Australia and the Japanese. The Kuri Bay venture was formed under the name, *Pearls Pty Ltd,* and consisted of *Male and Co, Broome Pearlers Brown and Dureau Ltd,* and the *Otto Gerdau Co of New York.* The Nippo Pearl Co. of Japan handled distribution and marketting. Tokiuchi Kuribayashi was the principal of this company and was the most powerful and influential man, since Mikimoto's death, in pearl culture. Kuri Bay was therefore named after him.

By 1981, there were five established pearl culture farms in the Broome region—*Kuri Bay, Port Smith, Cygnet Bay,* and two in *Roebuck Bay.*

It is well worth noting, at this point, a little about the history of pearl culture which began in Japan. The first known attempts were in the 13th century, where crude semi-round pearls were the result. Tatsuhei Mise (1880-1924) produced the first round cultured pearl before 1904. Tokichi Nishikawa (1874-1909) became the first person to produce a round cultivated pearl from planned experiments using tiny gold and silver nuclei.

Kokichi Mikimoto (1858-1954) is the man we most associate with Japanese cultured pearls. However his method

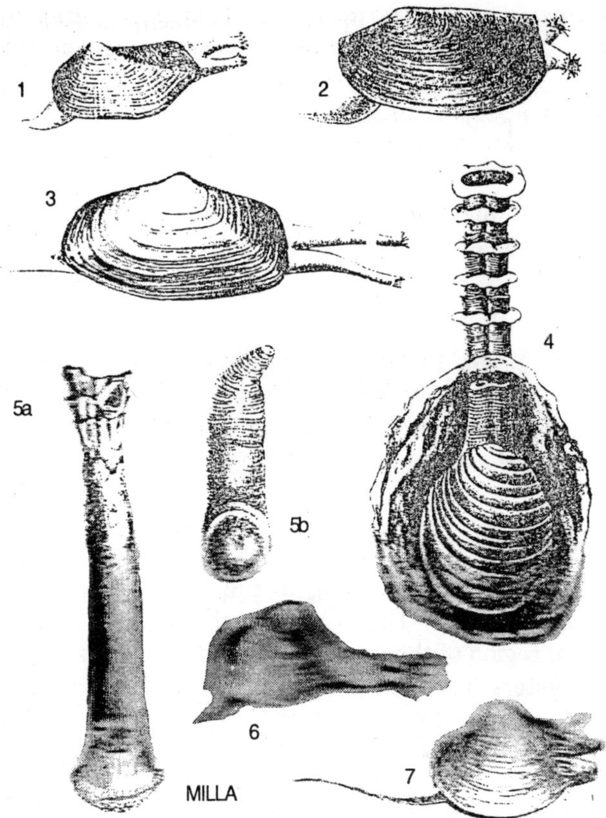

Anomalodesmacea: 1. Glassy lyonsia *(Lyonssia hyalina);* 2. Pandora clam *(Pandora inaequivalvis);* 3. *Thracia pubescens* (L 3 cm); 4. *Clavagella aperta;* 5. Watering-pot shell *(Penicillus vaginiferus;* L 12 cm), (a) Calcareous tube, (b) Soft body; 6. *Cuspidari cuspidata* (L 18 mm); 7. *Poromya granulata.*

Pearl producing molluscs

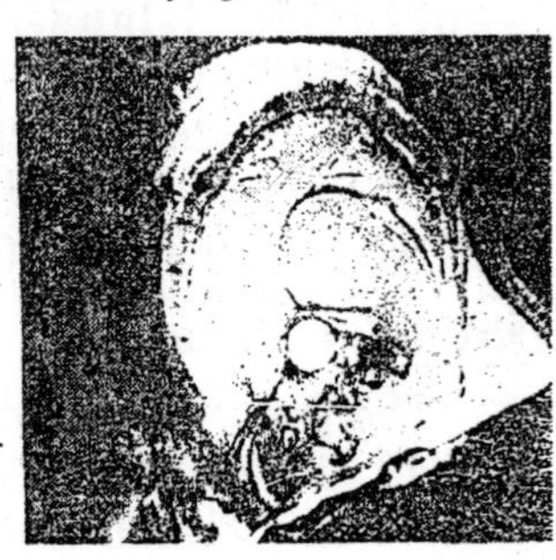

Pearl oyster with pearl in situ

of depositing nacre from graft mantle tissue has proved too delicate and difficult. The method used by the cultured pearl farmers today is based on T. Mise's basic requirements for the spherical pearl production which is:

(a) A spherical nucleus.

(b) The introduction of the nucleus into the connective tissue of the oyster.

(c) The nucleus must be inserted along with a piece of mantle epithelium.

In 1922, the Pearling Act inserted section 113 which prohibited anyone in Australia producing artificially cultivated pearl. Luckily, this was reappealed in 1949, thereby opening the doors for the development of today's pearl culture formed in W.A.

Resources and Distribution

The pearl oysters belong to the genus *Pinctada* under the family Pteriidae. They occur in the seas of tropical and subtropical region of the world. From Indian waters, six species of pearl oysters have been reported, namely *Pinctada fucata* (Gould), *P. margaritifera* (Linnaeus), *P. chemnitzii* (Philippi), *P. sugillata* (Reeve), *P. anomioides* (Reeve) and *P. atropurpurea* (Dunder). Among them P. *fucta,* called the Indian pearl oyster, is the most dominant and contributes to the pearl fisheries of the Gulf of Mannar and the Gulf of Kutch. This species also occurs in the Red Sea, Persian Gulf, China, Korea, Japan, Indonesia, Australia, Margarita and Cubagua (Venezuelan Islands). The black-lip pearl oyster, *P. margarilifera,* produces the famous steel black pearls. In India it is confined to the Andaman and Nicobar group of islands. Recently several specimens of *P. margarilifera* were collected from the Gulf of Mannar.

Distinctive Characters

The genus *Pinctada* oding is characterised by long and straight hinge. The long axis of the shell is not at right angle to the hinge. The left valve is usually deeper than the right. There is a byssal notch on each valve at the base of the anterior lobe. The colouration of the periostracum is variable and often brown.

In P. *fucata* the hinge is almost as broad as the width of the shell, anterior border of the shell does not extend or just extends in front of a perpendicular line drawn from the tip of the anterior ear, posterior ear well developed with moderately developed sinus, byssal notch slit-like, left valve highly convex, outer shell with 6-8 radial bands of reddish brown colour on a pale yellow background, nacreous layer iridescent and of golden, pink or ivory colour and the non-nacreous border on the inner surface of values has brownish or reddish patches.

In *P. margarilifera* the hinge is shorter than the width of the shell, anterior border of shell far in advance of a perpendicular line drawn from the tip of the anterior ear, posterior ear and sinus absent, byssal notch broad, left value moderately convex, outer shell dark brown with greenish tinge and radially distributed white spots, nacreous layer iridescent and of silvery sheen colour and the non-nacreous border is of dark colour; hence the name 'black-lip pearl oyster' for this species.

ECOLOGY, MORPHOLOGY, ANATOMY AND BIOLOGY

Ecology

As is the case with other species of the genus *Pinctada,* the Indian pearl oyster *P. fucata* is truly marine and does not occur in estuaries of backwaters. The preferred salinity range is 30-36 ppt. It occurs from the intertidal region upto 25 m depth. It prefers clear water and high turbidity is not conducive since it reduces the filteration and food intake.

Morphology

The shell, also called valve, is composed of three layers. The outer layer is formed by a horny substance called conchiolin. The middle layer is the prismatic layer and is composed of minute prisms of inorganic calcium carbonate in the form of calcite crystals separated from each other by thin layers of conchiolin.

The innermost layer is the nacreous layer or the 'mother-of-pearl' layer and is lustrous. It is composed of numerous line lamellae or aragonite crystals. These three layers are formed by the secretion of the epithelial cells of the mantle. The left valve is deeper and more concave than the right one. The shells are joined together on the adductor muscle is noticeable substantially. Thus musclc is attached to both the valves and controls the movements of the valves. Small sears numbering 12 to 15 are present inside the shell for the attachment of pallial muscles.

Ecology Morphology, Anatomy and Biology

The mantle is a fold of skin covering the soft body of the pearl oyster between the valves. It is formed of two lobes on either side and these are fused together dorsally below the hinge line. Seawater enters the mantle cavity through the inhalant aperture and leaves by the exhalent aperture. The mantle can be divided into three zones, namely, the edge or the marginal zone the pallial zone and the central zone. The mantle from the pallial zone is used for the graft tissue preparation.

The foot is a tongue-shaped organ capable of expansion and contraction. It arises in the anterior region of the visceral mass. The byssus gland is lodged at the proximal end of the foot and a bundle of stout laterally compressed bronze green fibre known as byssus originate from this gland.

The gills are paired, filamentous and are attached to the visceral mass in the front; the opposite end is free in the mantle cavity. They actively filter the water, absorb oxygen and retain organic matter useful as food. When the water is highly turbid the gill filaments become choked and filteration is stopped, affecting vital functions.

The heart, consisting of a ventricle and two auricles, is enclosed by the pericardium; it is ventral to the intestine. The auricles, after receiving oxygenated blood from the gills and mantle, pass it to the ventricles. From the ventricle arise two aorta, the posterior aorta supplying blood to the rectum, anus and adductor muscle and the anterior aorta to the rest of the body. The deoxygenated blood collected in the veins reaches the kidneys and gills where it is purified.

Two curved leaf-like structures known as the labial palps are located on the mouth opening. They select the food particles, pass them on to the mouth and reject the unwanted matter as pseudo-faeces. The selected food is passed through the mouth, oesophagus and the stomach where it is smashed and subjected to action by the enzymes, supplied from the crystalline style. The larger particles are directed towards the intestine; other nutrients are dispatched to the digestive diverticula which surround the stomach. The intestine extends through the visceral mass and forms a loop, ascends above the heart and opens into the long rectum, followed by the anus through which the faeces is ejected out.

The excretory system consists of paired nephridia and numerous small pericardial glands projecting from the walls of the auricles.

The nervous system consists of three pairs of ganglia, viz. cerebropleural, pedal and visceral, from which the nerves arise. The mantle is supplied by a complex network of nerves, the pallial plexus and is sensitive to slight stimulus. A branchial eye is present on the first left branchial filament and it can detect variation in light intensity.

Biology

The pearl oyster is a filter feeder. The food mostly consists of microalgae and particulate matter. The food brought inside the mantle gets entangled in mucus sheeth produced by the gills. The labial palps direct these entangled food into the mouth. The food sorting mechanism in the pearl oyster is considered to be not efficient since several organisms which cannot be digested find their way into the stomach and intestine. Alcyonarian and sponge spicules, lamellibranch larvae, gastropods, heteropods and crustacean appendages are some of the items, apparently not of food value, but found in the stomach of pearl oysters.

P. fucata grows fast in the first year. At the Krussadai Island the estimated length at the end of the first year is 45 mm, second year 55 mm, third year 60 mm, fourth year 65 mm and fifth year 70 mm. The corresponding estimated weights are 10 gm., 30 gm., 45 gm., 60 gm. and 70 gm. Hatchery produced oysters have been reported to attain model sizes of 47.5 mm, 64.5 mm. and 75 mm. at the end of the first, second and third years, with corresponding weights of 8.3 gm., 31.6 gm. and 54.4 gm. In the Gulf of Kutch the estimated lengths at ages 1 to 6 are 44.05 mm, 61.68 mm., 76.20 mm., 81.62 mm., 85.15 mm. and 86.65 mm. respectively.

The sexes are separate but male and female oysters cannot be distinguished by the shell characters. Change of sex has been reported in pearl oysters. The reproductive system consists of a pair of gonads which spreads superficially over the hepatopancreas and intestine in mature condition. It is pale yellow in colour in males and is of a deeper shade in females. *P. fucate* attains first sexual maturity at about 15.5 mm. at the age of 3-4 months. The eggs or sperms are released through the paired gonoducts ending in the genital openings located anterior to the gills. In the Gulf of Mannar *P. fucata* spawns twice a year during June-August and November-January coinciding with the southwest and northeast monsoons respectively. Between year variations are known to occur in the spawning of *P. fucata,* and also differences under the wild and farming conditions.

LOOKING AFTER OYSTERS

The shells which have been collected and transferred to the pearl farm are placed in baskets or panels which are attached to long lines connected to the floating rafts.

The line is dropped down into the ocean with the oyster securely inside the basket, where they remain until they become operated on for seedlings.

Seedling takes place when the temperatures become stable of which is usually around July. The oysters are very sensitive and can easily die if handled during changing temperatures. They are, therefore not touched, if possible, in May or September. The oyster can produce more than one pearl in its lifetime by taking good care of it. Regular cleaning of the shells to remove seaweed results in better pearls plus makes them easier to handle. The cleaning is done by a machine which uses water jets and brushes to clean off any seaweed or other growth from the outside of the oyster shell. It may also be cleaned using a tool similar to an axe.

It is most important to regularly turn the oyster shells through 180°, so that the nacre of the developing pearl inside the oysters, will be evenly distributed around the nucleus. The oysters need very tender loving cares so as to be productive when harvested.

THE PEARL FARM

Selecting a pearl farm site is extremely important as there are many important factors which must be considered, of which will determine the type of pearls produced, and the oyster survival rate. Some of these environmental factors are:

(a) The relief of the region: Natural features, such as mountains and reefs, are needed to protect the farm from winds, currents, storms, etc.

(b) Constant regularity of temperature

(c) Type of sea bed, such as rocky or sandy.

(d) Presence of gentle currents: Currents are essential for the survival of the oysters as they bring food and oxygen.

The whole pearl farm system is based on a series of floating wooden rafts. Ten basic units, each 20 ft. times 20 ft., are joined together in two rows of five, to complete one raft. 100 plastic covered wire mesh baskets, each of which house 10 oysters, are suspended from each unit. 8-10 empty metal drums are used for unit flotation and each raft is moved by iron anchors.

A typical work schedule for a pearl farm can be seen as follows:

January. Prepare for wild shell collection, organise dive crews, fishing gear, paperwork, and licence fees.

February. Fishing for 20,000 wild shell begins, linked to tide patterns. (Note: Tides in the area can vary by 10 m. per day)

March. Collected shell is 'dumped' on the seabed or site leased by the company and allowed to rest. Maintenance of dumped shells, turning and cleaning them. X-ray shells seeded last year to check if implanted nuclei have been rejected. Oysters which reject nuclei are usually re-seeded.

April. Water temperature begins to drop as winter approaches, rest period for the shells.

May. Ongoing farm work, turning and cleaning previous two years' seeded oysters kept suspended in wire panels in the water column.

June. Prepare for operating on oysters to implant nuclei. (Note: Some technicians may come from overseas, and some companies have boats fitted as mobile laboratories so seeding can be done on the pearling grounds.) Seeding and harvesting begin.

July. Normal operating time for pearls, seeding new oyster, re-seeding those which have rejected nuclei. Oysters which produce acceptable pearls are also re-seeded.

August. Harvest of previous year's seeded shell continues, then a two-month turning program follows operations. The oysters are turned over to encourage production of round pearls.

September. Turning operated shell.

October. Turning, cleaning, and change of areas.

November. Transportation of operated shell to grow-out areas.

December. Oysters into longline system. Clean gear.

SEEDING

Once oyster are two-three years old and are healthy, they are able to undergo surgical implanting, otherwise known as seeding. This is a very delicate operation and takes many years to perfect, as it is designed to minimise stress on the oyster. It involves three stages:

Preparation of the Graft

A donor oyster is sacrificed in order to obtain a piece of mantle needed so that the host oyster will accept the nucleus.

The mantle is located on the outer section of the oyster and produces the nacre forming pearls. Before a graft is taken from the mantle, the oysters are starved for several days. This allows the oysters metabolism to slow down, therefore decreasing the risk of core rejection as well as making it easier to open the oyster.

Attaching the Graft

The oyster is opened with special wedges and pliers, and a scalpel slit is made in the soft tissue near the reproductive organ. A graft of living mantle, four-five millimetres square, of which is obtained from another oyster, is then inserted into the scalpal slit.

Inserting the Core

A core/nucleus, usually a sphere made of Mississippi mussel shell is placed in the scalpal slit and the oyster is then returned back to the water. The inserted core irritates the oyster, provoking it to gradually coat the core with thin layers of mother of pearl nacre. After some time, the oysters are collected, and x-rayed to see whether the implants have been accepted. Oysters which have rejected the implant are returned to the water and are once again operated on the following year.

The seeder must be extremely careful not to harm the tiny pea-crab which lives unharmed within every healthy oyster. It is presumed that the crab assists the oyster by keeping it clean and by sharing the debris which the oyster sucks in.

HARVESTING

After 2-3 years, the oysters are harvested, as it takes at least this long for only 2 mm. of nacre to have formed around the core. This is a time of high anxiety for the pearl farmers as it will determine whether the seeding was successful. If so, thousands of dollars worth of fine quality pearls will have been produced.

The oysters are split open the and the pearl bag cut by the scalpel, and the pearl is removed. If the oyster has survived the past three years of gestation without any major problems, the harvester will place another core in the same position as the extracted pearl, without the need for another mantle. This occurs in about one-third of oysters. Approximately 35% of the oysters produce good pearls, 20-30% die, 20% reject the nucleus, and 10% are poor quality pearls.

Although these figures do not seem ideal, they are constantly improving. As time progresses, more successful methods and equipment will hopefully be introduced, which will reduce the number of deaths and increase the number of quality pearls produced.

The high mortality rate amongst oysters is thought to be associated with transportation, the difference in water temperature, and the increase in marine bacteria found in the water of the carrier tanks. Luckily, nothing is wasted in the pearling industry, as excellent prices are paid for the oyster meat as well as the *Pinctada maxima* mother of pearl shell, of which is used for buttons and inlay work.

SELLING PEARLS

The town of Broome has many pearling shops and boutiques offering pearls direct from the sea, as well as pearl jewellery, in a wide range of products, unmatched in the world for variety and quality. Retail sales in Broome totaled several million dollars in 1994 of which is expected to pass $10,000,000 a year before the end of the 1990's. This an extremely important contribution to the towns economy, and the development of pearl retail stores and businesses has promoted Broome as a tourist attraction as well as generating jobs.

Large-scale pearling companies, such as Paspaley Pearls, establish annual auctions which are held at various locations in the world, some major centres being Tokyo, Darwin, Brisbane, and Hong Kong. They are attended by top buyers from all over the world, who spend as long as 4-5 days examining pearls in minute detail, before the actual auction takes place. When bidding commences, sums in excess of US$ 1,000,000 are paid for exquisite pearls.

These auctions bring the world's best known jewellery agents and buyers together, assisting in widening international exposure to the splendour of Australia's South Sea pearls: In

Hydrabad, Bhuvneshwar, Cochin, Rameshwaram, Madras, pearl culture is very common.

The future of the pearl industry lies in breeding oysters so as to avoid overfishing and thus extinction. Mortality rates are too high and therefore, there must be a continuous supply of oysters to ensure the industry survives.

Pearl culture is being carried out only in countries such as Japan, Australia, China and Indonesia. If culture can be done, successfully to produce internationally qualifying grades of pearl, then it would be a great success. Hence the lack of thriving domestic market can be very well compensated for a promising international market.

The pearls are one of the most sought after materials in jewel industry. Thus it is a very important component in the jewelry export market and also in local market.

Normal Mussel gives 3½ ounce of raw edible portion 81 calories and 2.2 gm. fat.

Pearl has several importance

1. Jewellery (Pearl 1)
2. Food (Adductor Muscle-soup)
3. Poultry feed (shell)
4. Flooring chips (shell)
5. Medicinal
6. Button
7. Ornaments.

Mussel

1. Lime Industry (shell)
2. Cement Industry (shell)
3. Food (Meat)
4. Cancer Cure (Meat)
5. Biological filter (aquaculture)

Edible Oysters

In India six species of oysters, namely, Indian backwater oyster *Crassostrea madrasensis* (Preston), Chinese oyster

C. rivularis (Gould), west coast oyster *C,* gryphoides (schlotheim), India rock oyster *Saccostrea cucullata* (Born), Bombay oyster *Saxostrea cucullata* (Awati and Rai) and giant oyster *Hyostissa hyotis* (Linn) are found. The first four have commercial value.

The shape and size of oyster shell is highly variable. If grown individually on a soft bottom, shell tends to be smooth and elongate. If grown individually attached to a firm surface the lower value grows attached to surface ratherflat thus losing the normal cupped appearance. Proximity to other oysters when grown in a cluster causes distortion in shape of shell called Xenomorphoism shape determined by contours of substrate and is important in oyster culture.

Various general and species are identified based on shape, size and other characteristics of shell, anatomy, promyal chamber, gill ostia, heart and gut and breeding habit.

C. madrasensis: Shell valves irregular in shape, usually elongate, left valve deep and right one slightly concave, hinge narrow and elongated. *Adductor* muscle kidney shaped and dark purple, grows 22 cm. euryhaline, inhabits backwaters, creeks bays and lagoons, found from intertidal region to 17 m depth.

C. gryphoides: Shell valves elongate and thick. Left valve cuplike. Hinge are well developed and has a deep median groove with later elevations. Adductor muscale scar is broad, more or less oblong and striations on scar are absent or obscure. Grow 17 cm. it is also euryhaline occurring along north Karnataka, Goa, Maharashtra coasts. In Maharashtra it is exploited regularly from several creeks. It occurs in water upto 7 m. depth.

C. rivularis Shell valves large, roughly round, flat, thick and with a shallow shell activity. Left valve is thick and slightly concave and right one is about same size or slightly larger. Adductor muscle scar is oblong and white or smoky white Grow 15 cm. It found in Gujarat coast and also in Maharastra together with *C. gryphodies* in Mahim, Ratnagiri and Jaytapur creeks.

Saucostrea cucullata: Shell valves hand, stony, trigonal and pearshaped. Hinge straight, devoid of teeth and numbonal cavity well developed. Margin of both valves have well developed angular folds sculptured with laminae. Small tubercles present along inner margin of right valve and corresponding pits in left valve. Adductor scar is kidneyshaped and white or colourful. Occurs in all along the main land coast and Andaman and Lakshadweep Islands. In shallow water they attached to rocks and boulders and can withstand surf and wave action in sea.

Edible oyster food consists of organic detritus and phyto plankters such as diatoms and nanoplankters. The food parties are entrapped in mucus of gills and are passed in water currents towards the mouth by rapidly beating gill cilia (fine hairs). The labial palps, number four located close to mouth, surf the food before it enters the mouth. The unwanted food particles are rejected as pseudofaeces. Close to entrance to intestine is located the 2-3 cm. long crystalline style sac which contains a gelatinous rod (crystalline stole) known mucoprotein, contains digestive enzymes which convert starch into sugars. Apart from usual extracellar digestion in stomach intracellular digestion occurs in digestive gland.

The growth is related to food availability and also to environmental conditions, particularly to salinity and temperature.

Condition index of oysters denotes the quality of meat and it is useful to determine the best period for harvest. It is also helpful to assess the suitability of a locality for culture. High condition indicates greater proportion of meat in whole weight of oyster; those prime condition are tasty when compared to flacid and watery meats of oysters in poor conditions. Soft body of oysters changes according to the different stages of reproductive cycle. In maturation gonad size increases in weight causing increase of weight in soft body.

$$\text{Condition factor} = \frac{\text{weight of dry meat (ovendried at } 90-100^{\circ}\text{c)}}{\text{volume of shell cavity}}$$

In *C. madrasensis* the condition is considered as high if it is above 140 and poor if it is below 70.

In genes *Crassostrea* sexes are separate but occasionally hermaphrodite may found. Ovary and testes comprise a series of branching tubules, follicles on both sides of body. After spawning fertilization occurs externally. In nonbreeding season gonad is replaced by connective tissue called leydig tissue which mostly have glycogen. In this stage sex cannot be determined. The sex of oyster may change during breeding season. In *Crossostrea madrasensis* young oysters primarily function as males 60-75% and later change their sex. A single female measuring 80-90 mm. spawns 10-15 million eggs at a time. In Kakinada Bay, it spawns during January-June. In Madras harbour, under marine conditions year round spawning has been reported. In Adayar estuary it spawns during October-December and in March-April. In Muli estuary, major spawning activity is confined to April-June and minor in November.

As per survey conducted by CMFRI at some important centres the annual catch is estimated at less than 2,000t. *C. madrasensis* is dominant species from fisheries point of view. Oyster seed may be collected by clutch in water column at appropriate period. During spawning period the spat collectors are laid on bottom on racks or suspended from racks. If collectors are laid in advance of spatfall, they may be covered with silt or settlement of foulers making them unsuitable for oyster larvae to settle. The larval period in *C. madrasensis* is 15-20 days. The ideal time is 7-10 days after peak spawning. Strong currents interfere with larval settlement and may result in poor spat collection. Different methods are used for different culture method e.g. for ren method oyster shell strings are used.

The techniques for mass production of both attached and free seed of *C. madrasensis* has been developed by CMFRI at its shellfish Hatchery, Tuticorin which is described here. The creteria for site selection and inputs for establishment of hatchery of pearl oyster and edible oyster are similar with minor differences.

For brood selection oyster conditions and age are to be considered. Mixed stock from various area provides good results. Oysters of 60-90 mm, with 30% of them (males) in 60-75 mm. length are desirable. First of all clean them with wire brush to get rid of plants and animals adhering to shell. A batch of 25 oyster can be placed on a synthetic twince-knit PVC frame in 100-1 FRP tank, and raw (unfiltered) pre-cooled seawater (at 20-22°C) is filled in tank and aerated. About 2-3 l of mixed algal culture (1.5-2 million cells/ml) precooled at 20-22°C is given as feed twice a day and maintained at 10°C below ambient temperature.

Conditioned oyster are subjected to thermal stimulation by suddenly transferring them to sea with temperature 34-35°C. This sudden change causes spawning. The spawning oyster are immediately separated and placed in spawning trays having filtered seawater at ambient temperature. Remove oyster after spawning. Gametes are mixed and gentle aeration provided. Fertilization takes place immediately and first cleavage with 45 minutes. After four hours morulas stage is reached and swimming morula is siphoned and reared in filtered sea water. At the end of 20 hours the straight hinge or D-shaped larval stage is reached.

Stage	*Size (μ)m*	*Hours/Days*
Straight hinge	60-70	20 hours
Early umbo	100	3rd days
Mid umbo	150	7th day
Advance umbo	200-270	12th to 15th day
Eyed larva	280-290	13th to 17th day
Pedireliger	330-350	14th to 18th day

As larve grow, their density in rearing tanks is reduced and food ration increased. The schedule for rearing density and food requirements of larvae is given below:

Larval stage	*Density of larvae/ml*	*Algal cell concentration in ml/larval/day*
Straight-hinge	5	3000-4000
Umbo	3	4000-5000
Advanced umboto eyestage Pediveliger	2	5000-8000

Non toxic, hard, chemically, stable and clean spat collectors are used to produce attached spat oyster shells as spat collectors are widely used for the ren method of culture. Brush well the shell, washed in chlorinated water followed by soaking and repeated washing in seawater, so that the pH of water in rearing tanks is maintained. Shell collectors are spread uniformly on bottom of one tone FRP tanks containing filtered seawater. In addition several rows of shell rens are also suspended in settlement tanks to increase the surface area. The eyed oyster larvae are released in tanks at a concentration of 1 larvae/ml in well aerated seawater. In following few days larvae are set as spat on oyster shells. The concave side of shell usually has more settlement.

For production of clutchless spats (free spat or single spat) pretreated polythene sheet is spread as a lining on bottom and sides of a FRP tank. Oystershell grit of 0.5 mm size are wasted thoroughly sterilized in 10 ppm chlorine further washed in running filtered seawater and dried. These shell grit are spread uniformly at bottom of tank and larvae are released at setting stage. Rear spat for three weeks after setting in hatchery before they are transferred to field for nursery rearing.

Clam plays an important role in economy of coastal areas, their meat is cheap practised for culture in different countries.

The clam species belonging to families Arcidae, Veneridae, Corbiculidae, Solemidae, Mesodesmatidae Tellinidae and Donacidae, Arcid clam are called blood clams.

Paphia malabarica, Meretrix costa, M. meretrix Katelysia opima, Villorita cyprinoides, Tridacna crocea, T. maxima, T. squamosa. Hippopus hippopus, Anadra granosa, Tridacnid clams are well known. They filter feeder and their biology is similar to oyster as they are also a bivalve. Their bottom culture is quite popular.

The giant clam is potentially dangerous to man because the bivalve is extremely well camouflaged and have tremendous strength of single adductor muscle (which is considered a delicacy also). Once an object is caught between the two valves it can only be released again by either severing adductor muscle or breaking the shell.

In the early 1930s a crude pearl weighing 7 kg and measuring 23 by 15 by 14 cm was extracted from Tridacid clam in Philippines.

The Venus clams (family-Veneridae) are represented by some noteworthy forms in Europe and Americas. They are quite frequently jump a short distance with the aid of their foot. In the venus clams siphons are much longer and are partially fused, and externally these shells are quite distinct from cockles. Their shells are almost always weakly ribbed and lack spines and they are often decorated with dark brown bands. The watery venus (*Venus verrucosa* L-4cm) is one of the few ribbed species. European venus clam are characterized by distinct lamellar ribs which are arranged in concentrating the light brown coloration of shall values is often complemented by three or four interrupted bands which the concentric rings from inside to outside *Venus gallina* is somewhat smaller, numerous in North Sea. Where it is characteristics species of an entire animal ecosystem on the sandy ground at moderate depths. The edible *Callista chione* is equally abundant in Mediterranean. In contrast to true Venus clams, the Carpet shell (*Venerupis perforans*) is clearly more elongated and therefore its form is reminiscent of a blunt arcid bivalve.

The American Quahog (*Mercenaria mercenaria*) is another member of Venus clams. It was introduced to northwestern France from eastern coast of North America. The Algonguin tribes which formerly inhabited what is now the eastern USA, used to make beads out of shells which they pierced in middle and strung together into necklaces. These strings of shell beads also served as money which was known by Indian name of *Wampumpeag*. The Indians also strung these shell beads into very intricately designed belts known as wampums which were used as token to bind treaties.

The other candidate species with great economic importance are *Perna viridis* green mussel and *P. indica* brown mussel. The farming is done is coastal water beyond the surface zone at 10 to 15 meters depth in non-polluted area. The spawn of green mussels are reported in July-September. In July then juvenile (spat) form over the rocks and intertidial zones with density of 6550 per m^2. In one hour 18-20 kg spat can be collected. Spat collectors like coir ropes, any nylon string, roof tiles may also be used. in west coast of India during heavy monsoon in shallow water 10-15 m depth longline culture of mussel is practised. Loneline unit consist of 60 m long horizontal HPD rope of 20-40 mm thickness, anchored at both ends with 150 kg concrete blocks and a series of 100 litre capacity barrels as floats fixed at 3 m intervals. Vertical lines of 6 m length seeded with mussel spats are hung at a distance of 75 cm between two floats in main line. A longline unit of 60 × 60m can accommodate 12 horizontal ropes and 920 to 1,000 vertical ropes. The distance between two horizontal lines is 5 m. At every 20 m horizontal lines are connected using additional horizontal lines pole culture and stake culture are done in estuaries at a depth of 1.5-3 m. The spats of 15-20 mm. are wrapped around the poles or stake with cotton mosquito nettings. Spats get attached to poles in 72-96 hours and it is enough time for cotton net disintegration. Do periodic thining also. In long line culture two crops can be obtained.

Table Identification of *Paviridis* and *P. indica* taken within a year. The growth in open sea is very fast. The weight increases 6.0-7.4 gm. in one month and production reaches upto 10-13 kg. marketable size from one meter of rope. From a longline unit of 360m^2 a total production 54,720 kg shell on mussels can be obtained of which 40 per cent will be the meat (CMFRI Report TTS-I)

Bottom culture, stake culture, Rack bag culture Raft Culture.

Character	*P. viridis*	*P. indica*
1. External colour	Green	Dark brown
2. Mantle margin colour	Yellowish green	Brown
3. Ventral shell margin	Higher concave	Almost straight
4. Middle dorsal margin	Accurate	A distinct dorsal angle or lump present
5. Anterior end of shell	Pointed, break down turn	Pointed and straight
6. Number and size of hinge teeth	Two small teeth on left valve and one on the right	One large tooth on left value and corresponding depression one right value.

The mussels also have organ system like other bivalves. Its foot is like clams and small with little mobility. Mussel threads can be secreted more than one time. As mussels are filter feeder, phytoplankters are good for them. Sexes are separate and gonads located in body proliferate into mantle. They have plankto trophic larvae. Male gonads are creamy white in colour while in female it is pink or reddish.

At 4°C above the ambient gave results in inducing spawing, egg are brick red and measured 40-50 mm. morula size 58-60 (μm) in 4 hours, trochophore 65-70 (μm) in 7 hours. Early straight hinge 70-76 (μm) in 17-20 hours. Early umbo 120-140 (μm) 7 days, Eyed larvae 208-260 (μm) in 13-14 days. Prediveliger 280-320 (μm) 16 days onwards. Heavy spat settlement from 20 days onward.

Other molluses widely like cephalopods used for food purposes in fresh as well as processed form are cephalopods like octopus, sepia, loligo squids etc. Here only squid processing is described with cephalopods distribution (CMFRI report and MPEDA report).

IMPORT STANDARDS OF JAPAN

Form and shape. The form, trimmed or untrimmed, should be good, having no cuts, splits or any other wounds and having had the head and tentacled section, the cuttlebone, the fin and the outer skin removed.

Temperature. The temperature of the squid of cuttlefish should be -18°C.

Foreign substances. No foreign substance should be attached to or mixed with the fish.

Packaging. Packaging material and methods should sufficiently satisfy the quality and the use of foodstuff in question.

Sanitary inspection standards. Squid and cuttlefish should meet the following sanitary standards.

No. of bacteria	:	5 million or less per gram
E. Coli	:	Negative
Volatile basic nitrogen	:	26 mg or less per 100 grams

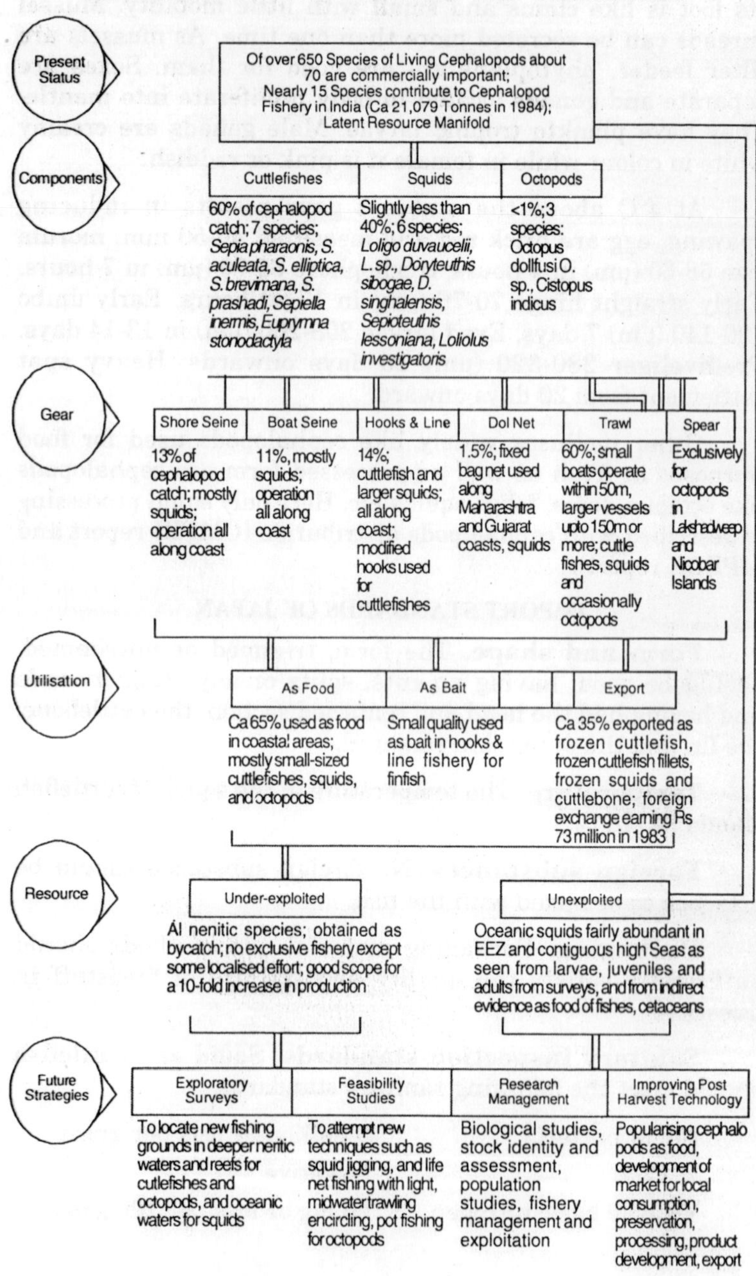
CEPHALOPODS: RESOURCES AND UTILISATION
Present Status
Of over 650 Species of Living Cephalopods about 70 are commercially important; Nearly 15 Species contribute to Cephalopod Fishery in India (Ca 21,079 Tonnes in 1984); Latent Resource Manifold
Components
Cuttlefishes
Squids
Octopods
60% of cephalopod catch; 7 species; Sepia pharaonis, S. aculeata, S. elliptica, S. brevimana, S. prashadi, Sepiella inermis, Euprymna stonodactyla
Slightly less than 40%; 6 species; Loligo duvaucelii, L. sp., Doryteuthis sibogae, D. singhalensis, Sepioteuthis lessoniana, Loliolus investigatoris
<1%; 3 species: Octopus dollfusi O. sp., Cistopus indicus
Gear
Shore Seine
Boat Seine
Hooks & Line
Dol Net
Trawl
Spear
13% of cephalopod catch; mostly squids; operation all along coast
11%, mostly squids; operation all along coast
14%; cuttlefish and larger squids; all along coast; modified hooks used for cuttlefishes
1.5%; fixed bag net used along Maharashtra and Gujarat coasts, squids
60%; small boats operate within 50m, larger vessels upto 150m or more; cuttle fishes, squids and occasionally octopods
Exclusively for octopods in Lakshadweep and Nicobar Islands
Utilisation
As Food
As Bait
Export
Ca 65% used as food in coastal areas; mostly small-sized cuttlefishes, squids, and octopods
Small quality used as bait in hooks & line fishery for finfish
Ca 35% exported as frozen cuttlefish, frozen cuttlefish fillets, frozen squids and cuttlebone; foreign exchange earning Rs 73 million in 1983
Resource
Under-exploited
Unexploited
Al nenitic species; obtained as bycatch; no exclusive fishery except some localised effort; good scope for a 10-fold increase in production
Oceanic squids fairly abundant in EEZ and contiguous high Seas as seen from larvae, juveniles and adults from surveys, and from indirect evidence as food of fishes, cetaceans
Future Strategies
Exploratory Surveys
Feasibility Studies
Research Management
Improving Post Harvest Technology
To locate new fishing grounds in deeper neritic waters and reefs for cutlefishes and octopods, and oceanic waters for squids
To attempt new techniques such as squid jigging, and life net fishing with light, midwater trawling encircling, pot fishing for octopods
Biological studies, stock identity and assessment, population studies, fishery management and exploitation
Popularising cephalo pods as food, development of market for local consumption, preservation, processing, product development, export

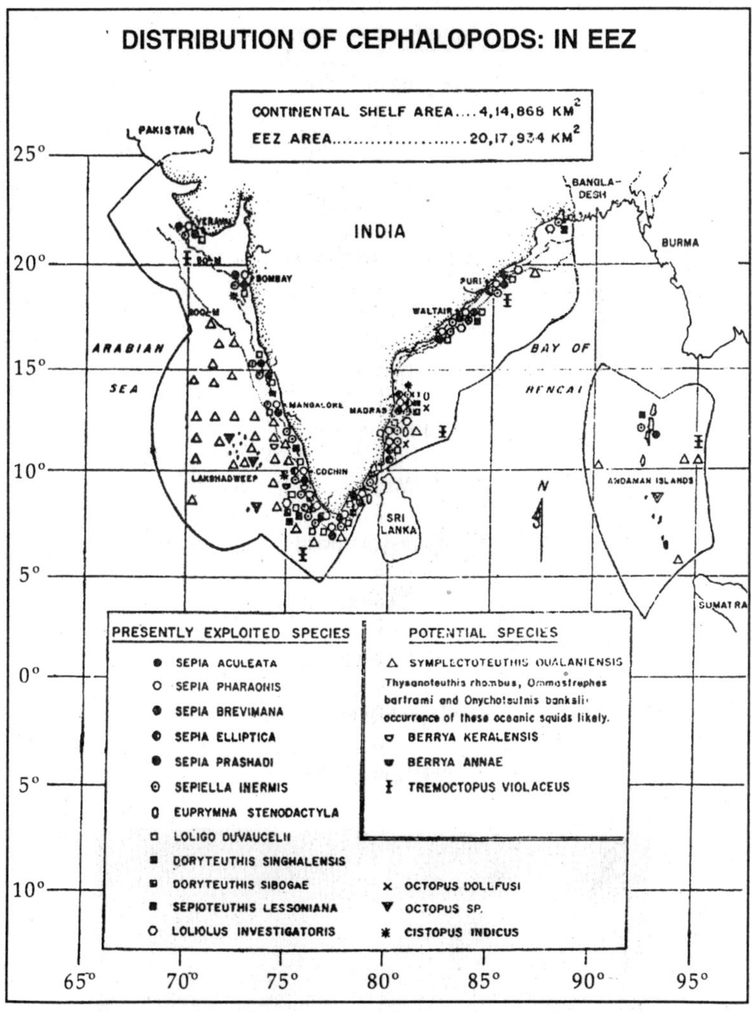

Occurrence of exploited and potentially important species of cephalopods in the Exclusive Economic Zone of India

Courtesy: CMFRI

Processing Flow Chart

Squid Head on (Whole)

Squid: Fresh or Frozen)

↙ ↓ ↙

Weighing ← Thawing

↓

Washing

↓

Cleaning
(Remove Viscera & Pen)

↓

Skinning Tube/Eye Out

↓

Washing

↓

Draining

↓

Packing in Trays

↓

Weighing

↓

Freezing (Air Blast)

↓

Sealing

↓

Labelling

↓

Packing in Master Cartons

↓

Storing at (–) 18ºC

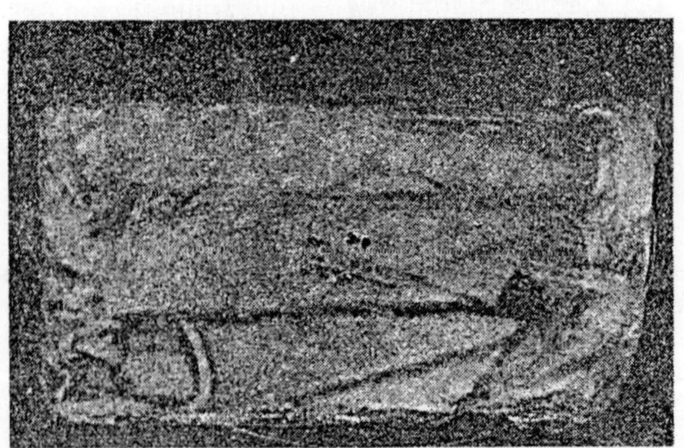

Squid Tubes

Squid: Fresh or Frozen)

↙ ↓ ↙

Weighing ← Thawing

↓

Washing

↓

Separation of Head
(Tentacles)

↓

Cleaning The Tube
(Remove Viscera and bone)

↓

Separation of Wings

↓

Skinning

↓

Washing

↓

Draining

↓

Packing in Tray

↓

Weighing

↓

Freezing (Air Blast)

↓

Sealing

↓

Labelling

↓

Packing in Master Cartons

↓

Storing (–) 18ºC

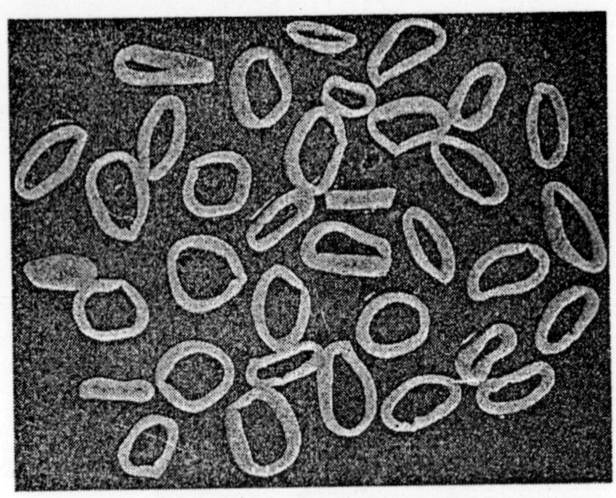

Squid: Fresh or Frozen

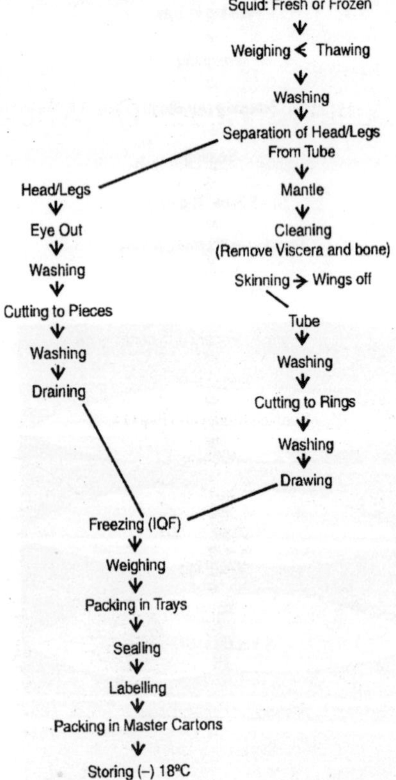

Stuffed Squid

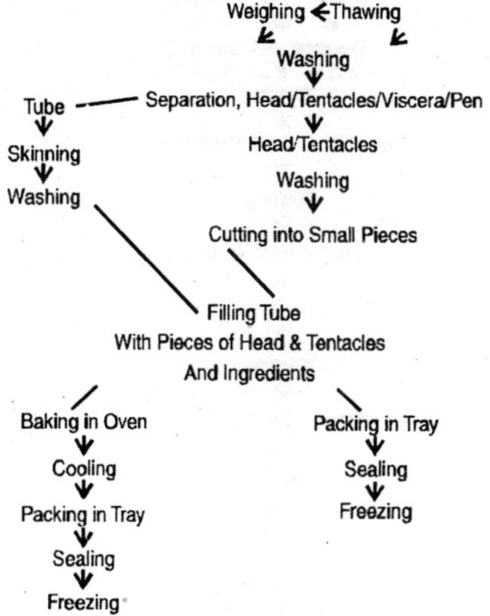

Cuttlefish Head - On Whole

Cuttlefish: Fresh or Frozen

↙ ↓ ↙

Grading ← Thawing

↓

Weighing

↓

Washing

↓

Cleaning

(Remove viscera and bone)

↓

Washing

↓

Skinning (Body) and Eye Out

↓

Washing

↓

Draining

↓

Packing in Trays

↓

Weighing

↓

Sealing

↓

Freezing (CPF)

↓

Labelling

↓

Packing in Master Cartons

↓

Storing at (–) 18°C

Cuttlefish Fillets

Cuttlefish: Fresh or Frozen
↓
Weighing ← Thawing
↓
Washing
↓
Separation of Head and Viscera
↓
Washing
↓
Separation of Wings
↓
Skinning
↓
Washing
↓
Filleting (Tube Opening)
↓
Washing
↓
Draining
↓
Packing in Trays
↓
Weighing
↓
Sealing
↓
Freezing
↓
Labelling
↓
Packing in Master Cartons
↓
Store at (–) 18ºC

EIA'S EXPORT SPECIFICATION FOR SQUID & CUTTLEFISH

Frozen Cuttlefish and Squid in any form shall be prepared by quick freezing from freshly caught wholesome cuttlefish and squid of a quality suitable for human consumption. The processing shall be carried out in a hygienic manner. Cuttlefish/ Squid of different varieties and products shall not be packed together. The freezing process shall be carried out in appropriate equipment in such a way that the range of temperatures of maximum crystallisation is passed quickly. The product shall be maintained under such conditions as to maintain the quality during storage, transportation and shipment. Prohibited chemical additives shall not be used at any stage of processing. Frozen cuttlefish and squid shall have the characteristic colour and odour. They shall not have any discolouration, objectionable odour and flavour. The material on thawing shall be clean, having an attractive appearance and shall be reasonably free from any defects and objectionable foreign matter.

Squid Products of Spain

(i) whole frozen squids

(ii) Block frozen tubes

(iii) IQF squids

(iv) Squid rings

(v) Frozen tentacles

Squid Products of Tapav

(i) Dried Squid product Surume

(ii) Seasoned Squid products Chinmi-ika

(iii) Fermented Squid Products ika-shiokara

(iv) Smoked Squid products ika kunsei

(v) Pickle squid meat sui-ika

Squids tops all other marine products in terms of ratio of edible to non-edible portions. The edible portion is 80-85 per cent of the weight. Nothing is wasted in processing such as viscera, liver, skin, pen, beak and eyes can be converted into squid meal. Its bone has demand in world market as its use in animal feeds, handicrafts, glass industry etc.

Bacteriological standards	*Cuttlefish / squid*	*Cooked and boiled*
1. Total plate count/gm.	Max. 5,00,000 Max. 2,00,000 (for products) meant for raw consumption)	Max. 1,00,000
2. E. coli count/gm	Max. 20	Nil
3. Coagulase positive staphylococcus/gm.	Max. 100	Max. 100
4. Salmonella & S. Arizona:	Negative	Negative

Beside this in class cephalopoda a beautiful silvery metallic coiled shelled animal *Nautieus* has great decorative and aesthelic importance.

REFERENCES

1. Advance Study of Molluscs by Jenkins. FEC (1992)
2. Algarswami, K and Dharmaraj (1984) Manual on Pearl Culture Techniques CMFRI Publication Cochin.
3. Algarswami K (1991) Production of Cultured Pearls 1-111 ICAR, New Delhi.
4. Anony mous, E (1991) Pearl Oyster farming and Pearl Culture Training Manual No. 8. Regional sea farming Development and Demonstration Project, Bankok.
5. Ballantine, D and Morton, J.E (1965) Filtering, feeding and digestion in the lamellibranch *Lasaea Rubra,* J. Marine, Biol. Assoc. U.K. 35, 241-274.
6. CMFRI Booklets on Oysters. (1995)
7. Elizabath Goling (1196) Bivalves Molluscs, Fishing News Books, Blackwell Science Ltd., Oxford, 200.
8. Eswaran C.R., K.R. Narayanan and M.S. Michael (1969) Pearl Fisheries of Gulf of Kutch, JBNHS 33(2); 338-344.
9. Hancock, D.A. (1973) Kuri Bay Pearls some of the finest in world Australian Fisheries 32(4); 11-12.

10. James and Narasimhan (1997).
11. Molluscs by P.S.B.R. James and K.A. Narasimham. (1997)
12. MPEDA House, Panampilly Avenue, P.B. No. 4272, Kochi-682035.
13. Narasimham, K.A. (1991) Present status of clam Fisheries of India. J. Mar. boil Ass. India 33(112) 77-88.
14. Pearl culture in Fresh Water Mussels Fresh Water Fisheries Research Station, Kalyani, W.B. (1994)
15. Ref: Malluscs by P.S.B.R. James and K.A. Narasimhan (1997)
16. Santhanam R, Ramanathan, N and Jegatheesan G (1990), Coastal aquaculture in India, CBH Publishers, New Delhi 92-107 and 117-124.
17. Squid and Cuttle fish—MPEDA Publications Cochin. (1982)
18. Thompson T.E (1958) The natural history embryology, larval biology and past larval development of gastropods. Phil. Trans. Roy. Soc B242 1-58.
19. Ward, F (1985) The Pearl, National Geographical 168(2); 1993-222.
20. Yonge, C.M. (1949) The Sea Shore, Collins London.

9

Phytomedicinal Research: Towards New Perspectives Based on Indigenous Knowledge System

Y.C. Tripathi and *Sushma Singh*

The medicinal uses of plants are probably one of the earliest of man's discoveries and the medicine man or witchdoctor was probably the first professional man in the development of human society. Since time immemorial, medicinal plants have been used as therapeutic agents for alleviation of human ailments throughout the world. The traditional medicine systems of all ancient cultures of the world had their root and origin in folk traditions and still continue to cater to the medicinal need of majority of population in the developing countries. The World Health Organization (WHO) estimated that 80% of the population of developing countries relies on traditional medicines, mostly plant drugs, for their primary health care requirements. More so, people are becoming disillusioned with many allopathic remedies, which cause side effects that are most often worse than the disease itself (Lalitha, 2000). Recently, there is a tendency towards traditional and more natural way of life in the entire world. Much awareness has been created towards the use of natural products and ecofriendly standards have become a part of consumers' demand. The resurgence of interest towards green medicines derived from the herbals throughout the world has led to the search for safe, sustainable and effective medicines for treating many dreaded diseases like asthma, rheumatism and gout, arthritis, paralysis, filaria, tuberculosis, diabetes, hepatitis, meningitis, cancer, AIDS etc. and is gaining momentum among the scientific community. Researchers and scientists are showing interest at global level in low priced, potential, sustainable herbal medicines for common people. Medicinal plants with promising traditional therapeutic efficacies (Table 1) represent an unparalleled source of molecular diversity to drug discovery and development, and are indeed complementary to those emerging sources.

HERBALISM: PAST AND PRESENT

The history of the use of plant in medicine systems can be traced back to the ancient civilizations or pre-Rigvedic times. The earliest written record of the preparation and use of plant-based medicines in India is the 'Rigveda', the earliest scripture of the Hindus (4500-1600 BC). The herbal of the Chinese emperor Sheng Noong dates from 2700 BC and it is known that in ancient Egypt, scores of herbal doctors made a living by treating their patients with measured doses of plant extracts. In '*Atharva Veda*', which was a later work in India followed by the monumental contribution like '*Charak Samhita*' (1000-800 BC) and Vagbhatta's '*Astanga Hridaya*' the uses of medicinal plants described are more varied. A number of books on Ayurveda, the traditional medicine system of India, were written by learned scholars such as Bhikshu Atreya, Patanjali, Nagarjun, Madhavakar, Chakradatta, Sarangadhar, Sankara and Bangasen (500-100 BC) who further expanded the vegetable materia medica of India. Another valuable contribution '*Kalpastanum*' or '*Vrikshayurveda*' gave detailed characteristics of vegetable drugs including their distribution, habitat, suitable soil for growth, season of collection, duration of efficacy and method of storage. In 400 BC, Hippocrates prescribed plant extracts for the treatment of sick. During the middle ages, the use of plant-derived medicines increased substantially and a number of them are still in use today. The remarkable progress of the Indian Medicine from Vedic to the period of Tantras and Siddhas declined with the invasions of Greeks Scythians, Huns, Moghals and Europeans. In addition, the dramatic rise of the synthetic chemical industry in the first quarter of the century (The Coal-tar era) caused an appreciable fall in these figures to 45% in 1936 (Mangelsdorf, 1972).

Table 1. Traditional therapeutic claims of medicinal plants

S.No.	Scientific name of the plants	Common name	Parts used	Traditional therapeutic claims
1	2	3	4	4
1.	*Abelmoscus moschatus*	Muskdana	Seed	Health tonic
2.	*Abrus precatorious*	Gunj	Leaves, Seed	Sore throat, Pneumonia
3.	*Acorus calamus*	Vacha	Rhizome	Nerve tonic
4.	*Acyranthus aspera*	Aglab	Whole plant	Scorpion bite, Toothache
5.	*Adhatoda vasica*	Vasaka	Leaves	Cough reliever
6.	*Aegle marmelos*	Bel	Fruit	Cooling, Ulcer, Stomach problem
7.	*Aloe vera*	Ghreet Kumari	Leaves	Burns, Cosmetic use
8.	*Andrographis peniculata*	Kalmegh	Whole plant	Liver tonic, Malaria
9.	*Asparagus racemosus*	Shatavar	Tuberous root	Female disorders
10.	*Azadirachta indica*	Neem	All parts	Fever, Skin problems
11.	*Bacopa monieri*	Brahmi	Whole plant	Memory improver
12.	*Boerhavia diffura*	Punannavan	Whole plant	Urinary problems, Blood purifier
13.	*Butea monosperma*	Palas	Flower, Gum	Cooling tonic, Post natal care
14.	*Calotropis gigantia*	Safed Aak	Leaves	Swellings
15.	*Cassia angustifolia*	Senna	Leaves, pods	Laxative, Purgative
16.	*Cassia fistula*	Amaltas	Fruit pulp	Stomach worm
17.	*Catharanthus roseus*	Sadabhar	Leaves, Flowers	Diabetes, Cancer
18.	*Centella asiatica*	Mandukapaini	Whole plant	Brain tonic
19.	*Chlorophytum borivelinum*	Safed Musli	Tuberous root	Aphrodisiac
20.	*Gloriosa superba*	Kalihari	Rhizome	Abortion, Joints problem, Fever

(Table Contd...)

Table 1. (Contd...)

1	2	3	4	4
21.	*Gymnema sylvestre*	Gurmar	Leaves	Diabetes
22.	*Mentha piperita*	Peppermint	Leaves	Anti-bacterial
23.	*Papaver somniferum*	Poppy	Flowers, Fruits	Sedative, Tranquilizer
24.	*Phyllanthus emblica*	Aonla	Fruit	Cough & Cold, Toothache, Tonic
25.	*Piper longum*	Pippali	Fruits, Root	Cough reliever
26.	*Plantago ovata*	Isabgol	Seed husk	Laxative
27.	*Plumbago zeylanica*	Chitrak	Root	Blood purifier, Skin problems
28.	*Pongamia pinnata*	Karanj	Seed	Skin diseases
29.	*Pogostemon cablin*	Ptchouli	Fragrant oil	Skin care
30.	*Pterocarpus marsupium*	Vijaysar	Heart wood	Diabetes
31.	*Rouwolfia serpentina*	Sarpgandha	Root	High blood pressure
32.	*Saraca asoca*	Ashok	Stem bark	Heart diseases
33.	*Solanum khasianum*	Kantkari	Fruit, Root	Conceptive, Stimulant
34.	*Terminalia arjuna*	Arjun	Stem bark	Fever
35.	*Terminalia bellerica*	Behera	Fruit, Seed	General tonic, Digestive
36.	*Terminalia chebula*	Harad	Fruit	Abdominal problems
37.	*Tinospora cordifolia*	Guduchi	Stem	Health tonic, Fever
38.	*Tribulus terrestris*	Gokhru	Whole plant	Renal problems, Stimulant
39.	*Valeriana officinallis*	Valerian	Root	Joints pain, Fever
40.	*Vitex negundo*	Nirgundi	Leaves	Leucoderma, Malaria
41.	*Withania somnifera*	Ashwagandha	Root	General Tonic, Inflammation

The traditional medicine system attracted attention of various scholars both in India and abroad during the 19th century. Based on the knowledge of plants gathered from various ancient medical writings, folklores, cultures and civilizations, observation during field exploration and consultation with folk medicinemen, research on medicinal plants has been undertaken all over the world for development of new herbal drugs. Modern pharmacopoeia contains atleast 25% drugs derived from plants and many others are synthetic analogues built on prototype compounds isolated from plants. As research activity on the isolation of the active principles from reputed medicinal plants was being carried out, synthetic drugs were made available with the advent of synthetic organic chemistry during 1940-1950. Plant based drugs were included in the pharmacopoeias of Russia, Japan and Germany. By 1820, the US Pharmacopoea listed 650 drugs of which 455 (70%) were of plant origin. In USA, 25 % of the total prescriptions dispensed from the community pharmacies were found containing one or more products of natural origin as therapeutic agents during 1959-1980. During the year 1980, consumers in USA paid more than 8 billion USD for prescriptions containing active constituents from medicinal plants. Japanese Pharmacopoeia (1986) contains 123 plant drugs, both crude and active principles, of which only 29 are used in the Western medicine. During the period 1950-1970, about 100 new basic drugs were introduced in USA, which contained only five drugs, namely, reserpine, deserpidine, rescinnamine, vinblastin and vincristin derived from higher plants. Again, during the next 20 years i.e. from 1971-1990, over 600 new drugs were introduced throughout the world but the plant derived drugs including those obtained by chemical modification of the structural moieties were not again more than 2%; important drugs were teneposide, etoposide, d-9-tetrahydrocannabinol, nabilane, artemisinine, (Z-) and (E-) guggulsterones and ginkogalides (Sukhdev, 1997, Camido *et al.*, 2000). Subsequently during the period 1991-1995, approximately 200 chemical compounds were introduced as drug out of which only four viz.; paclitaxel (Taxol), irenotecan, topotecan and gomishin (Farnsworth, 1990) were from higher plants. As per the data from global survey, it has been found that at least 130 chemicals of definite structure, all single entities extracted from higher plants or modified further synthetically are currently in use (Sukhdev, 1997, Farnsworth, 1990). A recent study indicates that approximately 60% of the anti-tumour and anti-infective agents that are commercially available or in late stages of clinical trails have plant origin. Other studies carried in the last 10-15 years have also demonstrated that plants represent an unparallel source of molecular diversity for drug development. India's contribution has attained some global visibility with respect to its self-reliant technical status. The prioritization of diseases related search for new therapeutic agent might be different for developed and developing countries.

DEVELOPMENT OF HERBAL DRUGS

The biological activities of plants are mainly due to the presence of a number of chemical constituents known as secondary metabolites, biosynthesized in plant tissues and utilized for their own growth and development. Such constituents, individually or in groups, can have specific action on the different body organs. Attempts have been made to analyze a number of medicinal species for their active constituents. Different methods have, therefore, been devised for extraction, structure determination, pharmacological testing for verification of their traditional therapeutic claims and synthesis. Plant chemical constituents play a dominant role in pharmaceutical sector and systematic investigation of plant resources for the discovery of new drug molecules has been a primary objective of modern pharmaceutical and clinical programmes (Gupta, 1994). The natural product research as a part of drug discovery efforts faces increasing challenge of improving its competitiveness with those synthetically produced. The search of new molecules has obviously led to drug development and a better understanding of unique and complex metabolic phenomena of living organisms. The discovery and introduction of reserpine, an alkaloid obtained from the roots of the Indian plant *Rauwolfia serpentina,* as drug for the treatment of hypertension and as a tranquilizer (Sukhdev, 1997) was considered as a revolutionary event. Similarly other bioactive plants with extensive folklore histories have yielded the cardiac (*Digitalis*) glycosides, the cannabinoids, the opiates (codeine, morphine), the cinchona alkaloids (quinine, quinidine), the belladonna alkaloids (atropine), pilocarpine, nicotine, ephedrine, cocaine, theophylline, vincristine, vinblastin and other useful drugs (Tyler *et al.*, 1986). Certain biologically active plant chemical constituents have served as a model compounds for drug synthesis and semi-synthesis. Natural compounds of pharmaceutical importance that were once obtained from plant sources, but now produced synthetically on large scale include caffeine, theophylline, theobromine, ephedrine, pseudoephedrine, emetine, papaverine, L-dopa, salicylic acid and tetrahydrocannabinol. Natural plant products serve additionally as chemical models for the design and total synthesis of new drug entities viz.; the belladonna alkaloids (atropine), physostigmine, quinine, cocaine, the opiates (codeine and morphine) and salicylic acid have served as model for the design and synthesis of anticholinergics, anticholinesterases, antimalarial drugs, benzocaine, procaine, lidocaine (xylocaine) and other anaesthetics, pentazocine (Talwin), propoxyphene (Darvon), methadone, meperidine (Demerol) and aspirin (acetylsalicylic acid) respectively (Goldstein *et al.*, 1974, Gund *et al.*, 1980). Important plant derived active principles plants (Table 2) and used in modern medicine include quinine atropine, caffeine, morphine, chymopapain, nicotine, morphine, codeine, hyoscyamin, papaverine, berberine, colchicoside, digitoxin, emetin, ephedrine, strychnine, gomishin, quinidine, quinine, reserpine, sennoside, taxol, vinblastine, vincristin, tenoposide (Tripathi, 1998). These drugs played an inestimable role in the treatment of human diseases. Today more than 20 best selling pharmaceuticals all over the world are derived from plants. However, the number of new chemical compounds from higher plants, which found entry as drug, has been extremely low.

CHEMO-PHARMACOLOGICAL EVALUATION

The recent discovery of remarkable medicinal properties of several plant species gave a new impetus to ethnomedicines and during the last three decades, considerable works have been done on folk medicines. Though a number of medicinal plants are being used to bring about traditional cure of various diseases, many of them are yet to be properly investigated

Table 2. Important active constituents and their biological activities

S.No.	Active constituents	Source plant	Biological activity
1	2	3	4
1.	Ajmalicine	*Catharanthus roseus*	Hypotensive
2.	Allin	*Allium sativum*	Anti-ulcerogenic
3.	Andrographalide	*Andrographis peniculata*	Hepatoprotective
4.	Artemisinin (Arteether)	*Artemisia annua*	Anti-malarial
5.	Asiatic acid, Asiaticoside	*Centella asiatica*	Braintonic, Psychotropic
6.	Atropine	*Atropa belladonna*	Parasympatholenic
7.	Baccosides	*Bacopa monneri*	Memory improver
8.	Berberine	*Coptis japonica*	Antifertility
9.	Boswellic acid	*Boswellia serrata*	Arthritis
10.	Camptothecein	*Mappia foetida*	Anti-cancer
11.	Cocaine	*Erythroxylin coca*	Local anaesthetic
12.	Codeine, Morphine	*Papaver somniferum*	Sedative, Analgesic
13.	Conescine	*Holorrhena anti-dysentrica*	Anti-dysentric
14.	Curcumin	*Curcuma longa*	Anti-inflammatory
15.	Digoxine, Lanatoside	*Digitalis lantana*	Cardiotonic
16.	Diosgenin	*Dioscorea deltoides*	Antifertility
17.	Etoposide, Teneposide	*Podophyllum emodi*	Anti-cancer
18.	Ginseniside	*Panax Ginseng*	Adaptogenic
19.	Gingkolides	*Gingko biloba*	Circulatory disorders
20.	Gomoshin	*Schizandra chininsis*	Hypatoprotective

(Table Contd...)

Table 2. (Contd...)

1	2	3	4
21.	Gossypol	*Gossypium spp.*	Male contraceptive
22.	Glycyrrhotic acid	*Glycyrrhiza glabra*	Anti-inflammatory
23.	Guggulsterone	*Commiphora mukul*	Hypocholesterolenic
24.	Gymnemic acid	*Gymnema Sylvester*	Anti-diabetic
25.	Hydroxycitric acid	*Garcinia cambogia*	Obesity treatment
26.	Hyosyamine	*Hyoscyamus muticus, D. metel*	Parasympatholenic
27.	Irenotecan, Topotecna	*Nathopodytis nimmoniana*	Anti-cancer
28.	Paclitaxel, Texatere	*Taxas brevifolia, T. wallicihina*	Anti-cancer
29.	Picrosides	*Picrorhia kurroa*	Hepatoprotective
30.	Psoralin	*Psoralia corylifolia*	Anti-bacterial
31.	Reserpine	*Rauwolfia serpentina*	Hypotensive
32.	Sennoside	*Cassia angustifolia*	Laxative
33.	Shatavarin I-IV	*Asparagus racemosus*	Abortifacient
34.	Scillin A & B	*Urginia indica*	Cardiotonic
35.	Silymarin	*Silybum marianum*	Hepatoprotective
36.	Taxol	*Taxus brevifolia*	Breast cancer
37.	Tenosporin	*Tinospora cordifolia*	Circulatory disorders
38.	Vinblastin, Vincristin	*Catharanthus roseus*	Anti-cancer
39.	Valepotriates	*Valeriana wallachi*	Sedative, Tranquilizer
40.	Vascine	*Adhatoda vasica*	Bronchodilator, Oxytocic
41.	Withanolodes	*Withania somnifera*	Adaptogenic
42.	Yuechukene	*Murraya paniculata*	Anti-fertility agent

and scientifically validated. In spite of these numerous past success, it has been estimated that only 5-15% of the 250000-500000 existing species of higher plant have been surveyed for biological activity (Spjut, 1985). In recent years, interest of scientific community in pharmacognosy and natural product research has increased rapidly in search of new molecules leading to the development of new herbal remedies and for validation of traditional medicines. Rapid technological advancement and increasing demand of traditional plant based drugs; phytochemical, pharmacological studies have been conducted on some of the medicinal species to elaborate new scientifically tested recipes for more effective treatment of specific ailments. The studies not only scientifically verified their traditional medicinal claims but also discovered newer biological activities not claimed traditionally.

Chemical analysis of *Adhatoda vasica* commonly known as Vasaka, traditionally regarded as an expectorant, bronchodilator, and used in relieving cough yielded active constituents Vasicine, vasicinone and 6-oxyvasicine. In a study, it has been shown that antiasthmaKada prepared by mixture of *Adhatoda vasica*, *Glycyrrhiza glabra*, *Solanum xanthocarpum*, bark of *Albizzia lebbeck* and roots of *Picrorhiza kurroa* protects asthma patients by its immunomodulatory activity and mast cell stabilizing property (Iyenger *et al.*, 1994). Recent pharmacological studies revealed that the active constituent vasicine possess oxytocic properties similar to oxytocine and methyl ergometrine (Atal, 1980). *Aegle marmelos* (Bel) investigated for a variety of chemical constituents viz., glycoprotein, alkaloids, coumarin, carbohydrates, sterols, fatty acids etc. Furanocoumarin obtained from the plant exhibited tyrosinase accelerating and tryptophan inhibiting effects (Chakraborty, 1978). The seed oil showed antibacterial activity (Banerjee and Kumar, 1980). Its root yields marmin, which exhibited anti-inflammatory effect in carcinogenic induced paw oedema in rats (Pitre and Srivastava, 1987). *Albizia lebbeck* has been used in traditional system of medicine for various ailments. The use of plant for allergy, cough, bronchitis, snakebite etc. has been described in Ayurvedic system of medicine. The antiallergic and antianaphylactic activities of bark decoction of A. leebeck have been established pharmacologically (Tripathi and Das, 1977). It also showed abortifacient action through the release of prostaglandins. The dried alcoholic extract of the plant *Allium cepa* prepared at low temperature showed anti-asthmatic action (Handa *et al.*, 1983). *Asparagus racemosus* is being employed as a general tonic, nutritive, antiulcerogenic, antispasmodic, galactegogue and in the treatment of jaundice in Ayurveda, Siddha and Unani Systems of Medicines. Some of the claims have already been validated by pharmacological screening (De *et al.*, 1997, Mandal *et al.*, 1998, Rege *et al.*, 1999). Root extract of the plant found to alter the function of macrophages, indicating its immunomodulatory properties (Dhuley, 1997). The drug has also galactogogue property and its petroleum ether extract showed diuretic activity. Roots of *Aristolochia indica* has been found effective postcoital contraceptive in both rats and hamsters (Che *et al.*, 1984). Aristolic acid isolated from the plant is an antifertility agent, which is non-mutagenic to bacterial system (Chakraborty *et al.*, 1987). The bitter principles (limonoids) of *Azadirachtra indica* (Neem) exhibit a variety of biological activity. Nimbin found to inhibit the stimulation produced by histamine. Nimbidine is reported to be active against eosinophilia with marked symptomatic relief. It has also shown potent anti-inflammatory activity (Pillai and Santhakumari, 1984). Recent studies revealed that neem oil constituents not only reduce fertility but also prevent implantation of embryo (Sinha *et al.*, 1984, Upadhyay *et al.*, 1990). Root of *Boerhavia diffusa* is regarded as diuretic, laxative, expectorant, stomachic and is used in asthma, oedema, anaemia, jaundice, ascites, anasarca etc. The aqueous extract of root showed broad spectrum antiviral activity (Verma and Awasthi, 1979). The root extract reduced the inflammatory changes as well as abscess formation in the kidney of animals infected with *Escherichia coli*. It also reduced the bacterial count in the urine samples of infected animals (Singh *et al.*, 1988). Isobutrin and butrin, the antihepatotoxic principles isolated from the flowers of *Butea monosperma* were found effective in various models of liver damage (Wagner *et al.*, 1986). Butin isolated from the seeds showed antiimplantation activity in pregnant rats (Bhargava, 1986). The glycosidic principles, Sennoside A & B isolated from the leaves of *Cassia angustifolia* have been found responsible for laxative action of the plant (Kobashi *et al.*, 1980). Juice of the leaf contains β–sitosterol, which showed anticancer properties (Perdue and Hartwell, 1976). Another leaf constituent, rhein has been found to be an antibiotic agent against Staphylococus aureus (Perry, 1980). *Cassia fistula* has been reported to possess hepatoprotective (Bhakta *et al.*, 1999a), antitussive (Bhakta *et al.*, 1998) and anti-inflammatory (Bhakta *et al.*, 1999b) properties. *Centella asiatica* (Brahmi) is a well-known brain tonic. A triterpene saponin has been isolated from the plant and is found to be effective in leprosy and different types of skin troubles. The saponin constituents, asiaticiside, madecassoside, thakunoside and isothakunoside have been reported from the plant. Ethanolic extract of the plant *Calotropis procera* showed 60% growth regression when applied to cancer ulcers (Bhatnagar and Verma, 1986). An ethanolic extract of the flowers showed anti-inflammatory, antipyretic, analgesic and antimicrobial activities (Moscalo *et al.*, 1988). The oleo gum resin obtained from *Commiphora wightii* is regarded as astringent, antiseptic, expectorant, demulcent, aperient, carminative, antispasmodic, emmenagogue and is used in leprosy, rheumatism, syphilitic disorders, nervous diseases. The steroidal constituent Z-gugglusterone isolated from the oleo gum resin showed a strong thyroid stimulating action in albino rats (Tripathy *et al.*, 1984). Essential oil from gum resin showed a depressant action on the isolated frog heart (Bagi *et al.*, 1985). The active constituent gangetin (pterocarpen) isolated from the roots of *Desmodium gangeticum*, one of the ingredients of a very popular traditional drug 'Dasmoolarishta' has been found to possess promising anti-inflammatory activity. Chemical analysis of leaves of *Ginko biloba*, known as 'living fossil' showed the presence of biflavonoids, ginkgetin, amentoflavone and some other simpler flavonoids. The biflavone amentoflavone showed anti-bradykinin action and found to increase the blood flow in vessels from 20-30%. *Fumaria indica* (Pitpapra), a well-known medicinal herb, has been chemically investigated for a number of alkaloids. The alkaoidal constituents were reported to possess central nervous system depressant and anti-inflammatory activities (Tripathi and Dwivedi, 1990). *Panax schinginseng* (Ginseng), a well-known drug of East Asian countries, is known for its properties of longevity. The active constituent ginsenosides belonging to two sapogenins 20-S-protopanaxadiol, 20-S-panaxatriol and of oleanosides have been isolated from this plant. Recent pharmacological studies confirmed its tranquilizer, sedative and stimulant properties. It is a tonic, reduces blood sugar and acts as adaptogen. Glycosides namely picroside I & II, kutkosides and another constituent apocyanin have been isolated from the plant. In recent pharmacological investigations and clinical trials, the drug has been found effective in different types of jaundice and especially in hepatitis. It exhibits cholagogue and choeretic action and is useful in infective and amoebic type of inflammation in liver. From the benzene extract of *Pueraria tuberosa* (Vidarikand) a new pterocarpen, tuberosin was isolated which was found to possess anti-staphylococal, anti-tubercular and anti-fungal activities. Petroleum ether extract and its sub fraction showed significant oestrogenic activity (CCRIMH, 1974-75). Its anti-fertility action claimed in folk tradition has also been confirmed. Chemical analysis *Salacia* spp (Saptrangi) resulted in the isolation of six closely related triterpenes of friedelane derivatives from *S. prinoides*. Alcoholic extract of the drug produced hypoglycemic effect in experimental animals and man. Betuline aldehyde isolated from *Tectona grandis* has been reported to have antitumour activity (Pathak *et al.,* 1988). The cardiac glycoside named Peruviside (Arora and Rangaswamy, 1972) isolated from the plant *Thevetia nerifolia* having high degree of therapeutic index as compared to digoxin has been found to be useful in insufficiency of heart or weak heart. Asclepin, another cardiac glycoside isolated from *Asclepias curassavica,* has been found to possess cardiotinic property. Leaves of *Tylophora asthamatica* have been found to contain alkaloidal constituents viz., tylophorine, tylocrebine and tylophoridine. All these three alkaloids showed significant anti-cancer activity. The total alkaloids as well as pure constituent tylophorine also showed beneficial effects in giving relief to patient in allergic arthritis and asthma. *Nelumbo nucifera* has been reported to possess hypoglycaemic antidiuretic, antidiarrhoel, antibacterial and anti fungal properties (Mukherjee *et al.*, 1995). *Tinospora cordifolia* has been traditionally used in general debility, fever, jaundice, skin disorders, rheumatism, dyspepsia and urinary diseases. A compound 1, 2 substituted pyrrolidine obtained from the ethol extract of stem showed central nervous system depressant and hypoglycemic activities in rabbit (Mahajan and Jolly, 1985). *Trichopus zeylanicus*, regarded as health tonic and had been used by Kani tribes of Kerala for instant stamina and vitality. It is also claimed by the tribals that a regular intake of the kernels and certain other preparations made of the leaf, stem and rhizomes would enable one to enjoy youthful health. Pharmacological studies have established the antifatigue, stimulant, Hepatoprotective and antistress properties of the herb (Pushpangadan *et al.*, 1988, Sharma *et al.*, 1989). Alcoholic extract of *Vitex negundo* contributed significant anti-inflammatory activity on acute inflammatory reaction in albino rats (Varshneya and Teleng, 1995). Withaferine A, the major withanolide constituent, has been found responsible for the bacteriostatic and anti-tumour properties of the plant *Withania somnifera* (Ashwagandha). Withaferin A and Withanolide D showed significant immunosuppressive activity. The new acyl steryl glucosides, sitoindoside VII and sitoindoside VIII isolated from the roots of the plant found to exhibit anti-stress activity, which validates the claim of the plant widely reputed as the Indian Ginseng (Bhattacharya *et al.*, 1987). Ethanolic extracts of *Pongamia pinnata* (seeds), *Delonix regia* (bark and flowers), *Psidium guavajava* (leaves) and *Aegle marmelos* (bark) possess significant anti-inflammatory activity. *P. pinnata* (seeds) and *Delonix regia* (bark and flowers) possess analgesic activity (Muruganandan *et al.*, 2000). Saponins were isolated from the bark of *Zizyphus rugosa* as a result of chemical screening which have been found to have cardiovascular properties (Acharya *et al.*, 1988).

Some of the pure individual constituents isolated from more than one plant and have wider distribution. They have been isolated from several species either belonging to the same family or different families. Their pharmacological studies have also been done and some of them are being used in the production of drug.

Catechin, the flavonoidal constituents of a number of species like *Pterocarpus marsupium*, *Acacia catechu*, *Artocarpus integra*, Oak and beech trees are considered as promising anti-ulcer agent. The active constituents, namely, ailanthone from *Ailanthus glandulosa*, 18-hydroxy quassin from *Quassia amara* have shown to be potent anti-amoebic. These constituents and some other compounds, namely, bruceantin, glucarubinone, dehydroailanthion and 2-acetyl glucarubinone have been found to possess anti-cancer activity. Digitoxin and Digoxin are the cardiotonic steroidal glycosides derived from *Digitallis* spp, *D. purpurea* and *D. lantana*. Vernolepin isolated from *Vernonia hymenolepsis* and *V. guinesis* is a sesquiterpene lactone, which showed *in vitro* anti-tumour activity against the Walker 256 sarcoma. Gossypol, the well-known polyphenolic chemical constituent of cotton seed which is also found in the plant *Montezuma speciosissima,* showed in vivo anti-cancer activity. It has also been investigated as an anti-spermatogenic drug for fertility control. Camptothecine, isolated from the plant *Camptotheca acuminata* and found in Indian Species *Mappa foetida* has marked anti-leukemic and anti-tumour activity. The other constituents of *M. foetida* are 9-methyl camptothecine and mappicine. 9-methoxy-camptothecine showed anti-cancer activity. Maytansine and other related maytanoid esters obtained from *Maytenus ovatus* and *M. bushmanii* and *Colubrina teonensis* have been found to exhibit strong anti-leukemic property.

OUTSTANDING CONTRIBUTIONS

Many plant-based drugs, which are introduced in modern medicine, have been considered as outstanding contribution to drug discovery. Some of the important drugs of this category alongwith their biological activity are mentioned here.

Anti-cancer Drugs

A wide range of higher plant yield compounds exhibiting anticancer activity. Until 1960, only a few active and well characterized compounds had been isolated from plants. These include podophyllotoxin and related lignans from *Podophyllum peltatum* (mandrake), colchicines derivatives from *Colchicum autumnale* (autumn crops) and vinblastin and vincristin from *Catharanthus roseus* (*Vinca rosea*). The bisindole alkaloids vinblastine and vincristine obtained from the leaves of *Catharanthus roseus* are potent anti-cancer agents used for the treatment of Hodgkin's diseases, lymphosarcoma and leukemia in children (De Conti and Creasey, 1975). Another anti-cancer compound obtained from *Podophyllum emodi and P. peltatum* is podophyllotoxin, a lignan (Camido *et al.*, 2000). Teniposide and etoposide are semisynthetic compounds developed from podophyllotoxin. They are currently being used for the treatment of testicular cancer, small cell lung cancer and lymphomas (Sukhdev, 1997, Camido *et al.*, 2000). Paclitaxel, a diterpenoid constituent from several *Taxus* species, including *T. wallichiana* growing in some regions in India is a valuable anti-eoplastic compound for the treatment of metastatic ovarian cancer, metastatic breast cancer and malignant melanoma (Sukhdev, 1997). Taxol, a toxic diterpene alkaloid from *Taxas brevifolia,* has been shown to prevent replication in mammalian cell. It has also been reported to have anti-cancer activity (Rowinskey *et al.*, 1990). Other important anti-cancer drugs are irenotecon and topotecon, which are semisynthetic, compounds obtained from the quinoline alkaloid camptothecine, a constituent of the Chinese tree *Camptotheca acuminata* (Sukhdev, 1997). Comptothecine is now mostly obtained from the Indian tree *Nothapodytes nimmoniana* Syn *Mappia foetida* (Govindachari and Viswanathan, 1972). Both irenotecan and topotecan were introduced as drug in Japan for the treatment of various types of cancer such as lung, ovarian and cervical cancers (Sukhdev, 1997).

Anti-malarials

Malaria continues to be major health problem in many areas of the world. Parasite resistance to chloroquine and mefloquine has been rapidly increasing in both degree and prevalence throughout the world. Presently artemisinine and its derivatives are receiving considerable attention because of their activity against resistant stains of *Plasmodium falciparum* and efficacy against cerebral malaria. Artemisinin (Quinghaous), a sesquiterpene peroxide obtained from the Chinese medicinal plant *Artemisia annua,* is a potent anti-malarial active against both chloroquine sensitive and chloroquine resistant strains of *Plasmodium falsiparum* and *P. vivax.* It is also effective against cerebral malaria (Ziffer *et al.,* 1997).

Hepato Protective

Gomishin, a lignan obtained from the fruits of the Chinese herb *Schiozandra chinansis,* was found to have hepatoprotective activity and used for the treatment of chronic hepatitis (Sukhdev, 1997). Another hepatoprotective drug is silymarin, a mixture of three flavanolignans obtained from the fruits of *Silybum marianum* (Shah, 1982). Silymarin has emerged as one of the effective hepatoprotective agents used for the treatment of various liver disorders and there is considerable demand for this drug in USA and European countries. Although there are several plants used in traditional medicine for treatment of liver disorders like *Phyllanthus amarus* (Mehrotra *et al.*, 1991), *Picorrhiza kurroa* etc (Dwivedi *et al.,* 1990) and several phytoconstituents were found to have anti-hepato activity, there appears to be no approved drug from these plant sources. *Sida cordifolia* is believed to ameliorate bilious and liver disorders (Rao and Mishra, 1998).

Hypocholesterolemic

(Z-) and (E-) guggulsterones obtained from the gum exudates of the Indian medicinal plant *Commiphora mukul* were found to have significant hypocholesterolemic activity (Nityanand and Kapoor, 1973, Patil *et al.,* 1972). A standardized ethyl acetate extract of the resin containing at least 4% guggulsterones, code named guggulipid was introduced in medicine (Sukhdev, 1997). The drug has been manufactured and marketed in India since 1987 and is being sold in the international market now.

Anti-fertility Drugs

In the ancient Indian literature, several plant preparations have been reported to possess fertility regulating properties. A large number of such plants have been tested for their anti-fertility activity in laboratory animals (Hiremath and Hanumantha Rao, 1999, Hiremath *et al.,* 1999). However, to date no single plant product is acceptable as a potent anti-ovulatory agent. The anti-fertility effect may be due to anti-ovulatory, anti-implantational or abortifacient activities. Anti-ovulatory activities of the seeds of *Nelumbo nucifera* (Mazumdar *et al.*, 1992), flowers of *Hibiscus rosa-sinensis* (Murthy *et al.*, 1997) and many other plants have been reported. Similarly, the leaves of *Ocimum sanctum* (Moya *et al.*, 1976), whole plant of *Striga densiflora* (Hiremath *et al.*, 1996) and *Striga orabanchioides* (Hiremath *et al.*, 1999) have shown anti-implantational activities. Abortificient activities of entire plants of *Acyranthes aspera* (Garg, 1976) and *Rubus* spp. have been investigated (Dhanbal *et al.*, 1999).

Thyroid Dysfunction

Thyroid hormones primarily regulate the body metabolism by managing energy production. Therefore, any drug of plant origin regulating thyroid function is likely to ameliorate thyroid dysfunction, both hypo- and hyper-thyroidism. Ethanolic extract of *Convolvulus pluricaulis* (Panda and Kar, 1997), *Withania somnifera*, *Bauhinia purpurea* (Panda and Kar, 1999), *Achyranthes aspera* (Tahiliani and Kar, 2000a) and *Moringa oleifera* Tahiliani and Kar, 2000b) have shown thyroid regulatory properties in pharmacological studies.

Anti-microbials

The emergence of multiple drug resistant bacteria has created immense clinical problems in the treatment of infectious diseases in normal, immuno-compromized and cancer patients. Multiple drug resistant *Staphylococcus aureus* continues to be a major cause of serious infection in humans both in hospitals and in community (Mehmood, 1999). Candidiasis and dermatophytosis are the most common forms of fungal infections in many countries (Odds, 1988, Ribbon, 1988). During the last two decades, the incidence of candidiasis and other fungal infections has increased, especially in immuno-comprosized hosts (Cohen *et al.*, 1991, Diamond, 1991, Rindal, 1991).

Recent advances in the field of pharmacology, especially in respect of anti-microbial properties of medicinal plants have aroused hopes for an alternative source of anti-microbial drugs

(Desta, 1993). A number Indian medicinal plants in common use have been screened for their antimicrobial properties (Ahmed *et al.*, 1998). Essential oil of plant origin and their components have been claimed to possess antibacterial (Jain *et al.*, 1974, Baratta *et al.*, 1998) and other antimicriobial properties (Singh *et al.*, 1992). Essential oil from leaves and rhizomes of *Zingiber chrysabthum* has been found to be antimicrobial (Hammer *et al.*, 1999). Leaves and barks of *Azadirachta indica* are reported to have antimicrobial properties (Ahmed, 1995). Antibacterial property of medicinal plants like *Curcuma longa* (Iyenge *et al.*, 1995) and *Pseudomonas solanacearum* (Poswal and Werner, 1997) has also been reported. Aqueous or ethanolic extract of leaves *Aloe vera* showed antibacterial properties (Levin *et al.*, 1988). Benzene extract of *Echinops echinatus* showed significant bacterial and fungicidal effects (Sharma and Mehta, 1989). Dimethoxy benzoquinone and aloe-emodine isolated from roots; questin, isotoralactone and torasachrysone from seeds of *Cassia obtusifolia* (Kitanaka and Takido, 1986) and triterpenoids isolated from *Terminalia belerica* (Nandy, *et al.*, 1994) showed antimicrobial activities. The alkaloidal constituent, tiliacorinine isolated from *Tiliacora racemosa* has showed antifungal activity in pathological studies (Tripathi and Dwivedi, 1989).

It is hoped that scientific evaluation of many potential medicinal plants extracts in this direction may provide safe and broad spectrum herbal formulations or new anti-microbial compounds, which may be exploited in the management of infectious diseases in both modern and traditional systems of medicines. Therefore, new drugs and their formulations have become a vital need of the time.

Protective Medicines

A delicate balance of cellular anti-oxidant defense systems is essential to protect the body against oxidative stress. Oxygen free radicals induced cellular damage have been implicated in many pathobiological conditions including malignancies, neurodegenerative diseases, diabetes mellitus, cardiovascular diseases, cataract, liver damage, bacterial and parasitic infections, lung damage, inflammatory diseases, reperfusion injury, cerebrovascular damage, atherosclerosis and aging

(Nadkarni *et al.*, 1991). Several studies have been carried out on anti-oxidant and the mechanism by which they render protection from oxidative damage. Extracts of a number of plants have shown to have anti-oxidant activities. Ascorbic acid (Vitamin C) is a strong antioxidant, which is protective against oxidative damage in Tissures (Sies *et al.*, 1992). Plants like *Azadirachta indica, Ocimum sanctum, Allium sativum* and *Adhatoda vasica* have been found to be good source of ascorbic acid. The therapeutic efficacy of herbal formulations may be due to the anti-oxidant principles of the constituent herbs. *Ocimum sanctum* (Tulsi), *Ficus bengaklensis* (Vat), *Aegle marmelos* (bel), *Allium sativum* (Garlic), *Saraca Asoka* (Ashok), *Cynodon dactylon* (Durva grass) and many other species have been recently described as plants which render several protective health benefits. *Ocimum sanctum* has strong free radical scavenging activity and after protection against free radical-induced lipid peroxidation and decreased level of GSH in mice (Devi *et al.*, 1998). Banerjee *et al.*, (1996) observed that alcoholic extract of the plant leaves has modulatory influence on the activities of glutathione-3-transferase and reduced glutathione level in the liver, lung and stomach of the mouse. *Spirula furiformis* is rich in antioxidants, provitamins (β-catotene) and SOD enzyme (Henrikson, 1989) and has been used in experimental, clinical and epidemiological studies for cancer chemotherapy and as a preventive medicine (Chen and Zhang, 1995). Extracts of roots of *Caesalpinia pulcherrima* yielded four novel dibenzoate diterpenes, out of which Pulcherrimins A & B were found to be active against DNA-repair deficient *Sacharomyces cerevisiae* mutants, suggesting their anti-mutagenic property (Patil *et al.*, 1997). The flowers of *C. pulcherrima* have been recognized as remedies for disorders like bronchial asthma and malarial fever. Flowers have a potential of combating the myriad disorders and diseases resulting from oxidative damage.

Radioprotective Agents

Radiation-induced hematological alterations have been studied extensively. Haematopoietic tissues are among the tissues most sensitive to ionizing radiation in living beings. After whole body exposure, manifestations of injury to mammalian tissues are well reflected in peripheral blood. The major effect of whole body irradiation are more obvious in the number than in morphology of circulating cells. Extensive research has been carried out on finding a suitable chemical radioprotective agent that can be administrated safely before radiation exposure. Recently increased interest has been developed on search of potential drugs of plant origin that are capable of modifying immune responses (immuno-modulators) and radiation responses (radio protectors/sensitizers) with minimum side effects. Plants like *Panax ginseng* (Pande *et al.,* 1988), *Allium sativum* (Garlic) (Gupta, 1988) and *Ocimum sanctum* (Uma Devi and Ganasoundari, 1999) have been found to have radioprotective effects.

STRUCTURAL MODIFICATION FOR NEW POTENT DRUGS

Structural modification of natural products to devise new useful drugs has been an interesting and important feature of research on phytomedicines. Semisynthetic steroidal hormones or corticosteroids provide an excellent example that is produced mainly from the steroidal sapogenins diosgenin, solasidine, stigmasterol, hecogenin etc (Ziffer *et al.,* 1997). Many biologically active molecules from higher plants have not been found useful as drug, but they serve as models for the discovery of new drugs. Khellin obtained from the leaves of *Ammi visanga* which was for the treatment of angina and as a bronchodilator, was later found to be toxic and produce side effect like nausea and vomiting. Subsequently using khellin as a model, a more useful and less toxic drug cromolyne was produced by synthesis. *Disodium cromoglycate* is marketed as a drug for the treatment of asthma (Farnsworth, 1990). Saphoradin, a chalkone obtained from the roots of the Chinese plant *Sophora substrata* was found to have significant anti-gastric activity (Sukhdev, 1997). Irenotecan and topotecan use the products of the structural modification of the anti-cancer alkaloild camptothecin (Youvraj *et al.,* 1995, Kingston *et al.,* 1990). Taxotere, a semisynthetic analogue of paclitaxel proved to be better and more potent than paclitaxel (Sukhdev, 1997). Similarly, arteether A & B derived from artemisinine has shown promise in clinical trials as better anti-malarial than artemisinine (Bhakuni, 1995).

QUALITY ASSURANCE OF HERBAL MEDICINES

Medicinal plant based therapeutics plays a vital role in primary health care and has the potential of leading the discovery of new herbal drugs. An analysis of the genesis of their induction in the modern medicine would show that invariably the starting point has been the therapeutic ingredients of plant origin (Cragg *et al.*, 1997). In fact, plant derived compounds are showing promise in the treatment of diseases like malaria, filariasis, leishmaniasis, tuberculosis, fertility regulation, aging related disorders, memory loss, osteroporosis, hypertension, diabetes, hepatitis, infectious diseases, cancer, human immunodeficiency virus (HIV) etc (Farnsworth, 1988; Camido *et al.*, 2000). In fact, plants are almost the exclusive source of drugs for a majority of the world population even today.

Global estimate indicates that over three-fourths of the world population can't afford the products of the western pharmacological industries and have to rely upon the use of traditional medicines, which are mainly derived from plants. As a part of the strategy of reducing financial burden on the developing countries, which spend 40-50% of their total health budget on medicines, WHO recommends inclusion and promotion of herbal drugs in national health care programmes (Farnsworth *et al.*, 1985). These drugs are time tested, thus considered to be much safer than the modern synthetic drugs. This has generated worldwide resurgence of interest in herbal medicine thereby increasing commercial value and market demand of plant based drugs leading to the increased use of adulterated and spurious raw material in drug production (Hafeel and Shankar, 1991). Increasing need of the population and the chronic shortage of the authentic raw material have made it incumbent to bring about some sort of uniformity in the manufacture of herbal drugs for quality control and quality assurance. These days, when quality has become the order of the day, pharmaceutical products need to be strictly scrutinized in terms of quality. The quality of medicine can be guaranteed only through the use of correct raw material and implementing a quality control system.

It has been amply demonstrated that the nature and amount of bioactive plant metabolites are responsible for the effectiveness of the medicine prepared out of medicinal plants. The pre-requisition of optimum drug effect is dependent on purity of the plant material used for drug preparation. Hence, authentication of medicinal plant raw materials by standardizing certain quality control parameters may lead to the isolation of desired ingredients in required quantities and will ensure effectiveness of the drug as well. It will also enhance their acceptability in the populace.

ADULTERATIONN: A SERIOUS PROBLEM

In early days of yore, Vidyas or folk medicinemen used to collect plants from the nearby forests and prepare various drugs by themselves. Therefore, there was a least chance of using material other than the genuine one for the medicine. However, now-a-days the collection and distribution of medicinal plants are mainly done by unskilled labours, crude drug dealers and inexperienced and profiteering community of the traders. Since there is no government check over these traders, unauthentic crude drugs may be marketed for the use. It has been quoted in the report of the task force on conservation and sustainable use of medicinal plants that no system of medicine can achieve any credibility and mass acceptance unless some degree of quality control or quality assurance is maintained (Tewari, 2000). In fact, the absence of proper regulations and measures governing the identity and quality of herb in the national and international trade inducted substitutes, adulterants as well as fake plants in the market.

Gum Arabic obtained from *Acacia senegal* (Kumbat) is adulterated with much cheaper gums of other *Acacia* spp., Neem and *Albizia* spp. Rhizomes of *Acorus calamus* (Vach) is used as carminative, anthelmintic and antispasmodic. Sometimes the officinal part of the plant is substituted with rhizomes of *Alipinia galanga* and *Aconitum* spp. The rhizomes are used in the form of powder, balms, enemas and pills and in ghee preparations. The powdered drug is reported to be adulterated with siliceous earth, ground root of marsh mallow *Althaea officinalis* and cereal flours. The market samples of the raw material of the diuretic plant *Boerhavia diffusa*

(Punarnava) are said to be adulterated with the plant materials of *Trianthema porulacastrum* and *T. monoguna* (Varshabhu). The chief adulterants of the laxative drug *Cassia angustifolia* (Senna) are the leaf stalks and stem of the plant itself, leaflets fruits of other *Cassia* spp. e.g. *C. obovata*. Leaves of *Momordica charantia* (Karela) is regarded as anthelmintic, vermifue and galactegogue and is used in jaundice, piles and and diabetes mellitus. Leaf powder procured from the market is often not genuine and is found mixed with some spurious material and hence the therapeutic effects are missing. Seeds of *Plantago ovata* (Isabgol) used as purgative and for other abdominal disorders are frequently mixed with seeds of *Salvia aegyptiaca*, yielding copious mucilage similar to *P. ovata*. Commercial samples of isabgol may consist of seeds of other *Plantago* spp. like *Plantago arenaria*, *P. lanceolata* and *P. major*. Roots of *Rauvolfia serpentina* (Sarpgandha) find variety of medicinal application in traditional as well as modern system of medicines. Due to its great industrial demand, roots of other lesser effective *Rouvolfia* spp. *like R. perakensis, R. beddmei, R. densiflora, R. micrantha* and *R. tetraphylla* and those of *Ophiorrhiza mungos* and white and red flower variety of *Clerodendron* spp. are found as adulterant in the market samples of *R. serpentina* roots. The medicinal plant *Ventilago maderaspatana* commonly known as Vembadam pattai or Surul pattai is attributed with several medicinal properties. The root bark of the plant (the officinal part) sold in the market is not the original bark of *V. maderaspatana* but is fake. The well-known general tonic and anti-inflammatory drug *Withania somnifera* (Ashwagandha) is adulterated with *Withania coagulans* (Vidya, 1979, WOI, 1988).

In addition some of the medicinal plants are naturally grown as a pervasive weed and many of them are cultivated with other simulating crops, incidental or intentional adulteration is quite possible while untrained collectors are involved in procuring the drug from the field.

Use of chemical fertilizers and pesticides during the cultivation of medicinal crops also contaminates the plant drug with toxic compounds resulting in the modification of their chemical constituents responsible for their therapeutic effect.

Recent investigations have shown the existence of anti-active compounds in plants employed in preparation of medicines, which may interfere the normal healing phenomenon in the system. This situation has created a lot of controversy and confusion about genuineness of these drugs.

CONCERN OVER QUALITY ASSURANCE

Over the last few decades, there has been a growing interest in plant drugs and the consumption of medicinal plant has almost doubled. The world demand for medicinal plants is growing by 7 % per annum. The trade and demand projection in medicinal plants globally indicates a steep upward trend in future. According to the World Bank Report 1996, the trade in medicinal plant is going to touch US $ 5 trillion by 2050 A.D. (Farnaworth and Soejarto, 1991). The phenomenal increase in the demand for herbal medicines is due to the greater consumers' preference and tendency towards increased dependence on ready-made preparation. In addition, the importance of medicinal plants in search of new potential herbal drugs is gaining momentum all over the world. The enhanced consumption of medicinal plant in pharmaceutical sector led to the adulteration of high valued plant material of medicinal importance that necessitates greater attention on ensuring safety and efficacy of plant-based products through adopting adequate quality control norms, exercising closer supervision and implementing proper regulatory measures.

The efficacies of plant-derived medicines are sometimes confusing; one lot of the drug may be effective and other lot of the same drug may be less effective or sometimes ineffective. This must be related to chemical viability of drugs collected from natural habitat as well as use of unauthentic and/or adulterated raw material in drug production. The problem of ensuring safety and efficacy of plant derived remedies prior to their introduction and utilization in health care has attracted a worldwide attention. However, lack of sufficient scientifically validated information and guidelines is one of the major problems in this context. A number of medicinal species are still being collected unscientifically by various collectors and user groups, which not only cause damage to the natural population of the medicinal plants but also lessen the efficacy of the drug prepared out of it. The use of ingenuine raw material in drug production and consequent hazards to the human health has led to the global concern over safety and efficacy of herbal drugs thus presenting a major challenge with regard to their quality assurance. So, standardization of parameters for authentication of plant material becomes imperative for pharmacological analysis and pharmaceutical production and investigation of a large number of plant based remedies by modern methods has become a formidable task.

VALIDATION AND QUALITY CONTROL OF MEDICINAL PLANTS

The market drive for production of phytopharmaceuticals and gradually evolving stringent quality standard for herbal medicines necessitates to ensure the sustained supply and use of genuine raw material in the production of curative and health care medicines (Stix, 1993). Hence, validation of plant drugs is now considered one of the top priority researches in the field of medicine. This can only be achieved by application of various methods available in modern sciences to scientifically assess, determine and identify the medicinal plants/crude drugs (Purohit, 1997). Strict quality control measures should be observed at every step of the extraction of drug involving quality control of raw material, refinement in their processing techniques for production of reliable and safe herbal medicines suitable for human consumption and meeting the quality norms of the world market (WHO, 1993). Analysis of plant materials including special handling procedures for sampling, macroscopic and microscopic examination is important towards establishing the identity and degree of purity of plant materials. Detection of adulterants and verification of authenticity of the medicinal plant raw materials can be done by standardizing following parameters (Nair *et al.*, 1997):

1. Organoleptic characters
2. Macroscopy and microscopy for botanical identity
3. Physicochemical evaluation
4. Fluorimetric and UV-VIS Spectroscopy
5. Chemical profile and finger printing

Test procedures have been recommended for assessing the purity and content bioactive constituents of medicinal plant materials to assist national laboratories engaged in drug quality control. Modern techniques like phytochemical standardization, microscopic examination and various chromatographic techniques prove to be helpful in verification of authenticity of medicinal plant raw material to be used in drug production.

Taxonomic Method

In this method, the exomorphological characters of the medicinal plants are examined. Efforts are made to trace out and gather the different parts of the plant from the crude drug material. The exomorphology of these parts are studied and analyzed as per a prevalent system of botanical classification, which helps in the identification of the drug.

Pharmacognostic Method

This method involves identifying the crude drugs by studying their macro and microscopic characters. Since most of the characters are common to groups of drugs, it may not be possible to arrive at the species level solely based on such analysis. However, the crude drug may be identified to the level of family and genus in most cases. To adopt this method, it is essential to know the tissue system in general and the different specific observations to be made for identifying the root and rhizome, bark, wood, leaf, flower in particular (Tyler *et al.,* 1981). The following characters are observed in the microscopic sections:

Root and Rhizomes Drugs

(i) Arrangement and types of vascular bundles.

(ii) Size, type and thickenings of vessels and fibres.

(iii) Width and nature of medullary rays

(iv) Presence or absence of stone cells

(v) Type of Cork cells

(vi) Endodermis and nature of the cells

(vii) Type and size of starch grain if present

(viii) Type and form of Calcium oxalate crystals

(ix) Crystoliths and latex cells as present in Moraceae.

Bark Drugs

The bark may be derived from trunk, branches and roots. The characters of outer and inner surface like lenticles, cracks or furrows, colour, striations must be observed. In the transverse section, the details of periderm including phellem, phellogen, phelloderm, cortex and cellular contents, phloem fibres, size and shape of stone cells, types of oxalate crystals, cells containing essential oil etc., and width of medullary ray must be observed.

Wood Drugs

Wood consists of the secondary tissues produced by the Cambium on its inner surface. They consist of vessels, tracheids, wood fibres, xylem parenchyma. They contain resin, tannin and colouring matter. The central region is known as the Heart wood and the outer the Sapwood. They exhibit annual rings and medullary rays. The characters to be observed are type of vessel and various thickenings, height and width of medullary ray cells, types of fibres, type of crystals present.

Leaf Drugs

The reticulate venation and parallel venation in a leaf determines whether it belongs to a dicot or monocot respectively. Several morphological characters like the stipulate, exstipulate, nature, petiolate or sessile, type of lamina (simple, compound, paripinnate or imparipinnate) shape, apex, texture and the surface are important in identification of leaf drugs. In addition, type and form of trichomes, epidermal palisade and spongy parenchyma details of midrib and secretory products should also be observed.

Flower Drugs

In case of flower drugs, characters like type of flower, calyx, corolla, androceium, gynoceium, ovules, pollen grains and transverse section of salk, help in determination of authenticity.

Fruit and Seed Drugs

The classification of drugs may be done by observing fruits if it belongs to the simple aggregate or collective type. Microscopically, the fruit may contain fibre or stone cells. The epidermis of the testa exhibit characteristic trichomas or hair. The perisperm and endosperm contain cell contents, like starch, calcium oxalate crystals, fixed oil, volatile oil etc. Sometimes scelereids are also present in the pericarp and testa. In addition to the above the size, shape, color, texture, ornamentation and presence of hairs are also important diagnostic characters.

PHYTOCHEMICAL TECHNIQUES: A RELIABLE TOOL

Phytochemical standardization involving chemical profiling of the raw materials and laying down specification for quality assurance of the finished products has nowadays emerged as a significant tool to ascertain the authenticity and purity of medicinal plant raw material. In this method, the organic constituents of the drugs are studied and classified depending upon the nature of the compound present like alkaloid, steroids, glycosides etc. Certain families characteristically contain specific chemical constituents, sometimes some genera also. Some of the major steps involved in this method are fractionation, colour test, chromatography, isolation and characterization of active ingredients using various spectroscopic techniques (Harbone, 1973). Plant samples are examined for organoleptic characters such as colour, odour and taste (Sarin, 1996). Determination of foreign matter, ash, maximum extractive value, water and volatile matter, volatile oils, bitterness value, haemolytic activity, tannins, swelling index, and foaming index are done by standard test methods (IP, 1996, Sarin, 1996).

Application of chromatographic techniques like thin-layer chromatography, paper chromatography etc. for the qualitative determination of impurities may prove useful to the pharmaceutical industry and pharmacists working with medicinal plant materials. Quantitative chromatography of the extracts of the plants have shown a considerable variation in concentration of active ingredients in plant material of different origin and collected in different season and developmental stages (Kappal Rao *et al.*, 1993). Absence of certain important biological ingredients and presence of certain unwanted constituents was also recorded in adulterated plant materials. Chemical profile of the drug is examined by TLC and HPTLC finger-printing (Bhutani, 2000).

The powdered drugs are studied through ultra violet method. The differences in fluorescence of the drug when exposed to UV are observed and compared with the authentic samples (Chase and Patt, 1949). Unorganized drugs of plant origin like gums, resins, latex or dried juice are identified by specific chemical tests. Test methods for the determination of important contaminants as pesticide residues, arsenic and heavy metals, and microorganisms, including aflatoxins.

CONCLUSIONS

Innumerable attempts have been made by the researchers to revitalize the traditional system of medicine. The study of phytotherapy is rapidly gaining importance throughout the world. As a result, the alternative treatments that make use of medicinal plants are gaining worldwide acceptance. These drugs are preferred over the synthetic drugs due to their natural affinity that allows total assimilation of a wide range of essential elements, developed by the growth organs of plants that are indispensable to human life. Another advantage of plant remedies is that they are harmless provided they are selected carefully and taken under medical guidance. New commercial organizations are emerging to study hitherto unknown medicinal plants adding to the information to elaborate new scientifically tested recipes for the more effective treatment of specific illness. Although a certain mystique still surrounds herbs and their medicinal uses today, progress in scientific research has greatly clarified the subject. Inspite of the development of a number of modern drugs, there is still genuine and urgent need to discover and develop new therapeutic agents especially for those diseases for which no satisfactory cure is available in modern system of medicine. Areas where new drugs are urgently needed include viral infections, liver diseases, bronchial asthma, diabetes, immunomodulators, adaptogens, anti-AIDS and anti-cancer agents.

The use of spurious or adulterated raw material in place of genuine one for manufacturing medicines may not only affect the efficacy of the finished products, but may cause deleterious effect on human health. Therefore, before using the crude drugs collected from the market their correct identity should be ensured. The quality control measures need to be followed strictly for authentication of the raw materials. All the crude drugs of standard quality need to be identified and preserved as the reference standard in a national repository of crude drugs with an excellent herbarium having authentic reference samples. Drug development may take advantage of modern sophisticated chromatographic tools like HPLC, HPTLC, GLC and spectroscopic techniques like IR, UV, NMR and MS to check and recheck the every batch of the final product for their efficacy.

Plant derived remedies are increasingly used in both developing and developed countries. In order to promote safe and appropriate use of such medicines and practices, national regulations are vital. Establishing national policies on drug production and herbal therapeutics should therefore include creation of legal framework. Verification of authenticity of raw material and observation of proper quality assurance norm should be mandatory for the bulk user of the crude drugs. These sectors should be cautious while procuring the medicinal plant raw material from the market. Otherwise, it may effect their reputation, market, human health and ultimately faith of the people in herbal medicines

REFERENCES

Acharya, S.B., Tripathi, S.K., Tripathi, Y.C. and Pandey, V.B. (1988). Some pharmacological studies on *Zizyphus rugosa* saponins, *Ind. J. Pharmacol.*, 20: 200-202.

Ahmed, I., Ahmed, F. and Hussain S. (1995). In vitro antimicrobial activity of leaf and bark extract of *Azadirachta indica* A. Juss. *Ind. Vet. Med. J.*, 19: 204-206.

Ahmed, I., Mehmood, Z. and Mohammad, F. (1998). Screening of some Indian medicinal plant for their antimicrobial properties. *J. Ethnopharmacol.*, 62: 183-193.

Anon (1974-75). Annual Report, CCRIMH, New Delhi, p. 47.

Arora, R.B. and Rangaswamy, S. (1972). Peruvosis and other cardiac glycosides of *Thevetia nerifolia* (Monograph), CCRIMH, New Delhi.

Atal, C.K. (1980). Chemistry and pharmacology of Vasicine—a new oxyticic and abortifacient, Regional Research Laboratory, Jammu-Tawi, p. 155.

Bagi. M.K.. Kakrani, H.K., Kalyani G.A., Datyanarayana, D. and Manvi, F.V. (1985). Preliminary pharmacological studies of essential oil from *Commiphora mukul*. *Fitoterapia,* 56: 245-248.

Banerjee, N. and Kumar, R. (1980). Studies on the seed oil of *Aegle marmelos* and its effect on some bacterial species. *J. & Proc. of Institution of Chemists*, India, 52: 59-60.

Banerjee, S., Prashar, R., Kumar, A. and Rao, A.R. (1996). Modulatory influence of alcohol extract of *Ocimum* leaves on carcinogen-metabolizing enzyme activities and reduced glutathione level in mouse. *Nutr. Cancer*, 25: 205-217.

Baratta, M.T., Dorman, H.J.D., Deans, S.G., Figueiredo, A.C., Barroso, J.G. (1998). Antibacterial and antioxidant properties of some commercial essential oils. *Flav. Frag. J.,* 13: 235-244.

Bhakta, T., Mukherjee, P.K., Saha, K. Pal, M. and Sahja, B.P. (1998) Studies on antitussive activity of *Cassia fistula* (Leguminosae) leaf extract. *Pharma. Biol., 36*: 140-143.

Bhakta, T. Mukherjee, P.K., Mukherjee, K., Banerjee, S. Mandal, S.C. Murthy, T.K., Pal., M. and Saha, B.P. (1999a). Evaluation of hepatoprotective activity of *Cassia fistula* leaf extract. *J. Ethnopharmacol.* 66: 277-282.

Bhakta, T., Mukherjee, P.K., Saha, K., Pal, M., Saha, B.P. and Mandal, S.C. (1999b). Evaluation of anti-inflammatory effects of *Cassia fistula* (Leguminoseae) Leaf extract on rats . *J. Spices. Med. Pl.,* 6: 67-72.

Bhakuni, D.S., Jain, D.C. and Sharma, R.P. (1995). An improved procedure for the synthesis of ethers of dihydroartemisinin. *Ind. J. Chem.,* 34 B: 529-530.

Bhargava, S.K. (1986). Estrogenic and postcoital anticonceptive activity in rats of Butin isolated *Butea monosperma* seed. *J. Ethnopharmacol.,* 18: 95-101.

Bhatnagar, S.K. and Verma, S.K. (1986). Effect of 50% extract of *Calotropis procera* Ait.f. on ulcers casual by assorted types of carcinoma. *J. Eco. Taxon. Bot.,* 8: 489-490.

Bhattacharya, S.K., Goel, R.K., Kaur, R. and Ghosal, S. (1987). Antistress activity of sitoindosides VII and VIII, new acylsteryl glucosides from *Withania somnifera Phytotherapy Res.* 1: 32-37.

Bhutani, K.K. (2000). Finger—printing of Ayurvedic Drugs. *Eastern Pharmist.*, 63: 21-26.

Camido, C., Morae, R.M., Dayan, F.E. and Fessriro, D. (2000). Molecules of interest. *Podophyllotovir,* 54: 11-120.

Chakraborty, D.P., Bhattacharya, P. and Chattopadhyay, K.K. (1978). Marmelide, a tyrosinase accelerating and tryptophan pyrrolase inhibitory furanocoumarin from *Aegle marmelos* Corr. *Chemistry and Industry*, 21: 848.

Chakraborty, S.N., Prakash, A. and Siddique, K.A.I. (1987). Aristolic acid an antifertility agent is non mutagenic to bacterial system. *Ind. J. Pharmacol.,* 19: 26-31.

Chase, C.R. and Patt, R. (1949). Fluorescense of powdered vegetable drugs with particular reference to development of a system of identification. *J. Am. Pharm. Assoc. Sci.,* 38: 324-331.

Che, C.T., Ahmad, M.S., Kang, S.S., Waller, D.P., Bingal, A.S., Martin, A., Rajamahendran, P., Bunyapraphatsara, N. and Lankin, D.C. (1984). Studies on Aristolochia III. Isolation and biological evaluation of constituents of *Aristolochia indica* roots for fertility regulating activity. *J. Nat. Prod.,* 47: 331-341.

Chen, F. and Zhang, Q (1995). Inhibition effect of *Spirulina* on aberract crypts in colon induced by dimethyl hydrazine. Chung. Hua. Yu, Fang, I., Husch, Tsa. Chin., Jan. 29 (1): 13-17.

Cragg, G.M., Newman, D.J. and Snadder, K.M. (1997). Natural products in drug discovery and development. *J. Nat. Prod.,* 60: 52-60.

Cohen, M.S., Insturiz, R.E., Malech, H.L., Root, R.K., Wilfert, C.M., Gutman, L. and Buckley, R.H. (1991). Fungal infection in chronic granulomatous diseases. *Amer. J. Med.,* 71: 59-66.

De Conti, R.C. and Creasey, W.A. (1975). Clinical aspects of the dimeric Catharanthus alkaloids. In *Catharanthus*: Alkaoilds, Botany, Pharmacology & Clinical Uses (Eds. Tayler, W. and Forasworth, N..) Marcel Dakker, New York Inc., pp. 237-278.

De, B., Maiti, R.N., Joshi, V.K., Agrawal, V.K., Goel, R.K. (1997). Effect of some sitavirya drugs on gastric secretion and ulceration. *Ind. J. Exp. Biol.,* 35: 1084-1087.

Desta, B. (1993). Ethiopian traditional herbal drugs. Part II: Antimicrobial activity of 63 medicinal plants. *J. Ethnopharmacol,* 39: 129-139.

Devi, P.U., Bist, K.S. and Vinitha, M. (1998). A comparative study of radioprotection by *Ocimum* flavonoids and synthetic aminothiol protectors in the mouse. *Br. J. Radiol.,* 71: 782-784.

Dhanbal, S.P., Shabu, P. Ramanathan, M., Elango, A.L., Suresh, B. (1999). Validation of antifertility activity of various *Rubus* species in female albino rats. *Ind. J. Pharm. Sci.*, 61: 58.

Dhuley, J.N. (1997). Effect of some Indian herbs on macrophase function in ochratoxin A-treated mice. *J. Ethnopharmacol.,* 58: 15-20.

Diamond, R.D. (1991) The growing problem of mycoses in patients infested with HIV. *Rev. Infect. Dis.*, 13: 480-486.

Dwivedi, Y., Rstogi, R., Chander, R., Sharma, S.K., Kapoor, N.K., Gerg, N.K. and Dhawan, B.N. (1990). Hepatoprotective activity of picrolin against carbon tetra chloride induced liver damage in rats. *Ind. J. Med. Res.,* 92: 195-200.

Farnsworth, N.R., Akerele, O. and Bingel, A.S. (1985). Medicinal plants in therapy. *Bull. Wld. Hlth. Org.* 63: 964-981.

Farnsworth, J.D. (1988). Screening plants for new medicines, In Biodiversity (Ed. Wilson, E.O.) National Academy Press, Washtington, pp. 83-97.

Farnsworth, N.R. (1990). The role of ethnopharma Cology in drug development. In Bioactive Compounds from Plants, John Wiley and Sons Chichester (Ciba Foundation Symposium 154), pp. 2-21.

Farnsworth, N.R. and Soejarto, D.D. (1991). Global importance of Medicinal plants, In: Conservation of Medicinal Plants (Eds. Akerele, O., Heywood, V. and Synge, H.), Cambridge Press, Cambridge, NY, pp. 79-124.

Garg, S.K. (1976). Effect of our indigenous plants in early pregnancy in female albino rats. *Ind. J. Med. Res.,* 64: 1077.

Goldstein, A., Aronow, L., Kalman, S.M. (1974). Principles of drug action: the basis of pharmacology, 2nd Edn., John Wiley & Sons, New York, pp. 741-761.

Govindachari, T.R. and Viswanathan, N. (1972). Alkaloids of *Mappa foetida*. *Phytochemistry*, 30: 3849-3854.

Gund, P., Andose, J.D., Rhodes, J.B., Smith, G.M. (1980). Three dimensional molecular modeling and design, *Science*, 208: 1425-1431.

Gupta, N.K. (1988). Hypolipidemic action of garlic unsaturated oils in irradiated mice. *Nat. Acad. Sci. Lett.* 11: 410-403.

Gupta, S.S. (1994). Prospects and perspectives of natural plant products in medicine. *Ind. J. Pharmacol.,* 26: 1-2.

Hafeel, A. and Shankar, D. (1999). Revitalizing indigenous health practices. *COMPAS Newslett.* (Feb., 1999): 28-29.

Hammer, K.A., Carson, C.F. and Riley, T.V. (1999). Antimicrobial activity of essential oil and other plant extracts. *J. Applied. Microbiol.,* 86: 985-990.

Handa, G. Singh, J. and Atal, C.K. (1983). Antiasthmatic principle of *Allium cepa. Indian Drugs,* 20(6): 239.

Harbone, J. B. (1973). Phytochemical Methods, Jackman Hall, London, p. 90.

Henrikson, R. (1989). Earth food Spirulina, Recolina Ltd., Ronore Enterprises Inc., Launa Beach, California, pp. 27-65.

Hiremath, S.P., Swamy, H.K.S., Shrishaillapa, B. Patil, S.B. and Londonker, R.L. (1996). Postcoital antifertility of the plant *Striga densiflora* (Scrophulariaceae) on female albino rats. *Ind. J. Pharmacology*, 34: 48.

Hiremath, S.P. and Hanumantha Rao, S.P. (1999). Antifertility efficacy of the plant *Striga lutea* (Scrophulariaceae). *Contraception,* 22: 467.

Hiremath, S.P., Rudresh, K., Badami, S., Patil, S.B. and Patil, S.R. (1999). Postcoital antifertility of *Acalypha indica* L. *J. Enthnopharmacol.,* 67: 253.

Indian Pharmacopoeia, (1996). 4th edition, Vol.2, Controller of Publications, Govt. of India, New Delhi, pp. A53–A54.

Iyenger, M.A., Jambiah, K., Rao, G.M., Kamath, M.S. (1994). Studies on anti-asthma Kada: a proprietary herbal combination Part II, Pharmacological studies. *Indian Drugs*, 31: 187-191.

Iyenge, M.A., Rao, M.P.R., Bairy, I. and Kamath, M.S. (1995). Antimicrobioal activity of essential oil of *Curcuma longa* leaves. *Indian Drugs*, 32: 249-250.

Jain, S.R., Jain, P.R. and Jain, M.R. (1974). Antibacterial evaluation of some indigenous volatile oils. *Planta Medica*, 26: 196-199.

Kappal Rao, H. and Sasibhushan, S. (1993). Quantitative assessment of medicinal plants found in Vishakhapatanam district of Andhra Pradesh, *Bull. Medico-Ethno-Bot. Res.*, XIV(1-2): 26-35.

Kingston, D.G.I., Samaranaya, G. and Iray, C.A. (1990). The chemistry of taxol, a clinically useful anticancer agent. *J. Nat. Prod..*, 53: 1-12.

Kitanaka, S. and Takido, M. (1986). Studies on the constituents in the roots of *Cassia obtusifolia* L. and the antimicrobial activities of constituents of the roots and the seeds. *Yakugaku Zasshi*, 106: 302-306.

Kobashi, K., Nishimura, T., Kusaka, M., Hattori, M. and Namba, T. (1980). Metabolism of sennosides by human intestinal bacteria. *Planta Medica*, 40: 225-236.

Lalitha, R. (2000). Awaken those energy auras, Folio, A Supplement to *The Hindu*, March, 2000, pp. 22-23.

Levin, H., Hazenfratz, R., Friedman, J., Palevitch, D. and Peri, M. (1988). Partial purification and some properties of an antibacterial compound from *Aloe vera*. *Phytother Res.*, 2: 67-69.

Magelsdorf, P.C. (1972). Introduction. In Plants in the Development of Modern Medicine (Ed. Swain, T.) XI-XII, Harvard University Press, Cambridge, MASS.

Mahajan, V.R. and Jolly, C.L. (1985). A new hypoglycemic agent from *Tinospora cordifolia* Miers. *Indian Drugs*, 23: 119-120.

Mandal, S.C., Maiti, B.C., Maiti, T.K., Pal, M., Saha, B.P.C. (1998) Evaluation of anti-inflammatory activity of *Asparagus racemosus* Willd. (Liliaceae) root extract. *Nat. Prod. Sci.*, 4: 230-233.

Mazumdar, V.K., Gupta, M.P., Pramanik, G., Mukhopadhyay, P.R. and Sarkar, S. (1992). Antifertility activity of seed of Nelumbo nucifera in mice. *Ind. J. Exp. Biol.*, 30: 533.

Mehmood, Z., Ahmed, I., Mohammad, F. and Ahmed, S. (1999). Indian medicinal plant: A potential source for anticandidal drugs. *Pharm. Biol.*, 37: 1-6.

Mehrotra, R., Rawat, S., Kulshshtra, B.K., Goyal, P., Patnaik, G.K. and Dhawan, B.N. (1991). In vitro effect of *Phyllanthus amaras* on hepatitis B virus. *Ind. J. Med. Res.*, 93: 71-74.

Moscolo, N., Sharma, R., Jain, S.C. and Capasso, F. (1988). Ethnopharmacology of *Calotropis procera* flowers. *J. Ethnopharmacol.*, 22: 211-221.

Moya, M. Gonzalev, C. and Forero (1976). Abstract, 24th Meeting of Intl. Congress for Research in Medicinal Plants, p. 6.

Mukherjee, P.K., Pal, S.K., Sahha, K. and Saha, B.P. (1995). Hypoglycaemic activity of *Nelumbo nucifera* Gaertn. (Family: Nymphaeaceae) rhizome (methanolic extract) in streptozotocin-induced diabetic rats. *Phytother. Res.,* 9: 522-524.

Murthy, D.R.K., Reddy, C.M. and Patil, S.B. (1997). Effect of benzene extract of Hibiscus rosa sinensis on the estrous cycle and ovarian activity in albino rats. *Biol. Pharm. Bull.,* 20: 756.

Muruganandan, S., Srinivasan, K., Tandan, S.K., Lal, J., Chandra, S. and Raviprakash, V. (2000). Antiinflammatory and analgesic activities of some medicinal plants. *J. Med. Aroma. Pl.,* 22/4A & 23/1A: 36-58.

Nadkarni, G.D., Mitra, A.G., Despande, U.R. and Pahuja, D.N. (1991). Liver anti-oxidant defense and lipid peroxidation in vitamin D deprived rats. *Ind. J. Biochem. Biophys.* 28: 224-225.

Nair, K.V., Nair, A.R. and Nair, C.P.R. (1997). Standardization of Ayurvedic drugs. *Bull. Medico-Ethno-Bot. Res.,* 18: 151-156.

Nandy, A.K., Chakarborthy, A. and Poddar, G.I. (1994) Antimicrobial activity of *Terminalia belerica* triterpenoids. *Fitoterapia*, 68: 178-180.

Nityanand, S. and Kapoor, N. (1973). Cholesterol lowering activity of the various fractions of guggal. *Ind. J. Exp. Biol.*, 11: 395-396.

Odds, F.C. (1988). *Candida and Candidiasis*, Bailliere Tindal, London, p. 42.

Panda, S. and Kar, A. (1997). Antithyroid property of root extract of Shankhpuspi (*Convolvulus plaricaulis*). *Med. Sci. Res.,* 25: 677-678.

Panda, S. and Kar, A. (1999). *Withania somnifera* and *Bauhinia purpurea* in the regulation of circulating thyroid hormone concentration in female mice. *J. Ethnopharmacol.*, 67: 233-239.

Pande, S. Kumar, M. and Kumar, A. (1988). Evaluation of radiomodifying effects of root extract of *Panax ginseng. Phytother. Res.*, 12: 13.

Pathak, N.K.R., Neogi, P., Biswas, M, Tripathi, Y.C. and Pandey, V.B. (1988). Betulin aldehyde, an anti-tumour agent from the bark of *Tectona grandis, Ind. J. Pharm. Sci.*, 50(2): 124-125.

Patil, V.D., Nayak, R.D. and Dev, S. (1972). Chemistry of Ayurvedic crude drugs 1. Guggulu (resin from *Commiphora mukul*), steroidal constituents. *Tetrahedron,* 28: 2341-2352.

Patil, A.D., Preyer, A.J., Webb, R.L., Zuber, G.M., Reichwein, R., Bean, M.F., Faucette, L. and Johnson, R.K. (1997). Pulcherrimine

A-D, novel diterpene dibenzoates from Caesalpinia pulcherrima with selective activity against DNA repair-deficient yeast mutants. *Tetrahedron,* 53: 1583-1592.

Perdue, R.E. and Hartwell, J.L. (eds) (1976). Plant and Cancer. Proc. of the 16th Annual Meeting of the Society of Economic Botany. *Cancer Treatment Report,* 60: 973.

Perry, M. (1980). Medicinal Plants of East and South East Asia, MIT Press, Cambridge, USA, p. 620.

Pillai, N.R. and Santhakumari, G. (1984). Antiarthritic and anti-inflammatory action of nimbin. *Planta Medica*, 43: 59-63.

Pitre, S. and Srivastava, S.K. (1987). Pharmacological, microbiological and phytochemical studies on roots of *Aegle marmelos*. *Fitoterapia,* 58: 194-197.

Poswal, M.A.T. and Werner, W. (1997). Antibacterial properties of essential oils on *Pseudomonas solanacearum*. *Plant Pathol.*, 9: 606-610.

Purohit, A.N. (1997). Medicinal plants: Need for upgrading technology for trading the traditions. *Hervesting Herbs*-2000, pp. 49-75.

Pushpangadan, P., Rajasekharan, S., Rathesh Kumar, P.K., Jawahar, C.R., Velayudan Nair, V., Lakshmi, N. and Sarada Amma, L. (1988). Arogyappacha (*Trichopus zeylanicus* Gaertn) The Ginseng of Kani Tribes of Asasthyar Hills (Kerala) for ever green health and vitality. *Ancient Science of Life*, VII (1): 13-16.

Rao, K.S. and Mishra, S.H. (1998). Antihepatoprotective activity of *Sida cordifolia* whole plant. *Fitoterapia*, 69: 20-23.

Rege, N.N., Thatte, U.M., Dahanukar, S.A. (1999). Adaptogenic properties of six rasayana herbs used in Ayurvedic medicine. *Phytother. Res.,* 13: 275-291.

Ribbon, J.W. (1988). The pathogenic fungi and pathogenic actinomycetes. In Medi Mycology (2nd Edn.) W.B. Sanders Co., Philadelphia, London, pp 21-32.

Rindal, M.G. (1991). Problems in the diagnosis of invasive fungal diseases. *Rev. Infect Dis.,* 13: 493-495.

Rowinsky, E.K., Mc Guire, M.P. and Donchower, R.S. (1990). In Workshop on Taxol and Taxas: Current and Furure Pespectives, Bethesda, MD.

Sarin, Y.K. (1996). Illustrated Manual of Herbal Drugs used in Ayurveda, Council of Scientific and Industrial Research and Indian Council of Medicinal Research, New Delhi.

Shah, C.S. (1982). Recent developments of natural products. In Cultivation and Utilization of Medicinal Plants, (Eds. Atal, C.K. and Kapoor, B.M.) RRL, Jammu, pp. 284-287.

Sharma, A.K., Pushpangadan, P., Chopra, C.L., Rajasekharan, S. and Sarada Amma, L. (1989). Adaptogenic activity of seeds of *Trichopus zeylanicus* Gaertn.. The Ginseng of Kerala. *Ancient Science of Life* 8(3-4): 212-219.

Sharma, S. and Mehta, B.K. (1989). Antimicrobial activity of *Echinops echinatus* root extract. *Fitoterapia*, 60: 82-83.

Sies, H., Stache, W. and Surdquist, A.R. (1992). Antioxidant functions of vitamins, vitamin E,C, β-carotene and other carotenoids. *Ann. NY. Acad. Sci.*, 669: 7-20.

Singh, S.P., Negi, S., Laxmi C., Singh, A.K. and Chand, L. (1992). Antimicrobial properties of essential oil from *Zinziber chrysabthum* leaves and rhizomes. *Fitoterapia,* 63: 73-75.

Singh, A., Singh, R.H., Singh, R.G., Misra, N., Vrat, S., Prakash, M. and Singh, N. (1988). Effect of *Boerhavia diffusa* (Punarnava) in experimental acute pyelonephitis in albino rats. *Indian Drugs*, 26: 10-13.

Sinha, K.C., Riar, Bardhan, J., Thomas, P., Kain, A.K. and Jain, R.K. (1984). Antiimplantation effect of Neem Oil. *Ind. J. Med. Res*, 80: 708-710.

Spjut, R.W. (1985). Limitations of a random screen: Search for new anticancer drugs in higher plants. *Econ. Bot.*, 39: 266-288.

Stix, G. (1993). Back to roots: Drug companies forage for new treatments (plant-derived pharmaceutics). *Scient. Amer.*, 268: 142-144.

Sukhdev, (1997). Ethnotherapeutics and modern drug development, The potential of Ayurveda. *Curr. Sci.*, 73: 909-928.

The Wealth of India – Raw Materials (1988) Publication and Information Directorate, Council of Scientific and Industrial Research, New Delhi.

Tahiliani, P. and Kar, A. (2000a). *Acyranthes aspera* elevates thyroid hormone levels and decreases heptic lipid peroxidation in male rats. *J. Ethnopharmacol.*, 71: 527-532.

Tahiliani, P. and Kar, A. (2000b). Role of *Moringa oleifera* leaf extract in the regulation of thyroid hormone status in adult male and female rats. *Pharmacol. Res.*, 41: 319-323.

Tiwari, D.N. (2000). Report of the Task Force on Conservation and Sustainable Use of Medicinal Plants, Planning Commission, Govt. of India, New Delhi.

Tripathy, Y.B., Malhotra, O.P. and Tripathi, S.N. (1984). Thyroid stimulating action of Z-guggusterone obtained from *Commiphora mukul*. *Planta Medica*, 50: 78-80.

Tripathi, Y.C. and Dwivedi, R.K. (1989). Antifungal activity of alkaloids of *Tiliacora racemosa*, *Nat. Acad. Sci. Lett.*, 12(3): 69-71.

Tripathi, Y.C. and Dwivedi, R.K. (1990). Central nervous system and anti-inflammatory activities of alkaloids of *Fumaria indica*, *Nat. Acad. Sci. Lett.*, 13 (6): 231-233.

Tripathi, Y.C. (1998). Advances in phytotherapeutic research. *J. of Non-Timber For. Prod.*, 5(1/2): 37-43.

Tripathi, R.M. and Das, P.K. (1977). Studies on anti-asthamatic and anaphylactic activity of *Albizzia lebbeck*. *Ind. J. Pharmacol.*, 9: 189-194.

Tyler, V.E., Brady, L.R. and Robbers, J.E. (1981). Pharmacognosy, 8th Edn., Lea & Febiger, Philadelphia.

Tyler, V.E. (1986). Plant drugs in the twenty-first century, *Econ. Bot.*, 40: 279-288.

Uma Devi, P. and Ganasoundari, A. (1999). Modulation of glutathione and anti-oxidant enzyme by *Ocimum sanctum* and its role in protection against radiation injury. *Ind. J. Exp. Biol.*, 37: 262-268.

Upadhyay, S.N., Kaushik, C. and Talwar, G.P. (1990). Antifertility effects of neem (*Azadirachta indica*) oil by single intraurine administration, a novel method for contraception. *Proc. R. Soc. Lond. B. Biol. Sci.*, 242: 175-179.

Varsheya and Teleng, R.S. (1995). Studies on anti-inflammatory and analgesic activities of *Vitex negundo*. National Convention Vety. Pharmacology and Toxicology, Dec. p. 59.

Verma, H.N. and Awasthi, L.P. (1979). Antiviral activity of *Boerhavia diffusa* root extract and the physical properties of the virus inhibitor. *Canadian J. Bot.*, 57: 926-932.

Vidya, B.G. (1972). Some controversial drugs of Indian Medicine VI. *J. Res. Ind. Med.*, 7(3): 69-75.

Wagner, H., Geyer, B., Feibig, M. Kiso, Y., Hikino, H. (1986). Liver protective drugs. Part 30. Drug for liver therapy. Part 12.

Isobutrin and butrin, the antihepatoprotective principle of *Butea monosperma* flowers. *Planta Medica*, 52: 77-79.

WHO Regional Office for Western Pacific (1993). Research guidelines for evaluating the safety and efficacy of herbal medicines, Manila.

Youvraj, R.S., Padmaja, Kaimal, A.V., Ranjan, M.B. (1995). Prophylatic therapy of Salmonella typhi septicemia in mice with traditionally prescribed drug for mutations. *J. Ethnopharmacol*, 45: 141-147.

Ziffer, H. Highet, R.J. and Klaymann, D.L. (1997). Artemisinin, Andoperoxidic antimalarial from *Artemisia annua* L. In Progress in Chemistry of Organic Natural Products, Vol. 72 (eds. Herz, W., Kirby, C.W., Moore, R.E., Steglich, W. and Amm, C.T.) Springer-Verleg, Wien, New York, pp. 121-124.

10

Ethnomedicinal Plants of Jharkhand

Prashant K. Mishra

The wealth of living resource of any region is the result of hundreds of millions of years of evolutionary history. Similarly tradition and culture of any takes its shape in several years. Conservation of both therefore is essential. Our mother planet earth is endowed with an immensely rich variety of life forms which includes 3,00,000 green plants, 8,00,000 insects, 23,000 fishes, 9,000 birds, 6,500 reptiles and 4,100 mammals. All the life forms are nicely woven in biosphere to constitute the ecosystem. Although each form of life has its own importance, plants definitely plays the pivotal role because their capacity of photosynthesis makes them only producer of any ecosystem. Man, from time immemorial, has utilized various plants for his various needs. Diseases and other health problems, on the other hand, are as old as human civilization. Human being has used, altered and nurtured living resource for various purposes; therapeutic need is obviously the most important.

Jharkhand state lying between 22^0 N Latitude to $25^0 5'$ N Latitude to 83^0 E Longitude to 88^0 5' E Longitude has recently got its place on the map of India although it was earlier popular by name Chotanagpur. God has been benevolent enough to bless this area with variety of plants, so is the name of this state Jharkhand meaning the place of plants. Because of predominance of tribal, the region has its representative culture, which has developed in thousands of years. Herbal therapeutic system is one of the important aspects of Jharkhandi culture which is age old and which has percolated in long run. During last century, fever of modernism has affected this state also and the arena of health care has not left unaffected. Synthetic drugs, antibiotics and steroids made their presence felt and traditional herbal treatment was pushed to the back seat. But soon after people became aware of the side effect of these drugs and importance of medicinal plants was reestablished. It will not be out of place to mention that western countries, where modern drugs were discovered, also turned towards herbal treatments. This pleasant transformation has led to a new trend of research and study of medicinal plants.

Rapid industrialization and unabated mining activities have considerably altered the land use pattern and have taken a heavy toll of biodiversity. Forest cover has reduced to meager 21%, canopy cover has diminished and a large number of medicinal plants have disappeared. Our colonial attitude towards forest has completely changed the ecology of Jharkhand. The region was once both beautiful as well as useful. At present this region is only useful as it has lost its beauty and if not taken care of, tomorrow it will be neither beautiful nor useful. Keeping these facts in mind a survey of medicinal plants of Jharkhand was made and some commonly used plans are described in following pages. Information was gathered by personal interview of elderly tribals.

Acacia catechu: Locally known as Kattha or Khair the plant belongs to family Mimosaceae.The plant contains alkaloids Catachin, Catachutannine and Cpicatachin. The plant is used as astringent as it is capable of causing contraction of body tissue. The juice of bark is used to treat sore throat.

Acacia arabica: Locally known as kikar the plant also belong to family Mimosaceae. Gum of this plant is used to treat diarrhea, dysentery and diabetes malltis. Bark of the plant has got astringent and demulcent properties.

Acacia concinna: Local people call this plant as Ritha. The plant belonging to the family Mimosaceae possesses alkaloid saponin. The pod has got aperients properties and is also used to cure cough because it is a good expectorant. The leaves are cathartic in nature and people of Jharkhand use it to cure disorder of bile formation. This plant is a common medicine for treating jaundice.

Acharas sapota: People of Jharkhand call it Sapota and the plant belongs to family Sapotaceae.Fruits are used against excessive bile formation. Local people use this plant to cure intermittent fever. The plant is also useful as febrifuge and tonic. Seeds have got diuratic properties hence used in kidney disorder. The plant contains Saponin and Sapotinin alkaloids.

Adhatoda vasaka: The plant is commonly known as vasak or bakas and belongs to family Acanthaceae. Juice of leaves and roots are administered in chronic bronchitis, cough and asthma. Paste of root is applied in rheumatism and is equally helpful in curing wounds as it has got antiseptic properties. Flower root and leaves have got insecticidal properties. Plant parts contain alkaloids vasicin and peganin.

Adina cordifolia: This plant belongs to family Rubiaceae and is commonly known as Haldu or Karam. The plant is used as febrifuge. Juice of leaves is given to kill worms and is also used as antiseptic agent. Crushed leaves are applied on sores and wound.

Aerva aspera: The plant is locally known as Chirchiri or Chaya and it belongs to family Amaranthaceae. Local practitioner uses this plant to treat headache. This plant is also popular for its anthelminthic, diuratic and demulcent properties.

Ageratum sp.: This plant belongs to family Asteraceae and is commonly known as Uchunti. The plant parts have got quick healing properties hence local people use it for healing wounds, burns and ulcers.

Ailanthus excelsa: The plant belongs to family Simarubaceae and is popularly known as Maharukhi.The plant is used in dyspeptic complains. The plant parts are also recommended as tonic, febrifuge and antiseptic. Juice of leaves is used as expectorant and administered in chronic bronchitis, asthma and common cold. The plant is also used as astringent in diarrhea and dysentery.

Albizzia sp.: Three varieties of this plant are found in this region and are commonly known as Siris, Kala Siris and Safed Siris. The plant belongs to family Caeselpinae.The plant is used as antidote in snakebite and scorpion bite. Bark of plant is used as tonic and as astringent in diarrhea and dysentery. Powered roots are used for strengthening gums. Local people use leaves for curing night blindness. Paste of leaves is applied externally in leprosy, ulcer and burns. Leaves are boiled with ghee and used to cure cough.

Alstonia sp.: The plant belongs to family Apocyanaceae and is popularly known as Chatiun. The plant is commonly used as tonic, alterative and febrifuge. Local vaidya uses the plant parts to cure malaria. Leaves are also administered in diarrhea and dysentery because of its astringent properties. Poultice of root is used on ulcer. The plant contains alkaloids echidnas, vitamins and echitemidin.

Anacardium sp.: The plant belongs to family Anacardiaceae and is locally known as Bhela or Bhelwa. The bark has got alterative and astringent properties. Poultice of bark is applied in leprosy, ulcer and corns as it effectively controls irritation. Root is purgative in nature and is recommended in constipation. Kernel is nutritious, demulcent and emollient. Extract of bark is given orally to lower B.P.

Artocarpus heterophyllum: Commonly known as Kathal and belonging to family Moraceae, fruits of this plant are a popular vegetable of Jharkhand. Paste of leaf is used in snakebite and ulcer. Plant extract is applied on glandular swelling and abscesses to promote suppuration. Root is used in dysentery. The plant contains alkaloids Morin, Cyanomachurine and Artosteron.

Asperagus sp.: Commonly known as Satawar or Satmuli the plant belongs to family Liliaceae. Root of this plant is used for curing various ailments, as it is refrigerent, demulcent, diuratic, aphrodisiac and astringent in nature. The root is administered to feeding females as it enhances lactation.

Barleria prionitis: The plant is locally known as Kantajati and it belongs to family Acanthaceae. Juice of leaf is useful catarrhal and other type of inflammation of mucous membrane, fever and common infection of children. Leaf of this plant is chewed to relief toothache. Dried bark is administered in cough. Root is crushed with mustard oil and made into paste, which is applied on boils and other glandular swelling. Decoction of whole plant is used in dropsy to wash the body.

Bauhinia sp.: Different varieties of this plant are locally known as Mahar champa, Jamlar or Patsingra. The plant belongs to family Caesalpinieae. Bark of stem is astringent in nature so it is used to cure diarrhea and dysentery. Paste of bark is used as antidote in snakebite and scorpion bite. Both leaf and bark is used in various types of skin diseases. Root of this plant is carminative in nature.

Boerhaavia diffusa: The plant is commonly known as Punerneva and belongs to family Nyctaginaceae. Root of the plant is diurative, laxative and expectorant in nature. Whole plant is used in various types of ailments like stomach disorder, anemia, odema, jaundice, scanty urine and infection of kidney. The plant also acts as antidote in snakebite. Active principle found in the plant Punernavine is highly effective in lowering the B.P.

Boswellia sersata: The plant is commonly known as Salai or Luban and it belongs to family Burseraceae. Gum of this plant is of therapeutic value and is diaphoratic, diuratic, and astringent in nature. It is administered in rheumatism, nervous and skin diseases.

Butea monosperma: Popularly known as Palash the plant belongs to family Papilionaceae. Seeds of this plant is anthelminthic in nature and is especially effective in children. Gum and leaves are astringent as well as used as tonic. Flower is astringent, diuretic, depurative and aphrodisiac in nature.

Casearia esculanta: The plant is commonly known as Churchuri and belongs to family Samydaceae. Root is astringent in nature and is also used as liver tonic. Decoction of root is recommended as medicine in diabetes and piles. Bark is boiled in water and is used to wash body in dropsy.

Cassia fistula: The plant belongs to family Caesalpinieae and is locally known as Amaltas. Root, bark, seeds and leaves are laxative in nature. Crushed fruit is applied in rheumatism and snakebite. Juice of leaf is externally applied in skin diseases. Root is astringent, febrifuge, tonic and purgative in nature.

Cleome viscose: The plant belongs to family Capparidaceae and is popularly known as Hurhur. Root of the plant is rubefacient, vesicant and sudorific in nature. Crushed leaf is externally used in wound and ulcer as it helps in healing. Seed is vesicant, carminative and rubefacient in nature. Juice of leaf is used to cure earache especially in children.

Corchorus capsularis: This plant is commonly known as Titpat and it belongs to family Tiliaceae. Infusion of leaf has got demulcent, stomachic, laxative and carminative properties. Leaf is also used as stimulant, appetizer and tonic. Decoction of root is administered in dysentery, fever, diarrhea and liver disorder.

Cordia dichotoma: This plant is locally called Buch and it belongs to family Boraginaceae. Fruit is astringent, anathematic, diuretic, demulcent and expectorant in nature. Dried fruit is used in infection of urinary passage and diseases of lung and spleen. Decoction of bark is used in fever and dyspepsia. Kernel is externally applied in ringworm. Paste of leaf is externally applied in ulcer, headache and snakebite.

Desmodium trifolium: The plant is popularly known as Kudalis and it belongs to family Cruciferae. Leaf of this plant is used in diarrhea and dysentery. Leaf is also used to check convulsion. Crushed leaf is applied to wound.

Dioscoria bulbifera: Commonly known as Ratalu, the plant belongs to family Dioscoriaceae. Tuber of this plant is boiled with ghee and applied in piles, syphilis and ulcer.

Diospyros melanoxylon: This plant belongs to family Ebenaceae and is commonly known as Tendu. Bark of plant is astringent in nature. Decoction of bark is used as tonic and to cure diarrhea as well as dysentery. Boiled bark is also used as eye lotion. Leaf is diuretic, carminative, laxative and styptic in nature. Dried flower is used to cure urinary disease, skin disease and blood disease.

Echinocloa sp.: The plant belongs to family Poaceae and is locally known as Sanwak. The plant is used to cure diseases of spleen and to check hemorrhage. Leaves are used as purgative in constipation.

Emblica officinalis: Popularly known as Amla, the plant belongs to family Euphorbiaceae. Fruit of this tree is cooling refrigerant, diuretic and laxative in nature. Fruits are used in various ailments like diarrhea, dysentery, cough, anemia and jaundice. Exudation of fruit is used externally in inflammation of eyes.

Euphorbia hirta: This plant is locally known as Dudhi and it belongs to family Euphorbiaceae. The plant is recommended to children in cough, bowel trouble and worm of stomach. Juice of plant is administered in dysentery. Decoction of plant is given in bronchial infection and asthma. Latex of plant is applied on warts.

Evolvulus alsinoides: The plant is locally known as Shankhpushpi and it belongs to family Convolvulaceae. Plant is febrifuge, vermifuge and a good brain tonic.

Holarrhena antidysentrica: Plant is commonly known as Kurchi and it belongs to family Apocyanaceae. Bark of the plant is used to control dysentery. Dried bark is boiled in water and used to wash body in dropsy. Seeds are astringent and febrifuge and administered in fever and intestinal worms.

Lagerstroemia parviflora: This is a tree belonging to family Lytheraceae and is commonly known as Sidha. Bark and leaf is purgative in nature. Root is astringent, stimulant and febrifuge. Crushed fruit is locally applied in insect bite.

Madhuca indica: Locally known as Mahua the plant belongs to family Sapotaceae. Flower is astringent, tonic, appetizer, cooling, nutritive and is used in cough in the form of decoction. Dried flower is used to foment in orchitis for sedative action. Flowers are fried in ghee and eaten by patients of piles.

Melia azedarach: The plant is locally known as Bakain and it belongs to family Meliaceae. Almost all parts of plant is deobstruent, resolvent and alexifarming in nature. Flower and leaf are applied externally as poultice to relieve headache. Juice of leaf used internally as anathematic, antilithic, diuratic and emmenogouge. Seeds are prescribed in rheumatism. Fruit, leaf and bark are used externally in leprosy and scrofula.

Morinda citrifolia: Locally known as Cheli, the plant belongs to family Rubiaceae. Leaf is administered internally as tonic and febrifuge. The juice of leaf is applied on wound and ulcer. Fruit is baked in ghee and given to the patient of asthma. Charred fruit is mixed with salt and applied on spongy gum. Juice of leaf is applied on gout.

Swertia chirata: This plant belongs to family Gentianaceae and is locally known, as Chireta.Plant is bitter, tonic, stomachic, febrifuge and laxative. Decoction of plant is used as blood purifier.

Terminalia arjuna: Locally known as Arjun, the plant belongs to family Combrataceae. Bark of the plant is astringent, tonic and febrifuge. It is used in cardiac disorder as tonic. The leaves are used in bilious affection, for sores and antidote to poison. Juice of fruit and leaf are administered in earache. Ash of bark is used in scorpion bite.

Terminalia belerica: This plant is commonly known as Bahera. Fruit is bitter, astringent, tonic, and laxative and antipyretic in properties. Fruits are variously used in piles, dropsy, leprosy, biliousness and headache. Half ripe fruit is used as purgative.

Terminalia chibula: Locally known as Harre, fruits of this plant are astringent, laxative, alterative and applied externally on chronic ulcer, wound and as gargle to stomatitic. Powder of fruit is dentifrice and is helpful in curing teeth ache and bleeding gum.

SOME REMEDIAL PREPARATIONS OF JHARKHAND

Combination of different plants and their specific preparations are common to cure various ailments. Plants of antiseptic nature like neem, karang, mango, turmeric and tamarind are used even to avoid surgery. Leaves of these plants are boiled in water and the preparation is used to wash wound. Dry and powered kernel of mango is mixed with karang oil and the paste is applied several times a day on wound. This helps in healing of wound. Paste of turmeric and powder of tamarind leaf is heated in limewater and the paste is applied on limb with fractured bone. Twice or thrice application for three days joins the bone (Dey1980).

Similarly there are various popular preparations to cure bowel trouble. Powder of leaf of (*Trachyspermum ammi*) is heated in mustered oil and the lukewarm solution is massaged on stomach of children to relieve stomachache. Leaves of mango, jamun and guava are crushed and their juice is given in loose motion. Similarly bark of bahera is boiled with water and warm preparation is given in case of dysentery.

Being predominantly forest area, cases of snake bite and scorpion bite are common and patients are treated with the help of various indigenous plants. Leaves of arjun, sidha, buch and amaltas are made into paste and applied in snakebite or scorpion bite. The paste not only checks pain but also acts as antidote to poison.

Similarly several locally available plants are used to cure fever and common respiratory diseases. Root of Ghorabuch is crushed and boiled with water to make a paste. Lukewarm paste is applied on chest of children suffering from fever and cough (Singh and Khan, 1990). Gum of babul is dried powdered and mixed with water. One teaspoon of this mixture is taken for a month in case of dry cough.

Skin and other types of infection are common ailments of Jharkhand region. Various herbal medicines are used for therapeutic purpose. Oil extracted from seeds of karang is regularly applied on infected portion for faster healing. Likewise gum of Sal, resin of guggul and leaves of neem are fumigated in the room of person suffering from infectious and contagious diseases, generally in evening hour.

Like above-mentioned diseases native people of Jharkhand have their own herbal answer to problems of family planning. Leaves of neem are dried, fumigated and female expose their private parts before intercourse. This considerably reduces the chance of conceiving. Similarly powder of plant *Vicoa indica* commonly known as banjhauri is taken as decoction to prevent chances of conceiving. Local population has also got medicinal preparation to induce permanent sterility. One spoonful of fresh juice of *Lasia spinosa* locally known as kanta saru is mixed with half liter of cow milk and taken orally in the morning on the third day of delivery of a child for three consecutive days (Sinha, 1995). This course makes the women permanently sterile. Another plant Bridalia spinosa locally known as kaj is also used to induce permanent sterility. Bark of the leaf is made into decoction and half cup of this preparation is given to female during first three days of menstruation. *Alangium lamaruii* and *Reullia suffroticosa* are two other plants used for identical purposes.

Local people have their own traditional method for termination of unwanted pregnancy. Root of *Reullia sp.* or *Alangium sp.* is powdered and mixed with water and given to terminate early pregnancy. Likewise seeds of carrot, sesame seeds and jaggery are boiled and warm preparation is administered to terminate unwanted pregnancy.

CONCLUSION

The very name of the state Jharkhand means a place of plants and trees. Importance of plants is very much evident in the day-to-day life of people of this state. Almost all needs of people of Jharkhand is fulfilled through forest and forest products. The life style of tribal is so congenially associated with the forest that while drawing benefit from forest utmost care is taken to maintain the forest wealth. Although with rapid industrialization and influx of outsiders things have changed considerably. The situation is almost true with health care and therapeutic system of Jharkhand. Local physicians known as Ojhas have rich traditional knowledge about medicinal plants and they used to cure almost all types of diseases with the help of them. Recent popularity of synthetic drugs has considerably damaged traditional therapeutic system and the traditional treasure of knowledge is fading out rapidly. Industrialization and mining activities have taken their toll of forests. A large number of medicinal plants are either not available or collected with great difficulty. Lack of well documented description is another bottleneck in propagation of this system. In most of the cases local people are less willing to share their knowledge with other people. This is high time the time tested and in majority of cases scientifically tested herbal medicines are conserved. Certain conventions and procedures of using these medicinal plants look bizarre in the first instance but their scientific validity needs to be tested. For example there is clear instruction regarding the mode of collection of plants specially the time of plucking. Some plants are suggested to be plucked in the morning hours while others are advised to be collected in the evening. Although at presently scientific proof of this suggestion is presently not available but there are sufficient background for further investigation. The concept pf endogenous circadian rhythm commonly known as biological clock is well known and it clearly indicates that physiological activities of plants are at optimum level during certain period of day. It is just possible that the time suggested for plucking a particular plant coincides the time when maximum alkaloid is available. So a detailed and scientific study of different aspects of ethno medicinal plants of this state is desired. It is a matter of pity that no organized step has been taken in this direction so far. Another important aspect is ever depleting number of medicinal plants. Industries and mines have engulfed vast forest area and this has got its significant implications on number and volume of medicinal plants of Jharkhand. So all possible steps should be taken to preserve them. Some steps have been taken at both governmental as well non-governmental levels in this regard. Parks of medicinal plants are being set up and medicinal plants are being tried to reclaim wasteland. There is need for coordinated effort from all local- government, NGO, industries, general people and mine operators.

REFERENCES

Dey A.C. (1980). *Indian medicinal plants used in Ayurvedic preparation*. Publication New Connaught Place, Dehradoon (U.P.) p. 165.

Singh M.B. and Khan A. (1990). *Indian Medicinal Plants* Todays and Tomorrow Publications, New Delhi.

Sinha A. (1995). *Traditional Indigenous Heal h Care Practices in Ho Tribes* Ph.D. Thesis. Avinashilingam Deemed University, Coimbtoor.

11

Economic Benefits of the Acacia for the People of Dry Region

G. Singh and *S. Mutha*

Sustainable management for timber, fuel and non-wood products for the rural communities and to fulfill the demands by urban communities for fuelwood cannot be met from forest resources within the arid and semi-arid regions. Furthermore, in severely desertified areas there is very little hope of improvement. However, *Acacia* species may be beneficial in the form of their utilization for fuel, timber, forage, gum, tannins, fibre, medicine, food, handicrafts, domestic utensils, environmental protection, soil fertility, shade and shelter, amenity and ornamental plantings, and agroforestry. The four most widely distributed and utilized species of this genus are *A. nilotica, A. senegal, A. tortilis* and *Faidherbia albida*. Their distribution and role in tropical dry zone are briefly described. Some of the acacias are introduced. However, there are problems in the management of these *Acacia* communities because of anthropogenic pressures. Protection of local plant communities through sound management is recommended in order to form a base from which it is possible to start reclamation of the degraded regions and therefore to improve the economic status of the local people.

BACKGROUND

This article is an attempt to study the role played by *Acacia* species in the rural economy of the drier regions. There are about 1250 species of *Acacia*, of which 134 species are native of Africa, with 20 species extending into Asia and 6 species native to East Asia (Ross, 1979; Lock, 1989, 1991). *Faidherbia albida* (formerly known as *Acacia albida)* is also included here because of its role in the rural economy and soil fertility improvement similar to *P. cineraria* in western Rajasthan. Despite the richness of *Acacia* species, relatively few appear to have been investigated in detail. They are *Acacia nilotica, A. senegal, A. tortilis and Faidherbia albida* (Brenan, 1983). This does not imply that the other species are worthless. Indeed, some are extremely important locally; all species make some contribution to the environment and rural economy by way of shade, shelter, soil stabilization and fertility, browse, fuel, etc.

THE ECOSYSTEMS

Ecosystems represent a state of equilibrium between the rainfall and soils in terms of available moisture and anthropogenic factors such as land clearance, grazing, cultivation, fuelwood etc. Excessive anthropogenic pressure will eventually lead to desertification in arid and semi-arid areas and are ultimately more damaging than deforestation of the rainforests.

The semi-arid ecosystems of tropical Africa are, with few exceptions, either thorn scrub or thorn savanna, with *Acacia* as the dominant species, either forming pure stands, e.g. *Acacia mellifera* thorn scrub, *A. seyal* thorn savanna, *Faidherbia albida* riparian woodland, etc. These ecosystems are important rangelands for the livestock production, especially in those areas where *A. tortilis* is widespread. *Acacia* species plays an essential part in the dietary requirements of herbivores that requires an intake of 20-25% browse, of which 5% is consumed during the rainy season when herbage is relatively plentiful and 30% during the long dry season. Loss of these species would mean that herbivores would no longer be able to live through the dry season without costly and currently uneconomic food supplements. Livestock production is undoubtedly the most suitable industry for the utilization of the rangelands but full regard must be made for sustainability not for exploitation. It has been suggested that, if the present trend in overexploitation and desertification continues unhindered, most of the browse species would disappear within the next 20 to 25 years and the livestock, which is the main economic activity in dry regions, would be seriously jeopardized (Le Houérou, 1989). The major regions of acacias distribution in Africa are presented in Table 1. In India, the distribution of acacias ranged from arid to dry sub-humid regions.

Table 1. Vegetation regions of Africa and the major subdivisions containing dry zone *Acacia* species (White 1983)

Region	*Area (km^2)*	*Distribution*
Mediterranean/	473,000	*Argania spinosa* scrub forest and bushland
Sahara Regional		*Acacia gummifera-Ziziphus* lotus bushland
Transition Zone		Succulent sub-Mediterranean shrubland
		Sub-Mediterranean anthropic landscape
Sahara Regional	2,842,000	Atlantic coastal desert
Transition Zone		Red Sea coastal desert
		Sahel semi-desert grassland
		Sahel Acacia wooded grassland and deciduous bushland
Somalia-Masai Region	1,873,000	Somalia-Masai Acacia-Commiphora deciduous bushland and ticket
Karoo-Namib	661,000	Kalahari deciduous Acacia
Region		bushland and wooded grassland
		Namib Desert
Total area	5,849,000	

GROWTH AND PRODUCTIVITY

There are very few quantified studies of fodder production from the acacias. Results from *A. tortilis subsp. heteracantha* in Africa (Milton, 1983), provide an estimated yield of 1 ton ha^{-1} yr^{-1} shoot and leaf dry weight, a figure that agrees favourably with foliage production in India from *A. tortilis* of

2.5 kg tree^{-1} yr^{-1} from young plantations at the plant density of 400 trees ha^{-1}. Production of above ground dry biomass of *A. tortilis* in dune of Indian desert was estimated to be 3.3 kg tree^{-1} yr^{-1} at the plant density of 400 trees ha^{-1} with additional benefits of forage production from the surface vegetation (Table 2). Booth (1966), working in Kordofan Province, Sudan, also obtained an oven-dry weight of 2.5 kg of leaf litter from a seven-year old *A. senegal*. Tables 3 provide some data on foliage and pod production of *A. senegal* in Senegal, Africa.

Table 2. Biomass of the 50 months old tree seedlings (kg seedling^{-1}) of *A. tortilis* on a shifting dune in association of CA: *Cassia angustifolia*; CC: *Cenchrus ciliaris* and CL: control; F wt: fresh weight; D wt: dry weight. Source: Singh and Rathod (2002)

Tree species		*Above biomass (kg plant^{-1})*		*Biomass of surface vegetation (tones ha^{-1} fresh weight)*		
		F wt	D wt	1998	1999	2000
A. tortilis	*CA*	18.8	13.3	3.97	3.80	3.70
	CC	19.4	13.0	0.97	1.40	1.29
	CL	19.2	13.2	–	0.30	0.46

Table 3. Annual foliage and pod biomass production of *A. senegal* in Senegal (Source: Bille, 1980)

Rainfall mm/yr	*Diameter cm*	*Twigs kg*	*Leaves g*	*Pods g*
250	6.4	2.9	580	200
	10.5	4.3	860	300
	15.9	19.2	3840	1340

Faidherbia albida in Mali over a number of sites produced 300 kg dry forage ha^{-1} yr^{-1} (Cissé and Koné, 1992), while pod yields varied from 200 to 600 kg/ha/yr (Montgolfier-Kouevi and Le Houérou, 1983). Pellew (1983a) obtained dry weight forage productivity of 5000 kg ha^{-1} yr^{-1} for *A. xanthophloea*, and 1725 kg/ha/yr for *A. tortilis*. Leaf biomass production had significant correlation with trunk circumference and annual foliage production had an exponential relationship per species (Bille, 1983). It was also related to rainfall and ecological environment and ranges from 100 kg to 1000 kg of edible DM ha^{-1} yr^{-1}. The feed value of the foliage in terms of energy is generally low to medium, ranging from 0.25 to 0.40 FU per kg of DM, i.e. 400 to 700 Kcal per kg of DM, while digestible protein is usually high, normally above 4% and often above 10% and may even reach as high as 30% in the Capparaceae (Le Houérou, 1983).

WOOD PRODUCTS

Fuel wood

The present trend of a 2-3% annual population increases and its concomitant increase in demands for more fuel and land for food has placed an impossible strain on existing forest resources, especially in the drier regions. Some of the countries are forced to reduce their fuel consumption; great log fires no longer burn overnight in the compounds (Riley and Brokensha, 1988). Similar economies must surely be practiced, where an exponential population explosion creates a demand that exceeds natural regeneration of the fuel source.

It indicates that the current use of trees for fuel is exceeding incremental growth. The standing volume obtainable from some of the best 20-year-old *Acacia* woodlands is only 8.5 m^3 ha, equivalent to about 0.42m^3/ha of mean annual increment. Wilson (1983) obtained annual values per capita of 0.50-0.62 m^3 of fuelwood for urban use in Niono, Mali, 0.62 m^3 in northern Ethiopia and 0.63 m^3 in northern Nigeria. In those areas of Ethiopia where wood is extremely limited, cattle dung has now become the principal fuel with annual fuelwood consumption per capita of 0.15 m^3 and, in areas where wood is still abundant, it is as high as 1.5 m^3. The annual fuelwood requirements for one person ranged from 0.7 m^3 (Maydell, 1987) to 1.5 m^3 in the Sahel and 2.1 m^3 (Hines and Eckman, 1993) in Tanzania. For the higher ranges it is uncertain whether fuel consumption is for domestic use only or includes bakeries, brewing, brick kilns, potteries, blacksmiths, etc. Since the degree of drying and efficiency as a fuel also affects fuelwood consumption, any conclusions based on the above figures would be inconclusive apart from obvious deduction that fuelwood consumption is exceeding supply and must inevitably affect other forms of tree usage too.

Firewood

The criteria for the selection of firewood species are (Riley and Brokensha, 1988; Hines and Eckman, 1993):

- Rapid growth, high volume production, ability to coppice or sprout readily when cut and that requirs minimum management.
- Wood would dense with a low moisture content, relatively easy to cut, easy to handle, with few or no thorns, splitting easily and readily transportable; wood collected during the rainy season should not absorb moisture.
- Slow burning with high calorific value, producing very little smoke without objectionable toxic fumes, and without spits and sparks.

Because of their high calorific value, preferred acacia species are *Acacia ehrenbergian, A. tortilis subsp. raddiana* and *A. nilotica subspp. adstringens, nilotica* and *tomentosa,* while the wood of *A. macrostachya* and *A. erythrocalyx* in Africa (Guinko, 1991). *Acacia seyal* however played a special role in the supply of charcoal to growing cities in Sudano-Sahelian Africa. Because of decreasing availability there is an increasing gap between preference and actual use. The deciding factor is availability and ease of collection (Riley and Brokensha, 1988). The use of such an obnoxious fuel as *A. oerfota* in the Yemen can only be tolerated because all other fuelwood sources have already been exploited; its local abundance in the lowlands is attributable to overgrazing and the replacement of more palatable species by the less palatable (Al-Hubaish and Müller-Hohenstein, 1984).

Charcoal

Charcoal is often preferred in urban areas. The desirable criteria for charcoal are similar to those for fuelwood but requiring large, uniform stands producing wood of suitable dimensions and a high specific gravity (0.7-0.9) for efficiency and ease of manufacture. The characteristics of the finished product are: moisture <7%, ash <3 %, fixed carbon >75 % and apparent density >0.3 g/cc (FAO, 1962).

Timber

Acacias are also useful in producing valuable timber for its utilization in many industrial processes. The desirable criteria for construction and pole timber defined by Hines and Eckman (1993) are:

- Rapid growing, with straight stems of uniform size and small branches, naturally pruning and rapidly self-healing.
- Possess good physical, mechanical, seasoning, preserving, and processing properties and resistant to insect, fungus and rot.

For domestic use and handicrafts, such as platters, spoons, shuttles, mortars, tool handles, musical instruments, carvings, bows, arrows, etc. shape, strength, density, ability to peel, work or carve easily, freedom from splinters and, for such items as spoons and bowls, low permeability to liquids, are additional requirements. A number of species provide pole timber, which is widely used in the domestic economy for construction purposes. Commercial sawn timber is largely obtained from *A. nilotica*.

NON-WOOD PRODUCTS

Forage

Forage are the basic requirement for sustainable livestock productivity in dry areas. The desirable criteria (Hines and Eckman, 1993) for forage are:

- Provision of palatable, non-toxic, nutritious foliage and fruit.
- Ability to withstand browsing, lopping, pruning and coppicing
- Quick growing, especially during the early stages of growth.

Acacias provide a valuable browse for game and livestock as well as being a valuable source of highly nutritious pods, which can be stored as a dry season supplement for livestock. Depending on the species, a flush of growth is often available at the end of the dry season before the grasses have begun to grow following the first rains. Browse is essential to all herbivores in arid and semi-arid environments since grasses alone are unable to supply maintenance requirements for more than a few months of the year. The crude protein content of most browse is generally considerably higher than that for grasses except during the early growing season (Walker, 1983). In the arid, semi-arid and sub-humid zones of sub-Saharan Africa, woody species are an important source of fodder as well as exerting an influence on the seasonality and productivity of the grass cover growing beneath their canopy.

Nutritional Value

Calcium content is high in the bark of *A. tortilis* subsp. *spirocarpa* (5.68%) and *A. xanthophloea* (4.07%). Cyanogenic glucosides have been reported by Steyn and Rimington (1935) in the pods of *A. erioloba, A. hebeclada* subsp. *hebeclada, A. lasiopetala, A. tortilis* subsp. *heteracantha* and *A. robusta*. Tannins are also present in most vascular plants and in a number of acacia species. It acts as an anti-nutritional factor due to their ability to precipitate proteins from an aqueous solution. Condensed tannins are more effective in reducing digestibility whereas hydrolysable tannins cause various toxic manifestations. Dietary condensed tannins (2-3%) may even assist rumen digestion by forming a protein-tannin complex, thereby reducing wasteful protein degradation. Condensed tannins present in *A. nilotica* pods have been reported to lower growth rate in sheep due to their ability to reduce nitrogen and NDF digestibility; similar to the hydrolysable tannins in *A. sieberiana* pods (Kumar, 1992).

A number of species are regarded locally as a good source of pollen and nectar for bees. Wild honey is largely collected from natural cavities in trees or from crude hives suspended in the trees. Acacia honey is an underexploited food resource for both local and export markets and should be encouraged in benefits of the local people.

Gum

Gum arabic was traditionally defined as 'the gummy exudate from *Acacia senegal* or its related 18 other species. Gum arabic is used in the food industry to fix flavours and as an emulsifier, to prevent the crystalization of sugar in confectionery products, as a stabilizer in frozen dairy products; its viscosity and adhesive properties find use in bakery products, and as a foam stabilizer and clouding agent in beer. In the pharmaceutical industry gum arabic is used as a stabilizer for emulsions, binder and coating for tablets, and as an ingredient in cough drops and syrups. In cosmetics it finds use as an adhesive for facial masks and powders, and to give a smooth feel to lotions. Industrially, gum arabic is applied as an adhesive, as a protective colloid and safeguarding agent for inks, sensitizer for lithographic plates, coatings for special papers, sizing agent to give body to certain fabrics, and anti-corrosive coating for metals; it is also used in the manufacture of matches and ceramic pottery (Cossalter, 1991). It is a polysaccharide exudates. The peoples in Africa are able to survive on gum for days and harvest gum to survive on a daily ration of 170 g (Grieve, 1931). Gums are also widely used in traditional medicine since Pharaonic times as a soothing and softening agent, being taken internally for coughs, diarrhea, dysentery, haemorrhage, as well as being used externally to cover inflamed areas.

The increasing international pressure towards tighter trade specifications and labelling regulations, identity and purity has led to the Revised Specification (WHO, 1990a, b; FAO, 1990) where gum arabic is defined as originating from *A. senegal* or closely related species, with a specific optical rotation range of -26° to -34° and a Kjeldahl nitrogen content of 0.27-0.39%. This has limited the designation of gum arabic to members of the subgenus *aculeiferum* which, in addition to *A. senegal*, includes *A. laeta*, *A. mellifera*, *A. polyacantha* subsp. *campylacantha*, although gum is not available commercially from any of these closely related species. The gums containing tannins and with a positive optical rotation from species such as *A. seyal*, *A. xanthophloea*, *A. karroo* and *A. nilotica* are now excluded for use in food and consequently attract a lower price (Anderson, 1993); the presence of tannin in these gums is also considered carcenogenic. Exudates from species other than *A. senegal*, occur as small tears and dribblets; their collection is consequently extremely time-consuming. This, together with the low price, currently at $US 1000 per tonne, compared to $US 5000 per tonne for gum arabic, is likely to kill the export trade in acacia gums but is unlikely to have any serious effect on internal use as an edible commodity, adhesive and ingedient of traditional medicines.

In the past, the rural population of the Sudan "gum belt" practiced a tree fallow rotation system with *A. senegal*, forming practically pure stands during the fallow period, thereby increasing soil fertility and reducing soil erosion while, at the same time, providing a dry-season source of income from the gum harvest. With increasing human and livestock populations and pressure on the cultivatable and grazing lands, the length of the tree fallow was decreased or, in some instances, eliminated. The carrying capacity of the range land also decreased, resulting in the necessity to fell the young trees for forage. The poor prices paid to the farmers for their gum has also encouraged the cutting of trees for firewood and charcoal for an albeit short-term source of income.

Tannins

Pods of *Acacia nilotica subsp. nilotica* are intensively used for tannin exploitation under the name 'Nep Nep'; also in the Sudan as 'sunt grains'. Sunt grains are obtained mainly from the crushed pods of *A. nilotica subsp. nilotica* or subsp. *tomentosa*, after first removing the seeds; most of the pod cases are then removed by sifting. The residue contains 50-60% tannin, which gives a soft, plump, very light coloured leather. However, contamination of the pods with iron and mineral matter causes undesirable spotting of the leather. In Kano, Nigeria subsp. *adstringens* is reported to be the preferred subspecies for tanning goat skins for the Moroccan leather trade, the pod residues are afterwards fed to cattle. In Kenya the bark of the Australian *Acacia meansii* is processed for tannin at Thika.

Fibre

Acacia species are also used for root and stem bark fibre production. The fibres are used in the domestic economy and there appears to be no commercial potential. The general impression obtained is that such usages could also be carried out by other genera and that acacias are not an important fibre source.

Local Medicine

Gum, for example, has an emollient activity, resulting in a softening soothing action on the skin or irritated internal surfaces. The astringent activity of tannins causes a contraction of mucous surfaces, coagulates proteins and is useful in stopping bleeding of small wounds and other discharges (Brown, 1977). Hagos *et al.*, (1987) have demonstrated that the stem bark of *A. tortilis* subsp. *radiana*, which is used in Somalia against asthma, contains pharmaceutically active compounds that inhibit muscular contractions of the lower small intestine of the guinea pig. Further research will doubtless reveal the presence of other pharmaceutically active compounds within the genus. Acetic acid, alcohol and water extracts of the fruits of *A. dudgeoni*, *A. nilotica* subsp. *adstringens* and subsp. *nilotica* have been shown to have molluscocidal activity. The planting of the latter subsp. along waterways could prove beneficial in the control of schistosomiasis (Ayoub, 1982, 1985; Kloos and McCullough, 1987).

Food

The seeds of several African species are known to be eaten, either cooked or raw, but very little is known about their nutritional value. What little evidence is available suggests that *Acacia* seeds could be a very valuable and underexploited food resource. Certainly the Australian species have been found to be generally high in protein (17-25%), fat (416%) and carbohydrate (30-40%), i.e. comparing very favourably with such cereals as wheat and rice and even higher than some meats (Brand and Cherikoff, 1985; Thomson 1992). Edible seeds from *A. senegal* is an important ingredient of 'Panchkuta' usual vegetable in western Rajasthan.

Handicrafts

The uses of acacias in handicrafts are mainly domestic although there is a small commercial potential for sales to the tourist industry. The general impression is that such usages could be carried out equally well by other genera and that acacias are not essential sources of handicraft materials.

ENVIRONMENTAL IMPROVEMENT

Many species, especially in the drier regions, have extensive, subsurface lateral roots to take advantage of any light rainfall that might occur. Such subsurface root systems help to stabilize the soil. The aerial system too has a role in reducing wind erosion and ameliorating the microclimate although information on changes in humidity and temperature is generally lacking. Such species as *A. nilotica* subsp. *indica*, *A. senegal*, *A. tortiliis* and *F. albida* have been widely planted to fix and stabilize sand dunes and combat wind erosion (Singh and Rathod, 2002).

The shade effect by reducing air temperatures also reduces soil evaporation and lowers soil surface temperatures; factual information on shade and soil temperatures is poorly documented. Midafternoon soil temperatures beneath the canopy of *A. tortilis* subsp. *spirocarpa* were reduced by 3-12°C including other favouarble changes (Belsky, 1994; Belsky and Amundson, 1994; Belsky *et al.*, 1989, 1993). However, in the absence of shade the high surface temperatures attained by bare soils must have an adverse and possibly lethal effect on any dormant or germinating seed lying on the soil surface, discouraging revegetation and encouraging desertification.

The lower shade temperatures also encourage a more mesophytic and generally more nutritious ground cover, which is further encouraged by the higher soil moisture in the vicinity of the trunk as a result of stem flow. The rainfall collected as a result of stem flow is not immediately available to the lateral roots, whose root hairs are more able to take advantage of leaf drip around the periphery of the canopy. Stem flow infiltrates through the soil. Immediately the trees can eventually utilize by the surrounding lateral roots. Measurements of total soil water content under *F. albida* taken just before the start of the rainy season gave a value of 8% beneath the canopy and 4% outside (Radwanski and Wickens, 1967).

Soil Fertility

The direct contribution of acacias to soil fertility is two-fold. Firstly, there is the contribution through nitrogen fixation and, secondly, that due to litter fall. The contribution of the latter is from the recycling of the minerals extracted from the soil by the root system (Radwanski and Wickens, 1967). The contribution in terms of nitrogen is probably minimal since the foliage on the ground probably undergoes two periods of rapid degeneration. The first is the rapid dehydration of the litter and loss of any volatile compounds, so that by the end of the long dry season the second decomposition will be of the remaining fibrous mass and minerals.

Investigations (Singh *et al.*, 2002) on soil properties under four age classes of *Acacia tortilis* indicated increase in soil organic carbon, available nitrogen N (NH_4-N+ NO_3-N), PO_4-P and Ca including micronutrients. Accumulation was greater in 0-15 cm soil layer compared to 15-30 cm layer. The increased soil nutrients under *A.tortilis* plantation had direct impact on site improvement (Table 4).

The exception is that of *Faidherbia albida*, where leaf fall occurs at the start of the rainy season and the litter is consequently readily decomposed and the nutrients incorporated into the soil. It has been demonstrated by Giffard, 1964, that the leaf litter from an average stand of 50 trees of *A. albida* per hectare returns to the soil an annual equivalent of 75 kg N, 12 kg P, 13 kg K, 20 kg S, 25 kg Mg and 120 kg Ca per hectare (Giffard, 1964). The growing of millet *(Pennisetum glaucum)* beneath the tree canopy can result in a 2.5- to 3-fold increase in yield without additional fertilizers. There is also an indirect contribution from livestock browsing on the fallen litter, the decomposition of the dung following similar cycles to that of the litter. The quantitative effect of this contribution does not appear to have been evaluated.

Table 4. Soil nutrients of *A.tortilis* plantation under the influence of stand age. *Source:* **Singh *et al.*, (2002).**

Stand age	*Soil depth*	*Soil parameters* SOC	NH_4-N	NO_3-N	PO_4-P	Ca	Mg	K
		%	mg kg^{-1}	mg kg^{-1}	mg kg^{-1}	g kg^{-1}	g kg^{-1}	g kg^{-1}
Y0	0-15	0.24	1.32	0.44	4.34	18.18	0.67	0.68
	15-30	0.21	1.02	1.23	3.35	18.35	0.60	0.59
Y4	0-15	0.38	3.45	2.92	7.45	17.96	0.63	0.49
	15-30	0.33	1.45	1.61	3.89	18.00	0.56	0.47
Y9	0-15	0.52	6.09	6.76	12.56	18.43	0.58	0.53
	15-30	0.42	1.93	1.47	3.98	18.08	0.52	0.51
Y12	0-15	0.55	5.80	7.05	13.52	18.90	0.55	0.59
	15-30	0.45	1.62	2.34	4.52	18.33	0.50	0.49
Y19	0-15	0.85	4.28	9.03	11.26	18.48	0.47	0.65
	15-30	0.57	1.12	4.08	4.45	17.68	0.49	0.48

Hedges, Shade and Shelter

The acacias are also reported for live and brushwood hedges, shade and shelter (Singh and Gupta, 2000). Doubtless other species can also fulfill these functions equally well. The shade value of these trees is evident from the large number of livestock crowding beneath a tree canopy during the midday heat. However, the practice of pastoralists of cutting large branches of such thorny species as *A. nilotica*, *A. tortilis* subsp. *raddiana*, *A. seyal* and *Zyziphus* and *Prosopis* species to construct 'zeribas' in Africa and 'dhola' fencing in western Rajasthan (enclosures for livestock) is severely damaging. This practice is even more extensively used by agriculturists to keep livestock off their cultivated land. These zeribas/dhola fencing last in about two years before being destroyed by termites or even earlier by bush fires. The brushwood is either obtained by drastic mutilation of the trees or, as is often the case with the agriculturists, from trees felled during land clearance. In many areas of the Sahel, where large areas have been cleared in an attempt to counteract the low crop yields due to desertification and low rainfall, there are no longer any trees remaining to provide the brushwood hedges. Live fences using *A. mellifera*, *A. tortilis*, *A. senegal*, *P. juliflora*, and the species of *Euphorbias* and *Opuntias* would be a solution, although the eratic rainfall and protection during early establishment may affect the planting.

Wildlife Resource

Acacias are recorded as attractive to mammals, birds and edible trees are browsed by wild herbivores and are a good source of bee food. It requires a high standard of management, not only to ensure against over-grazing but also to ensure that a proper balance is maintained between the different species, especially predators and prey, as well as between the browsing animal and the vegetation strata utilized. Recruitment of acacias and associated bush encroachment may result from removal of grazing pressures. However, the degree to which browse is utilized by herbivores is species dependent, some are mainly grazers, but during the dry season balance their diet by browsing while other species can thrive satisfactorily on a purely browse diet. This can be explained in terms of the nutritive value and intake of browse by the two groups of herbivores. Since the maintenance requirements for cattle are 0.09 MJ/kg (0.65 FU/kg) of dry matter and the nutritional value of browse varies from 0.04-0.06 MJ/kg (0.25 to 0.40 FU) per kg of dry matter, browse alone cannot maintain cattle but can maintain sheep, who require only 0.05 MJ/kg (0.35 FU/kg) of dry matter, but this browse does not allow for production. However, goats, who only require 0.03 MJ/kg (0.19 FU/kg), can obtain maintenance and production from a purely browse diet. This explains why only goats, camels and some wild herbivores are able to survive in those arid and semi-arid where browse is the main dietary intake and also explains why these animals are less susceptible to catastrophic droughts than cattle and sheep. It also follows that for the yearlong productivity of sheep and cattle, a balance must be maintained between grazing and browsing, with browse forming 20 to 30% of the dry season diet. In West Africa a density of 25-50 trees per hectare is considered optimal in a *Faidherbia albida* - millet agricultural system with a canopy cover of 30-60% (Le Houérou, 1983).

There is a significant difference between the grazing of domestic livestock and that of indigenous game animals, the former being more selective in their grazing habits as well as being less well adapted to coping with the range of vegetation available. For these reasons, the encroachment area of *Acacia mellifera* subsp. *detiens* and *Dichrostachys cinerea* are encouraged for browsing by indigenous game (Erkkilä and Siiskonem, 1992).

Agroforestry

The traditional agricultural system of rotating cultivated land with a period under fallow until such time as the soil has recovered its fertility is effective but slow, and is largely dependent for its speed of recovery on the natural regeneration of indigenous woody species. However, with an ever increasing population demanding more and more land under cultivation to provide more food, a more certain and rapid means of restoring soil fertility is required. Agroforestry systems would appear to ensure a more rapid recovery and longer maintenance of soil fertility than under traditional fallow; with proper planning it should also offer greater protection against erosion in addition to providing a source of fuel, pole timber, wild food, browse, herbal medicines, etc.

The development of agroforestry systems have tended to be concentrated in the more humid tropics, where tree growth is relatively rapid and soil fertility is more speedily restored. There have been relatively few attempts to develop agroforestry in the drier regions. Kessler and Breman (1991) point out that water availability limits primary production in the northern Sahel of West Africa, with nutrient availability the critical factor in the southern Sahel and Sudanian zones. The woody species influence the water balance through rainfall inception, evapotranspiration and water infiltration as well as influencing nutrient redistribution by recycling minerals and increasing fertility. In a southwards direction the benefits of agroforestry increase more than proportionally with the rainfall. The potential to increase nutrient availability through agroforestry systems in the drier regions is limited and that windbreaks would be more beneficial. In fact, Le Houérou (1989) recommends a reticulated system of windbreaks for the Peanut Basin of Senegal where increasing human pressure on the land is destroying the existing millet-*Faidherbia*-millet rotation. His recommended solution is for an increased integration of agroforestry and animal production in order to increase draught power and restore fertility without reducing the area cropped. This involves a perimeter box of primary windbreaks of *A. senegal*, or *F. albida* subdivided internally by secondary windbreaks of *A. holosericea* and further subdivided by tertiary windbreaks of fodder species, according to rainfall, *Combretum aculeatum*, *Bauhinia rufescens* or *Feretia apodanthera* interplanted with *Ziziphus mauritiana* for protection. *F. albida* is an ideal tree for agroforestry in the higher rainfall regions of the semi-arid zone of Africa, the lack of foliage during the rainy season ensures a suitable light regime for the growing crops. The incorporation of leaf litter and the faces of browsing animals into the soil at the start of the dry season without excessive desiccation are conducive to improving soil fertility, an improvement that is manifested by the enhanced yields obtained from crops grown beneath the canopy. The study carried out in Sahelian region of Africa indicates significant influence of *F. albida* on crop yield (Table 5).

In the drier regions of India *Acacia nilotica* subsp. *indica* has been widely grown in agroforestry systems. There have been a number of recent reports of crop yields growing below the canopy are depressed. Root competition for moisture appears to be a factor but whether high tannin compounds in the soil could also be a factor has not been investigated (Adjers and Hadi, 1993). Comparative trials should also be carried out with subsp. *cupressiformis*, which has a narrow canopy, to see whether shade too could be a factor since both subspecies would be expected to more or less equally compete with crops for soil moisture.

MANAGEMENT AND PRODUCTIVITY ENHANCEMENT

Overgrazing causes stunted growth and prevents regeneration. It also leads the encroachment by the unwanted species. Large stands of *A. mellifera* that dominate the clay soils of the Nile and that had supplied charcoal for Khartoum

Table 5. Crop yield under and outside *Faidhebia albida* canopies

Crop	*Grain yield (kg/ ha)*				*Biomass yield (kg/ ha)*				*Source*
	1a	*2a*	*Diff.*	*P*	*1*	*2*	*diff*	*P*	
Millet	660	1668	+153	0.01					Charreau and Vidal, 1965
Millet	1008	1506	+49	–					Maiga, in Kessler, 1992
Sorghum	671	1674	+149	<0.05	2717	3855	+42	<0.05	Depommeir et al., 1992
Sorghum fertlizer	1340	1388	+3	Ns	15870	18140	+14	0.01	Dancett and Poulain, 1969
Sorghum	889	937	+5						Maiga, in Kessler, 1992
Maiz	1920	3390	+76e	<0.01	53g/pl	66	+24	ns	Poschen, 1986
Maiz	–	–	+27	<0.05	–	–	–	<0.05	Depommier, 1996a
Groundnut, mature	1289	992c	-23	Ns	1492	1459c	-2	ns	Louppe *et al.*, 1996
Groundnut	810	1108	+37	0.01	860	1266	+47	0.001	CTFT, 1988
Groundnut fertilizer	954	1136	+19	0.1	1091	1386	+27	0.01	CTFT, 1988

Where, a 1-open controls, 2-sub canopy plots; b-difference= sub canopy-open field/ open field x 100; c-average over the whole sub canopy area; d- total biomass.

Source: FAO, 1997.

have been clear felled for irrigation is the best example. Fuel wood demands of the urban populations must have a large impact on the tree resources of the rural areas, affecting not only the rural population but also the environment and wildlife, etc. It is possible that a wider implementation of agroforestry practices supplemented by selective thinning for fuel and timber from communal shelterbelts could meet the requirements of the rural agrarian communities. For such developments the cooperation of the rural populations is essential

Large scale reseeding and re-afforestation of rangelands is not an economic proposition when considered in the purely economic terms of carrying capacity. The growing of fodder banks as part of the agroforestry rotation can help carry the livestock through the difficult dry season period. Livestock watering points must be so distributed and regulated that overgrazing in the immediate vicinity is minimized, preferably by rotation. The management of *Acacia* communities for the benefit of the rural population in order to meet their needs for fuel, fodder, medicine, etc as well as maintain the environment and its associated wildlife is a tough task. Forestry alone cannot meet all the fuel requirements of the major urban areas of the drylands. Trees must be managed for sustainable production. However, it must be stressed that the most important role of acacias and other tree and shrub species is to stabilize the environment and thus prevent desertification, a role that safeguards both the national and rural economy.

FUTURE PROSPECTS

Over-cultivation, reduction in fallow period, over-grazing and deforestation lead to desertification and degradation of the environment. To reverse the process, to return the land to its pre-desertified condition, is a tough task. Rapid increase in human and livestock populations has created additional pressures which are unlikely to be controlled until too late. There is a finite, environmental limit to the carrying capacities of the drier regions in terms of human and livestock sustainability, even with fertilizer and irrigation inputs. The prospect of a climate change casts additional gloom on future prospects.

The strategy to rehabilitate this degradation should now be to prevent further desertification by concentrating on protecting the higher rainfall regions and later attempt reclamation by working from the protected area into the area of desertification. In this attempt, more thought should be given to preventive rather than remedial measures.

REFERENCES

Adjers, G. and Hadi, T.S (1993). Acacias in agroforestry. In: Awang, K. and Taylor, D.A. (eds.) Acacias for rural, industrial and environmental development. pp. 134-143. Winrock International Institute for Agricultural Research and Food and Agriculture Organization of the United Nations, Bangkok.

Al-Hubaish, A. and Müller-Hohenstein, K. (1984). An introduction to the vegetation of Yemen. Ecological basis, floristic composition, human influence. GTZ, Eschborn.

Anderson, D.M.W. (1993). Some factors influencing the demand for gum arabic (*Acacia senegal,* (L.) Willd.) and other water-soluble tree exudates. *Forest Ecology and Management,* 58: 1-18.

Ayoub, S.M.H. (1982). Molluscicidal properties of *Acacia nilotica. Planta Medica,* 46:181-183.

Ayoub, S.M.H. (1985). Molluscicidal properties of *Acacia nilotica* subsp. *adstringens. Journal of Tropical Medicine and Hygene,* 88:197-200.

Badi, K.H., Ahmed, A. El A. and Bayoumi, A.A.M.S. (1989). The Forests of the Sudan. Agricultural Research Council. Khartoun.

Belsky, A.J. (1994). Influences of trees on savanna productivity: Tests of shade, nutrients, and tree-grass competition. *Ecology,* 75 (4): 922-932.

Belsky, A.J. and Amundson, R.G. (1992). Effect of tree on under storey vegetation and soil at forest/savanna boundaries in East Africa. In: *The Nature and Dynamics of Forest-Savanna Boundaries* (Eds. Furley, P.A. and Proctor, J.).

Belsky, A.J., Amundson, R.G., Duxbury, J.M., Riha, S.J., Ali, A.R. and Mwonga, S.M. (1989). The effects of trees on their physical, chemical and biological environments in a semi-arid savanna in Kenya. *Journal of Applied Ecology,* 26: 1005-1024.

Belsky, A.J., Mwonga, S.M., Amundson, R.G., Duxbury, J.M. and Ali, A.R. (1993). Comparative effects of isolated trees on their undercanopy environments in high- and low-rainfall savannas. *Journal of Applied Ecology*, 30:143-155.

Bille, J.C. (1983). Measuring the primary palatable production of browse plants. In: Le Houérou, H.N. (Ed.), pp. 185-195.

Booth, G.A. (1966). Forestry Report. Study of the gum *"Acacia senegal,"* and the supply of other forestry produce in relation to land use planning. Land and Water Use Survey in Kordofan Province of the Republic of the Sudan. DOX-SUD-A45. Doxiades Associates, Athens.

Brand, J.C. and Cherikoff, V. (1985). The nutritional composition of Australian Aboriginal food plants of the desert regions. In: Wickens, G.E., Goodin, J.R. and Field, D.V. Plants for Arid Lands, pp. 53-69. George Allen and Unwin, London.

Brenan, J.P.M. (1983). Manual on taxonomy of Acacia species. FAO, Rome.

Brown, J.A.C. (1977). Pears medical encyclopaedia. (revised by Hastin-Bennett, A.M.). Sphere Books, London.

Cissé, M.I. and Koné, A.R. (1983). The fodder role of *Acacia albida* Del.: extent of knowledge and prospects. pp. 29-37. In: Vandenbeldt, R.J. (Ed.), Faidherbia albida in the west African semi-arid tropics. Proceedings of a workshop, 22-26 April 1991, Niamey, Niger. ICRISAT, Patancheru and ICRAF, Nairobi.

Charreau, C. and Vidal, P. (1965). Influence de *Acacia albida* del. sur le sol, nutrition minerale et rendements des mils *Pennisetum au* Senegal. *Agronomie Tropicale*, 20: 600-626.

Cossalter, C. (1991). *Acacia senegal,* gum tree with promise for agroforestry. NFT Highlights, NFTA 91-02. Nitrogen Fixing Tree Association, Hawaii.

CTFT (Centre technique forestier tropical). (1988). *Faidherbia albida* PeL) A. Chev, (Synonyme: Acacia albida De/.). *Monographie.* Nogent-sur-Marne, France, CTFT p. 72.

Dancette, C. and Poulain, J.F. (1969). Influence of *Acacia albida* on pedoclimatic factors and crop yields. *African Soils*, 14 (1-2): 143-184.

Depommier, D. (1996a). Structure, dynamique et fonctionnement des parcs a *Faidherbia albida* (Del.) A. Chev. Caracterisation et incidence des facteurs biophysiques et anthropiques sur l'amenagement et le devenir des parcs de Dossi et de Watinoma, Burkina Faso. These cle doctorat. Paris VI, Universite Pierre et Marie Curie.

Depommier, D., Janodet, E. and Oliver, R. (1992). *Faidherbia albida* parks and their influence on soils and crops at Watinoma,

Burkina Faso. *In* R.J. van den Beldt, ed. Faidherbia albida *in the West* African *semi-arid tropics,* pp. 111-115. Proceedings of a workshop, ICRISAT/ICRAF, 22-26 Apr. 1991, Niamey, Niger. ICRISAT, Patancheru, India and ICRAF, Nairobi.

Erkkilä, A. and Siiskonen, H. (1992). Forestry in Namibia. Silva Carelia 20. University of Joensuu, Finland.

FAO (1962). Charcoal for domestic and industrial use. FAO, Rome

FAO (1990). Specifications for identity and purity, pp. 23-25. In: Specifications for identity and purity of certain food additives. Emulsifiers, enzyme preparations, flavouring agents, food colours, thickening agents, miscellaneous food additives. FAO Food Nutrition Paper 49, Rome.

FAO, (1997). Amenagement des forets naturaelles des zone tropicales seches. Cahier FAO consevation 32.

Giffard, P.L. (1964). Les possibilités de . eboisement en Acacia albida au Senegal. Bois et forêts des tropiques 95:21-33.

Grieve, M.A. (1931). Modern Herbal. Jonathan Cape, London.

Guinko, S. (1977). Traitment traditionelle de quelques maladies en Pays Bisa (Republique de Haute-Volta. pp. 26-45. In: Troisième colloque du CAMES. Médecine traditionelle et pharmacopée africaines, Kigali, Rwanda 8-15 October 1977. Conseil Africain et Malagache pour l'Enseignement Superieur, Ouagadoougou, Haute-Volta.

Hagos, M., Samuelsson, G., Kenne, L. and Modawi, B.M. (1987). Isolation of smooth muscle relaxing 1,3-diaryl-propan-2-ol derivatives from *Acacia tortilis. Planta Medica,* 53:27-31.

Hines, D.A. and Eckman, K. (1993). Indigenous Multipurpose Trees of Tanzania: Uses and Economic Benefits for People. FO:Misc/ 93/9 Working Paper, FAO, Rome.

Kessler, J.J. (1992). The influence of karite (*Vitellaria paradoxa*) and nere (*Parkia biglobosa) trees* on sorghum production in Burkina Faso. *Agroforestry Systems*, 17: 97-118.

Kessler, J.J. and Breman, H. (1991). The potential of agroforestry to increase primary production in the Sahelian and Sudanian zone of West Africa. *Agroforestry Systems,* 13:41-62.

Kloos, H. and McCullough, F.S. (1987). Plants with recognized molluscicidal activity. pp 45-108 In: Mott, K E (Ed.) Plant molluscicides. John Wiley and Sons, Chichester.

Kumar, R. (1992). Anti-nutritional factors. The potential risks of toxicity and methods to alleviate them. *In:* Speedy, A. and Pugliese, P.L. (Eds), pp. 145-160.

Le Houérou, H.N. (1983). The role of browse in the management of natural grazing lands. In: Le Houérou, H.N. (Ed.), pp. 329-338.

Le Houérou, H.N. (1989). The Grazing Land Ecosystems of the African Sahel. Springer Verlag, Berlin.

Lock, J.M. (1989). Legumes of Africa: A Check-list. Royal Botanic Gardens, Kew.

Lock, J.M. and Simpson, K. (1991). Legumes of West Asia: A Check-list. Royal Botanic Gardens, Kew.

Louppe, D., N'Dour, B. and Samba, A.N'D. (1996). Influence cle Faidherbia *albida* sur I'arachide et le mil au S6n6gal. Methodologie de mesure et estimations des effets d'arbres emondes avec ou sans parcage d'animaux. *In:* Les parcs A *Faidherbia, Cah. Sci. du CIRAD-For6t,* 12:123-139. Paris, CIRAD-ORSTOM-CORAF.

Maydell, H.J. (1986). von Trees and shrubs of the Sahel - their characteristics and uses. G.T.Z., Eschborn.

Milton, S.J. (1983). *Acacia tortilis* subsp. *heteracantha* productivity in the Tugela dry valley bushveldt: preliminary results. *Bothalial,* 4:767-772.

Pellew, R.A. (1983a). The production and consumption of *Acacia* browse and its potential for animal protein production. In: Le Houérou,, H.N. (Ed.), pp. 223-231.

Poschen, P. (1986). An evaluation of the Acacia albida-based agroforestry practices in the Hararghe highlands of Eastern Ethiopia. *Agroforestry Systems,* 4:129-143.

Radwanski, S.A. and Wickens, G.E. (1967). The ecology of *Acacia albida* on mantle soils in Zalingei, Jebel Marra, Sudan. *Journal of Applied Ecology,* 4: 569-579.

Riley, B. W. and Brokensha, D. (1988). The Mbeere in Kenya. 2 vols. University Press of America, Lanham, MD.

Ross, J.H. (1979). A conspectus of the African *Acacia* species. Memoirs of the Botanical Survey of South Africa No. 44.

Singh, G. and Gupta, G.N. (2000). Living barriers in Indian arid zone and their ecological Benefits. *The Indian Forester,* 126: 257-268.

Singh, G. and Rathod T.R. (2002). Plant growth, biomass production and soil water dynamics in a shifting dune of Indian desert. *Forest Ecology and Management,* 171(3): 309-320.

Singh, G.; Singh, B.; Kuppusamy, V. and Bala, N. (2002). Variations in foliage and soil nutrient composition in different age classes *Acacia tortilis* plantation. *Indian Forester*, 128: 514-522.

Steyn, D.G. and Rimington, C. (1931). The occurrence of cyanogetic glucocides in South African species of *Acacia*. Onderstepoort *Journal of Veterinary Science and Animal Industry,* 4:51-63.

Thomson, L. (1992). Australia's subtropical dry-zone *Acacia* species with human food potential. *In:* House, A.P.N. and Harwood, C.E. (Eds), pp. 3-36.

Walker, B.H. (1980). A review of browse and its role in livestock production in southern Africa. *In:* Le Houérou, H.N. (Ed.), pp. 7-24.

White, F. (1983). The vegetation of Africa. A descriptive memoir to accompany the Unesco/Aetfat/UNSO vegetation map of Africa. UNESCO, Paris.

WHO (1990a). Technological evaluation of certain food additives and contaminants. Food Additive Series 26: 77-79. WHO, Geneva.

WHO (1990b). Evaluation of certain food additives and contaminants. *Technical Report Series,* 798: 24. WHO, Geneva.

Wilson, R.T. (1983). Fuelwood consumption in a central Malian town and its effects on browse availability. In Le Houérou H.N. (Ed.), pp. 473-476.

12

Bamboo—The Green Glod of Northeastern India

Y.C. Tripathi, Sushma Singh and *P. Hazarika*

Bamboos are one of the most important renewable forest produce having the capability to produce maximum biomass per unit area and time as compared with other forest species. It is the fastest growing plant in the world; some varieties grow by a meter a day. Its wide distribution, natural occurrence and easy availability have favoured millions' choice of this versatile material. Bamboo has played an integral role in social, cultural, and economic development, especially in Asia, throughout history. The range of traditional uses has varied from housing to music, from food to transport, from agricultural tools to weapons and from basketry to bridges. Local communities have adapted this versatile material to meet their needs in ingenious and innovative ways.

Unfortunately, the recent past has seen in many cases the slow dissolution of these traditions and uses and their replacement by often less sustainable and less culturally or physically appropriate substitutes. Where once bamboo was used, plastic, metals or wood is used now. This has resulted in obvious and direct problems in environmental terms, as deforestation and climate change reflected in obvious economic hardship for bamboo workers. Bamboo has seen as a backward material and unable to compete with other materials. It has also resulted in less discernible, but deeper, changes in terms of cultural and social affinities. This is especially notable in Asia, where bamboo was once known as the source of goodness.

Bamboo claims a significant status and role in the socio-economy of Northeastern India and national economy as well (Tewari, 1992, Mohan Ram and Tondon, 1997, Seethalaxmi and Kumar, 1998). Out of 23 genera and 126 species of bamboo available in India, 63 species of 20 genera are found in northeast region accounting for 28% of the total bamboo growing area of the country (Biswas, 1988, Rai and Chouhan, 1998). The general consumption pattern of bamboo in India (Table 1) indicates that 35% of bamboo is being consumed by pulp and paper industries while rest of the 65% is used by locals under cottage, construction, furniture and implement industries.

Yet, it is recognized by a small but growing community of people involved in promotion of bamboo that, with new technical inputs, marketing finesse and a renewed commitment, bamboo could take on the role as the "material of the future". This is a realistic expectation, given the physical qualities of bamboo and its long and rich history of use.

Table 1. Consumption of Bamboo in India

S.No.	*Uses*	*Percentage consumption*
1.	Pulp & Paper	35.0
2.	Housing	20.0
3.	Nonresidential	5.0
4.	Rural uses	20.0
5.	Fuel	8.5
6.	Packing including baskets	5.0
7.	Transport	1.5
8.	Furniture	1.0
9.	Other wood industries	1.0
10.	Other including ladder, Scaffolding	3.0

DISTRIBUTION OF BAMBOO IN NORTHEASTERN INDIA

Bamboos are found naturally in many types of forests. Their distribution in India is largely governed by rainfall, temperature, altitude and soil type. Most of the bamboos require a temperature from 8° to 36°C, a minimum of 100 cm annual rainfall and high humidity for good growth. They form an important constituent of many deciduous and evergreen forests from the tropical to mild temperate regions and from sea level and flat alluvial plains to high mountains, up to an altitude of 4000 meters. Bamboos are native to all the continents except Europe and most of them confined in South and Southeast Asia. The bamboo with *ca* 70 genera and about 1200 species are mostly distributed in tropical, subtropical and temperate zones of different parts of the world (Biswas, 1988). Worldwide distribution of bamboo species belonging to some important genera is presented in Table 2. Tropical Asia can be referred as center of origin of bamboo germplasm with as many as 45 genera and 750 species. Indo-Malaysian region has about 25 genera represented by about 160 species of bamboo. The total forest area of India is 76.52 million ha, out of which 8.96 million ha are covered by different species of bamboo (Rai and Chauhan, 1998). This estimate does not include bamboos growing in farmlands. According to Rai and Chauhan (1998), the states/

Table 2. Worldwide Distribution of Bamboo

Genera	*Number of species*		
	World	*India*	*N. E. India*
Arundinaria	150	10	10
Bambusa	70	22	17
Cephalostachyum	12	7	5
Chimonobambusa	14	9	6
Dendrocalamus	20	15	10
Thamnocalamus	5	4	4

UTs, which constitute a large proportion of the growing stock, of which Assam contributes 16%, Manipur (14%), Mizoram (14%), Arunachal Pradesh (12%), Madhya Pradesh (12%), Orrisa (7%), Meghalaya (6%), Nagaland (5%) and Maharastra (5%). Northeastern states comprise 34% of India's bamboo growing stock and rest 66% in other states of the country (Table 3). There is 22 genera of bamboos distributed in India, Nineteen (19) are indigenous and 3 are exotic (Mauria and Arora, 1988). The indigenous genera of Bamboo in India are *Arundinaria, Bambusa, Cephalostachyum, Chimonobambusa, Dendrocalamus, Dinochloa, Gigantochloa, Indocalamus, Melochanna,*

Neohouzeaga, Ochlandra, Oxytenanthera, Phyllostachys, Pseudostachyam, Schizostachyum, Seminarundinaria, Sinobambusa, Teinostachyum and *Thyrsotachystachys*. The three exotic species that introduced to cultivate occasionally in some parts of the country are *Guadua, Pseudosasa* and *Thyrsostachys*.

Table 3. Availability of Bamboo in India

S.No.	*State/Region*	*Area (%)*	*Growing stock (%)*
1.	Northeast	28.0	66
2.	M.P.	20.3	12
3.	Maharashtra	9.9	5
4.	Orissa	8.7	7
5.	A.P.	7.4	2
6.	Karnataka	5.5	3
7.	others	20.2	5

Northeastern region of India comprising States of Sikkim, North Bengal, Arunachal Pradesh, Assam, Manipur, Meghalaya, Mizoram, Nagaland and Tripura, has over 50% of bamboo species of total species reported from India *ca* 115. All Nineteen genera of bamboo found in India are available in N.E. region. The genera such as *Arundinaria, Bambusa, Cephalostachyum, Chimonobambusa, Dendrocalamus* and *Thamnocalamus* are rich in species diversity in this region (Biswas, 1988). The state wise distribution of genera and species of bamboo is presented in Table 4.

Bamboos are found to distribute almost all the forest types of Northeastern region of India. The common bamboo species in tropical forests of this region are *Dendrocalamus hamiltonii, Bambusa balcooa, B.tulda, B.pallida, Melocana baccifera* and *Neohouzeaua dullooa*. These bamboo species are met with other plant species such as *Shorea robusta, Gmelina, Bombux, Dillenia spp, Schima walichii, Adina, Kydia, Bauhinia, Cassia, Albizzia, Lannea, Tarminalia spp, Amoora, Dysoxylum, Chukrasia,Canarium, Phoebe, Cinnamomum, Castanopsis, Syzygium, Artocarpus, Michelia, Dipterocarpus, Garcinia, Lagerostroemia* etc. In sub-tropical forests of the region, the common bamboo species are *Chimonobambusa callosa,*

Table 4. Distribution of bamboo species in Northeastern region of India

State	*Taxa*	*Main genera*	*Main species*
1	**2**	**3**	**4**
Arunachal Pradseh	Genera (14) Species (41)	*Arundineria, Bambusa, Cephalostachyum, Chimonobambusa, Dedrocalamus*	*B. balcooa, B.tulda,B.pallida, B.nutans, Dendrocalamus giganteus, D. hamiltonii, Cephalostachyum capitatum, C. fuchianum, C. latifolium, C pallidum, Chimonobambusa. griffitiana, C. intermedia, Mecocanna baccifera, Neohouzeaua dullooa*
Assam	Genera (13) Species (33)	Bambusa, Dendrocalamus	*Bambusa arundinacea, B.khasiana, B. nutans, B. pallida, B. tulda, Dendrocalamus hamiltonii, D. patellaris, D. sikkimensis, Neohouzeaua dullooa, Pseudostachyum polymorphum*
Nagaland	Genera (11) Species (23)	*Cephalostachyum, Dendrocalamus*	*B. khasiana, B. tulda, Dendrocalamus hamiltonii, D. patellaris, D. sikkimensis,D. hookeri Neohouzeaua dullooa, Cephalostachyum capitatum, C. latifolium*
Manipur	Genera (9) Species (20)	*Bambusa,*	*Arundinaria clarki, B. khasiana, B. pallida, B. tulda, B. balcooa, Dendrocalamus hamiltonii, Cephalostachyum fuchsianum, C. latifolium,Chimonobambusa khasiana*

(Table Contd...)

Table 4. (Contd...)

1	2	3	4
Mizoram	Genera (7) Species (15)	*Bambusa, Dendrocalamus*	*Bambusa arundinacea, B. tulda, Cephalostachyum capitatum, Dendrocalamus hamiltonii, D. hookeri,, D. sikkimensis, Neohouzeaua dullooa, Pseudostachyum polymorphum*
Meghalaya	Genera (14) Species (36)	*Bambusa, Dendrocalamus, Arundineria, Chimonobambusa,*	*Arudinaria microphylla, A. subercta, Bambusa arundinaria, B. pallida, B. tulda, B. balcooa, B.khasiana, Dendrocalamus hamiltonii, D. hookeri, D.sikkimensis, Oxynenthara nigrociliata Cephalostachyum capitatum, Chimonobambusa khasiana,C. hookeriana, C. polystachya, Pseudostachyum polymorphum*
Tripura	Genera (5) Species (13)	*Bambusa,*	*Bambusa arundinaria, B. pallida, B. tulda, B. nutans, B. balcooa, Dendrocalamus hamiltonii, Neohouzeaua dullooa*

( ) Value in parentheses are number of either genera or species of Bamboo

C. khasiana, Neouhouzeaua dulla, Thamnocalamus prainii, Dendrocalamus hookeri, D.sikkimensis, Pseudostachyum polymorphum, Teinostachyum griffithii etc. In temperate forests of the region the common bamboo genera distributed as an admixture with the other plant species are *Arundinaria, Chimonobambusa, Thamnocalamus, Semiarundinaria* with their different species.

There are about 13 rare and endangered bamboo species reported so far from this region (Biswas, 1988). They are *Arundinaria clarki* (Manipur, Sikkim); *Arundinaria mannii* (Meghalaya); *Bambusa mastersii* (Assam); *Cephalostachyum capitatum* and *C. decomposita* (Sikkim); *Dendrocalamus sahnii* (Arunachal Pradesh); *Dinochloa maclellandii* (Assam), *Gigantochloa takserah* (Meghalaya), *Phyllostachys mannii (Nagaland), Seminarundinaria pantlingii* (North Bengal, A.P., Sikkim), *Sinobambus elegans* (Nagaland).

TRADITIONAL USES

Bamboos have been in use since time immemorial. It is perhaps the best known plant on the globe. It has over 1500 uses and has tremendous versatility. A list published in Japan in 1903 includes 1048 articles of practical use. Bamboo has been an integral part of the socioeconomic status and cultural heritage for the northeastern India. It has found manifold utilizations in the daily life of inhabitants of this region for a long time (Table 5). Some of the common uses include fences, props for fruit trees, laundry poles, tent poles, electricity poles, flag staffs, oyster raising sticks, ladders, traps for fish and crabs, fishing rods, sport articles, weapons, bows and arrows, umbrella handles, ladies bag, woven cases for lacquer ware, spoons, forks, brooms, sieves, fire-blowing tubes, brushes, water pipes, water buckets, fans, toys, cloth pins, gramophone needles, serving trays, chopsticks, cordage, hair clips, buttons, lampshades, musical instruments, highly prized artistic handicrafts etc. Bamboo rhizomes are used as ornamental borders to many bamboo articles. Bamboo jewellery such as earrings, bracelets or necklaces has a fresh, glistening appearance (Subba Rao, 1966, Chakraborty, 1988). Bamboo nails have a good holding power, the round form being superior to the square form and can replace steel fasteners in wooden

Table 5. Major timber and non-timber uses of bamboo in North eastern India

Major categories	*Use pattern*
House Construction	Roofing, walls, scaffolding for building construction, mine poles, huts, roller-binds for windows, floor, partition
Household goods	Walking stick, bamboo torch, cradle, furniture, comb, racks and brooms, bamboo (took) umbrella, filter, water carrying pots, water vessels
Agricultural uses	Bamboo stem support for agricultural field bunds and terraces and drainage, tools like yoke, food grain container, seed store, paddy storage bin, bamboo for packaging the agri-produce, ropes
Human Food	Bamboo shoots
Animal husbandry	Green fodder
Fishery	Fish traps
Water sports goods	Water conduits, boat masts, rafts
Musical instruments	Flute
Aesthetic use	Hedge, groove, flower (made up of fine fibrous peelings developed from green shoots)
Burning material	Firewood
Defense instruments	Bow and arrow, Spade
Community level uses	Palisade around the village, hanging bridge, domestic animal enclosures, bamboo plantation fence (Harvested bamboo), biofence (live bamboo), ladder
Handicrafts	Woven bamboo mats, winnowing fans, baskets
Religious	In making God (alter)/Temple during local festivals like Mopin, Dree and Nyokum (Adi, Apatani and Nishing tribe festivals, respectively in Arunachal Pradesh), Temple flag pole
Industrial	Pulp, bamboo dust for plywood, ice cream sticks, match sticks, Agarbatti sticks
Environmental	Soil and water conservation, carbon sink as major secondary plant species in abandoned jhum (shifting) cultivation areas.

constructions. Sometimes the clums are bleached to make them more pleasant and ornamental. Certain bamboo articles are almost indispensable to all kinds of farm works, such as irrigation wheels, tool handles, rakes, flails, pumps, threshing boards, etc. Sticks from bamboo are used for carrying heavy goods and their swaying eases the weight on the shoulders. Bowls made out of bamboo clums are used for storing liquors for a special taste. In Japan and China, hoofs made out of bamboo are used for large wooden tubes intended for picking and food processing industries. Bamboo leaves serve as fodder and thatch material. Dried mature leaves are used for deodorizing fish oil. With the rapid socio-economic transformation and industrialization of the country, it gained importance as a raw material not only for cottage industry but also for large-scale industries.

INDUSTRIAL UTILIZATION OF BAMBOO

Bamboo has been one of the first materials in the tropical countries used by mankind to increase comfort and well-being. Its wide distribution and availability, rapid growth rate, superior technological properties and its easy handling make bamboo an ideal material for countless uses. Bamboo is sometimes called the poor man's timber, which is not quite fair; in certain fields of application, bamboo has superior technological qualities. Besides its role as a material for consumer products, bamboo has great potential as a Industrial raw material and structural material with attractive physical properties for meeting the needs effectively and inexpensively. Pulp and paper industry and bamboo craft sector are the two major users of bamboo resources. However, its industrial potential in the manufacture of a number of other value added products like laminated boards, bamboo fiber-cement board, flooring, wood substitute, medicine, food products etc provides enough opportunity for employment and income generation and at the same time ensures efficient and sustained utilization of available resources.

In the recent past, considerable achievements have been gained in commercial exploration of bamboo resources that has opened new vistas for its industrial exploitation. A number of marketable products have been developed out of bamboos that are well appreciated by the consumers all over the world. Since 1970s, China has made great progress in the industrial processing and utilization of bamboo resources and increased the value of bamboo. It is important to note in this connection that bamboo species vary considerably in their properties that suggest their suitability for different products; so, different species of bamboo can be put to various uses based on proper identification of their properties for their manifold utilization. Hence, it is evidently clear that in view of the adequacy of bamboo as raw material for manufacture of numerous value added products; its industrial exploration is of paramount importance. It will lead to the efficient utilization of available bamboo resources needed for the economic development of bamboo growing areas and the country as well and also explore the means and ways for sustainable utilization of huge bamboo resource expected to be available consequent to gregarious flowering. Some of the major industrial utilization of bamboo now-a-days include pulp and paper, construction, engineering articles, various types of boards, panels, food products, medicines, bamboo charcoal, fabrication of equipments etc.

Pulp and Paper

The use of bamboo in papermaking had started in China as early as during Jin Dynasty (265-420). Pulp mills using bamboo as raw material exist in several Asian countries. Because of its anatomical structures, bamboo is a suitable material for pulping. The quality of bamboo pulp is considered to be fairly good due to its high fiber content ranging from 40% to 60%. The fiber length of bamboo varies from species to species and with internodal length. Mean values for species differ for fiber length from 1.5 – 4 mm, fiber diameter from 11-19 mm, lumen diameter from 2-4 mm and wall thickness from 4-6 mm. Bamboo fibers are thick walled as compared to other non-wood paper making raw materials. They are more slender than wood fibers, which contribute to the desired smoothness and flexibility. Bamboo constitutes as high as 70% of the total cellulosic material that implies suitability for dissolving pulp and for rayon production. The high content of cellulose make the paper made out of bamboo fine textured, water resistant, ink absorber and fuzzless. Since bamboo contains more impurity than wood, however, cooking is more costly and the pulp is lower (Singh and Guha, 1981). Bamboo pulp has a high tearing strength and compares well with softwood kraft pulp in this respect. Tensile and bursting strength are, however, considerably lower than those of wood pulp. For many purposes, bamboo pulp is mixed with pulp from grass, wood, rag or waste paper. It can be used in the production of newsprint by mixing with ground wood pulp in the ratio of 3:2. Bamboo pulp is suitable for a large variety of paper, e.g., writing, printing, wrapping, tissue etc. It is used for the production of bond paper, ledger-paper, art paper, facial tissue paper, paper for airmail, newsprints etc. The paper prepared from the bamboo pulp compares favourably with ordinary super-calendared paper, printing paper and other conventional paper made from wood pulp. It is capable of withstanding considerable wear and tear and retains its whiteness and brilliance for reasonably long periods. The unbleached pulp can be used for the manufacture of cable-insulations, industrial saturating and toweling papers. Bamboo is also used in rayon and cellophane production.

In India, besides local consumption substantial demand on bamboo resource comes from pulp and paper industry. Out of the estimated annual production of 9.5 million tons of bamboo, nearly half is consumed by the paper industry. In earlier days, uses of bamboo to produce paper and board were 50-70 %. However there is a drastic shift during recent past reducing its usages to 10%. Some mills are going ahead with total replacement of bamboo with hard wood to produce paper and board. Utilization of bamboo in paper industry has been declined during the last few years; wood and other non-traditional raw materials like rags, waste paper, agricultural waste, etc. partly replace bamboo. Currently Indian paper mills are using bamboo (8%), Wood (30%), Agricultural residues (36%) and waste paper (26%). Out of 28 raw material based mills, 10-12 paper and board mills (IPMA members) are using bamboo and reed as raw material. In all 1.394 million tons of bamboo is used alongwith 40 million tons of wood to produce 1.30 million ton of pulp per annum. The consumption of bamboo for pulping is steadily increasing particularly in India. At present approximately 2.5 million tons of air dried bamboo are being used for paper making in India. Demand is expected to go up further. With the expansion of Indian Paper Industry and falling supplies of bamboo, the industries are forced to mix hardwood pulp with bamboo pulp for the production of paper. Also, the consumption of bamboo for pulp and paper has not increased in some other countries due to the difficulty of collecting the necessary raw material due to the cost of transport and production.

As Engineering Material

The strength properties of bamboo are far more superior in many aspects to manmade engineering materials. Its tensile strength is greater than mild steel. Use of bamboo as engineering material aimed to develop the underutilized potential of bamboo as a structural material into its full range of housing and related applications. The activities under the programme should include the following activities:

Model development for predicting strength properties of bamboo: As the strength of a bamboo clum is affected by various general and species-specific aspects, the development of a reliable and practicable model so as to ascertain reliably its strength properties.

Development of testing manual for the use of bamboo in engineering applications: With a tensile and flexural strength greater than steel, the intrinsic and unique structural features and properties of bamboo make it an ideal engineering material for the developing countries in which it occurs and /or can be grown. However, the use of bamboo in engineering applications has been limited by the lack of awareness on method for testing bamboo.

Formulation of a bamboo building code: Formulation of a bamboo building code is essential that will facilitate transfer of knowledge on engineered bamboo housing from the laboratories to the field, and enable architects to incorporate and use bamboo in their building plans. This will ultimately help make improved bamboo houses more readily available.

Formulation of a code for disaster-resistant bamboo buildings: On several occasions, the high potential of bamboo for use in disaster-resistant housing has been proved. However, there has been no attempt to gather related data from different sources, analyze these and consolidate the results into a standardized format replicable anywhere in the world. It is, therefore, necessary to formulate a code for disaster-resistant bamboo buildings.

Development of improved bamboo scaffolding: In many parts of Asia, bamboo finds wide use as a material for scaffolding because of the suitability, low cost and easy availability of the material. This application, however, remains geographically limited owing mainly to bamboo's inherent low durability and the lack of more structured methods of use. As building construction is a continual activity around the world, the development of improved bamboo scaffolding holds immense potential for popularizing and globalizing the usages of bamboo.

Development of improved bamboo truss design: Trussing is an important use of bamboo in traditional house construction. However, better and more functional truss designs will be required before bamboo can enter the world of modern construction.

In Construction

Bamboo has been traditionally and now is extensively used as construction material in many countries with plentiful bamboo resources. Various bamboo species have widely different characteristics affecting its usefulness as a constructional material. The strength of bamboo culms, their straightness and lightness combined with hardness, range in sizes, hollowness, easy working qualities make them suitable for a variety of purposes. It constitutes the principal construction material for building structures, walls, fences and scaffolding in Southeast Asia. Different tribal communities of Northeastern India use bamboo for housing purposes (Laha, 2000). Its wide application as an appropriate building material has been greatly emphasized (Stulz, 1981; Bamboo, 1982; Schreckenbach and Abankwa, 1983). The mechanical properties of the stiff, hollow cylinder with dispersed fiber bundles of high flexibility are ideal for many technological purposes. In Burma and Bangladesh about 50% of all rural houses are made entirely out of bamboo. In Java the houses are often built with timber as frames and bamboo mats as walls. In Ethiopia, the traditional roundhouse patterns are followed. In Japan, a square form of clum is obtained by putting a wood mould over the developing shoot soon after they emerge from the ground. Such square clums are often used as alcove posts for a tokonoma in a Japanese house. With the advent of modern building materials, bamboo was forced to beat a retreat. However, there is a renewed interest on bamboo due to its ability to withstand earthquakes and its suitability for emergency housing. Most low-income houses are made of bamboo and a number of them have been standing for over half a century. Some well-finished houses with bamboo frame and foundation and walls covered with earth or cement survive for a much longer period. Bamboo's image as the poor man's timber still restricts its wider usage. Fighting against this mental block some architects have designed modern bamboo houses. Such houses owing to their high cost are not within the rich of majority of peoples. Nevertheless, these efforts have succeeded in removing the prejudice of public mind against bamboo as a construction material of the poor. Approval of bamboo as the building material of choice by the upper class is expected to boost its use in other classes of the society also. An increase in demand for bamboo houses will then bring about technological upgrading, which will in turn boost the demand further.

The use of bamboo for trusses has been elaborated by Janssen (1981). Trusses with a free span of 8 m or more could be built to support roofs of galvanized or fiber cement corrugated sheets. Different types of joints have been designed for the trusses. Bamboo is widely used for scaffolding of high buildings. In several countries, attempts have been made to use bamboo in reinforcement concrete in place of steel. Bamboo Crete is especially suitable for non-load bearing structures like wall panels. The estimated saving in cost in making a concrete slab reinforced with bamboo rather than steel was 30% in India (Purushotham, 1963; Masani *et al.*, 1977).

To realize the potential of bamboo in constructional applications, works need to be done to further examine the technical properties of several important species of bamboo and to identify which species merit further development. New industrial and building products that can be made with bamboo should also be searched out, particularly as a substitute for products that traditionally used timber. Bamboos used for constructing low-cost houses are hardly treated for preservation. Bamboo that goes into higher income houses gets a chemical treatment, usually using environmentally hazardous chemicals such as pentachlorophenol. The growing environmental concerns have spurred research into ecofriendly preservation methods. It is due to the reason that bamboo has also started to figure in projects aimed at environmental conservation and land reclamation. Bamboo advocates in the world hope that the movement will catch on and provide enough raw material to spark off a bamboo revolution that will change forever the way the society views this wonder plant.

Artificial Boards

Bamboo has emerged to stake a claim for timber substitute. Bamboo panels have proved their merit as ideal replacement for wood in several applications. Bamboo clums cut into strips of same size and after dried and milled, they are arranged horizontally or vertically and hot pressed. The processing of the hot pressed boards into final product is the same as with wood floorboards. The products that can be made from laminated bamboo strips include flooring boards, wallboards, dowels and moldings. These products have the advantage that they are harder wearing than many hardwoods, softwood and particleboard alternatives in the market. Of these products, flooring boards are becoming very popular worldwide, and growing markets have already been established in Japan, Europe and North America. In China alone, there are more than one hundred manufacturers of bamboo flooring boards, with a total output of approximately 10 million square meters. Bamboo flooring provides a unique and sensible solution for interiors—from homes to offices and even basketball courts. Besides the pleasing luster and trademark grain of the bamboo clum, bamboo flooring is resistant to moisture, pressure and damage. It is also flexible, acts as sound insulation, lasts longer than wood and is less expensive.

Floorboards: Bamboo flooring boards are praised for their strength, toughness, abrsion, resistance, graceful luster, beautiful appearance and novel structure. On accout of their durability and aesthetic effects, bamboo-flooring boards are expecting a very good market all over the world.

Laminated boards: Laminated bamboo boards have a high utilization rate of raw material that are bamboo stripes of 2 mm in thickness and 10-20 mm in width with varied length. The art and equipments involved are also simple but the major disadvantage in producing laminated boards is that it has a greater consumption of glue and demands high pressure. This type of board is mainly used as the floor of the trucks and train cars.

Bamboo Fiber-Cement Panels: There is a keen interest in utilization of bamboo as raw material in industrial production. Bamboo-fiber cement panel is one such product made by combining bamboo fibers and cement to form panels that take advantage of the strength, adhesiveness and anti-corrosiveness of bamboo have proved to be excellent construction material and partly fulfills its responsibility for protecting the global environment. After five years of research and trial in Java and Indonesia, Ask Corporation have developed the basic technology and the firm is now producing and exporting Ask boards. Japan is currently importing these boards where it is mainly used as a roofing material. Anticipating further demand in Asia, Ask Corp. has been set up in Thailand also to manufacture these panels. While *Gigantochloa apus* is employed in Indonesia, *Dandrocalamus asper* is being tried in Thailand with suitable modifications to the manufacturing process. It has been estimated that a plantation of 1500-200 ha will be needed to meet the raw material demand of 40 tons per day for a bamboo fiber-cement plant.

As Food

Traditionally, edible bamboos have formed a part of the human diet particularly of the tribal people in the northeastern parts of India. Tender shoots of *Dendrocalamus giganteus, D. hamiltonii, D. membranaceus* and *D. strictus* are consumed as vegetable or pickles. In Manipur, *Bambusa bambos* and *D. giganteus* are the two main species whose young shoots are used as vegetables. Bamboo shoots are an important daily food for many people in Asian countries especially in China, Japan, Thailand and northeastern India. The tender shoots of *B. bambos* are pickled and steeped in oil for direct consumption or made into curries. *B. polymorpha* shoot, with its distinct sweet taste, is considered to be the best in the world for producing quality edible shoots. Crooked tender shoots of *B. balcooa* are eaten, or preserved through fermentation and drying for later consumption. Bamboo seeds are pickled and used for making beer. The leaves of a number of bamboo species are much valued as fodder, particularly during scarcity. Young bamboo leaves and twigs are the favorite fodder of elephants. Fruits of *Melocanna baccifera* are devoured by cattle, elephants, pigs, etc. Seeds of *B. Bambos, Cephalostachyum pergracile, D. strictus* and *M. baccifera* are extensively consumed by tribal and poor people during famine. The protein content of seeds of *B. bambos* growing in Orissa was found comparable with wheat, although they were similar to rice in quality. The protein content of seeds of *B. bambos* growing in Orissa was found comparable with wheat, although they were similar to rice in quality (Mitra and Nayak, 1972).

Healthy shoots of bamboo, stunted or even crooked, are used for the preparation of edible products. The outer sheaths are removed and the tender partions are cut into rings or pieces of suitable sizes. Bamboo shoots contain cyanogenic glucoside, which on endogenic hydrolysis yields hydrocyanic acid. Hydrocyanic acid can be removed by steaming the pieces of bamboo shoots. Another toxic constituents, taxiphylin have been reported in the shoots of *B. guadua, B. vulgaris, B. arundinacea, D. giganteus* and *D. hamiltonii*. The toxin can be removed during cooking, by changing the water several times. Soaking thin pieces of shoots in different changes of 2% salt water for several hours will also remove the toxicity. Bamboo shoots are used for the preparation of many edible products like candy, chutney, curry and vinegar, lime and oil pickles.

The nutritive value of bamboo shoots has also been studied by various workers. The average values for various species from different regions for various species are Carbohydrate, 4.5%; protein, 2.6%; fat, 0.3% and ash 0.9%

(Young, 1984). Shoots are also a good source of thiamine and a fair source of niacin, two members of vitamin B complex. Analysis of 100 g of edible portion of *B. bambos* yielded moisture, 88.8 g; protein, 3.9 g; fat, 0.5 g; carbohydrates, 5.7 g and minerals, 1.1 g. Presence of Fe, Mn, Pb, Sn, Ga, Ti, Ni, Bi, Mo, V, Zr, Cu, Ag, Zn, Co and Sr in the shoots of *Phyllostachys pubescens* was established spectroscopically. The content of Zn and Fe was very high followed by Ni, V, Cu and Sn where as content of Pb was very low (Chen, 1990). Studies have also been done on edible shoots and seeds of *B. vulgaris* and *M. baccifera* and seeds of *B. arundinacea, B. nutans* and *D. strictus* from the point of view of human consumption. In addition, the leaves and shoots *of B. arundinaces, B. polymorpha, D. longispathus* and *M. baccifera* were analyzed for supplementing animal feed. The results showed substantial contents of vitamin C and protein. The mineral content sodium, potassium, calcium, magnesium and phosphorous – was also appreciable. On the basis of above studies, it can be inferred that edible bamboo, if promoted extensively, may help solve nutritional deficiency of the rural poor.

As Medicine

Bamboos find various medicinal applications in traditional system of medicines. Extraction of an important drug named *Talbashir* or *Banslochan* from the substance accumulated at the hollow internodes of bamboo is well known. Banslochan is reported to have many medicinal properties (Shukla and Das, 1981). It is valued as a stimulant and febrifuge and is said to be useful as a cooling tonic and as an aphrodisiac. It can also be used for the treatment of asthma, cough, paralytic complaints and other debilitating diseases. In China, a decoction of the leaves is fermented and used to treat fevers. When taken with congee (a rice preparation) it is thirst relieving and purifies blood. An extract of leaves is capable of controlling halitosis. Fresh leaves are given to cattle for post-natal cleansing of uterus. From the powder produced on the surface of young stems of a Chinese bamboo *Bambusa changii*, a crystalline compound similar to the female sex hormones has been isolated. An aqueous extract from bamboo shoots has been found to be superior to conventional media for the culture of certain pathogenic bacteria (Yoshida and Ikejiri, 1950). The extract, when added to tuberculin (a solution of T.B. germs), increases the intensity of responses of the skin test. From mosochiku bamboo, quinone derivatives having antibacterial and hair growth promoting activities have been isolated. They do not have acute toxicity or mutagenetic activities (Absuyshi, 1990). The sugar silica from the clums is used as a cooling tonic and an aphrodisiac.

Stems and leaves of *Bambusa arundinacea* are used in Ayurvedic system of medicine as blood-purifier, in leucoderma and inflammatory conditions. An infusion of the leaves is used as eyewash. It is also given internally for bronchitis, gonorrhoea and fever. The dealcoholized extract of leaves has shown anti-bacterial activity. Its brunt roots are applied to ringworm, bleeding gums, and to painful joints. The black soot deposited on the clums, mixed with lime or even finely powdered coke is applied to cuts and wounds. The bark is used as a cure for eruptions. Tender shoots of *Bambusa arundinacea* are pickled or steeped in oil for direct consumption or made into curries that are reported to promote the appetite and help in digestion. Buds of the species are reported to show estrogenic activity. Extracts of bud has shown antifertility activity in rats (Menachery and Chandran, 1984; Sethi *et al.*, 1989). In North Lakhimpur forest division of Assam, very soft shoots are used for birth control. A poultice of the young shoots is applied to ulcer for the dislodgement of worms. Leaves are emmenagogue, anthelmintic, astringent and febrifuge. They are mixed with black pepper and common salt and used to check diarrhoea in cattle. They are also given to horses as a remedy for cough and cold. The leaf bud of *Bambusa spinosa* is used in leprosy, fever and haemoptysis. A decoction of its root is administered in cases of anuria. The sap of *Bambusa vulgaris* is given as a remedy for phthisis in the Philippines. The astringent bark is used in haemorrhage, nausea and vomiting. The shoots and roots possess diuretic, diaphoretic and emollient properties. Fresh roots mixed with equal parts of tobacco and betel-leaves is macerated in oil and applied to hard tumours. Because of their wide therapeutic properties, various species of bamboos are used in pharmaceutical industries.

Bamboo Activated Carbon

Bamboo charcoal is generally used by goldsmiths. Being superior to other charcoals from other sources, it is used in batteries. In Japan, bamboo charcoal is used in gardening, as it is believed to preserve the moisture available to plants. Carbonized filaments made from bamboo can be used as light-giving source for certain purposes (Varmah and Bahadur, 1980). In order to make charcoal out of overmatured bamboo and bamboo of low quality, a portable oven was designed (Inoue, 1988).

Activated carbon from bamboo carbon is a new product developed in recent years. It possesses high absorptive capacity due to its special micro-porous structure. Bamboo carbon has an iodine absorption value of 1160 mg/g and the decolorizing capacity for methylene blue as high as 250 mg/g as indicated in various tests. For these properties activated bamboo carbon is widely used in foods, medicine, chemical, and metallurgical industries. Bamboo carbon can absorb pigment and foreign substances from solutions and can be used as catalyst, catalyst carrier or directly applied as medicines. Application of activated bamboo carbon in hi-technology areas needs special attention and adequate efforts.

Bamboo Vinegar Generator

Bamboo has been suitable for fabrication of vinegar generator for large scale manufacture of vinegar. In the production of desiccated coconut, one of the main waste products is coconut water that can be converted into marketable value added products, such as coconut vinegar through integrated processes. With the use of a vinegar generator made of bamboo clums and using a blend of coconut toddy and coconut water, Ceylon Institute of Scientific and Industrial Research (CISIR), Sri Lanka has successfully produced coconut vinegar.

Miscellaneous Products

A bamboo powder suitable for use as a deodorizer was prepared by dry distilling powdered bamboo of about 12-mesh size and drying the powder of the resultant product in a current of hot air (Hirosuke, 1990). It has been shown that chips of

Phyllostachys pubescens can be converted to a straw substitute, suitable for use as fodder by steaming the chips at a temperature of 184^{0}C at a pressure of 10 kg/sq. cm for 20 minutes (Mizobuchi, 1987). From the chopped off tips of *Oxytenanthera braunii*, a kind of wine is collected in Tanzania (Mgeni, 1983). A diesel like fuel has been prepared from bamboo culm (Piatti, 1947). Bambusa bambos (Dowga bamboo) reported to contain 0.25 % wax (Beri *et al.*, 1967), which can be used as a base for shoe polish, carbon paper, waterproofing kraft paper and as a cheap grade sealing wax.

ENVIRONMENTAL APPLICATIONS

The world as a whole and developing countries in particular are facing a serious problem of depletion in the natural resource base. The livelihoods of rural people are particularly vulnerable to depletion of soil and water resources, as millions of hectares of productive land become unproductive because of topsoil loss from erosion. Erosion of riverbanks exacerbated by the scouring power sediment-laden streams and hilly regions causes thousands more hectares to be lost annually. Moreover, global warming from excesses of atmospheric carbon will present a very serious problem for developing countries in years to come. There is a pressing need for the development sector to find ways to address these problems. About 17% of the world's total cultivated land is degraded, of which 62% is moderately to severely damaged. Rehabilitating this land will require policy and ground level efforts; bamboo can make a significant contribution to the latter. The root system of bamboo is ideal for soil stabilization—reventing erosion and landslides, protecting riverbanks and preventing floods. Its unique ground colonizing, horizontally spreading growth habit makes plantation expand in area automatically unlike trees, which have to be individually planted. Its growth characteristics and requirements help the environment both above the ground in terms of carbon sequestration, as well as below the ground in terms of soil rehabilitation. Bamboo can also grow well in poor soils, under low rainfall and moisture conditions. There have been a number of projects in recent years that utilize bamboo to reclaim degraded lands and wastelands. What needs to be done for the future, however, is to develop standardized information and techniques that can enable bamboo to be used on a much larger scale. This will require estimation of total reclaimable area across countries, matching of most appropriate species to various types of damaged site, and development of new technologies to increase the efficacy of rehabilitation plantations, while making the short-term economics attractive to landholders.

Besides recovering wastelands and degraded lands, the appropriate use of bamboo can mitigate the impact of unsustainable agricultural practices and deforestation on the soil. This is important in riverine and hilly areas, which are particularly vulnerable to water and wind erosion. As with land rehabilitation, large-scale soil conservation efforts must be based on research into appropriate technique and standardization of processes. Bamboo provides a viable economic alternative to forest exploitation and conservation, so that forest can be protected for their biodiversity and their diverse other benefits. On account of better strength properties of bamboo wood its entry into the wood market as the new environment friendly wood of the 21st century can help save tropical forest via timber substitutes.

Control of greenhouse gas emission around the world being a politically charged issue needs development of novel means of removing carbon from the atmosphere. Bamboo as a carbon sequestration agent can play a major role in offsetting the continued emissions of carbon dioxide into the environment. As companies and governments look for ways to balance their carbon producing activities with carbon absorbing projects, bamboo would be an ideal species for such carbon credits. Recent studies suggest that well-managed bamboo plantations could remove carbon dioxide from the atmosphere more effectively than any other species. In addition, bamboo's uniquely rapid growth and its multiple uses would ensure the economic as well as ecological value of these plantations.

SUSTAINABLE DEVELOPMENT AND BAMBOO

Economic upliftment of people and environmental conservation are the most stressed key components of the sustainable development. Bamboo is uniquely geared to address both the issues (Fig. 1). It is an ideal example for bridging the sustainable subsistence with the present days unsustainable industrialization for a sustainable future. Due to its unique characteristics and myriad applications ranging from traditional (subsistence) and contemporary (industrial) uses, the plant remains in the mainstay of traditional bamboo microenterprises as well as modern demand and supply driven large scale industrial sectors. The traditional bamboo based cottage industry including woven products, other handicraft typified by basketry, and incense sticks easily fit with the daily schedule of village women. It constitutes a very large and prosperous industry and if expanded oriented into a form of family business offers a safe framework for children to learn a trade thereby bringing their livelihood in future. It has an enormous employment potential if economic value is added to the wide range of subsistence level bamboo based applications. Bamboo can help bridge the gap between subsistence and cash economics. Even at current levels of conversion of traditional knowledge into economic opportunities, bamboo cultivation and utilization accounts for millions of jobs. It has also the advantage of preserving culture as well and indigenous knowledge base that are rapidly eroding today owing to globalization and cultural homogenization of people.

Bamboo finds an important role in several major economic sectors like housing, pulp and paper, textiles (rayon), wood substitutes, agriculture, fisheries and sericulture. Production of bamboo and many of its products is labour intensive thus creating huge job opportunities. Maintenance and selective harvesting of bamboo is mostly done by hands. Even at industrial level, the product manufacturing has substantial labour requirement. Value addition to the traditional processes and products can increase the cash earnings also. The cost and energy requirement for production of bamboo-based articles are low making it economically advantageous for small and large-scale industrial entrepreneurs. Some of the sectoral applications of bamboo have an obligate relationship in backstopping key livelihoods that are important not only for income generation but also for

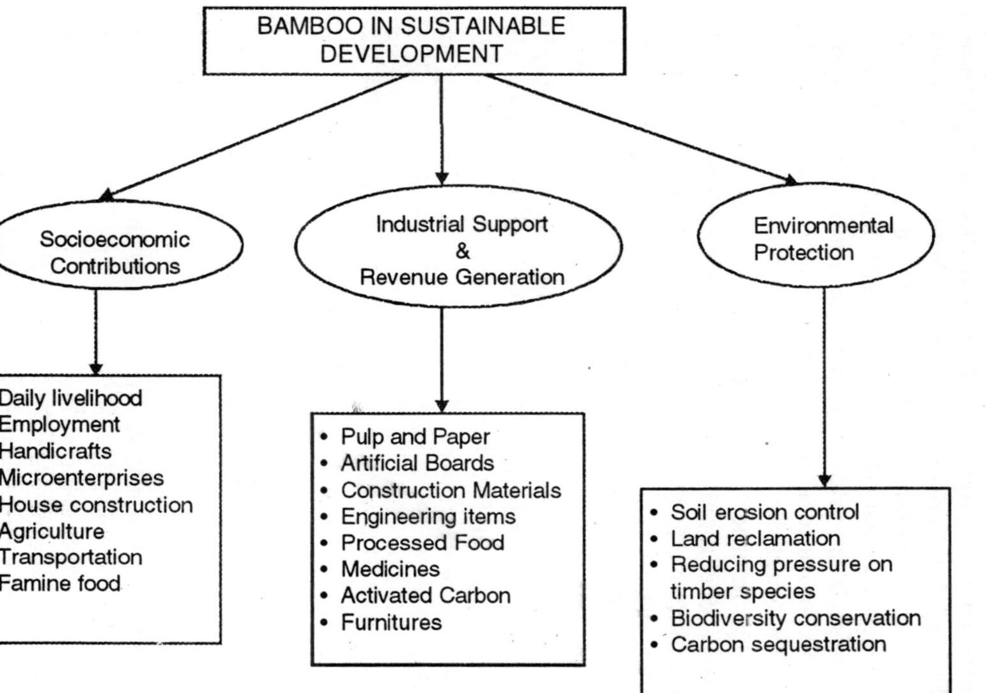

Fig. 1: Role of Bamboo in sustainable development

ensuring food security. For instance, bamboo is important in backstopping agriculture through land irrigation (used in treadle pumps, shallow tube wells and piping), rural transport, protection/fencing of agricultural crops and livestock and a host of other applications such as stakes and other supports. Agricultural implements for tilling and managing field, winnowing trays for cleaning grains and baskets for storing and transporting agricultural produce are also major uses of bamboo in agriculture. Another important use of bamboo is as fodder for livestock. In much of Southeast Asia and now increasingly in other parts of the world bamboo shoots are an important food source and a major economic industry. For bamboo to thrive in the cash economy, it is necessary that it register a presence across the market spectrum, particularly towards the higher end. Here, it has to win against some well-set prejudices. Considering bamboo as a prime means to achieve sustainable development, development of strategies, technologies and tools that blend environmental concerns with economic opportunities and transferring that knowledge worldwide is essential for enhancing human capacity and livelihoods.

BAMBOO FLOWERING: A CHALLENGING ENIGMA

Flowering of Bamboos in the intensity of sporadic and gregarious nature has always been looked upon as ill omen leading to significantly adverse social as well as environmental consequences. Most of the bamboo species have a definite schedule of flowering causing death of the flowered clum after seed setting. However, the exact physiological mechanism is yet not known. The consequent increase in seed production reported to increase in rat population resulting in the destruction of cultivated crop and creating a situation of climatic disaster like famine and flood. According to an estimate, flowering of Muli bamboo (*Melocanna baccifera*) with a flowering cycle of 48 years is expected in about 1.76 million ha area in the northeastern region affecting about 26 million growing bamboo stock by 2004-2007. Due to gregarious flowering, it is estimated that more than 10 million tons of bamboo would be available in accessible area that would need immediate disposal through effective utilization. Flowering of other bamboo species in future may also result a large loss in terms of resource and the environment. This alarming situation may create unprecedented problems in resource management and utilization.

Flowered bamboos require immediate attention so that timely harvest of bamboo and utilization are planned. In this context, industrial processing and utilization is an important part for efficient and sustained exploitation of bamboo resource and to save the huge economic loss in terms of resource destruction. An urgent and immediate policy decision with regard to establishing tie-up with industries engaged in value addition of bamboo could bring in lot of revenues to the northeastern states.

BAMBOO DEVELOPMENT PROGRAMMES IN INDIA

In India, the area under bamboo plantation is reducing gradually, according to the Ministry of Environment. Also, the use of bamboo has declined over the years and the government is concerned that the traditional skills of making bamboo handicrafts may also die out if not encouraged now. Bamboo has already been losing its popularity in parts of Uttar Pradesh, Northeast and Western Ghats. The government has convened a meeting with some industrial houses to build on bamboo's potential as a major export item. A Bangalore-based research organization, the Indian Plywood Research and Training Institute (IPRTI) has shown that building materials can be developed from bamboo. One such building material, the bamboo mat board has been found to be strong and durable. As a part of its attempt to popularize bamboo products, the government has also taken up the membership of INBAR. The Government of the state of Mizoram entered into an agreement with INBAR in March 2000 to advise and provide technical assistance to help develop an integrated bamboo processing industry in the State. With a bamboo resource base with a current productive potential of 3.2 million tons per annum, a small population of only 0.85 million, here are very significant social, economic and environmental benefits that can be realized from the development and utilization of the bamboo resource base in Mizoram.

On the World Environment Day in June 1999, the Prime Minister of India announced a major initiative, committing the Government of India to a comprehensive and integrated programme to develop the bamboo sector in the country. It involves a number of national government ministries and individual state governments and aims to give a major boost to the sector for economic development, income and employment; particularly in the Northeast, where the plentiful natural bamboo resources can be an important vehicle for substantial and widespread economic development. The Government of India, through the Technology Information, Forecasting and Assessment Council (TIFAC) of the Department of Science and Technology (DST) has launched the Mission on Bamboo Application as a special Action Plan under the Tenth Five Year Plan. This national mission would address the agrarian economy, especially in the remote regions of the country where development initiatives have so far not penetrated. Innovative techniques for value addition would immensely contribute to sustainable livelihood and also augment the employment generation initiative in the rural sector. Such value added utilization of bamboo would help reduce the strain on the environment and the fragile forest ecology. The National Mission on Bamboo Application would be an ideal platform for demonstrating the "technology with a human face". The mission will target a few major sectors namely building and construction, industrial and consumer products, food processing and packaging. The governments of Northeast states have greatest interest in developing the sector. Not only do they have the natural capital on which to base such development, but also they have the cultural and traditional associations with the material and crucially, a limited number of alternative options for regional development.

The International Network for Bamboo and Rattan (INBAR), a non-profit international organization established in 1997 by treaty with twenty one sovereign countries as members was recently contracted to provide expert assistance on the UNDP/UNIDO-funded Cane and Bamboo Technological Upgradation and Networking Project in Northeastern India. This large-scale project is being executed by the Department of Science and Technology and aims to provide support for developing and enlarging sustainable handicrafts industries in the Northeastern States by enhancing the access of craftspeople to improved technology and knowledge, employment and income generation opportunities.

CONCLUSIONS

Bamboo has great potential as wood substitute and can contribute to reducing the pressure on forest. The long history of the multiple uses of bamboo tends to give the impression that further research effort is needed. In spite of the large amount of processing and utilization information, a number of research areas require attention. To date, much of the processing of bamboo has followed traditional techniques, with little modification and improvement. With new demands and modern end-uses for products, there is now a need to improve the quality and quantity of outputs. One of the most pressing problems affecting the use of bamboo is the post-harvest damage, which greatly reduced the sale value of the processed material and limits their application to many end-uses. Research and testings are needed to develop better hand tools for primary processing, as currently there are few tools specifically tailored for bamboo. Despite the emphasis on modernity, there is also a need to examine traditional technologies to identify and revitalize those that have economic potential to enhance livelihood opportunities in rural areas.

The future of bamboo lies in their development for sale in urban and international markets. However, these markets have requirements for new designs with increased aesthetic and/or utility value. As the urban market develops, there will be an increasing competition between bamboo building and furniture products, other consumer goods and synthetic substitutes. To ensure the long-term viability of the sector and to increase profitability, there is a need to promote innovation and quality control at the micro-enterprise level. In particular, areas that need to be addressed include developing grading standards at various levels of production and processing, documenting and cataloging existing and traditional designs and developing novel designs that use bamboo. Because most bamboo products are hand-crafted, there is a strong need to establish and maintain uniform production norms, especially at high production volumes. Hence, implementable strategies should be evolved and in this context following points need immediate attention:

- Designing of marketable product lines keeping in view the demand and trend of global market.
- Technological innovations for production of such items efficiently at the market-friendly price.
- Selling of bamboo products to sustain the production.
- Assuring sustained supply of good quality raw material through improved raw material management.
- Provisions of round the year transportation system.
- Better integration of pre-processing, processing and ultimate use.
- Manufacturing of higher-end furniture, house-ware and gift items at industrial and village craft centers.
- A workforce capability development and infrastructural support.

The revolution of bamboo products as a substitute for wood is an area of great strategic importance and potential since this will open up the huge wood market to bamboo. Special attention should be paid to the use of bamboo in developing low-cost and quality products, which can play a significant role in the livelihoods and economic development of rural populations and the nation as a whole. At the same time, emphasis on high-end applications is critical, since experience has shown that this plays a major role in changing the social perception of the material and increases its acceptance.

REFERENCES

Absuyshi, N. (1990). Antibacterial activity of Mosochiko Bamboo extract. *Bamboo Abstracts* (1991) 4(2): 49.

Anon (1988). *The Wealth of India – Raw Materials,* Vol 2: B, Publication and Information Directorate, CSIR, New Delhi, pp. 7-38.

Beri, R.M., Kamik, M.G. and Pharasi, S.C. (1967). Note on Dowga Bamboo wax. *Indian Forester*, 93(3): 188-189.

Biswas, S. (1988). Studies on bamboo distribution in Northeastern region. *Indian Forester*, 114: 514-553.

Chakraborty, D. (1988). Utilization of bamboo as raw material in the handicraft industries in Tripura. *Indian Forester*, 114(10): 635-636.

Chang, F.C. (1938). Crystalline compound from the white powder found on *Bambusa changii*. *Lingman Science Journal*, 17:617-622.

Hirosuke, M. (1990). Manufacture of bamboo powders for deodourization. *Bamboo Abstract* (1991), 4(2): 47.

Inoue, M. (1988). Bamboo charcoal production by means of removable carbonization oven. *Bamboo Journal*, 6: 73-83.

Janssen, J.A. (1981). Bamboo trusses, In Bamboo Production and Utilization (Ed. Higuchi, T.) *Proc. XVII IUFRO Congress Group*, 5.3, Kyoto, pp. 158-163.

Janssen, J.A. (ed.) (1982). Bamboo: A series of articles on the use of bamboo in building construction, University of Techn., Eindhoven, p. 125.

Laha, R. (2000). Bamboo uses for housing by the different tribes of Northeast India. *Bamboo Science and Culture*, 14: 10-14.

Masani, N.J. Dhamni, B.C. and Singh Bachan (1977). Studies on bamboo concrete composite construction. Forest Research Institute, Dehra Dun, p. 39.

Menachary, M.D. and Chandran, K. (1984). Estrogenic activity of Bamboo (*Bambusa arundinacea*) Buds, *Kerala J. Vet. Sci.*, 15(1): 38-44.

Mgeni, A.S.M. (1983). Bamboo wine from *Oxytenanthera braunii*. *Indian Forester*, 109(5): 306-308.

Mitra, G.N. and Nayak, Y. (1972). Chemical composition of bamboo seeds (*Bambusa arundinacea* Willd). *Indian Forester*, 98(8): 479-481.

Mizobuchi, K. (1987). Bamboo Use as Fodder, Research on main bamboo species from Japan., Read Publishing House, Fukuoka, Japan, p. 291.

Mohan Ram, H.Y. and Tondon, R. (1997). Bamboo and Rattansd: From riches to rags. *Proc. Ind. Nat. Sci. Acad.*, 63: 245-267.

Piatti, L. (1947). Flussige B. enn-stoff aus Bambus Schweiz Arch Agew. *Wirs Tech.*, 13: 370-376.

Purushotham, A. (1963). Preliminary note on some experiments using bamboo as reinforcement in cement concrete. *J. Timb. Dryer's Pres. Assoc. India 9*, 4: 3-14.

Rai, S.N. and Chouhan, K.V.S. (1998). Distribution and growing stock of bamboo in India. *Indian Forester*, 124(2): 89-98.

Seethalaxmi, K.K. and Kumar, M.S.M. (1998). Bamboos of India: A compendium, BIC India; KFRI, Peech and INBAR, New Delhi.

Sethi, N., Nath, D. and Singh, R.K. (1989). Teratological evaluation of some commonly used indigenous antifertility plants in rats. *Ind. J. Crude Drug Res.* 17(2): 118-120.

Schreckenbach, H. and Abankwa, J.G.K. (1983). Construction technology for a tropical developing country, GTZ, Eschborn, p. 338.

Shukla, U. and Das, D. (1981). On the occurrence of some medicinal grasses in Eastern India. *Ind. J. Forestry*, 4(4): 273-279.

Singh, S.V. and Guha, S.R.D. (1981). Indian experience of papermaking from bamboo. In Bamboo Production and Utilization (Ed. Higuchi, T.), *Proc. XVII IUFRO Congress Group*, 5.3, Kyoto, pp. 33-38.

Sethi, N., Nath, D. and Singh, R.K. (1989). Teratological evaluation of some commonly used indigenous antifertility plants in rats. *Ind. J. Crude Drug Res.* 17 (2): 118-120.

Stulz, R. (1983). Appropriate building materials. Skat Publ., St. Gallen, 12, p. 324.

Subba Rao, T.V. (1966). Bamboo and its utilization. *Indian Forester*, 92(3): 186-190.

Tewari, D.N. (1992). A Monograph on Bamboo, International Book Distributor, Dehra Dun, pp. 207-251.

Varmah, J.C. and Bahadur, K.N. (1980). Country Report – India. In: Bamboo Research in Asia (Eds. Lessard, G. and Chouinard, A.) IDRC, Canada, pp. 19-46.

Yoshida, N. and Ikejiri, S. (1950). Bamboo-sprouts extract as growth promoting substance for microorganisms Part-I, *Shikoku Acta Medica*, 1: 25-28.

13

Eco-Restoration and the Stabilisation of Degraded Minespoil and Landslides Areas

K.S Dadhwal and *G.P. Juyal*

Mining is second only to agriculture as the Word's oldest and important industry. The dependence of primitive societies upon mined products is illustrated by the nomenclature viz. Stone age, Bronze age and Iron age, i.e. the sequence which also shows the increasing complexity of Society's relationship with mining. In a sense, the history of mining is the history of civilization itself (Khoshoo, 1984). Minerals are important raw materials for the economic development of any nation but unscientific mining of minerals poses a serious threat to the environment because of disturbed ecology. Large scale denudation occurs due to mining and the developmental activities. Due to excessive use of explosives in mining, shear resistance of mass is reduced and shear stresses are increased. Due to unscientific mining in hilly terrains slope height increases, resulting in slope instability. Due to this during heavy rains movements are caused by mechanical activation of overland flow and sub-surface water (Haruyama, 1980). Land degradation, pollution of water and air, droughts, floods, landslides, and disruption of communications are some important environmental hazards in open-cast mining.

Therefore, public awareness about environmental concerns in the recent times has compelled several developed and developing countries to launch extensive programmes for reclamation of mined areas and for control of environmental pollution caused by whole gamut of mining operations. Though not measured precisely, the mine overburden dumped annually the World over, is estimated at 3000 billion tones. The area of land currently disturbed by mining is computed to nearly 386,000 ha per year by 2005 A.D., this figure is projected to go up as high as 1000,000 ha per year. Therefore, in the last quarter of the century alone, the mine-degraded land amounted to nearly 24 million ha.

The Mining Industry of India

The mining industry in India is a major development sector. The total value of mineral production in the country during 1982 was Rs. 59430 million. The export projection for 1978-83 at 1976-77 prices are iron ore Rs. 17820 million and Coal Rs. 1150 million. Fortunately, our country has huge coal reserves of the order of 85700 million tones. Iron ore reserves have been estimated by GSI to be of the order of 1350 million tones (Katiyar *et al,* 1988). Value of mineral produced in India, which was Rs. 5414 million in 1972 spurted to Rs. 51290 million in 1982. More than 90% of our country's minerals production is from the public sector. According to Indian Bureau of Mines, out of about 60 minerals mined in India, 38 come from small mines, 17 from small and large mines and 5 from comparatively large mines confined to non-ferrous sector. In India mining is spread over an area of 9,43,380 hectares under 7365 mine leases (Anonymous, 1992). Number of working mines in India from 1993-96 are listed in Table 1.

Landslide Scenario

About 25% of the country's geographical area is under mountainous terrain. The southern, central and western mountains namely the western Ghats, Satpura and Vindhyan ranges and Aravalis are geologically very old and are stable formations as compared to the Himalayas and Shiwalik ranges in the north. These recent formations are geologically unstable, are in the seismic zone and are in the upheaval stage (Valdiya, 1988). The Himalayan and Shiwalik ranges extend from the north-west (Jammu & Kashmir) to north-east (Arunachal Pradesh and Nagaland). The elevation varies from 5000 to 9000m. The terrain is usually steep. Recently, there has been a very rapid increase in the developmental activities in the Himalayas and Shiwaliks comprising large scale construction

Table 1. Number of working mines in India

State / Union Territories	*Year*		
	1993-94	*1994-95*	*1995-96*
Andhra Pradesh	461	434	403
Assam	10	9	9
Goa	81	83	71
Haryana	27	23	26
Jammu & Kashmir	8	6	5
Kerala	48	42	48
Maharashtra	144	146	141
Meghalaya	2	2	2
Rajasthan	680	674	585
Tamil Nadu	132	127	132
West Bengal	133	127	127
Arunchal Pradesh	1	1	1
Bihar	389	369	364
Gujarat	513	524	496
Himachal Pradesh	32	35	34
Karnataka	244	241	229
Madhya Pradesh	594	567	510
Manipur	1	1	1
Orissa	282	277	257
Sikkim	1	1	1
Uttar Pradesh	57	54	48
Total	3,840	3,743	3,490

of mountain roads, mining, over-grazing, deforestation and opening up of steep land for agriculture. These activities have unleashed severe strain on the fragile ecosystem. Construction of a hill road involves clear felling of existing productive vegetation, cutting and blasting etc. These activities cause severe erosion resulting in extensive slope failures and are termed as landslides or slips. In the hills of Peninsular India annual occurrence of landslides and landslips is common during south-west and north-east monsoon periods in the Nilgiris and other high hills. Clearance of new areas for cultivation, industry, hydro-electric projects, townships, roads, bridges, coffee and tea plantations and absence of proper planning as regards to the proper disposal of rain-water results in severe erosion and finally result in landslides and landslips (Chinnamani *et at,* 1980). Krishnaswami and Jain (1975) identified 31 major landslides in northern and north-western Himalayas spread over an area of 1,20,000 km^2 and reviewed the causes of formation and problems associated with them. As many as 369 landslides have been recorded in the Alaknanda, Bhagirathi and Ganga basins of Uttaranchal Himalaya by Geological Survery of India (GSI). On an average the major road blocks in the country due to mine spoil deposition and landslides cause an annual loss of approximately 50,000 man hours and 5000 vehicle hours due to disruption of communication alone (Bansal and Mathur, 1976). Landslides due to hill slope materials and processes are given in Table 2.

Table 2. Landslides due to hill slope materials and processes (Selby, 1982)

Sl. No.	*Landslides*			
	Transitional slides	*Rotational slips*	*Falls*	*Sub-aqueous slides*
1.	Rockslides, block glides	Single rotational slips	Stone and boulder falls	Flow slides
2.	Slab, or flake slides	Multiple rotational slips (*a*) in stiff, fissured clays (*b*) In soft, extra sensitive clays	Rock and soil falls	Under consolidated clay slides
3.	Detritus or debris slides	Successive, or stepped rotational slips	–	–
4.	Mudflows (*a*) Climatic (*b*) Volcanic	–	–	–
5.	Bog flows; bog bursts	–	–	–
6.	Flow failures (*a*) Loess flows (*b*) flow slides	–	–	–

Factors Contributing to Landsliding in Soils

Many factors influence the development of landslides, and a particular slide can seldom be attributed to a single definite cause although it may be possible to identify a dominant or a triggering effect. Possible causes and contributing factors conducive to slides are listed in Table 3.

Factors causing landslides in India have been broadly classified by Deb *et al.*, (1969) (Table 4). Major landslides occurred in India in past are given in table 5.

Factors conducive for development of minespoils

Blasting of more and more areas unscientifically for mineral exploration and lack of stringent regulations governing environmental controls and precautions are the main factors conducive for the development of minespoils in India. This problem has resulted in shifting mass of sand and rubble, denudation of forests, reduction in water holding capacity of reservoirs, rivers, streams etc. (Dadhwal, *et al.*, 1988). Erosion on mine sites not only reduces the productivity but also causes offsite damage by depositing sediments in road ditches, stream channels, water reservoirs and in some of the most productive agricultural lands in the area. In coal strip-mining strata in the overburden contain iron disulfide minerals. These minerals when exposed to air and moisture, oxidize to produce acid and soluble salts. The end result is a spoil material that is far too acid for plant growth (Struthers, 1964). Apart from this, measures like not installing diversions, vegetative waterways etc. well ahead of the actual operation of mining is also a causative factor for development of mine spoils.

Deteriorating Effects of Mine Spoils

- It is a heterogeneous mixture of the overburden strata. Compaction of spoils adversely affects plant establishment and growth by increasing resistance to penetration by roots and to the emergence of seedlings (Byrnes *et al.*, 1982),
- Limited plant growth due to lack of available moisture.

Table 3. Factors contributing to landsliding in soils

Sl. No.	Types		Major mechanisms
A.	**High shear stress**		
1.	Removal of lateral support	*(i)*	Stream, water, or glacial erosion
		(ii)	Sub-aerial weathering, wetting, drying and frost action
		(iii)	Slope steepness increased by mass movement
		(ii)	Manmade quarries and pits, or removal of toe slopes
2.	Overloading	*(i)*	Weight of rain, snow, talus
		(ii)	Fills, waste piles, structures
3.	Transitory stresses	*(i)*	Earthquakes-ground motion and tilts
		(ii)	Man-made vibrations
4.	Removal of underlying support	*(i)*	Undercutting by running water
		(ii)	Sub-aerial weathering, wetting, drying and frost action
		(iii)	Subterranean erosion
		(iv)	Mining activities
5	Lateral pressure	*(i)*	Water in interstices
		(ii)	Freezing of water
		(iii)	Swelling by hydration clay
6.	Seismicity and other vibrations	*(i)*	Traffic
		(ii)	Machinery explosives, etc.
B.	**Low shear stress**		
1.	Composition and texture	*(i)*	Weak materials such as volcanic tuff and sedimentary clays
		(ii)	Loosely packed materials
		(iii)	Smooth grain shape
		(iv)	Uniform grain sizes
2.	Physico-chemical reactions	*(i)*	Cation exchange
		(ii)	Hydration of clay
		(iii)	Drying of clays
3.	Effects of pore water	*(i)*	Buoyancy effects
		(ii)	Reduction of capillary tension
		(iii)	Viscous drag of moving water on soil grains
4.	Changes in structure	*(i)*	Spontaneous liquefaction
5.	Vegetation	*(i)*	Removal of trees *(a)* Reducing normal loads *(b)* Removing apparent cohesion of tree roots

Table 4. Factors responsible causing landslides in India

Sl.No.	Factor	Place of occurrence in India
1.	Geological	Himalayas, Shiwaliks (Example : Nalota Nala, Kalagarh Nala in Dehradun District)
2.	Hydrological	Mostly in many places in India (caused by water seeping through disturbed slopes, toe cutting by rivers and torrents and pore water pressure)
3.	Seismic	Himalayan region (Examples: blasting for limestone, quarrying, road making, etc.)
4.	Land use	Many places in country (caused by : indiscriminate felling of trees, overgrazing, improper agriculture on steep slopes, blasting for mining and road making, unscientific minespoil disposal, undercutting of the base of existing slope, etc.)

Table 5. Major landslides in India

	Region	Name of slides
A.	(Northern and Himalayas north-western spread over 1,20,000 sq.km)	Nashri, Khuni-nala, Paithal nala, (on National Highway No. 1 Jammu Srinagar road), Rajour and Jhanammandi slides on Rajouri-Punch road; Nand Prayag, Kalinsor, Sial Phakhi and Tangi, Reni on the Rishikesh-Badrinath road, Narendranagar, Jamoli on Rishikesh-Tehri road; Kalagarh Nalota nala, Signikhala on Dehradun-Mussorie road; Laldhang, Chibru, Koti along Haripur-Koti road, Brik, Lepehera Jhora, Meli Bazar, Sangre, 9th Mile, 10th mile slide, etc. on National Highway No. 31A (Rampur-Rangpur) in Sikkim etc.
B.	Central India	No serious landslide
C.	Peninsular India	Slides in Western Ghats, Eastern Ghats, Vidhayas, Satpuras, etc.

- Reduced water holding capacities as compared to unmined soils.
- Water pollution and environmental hazards.
- Slope aspect, soil moisture, and shading by plants influences temperature of spoil material. For example, high temperature injury to plant seedlings is serious on black bituminous coal and black organic shales, put less on light coloured mineral shales and sand stones (Deely and Bordon, 1973).

- Spoil pH below 4.0 and above 9.5 is usually toxic for plant growth.
- Overburden materials are low in organic matter content.
- Due to low organic matter content of most minespoils, N is the primary limiting nutrient for plant growth. However, in most cases a complete fertilizer (N P and K) is needed for plant establishment.
- Reduced microbial growth and activity.
- Respiration and enzyme activities are hampered in minespoil as compared to undisturbed soil.
- Unreclaimed or poorly reclaimed lands often are barren because soil materials left on the surface after mining and reclamation does not support plant growth. This causes excessive erosion due to inadequate cover. Approximately 100 times more erosion has been reported for the abandoned mined lands compared to similarly located forest land (USEPA, 1976).

Mining Policy

A conscious strategy for the development of mineral resources constitutes the core of the mineral policy of India (1993), which clearly lays down the objectives of mineral development with due emphasis on their conservation and protection of the environment. The Mineral Conservation and Development Rules (MCDR), 1988, provide for the preparation of an Environmental Management Plan (EMP). The EMP adequately provides for controlling the environmental damage done, restoration of mined out areas and plantation of trees according to prescribed norms.

Rehabilitation Strategies for Mine Spoils

The mine spoil areas may be ameliorated by addition of soil amendments such as liming material for acidic mine spoils, sevage sludge, pepermill sludge, power plant flyash, normal soil, mulches, manures/compost etc. so that plant growth can be established (Apel, 1983). Once the abandoned mined lands has vegetation growing on the surface, the regeneration of these areas for productive use begins and offsite damages are minimized. In addition, establishment of vegetation on the abandoned mined lands also improves the aesthetics, location and objectives which may include agriculture, agroforestry, forestry, wildlife, and recreation. The rehabilitation strategies and technologies evolved in the country and elsewhere are summarized as follows:

Site Preparation and Overburden Placement

- Soil and overburden overlying the deposit is to be stripped out separately and can be stored during excavation for the final restoration.
- For site preparation seed, fertilizer/manure, mulch, normal soil, sludge and other amendments should be added.
- Water and sediment management must be considered from the very beginning. Water control facilities like channels, diversions and culverts be properly designed and maintained (Barfied *et al.,* 1981).
- Correction of drainage and long-term monitoring and maintenance is required for underground abandoned mines.
- Characteristics of the overburden must be considered in planning mining and reclamation programmes.
- Reclamation options often depend on how the overburden is handled during mining. Planned placement of overburden in mountainous terrain is widely accepted, because it maximizes the physical stability of the spoil pile, make vegetation establishment easier and create configuration more compatible with the natural topography.
- In highly degraded areas, where establishment of vegetation is difficult, mechanical measures are must for controlling erosion in the initial stages (Dhruvanarayana *et al.,* 1987).
- The spoil slopes which are highly friable should be moderated and can be reclaimed in stages.
- Mechanical measures to be adopted for the rehabilitation of limestone, lignite, coal mine spoils

etc. from India comprise rough levelling, making of plots depending upon the topography, gabion structures, loose check dams, toe walls, contour trenches, stone walls, grade stabilization structures, slope stabilization materials like geo-jute, formation of roads to transport plants, manure, water etc. (Narayana, 1986; Dadhwal *et al.,* 1998).

Top Soiling/Placing and Mixing

- Requirements regarding the depth of material for top soiling should be specific.
- Top soiling involves removal and storage of A, B, C, horizons—individually or in combination.
- The effects of storage on the physical, chemical and biological properties of top soil must be considered and documented.
- An alternative strategy to B-horizon replacement for the construction of post mine soils is sub-surface horizon blending, which may evaluate various combinations of substratum and B-horizons materials, as sub surface rooting media (McSweeney *et al.,* 1981).
- Top placing and mixing of normal soil upto 30 cm depth in limestone minespoil (Dadhwal *et al.,* 1991a); replacing top soil and subsoil over sedic spoil upto 70 cm, selected placement of various soil material over regarded lignite minespoil (Chichester, 1983); top soil placement upto 40-60 cm; for revegetation of coal mined land (Schumman *et al.,* 1985); top soil addition to an exposed 'C' horizon of a loess soil upto 10 cm (Mielke and Schephers, 1986) etc. have helped in revegetation of degraded minespoil areas.
- Minespoil amended with topsoil stimulates microbial activity.

Soil Amendments and Mulching

- Addition of N and P to establish and maintain vegetation alone or in combination.

- Use of legumes in seed mixtures for long term supply of N.
- Use of soil amendments, mulches and soil stabilizers.
- Minespoil pH has to be rectified and mulch materials aid in the establishment of vegetation, reduce soil erosion, conserve soil moisture and add organic matter.
- Use of sludge and sewage sludge tablets (Berry, 1983).
- Addition of dried sewage sludge in combination with NPK and CaO for *Pinus* spp. (McNab and Berry, 1985).
- Use of different synthetic conditioners viz, hygromull (a urea formaldehyde), agrosil LR and agrosil S (amorphos sodium hydrosilicates) which are helpful in modifying soil water relationships (Nimah *et al.*, 1983).
- Addition of sludge or hay increase number of microorganisms, enzyme activity, fungal generate distribution in non rhizosphere minespoil and enhance mycorrhizal infection (Lindemaan *et al.*, 1984).
- Minespoils treated with flyash favour establishment of grasses and legumes (Plass and Capp, 1974).
- Stubble mulch exhibit greater infiltration than straw mulch (Schuman *et al*, 1985).
- Combination of grass with legumes and nitrogen on coal minespoils produces optimum biomas (Yamanaka and Holl, 1984).
- For fertility restoration in limestone minespoil from outer Himalayas of India, N@50 kg ha^{-1} and FYM @ 20 t ha^{-1} for *Eulalioppsis binata* was found to be the ideal dose (Dadhwal *et al.*, 1991b; Dadhwal and Katiyar, 1989).
- Among sawdust, perlite and vermiculite, saw dust was found promising for promoting growth of forests, shrubs and trees on bentonite minespoil (Uresk and Yamamoto, 1986).

- While amending minespoils, the main points to be looked into seriously are moisture, nutrition, toxicity, presence of heavy metals, structure, acidity and salinity problems.

Revegetation Measures

- For revegetation, selection of plant species is generally determined by the physical and chemical properties of mine spoil and as per its classification (Juyal *et al.,* 1995).
- From soil conservation viewpoint length and thickness of plant roots in binding the soil particles is very important. Fibrous root system have greater soil binding capacity. Therefore, plant species which are native of the place, have ability to fix Nitrogen, hardy, have aesthetics, good in root soil binding characteristics, provide immediate coverage initially, and are sound ecologically and economically should be preferred and chosen (Dadhwal and Singh, 1993).
- For fluorspar mine tailings the use of legumes in seed mixtures has given promising results for the longterm supply of N and development of a low maintenance cover of vegetation (Johnson, 1976).
- For bauxite mined area of Madhya Pradesh, plant species viz, *Grevillea pteridifolia, Eucalyptus camaldulensis, Pinus, Shorea robusta* gave good performance (Chaturvedi, 1983).
- For coal mined spoil of Madhya Pradesh species like *Eucalyptus* hybrid, *Eucalyptus camaldulensis, Acacia auriculiformis, Acacia nilotica, Dalbergia sissoo, Pongamia pinnata* etc. have given promising results (Ramprasad & Shukla, 1985).
- for limestone minespoil reclamation from outer Himalayas of Uttaranchal plant species which gave promising results are: *Salix tetrasperma, Leucaena leucocephala, Bauhinia retusa, Acacia catechu, Ipomoea carnea, Eulaliopsis binata, Saccharum spontaneum, Chrysopogon fulvus, Arundo donax, Agave Americana, Pennisetum purpureum, Erythrina*

suberosa, Ficus rumphii, Pueraria hirsuta, etc. (Dhyana, *et al,* 1988; Dadhwal Katiyar, 1990).

- For rock phosphate minespoils reclamation (having composition of limestone, chert, shale, silica, etc.) at Maldeota in Uttaranchal, the most promising and suitable species tested are: *Pennisetum purpureum, Saccharum spontaneum, Vitex negundo, Rumex hastatus, Mimosa himalayana, Buddlea asiatica, Dalbergia sissoo, Acacia catechu, Leucaena Leucocephala* and *Salix tetrasperma* (Soni *et al.,* 1986).
- For lignite mine spoil reclamation in Tamil Nadu, Narayana (1986) observed best performance of *Eucalytpus, Leucaena leucocephala, acacia, agave, etc.*

Revegetation with Geojute

The geotexile (geo-jute), a woven netting made from jute fibers has been successfully used for stabilization of mine spoil slopes at Sahastradhara near Dehradun (Juyal *et al.,* 1991). The specification of geojute used were grade -500 g/m^2, strand thickness -5 m and open area -65 per cent. Geojute helped establishment of vegetation on degraded mine spoil sloppy lands by providing mechanical support to it and by trapping fine soil and moisture for its establishment and growth (Juyal and Dadhwal, 1996). Slopes ranging from 30-70 per cent were revegetated with this technique within a period of 3 years. Initially planting of grasses and shrubs is recommended. Once the soil conditions improve by biomass incorporation, tree seedlings may be planted or seeds sown.

Environmental Impact Assessment of Conservation Measures

Environmental impact assessment of treated Sahastradhara (Dehradun) mine spoil watershed showed that as a result of treatment measures, the annual debris flow was reduced to 8 t ha^{-1} as against 550 t ha^{-1} earlier within a period of 6 years. With increased storage of runoff water into the soil profile by conservation measures, base flow increased during dry weather period (October-May) and new water sources/ springs regenerated in the watershed. The water quality also improved with reduction in Ca, Mg and SO_4 content. As a result of leaf and root decay incorporation of biomass from vegetation, the soil properties improved (Juyal *et al.*, 1998).

Rehabilitation Strategies for Controlling Landslides

Rehabilitation strategies for controlling landslides comprise mechanical and biological measures which are summarized in Table 6.

Gaps in Technological Development and Its Actual Adoption

- There is practically no involvement of local people during the planning, implementation and maintenance of rehabilitation works.
- The methods adopted for rehabilitation in many cases are not based on the kind of plant material available i.e. choice of species, locality and site conditions (Dhyani *et al.*, 1988).
- Transfer of technology to the farmers, contractors, foresters, mine professionals, highway engineers, public works departments, electricity board, etc. is an important aspect which can be achieved through training by organizing short courses, training and by creating awareness among the user agencies.
- Technologies developed for stabilization of mine spoils/landslides should be immediately passed on to the user agencies and vice-versa feed back is essential which is lacking.
- Biotic interference in several affected areas is to be stopped by creating awareness among people.
- The landslides and landslips can be tackled by good management, soil and water conservation measures, planning proper disposal system for rain water during storms and by protection measures which can be achieved by an Act of law, educating the general public through different publicity media and in extreme cases fencing the area.
- Before mining and any other developmental activity like diversions, drains, waterways, etc. necessary steps are generally not planned or properly implemented.

Table 6. Rehabilitation measures for controlling landslides in India.

Region	*Mechanical measures*	*Biological measures*
1	**2**	**3**
A. Northwest Outer Himalayas (Sastry *et al.*, 1981)		
Debris cone reach	(*a*) Retention barrier parallel to torrent (*b*) Drop structures with shallow foundation and aprons in the main torrent (*c*) Protect against scouring action	• Along the torrent beds—*Pennisetum purpureum*, *Arundo donax*, *Salix tetrasperma*, etc. • Species suited for different reaches and slide face stablization are: (as listed for lower reach)
Lower reach	(*a*) Gabion toe walls (60 cm deep and 60 cm top width) along the concave bank of torrent (*b*) Deflecting type spurs	• *Lannea grandis*,*Erythrina suberosa*, *Vitex negundo*, *Salix tetrasperma*, *Cassia fistula*, *Mallotus phillippinensis*, *Cocculus laurifolius Hamiltonia suaveolens*, *Ipomoea carnea*, *Acacia catechu*, *Dalbergia sissoo*, *Pueraria hirsuta*, *Chrysopogon fulvus*, *etc.*
Middle reach	(*a*) Construction of a series of gabion drop structures with adequate aprons, clear spillway and proper alignment	**Mulching** Inter wattle areas mulching with *Chrysopogon fulvus*
Upper reach	(*a*) A series of drop structures (*b*) Stone checkdams (*c*) Vertical posts check dams (*d*) Logwood crib checkdams	

(Table Contd...)

Table 6. (Contd...)

1	2	3
Slide face (having rills and gullies	(a) Contour wattling by breaking the slope length into shorter portions (b) Jute/coir geotextile	Plantation of grasses/shrubs/trees.
B. Peninsular India (Chinnamani, *et al.*, 1980)		
Pure soil landslides of Nilgiris, Western and Eastern Ghats	(a) Provision of longitudinal drain with drop pits at 1 m VI. (b) Continuous contour trench at 1 to 2 m V.I connecting the drain (c) Diversion bund or drain of trench 3-5 m on the uphill side of the slide	*Hybrid napier, Gautamala grass, Sorghum helepens, Cymbopogon spp., Arundo donax, Eragrostis curvula, Pennisetum clandestinum, Cynodon dactylon, Acacia dealbata, Eucalyptus globulus*, etc.
Landslides of soil and boulders (avalanches) common throughout the hills of Peninsular India	(a Contour trenches (b) Diversion drains	*Agave Americana, Eucalyptus globules, Acacia dealbata, Acacia melanoxylon, Cupressus macrocarpa*, etc.

(Table Contd...)

Table 6. (Contd...)

1	2	3
Landslides of pure boulders	(a) Diversion drains of boulders (b) Stone mulching	Allowing natural vegetation of *Lichens, Ferns* and *Mosses; Castor, Sorghum helepens, Eragrostis curvula,Bamboos, etc.*
Landslides with mudflows (Due to high surface and subsurface runoff)	(a) Provision of diversion drains not less than 3m deep at the top of slide (b) Side drains not less than 3m deep around the slip	*Hybrid napier, Gautemala* grass *Sorghum helepens, Arundo donax, Chrysopogon* spp., *Eragrostis curvula, Eucalyptus globulus, Acacia* spp. etc.
Landslips on terrace embankments, Urban lands and roads	(a) Ease the landslip to angle of response (b) Construct loose stone embankments (c) Construct embankments with stone masonary (d) Provide drain and inner drains	*Gautamala grass, Eragrostis curvula, Digitaria* spp., *Cymbopogon spp., Pennisetum* spp., sorghum *helepens, Doddonea viscosa, Agave Americana, Polygonum spp.,* etc.
Landslips on torrents, perennial stream and rivers (common in Kerala, Karnataka, T. Nadu and *Maharashtra*)	(a) Revetment with loose stones (b) Masonary revetments (c) Safe water disposal structures	*Napier, Arundo donax, sorghum helpens, tripsacum laxum, Eragrostis curvula, Mimosa spp,, digitaria spp., Cynodon dactylon, Eucalyptus globulus, E. grandis, E. hybrid, Acacia dealbata, A. mearnsii, Bambusa arundinacea, Dendrocalamus strictus, Acacia nilotica, Terminalia spp., etc.*

- Physical, chemical and biological properties of minespoils are not considered before taking the rehabilitation programmes because they define limitations for vegetation establishment.
- Mining activities particularly in unorganized sector are not carried out strictly according to the Rules and Regulations concerning environmental protection.
- Area disturbed is not rehabilitated immediately and aggressive maintenance measures are not adopted.
- In most of the cases, mechanical and biological measures adopted are not on scientific lines.
- Lack of public awareness for scientific mining.
- Heavy expenditure and works of rehabilitation of minespoils and landslides.

CONCLUSIONS

- Mining activities should be mandated by stringent rules and regulations like developed countries e.g. USA, UK, France, etc.
- Before mining all necessary steps are required to be planned to prevent soil erosion, interference to water courses and also that provision should be made for rehabilitation and monitoring. There must be pre-planning of water and sediment control and installation of diversions and vegetative waterways much ahead of the actual operation.
- Disturb as an small area as possible at one time so that minimum vegetation is disturbed. Companies/ contractors should not be allowed to disturb further areas until previous one is rehabilitated.
- Reclamation process of minespoils and landslide areas should be speedy, cheap, reliable and should not require subsequent maintenance.
- Approach for rehabilitation must consist proper mechanical measures and aggressive maintenance programme.

- Important characteristics of minespoil and landslide areas should be considered before adopting biological measures for rehabilitation.
- Develop site specific reclamation and land use planning practices.
- Choose only those plant species for reclamation which grow quickly, provide immediate ground cover initially, native of that place, have good soil binding characteristics, fixes N, have aesthetics and have economical and ecological security.
- For improving fertility status and moisture content of mined lands and landslide areas leaves/leaf litter of the adjoining vegetation can be used as mulch and compost.
- For the better establishment of vegetation measures like normal soil placement and mixing, use of Farm Yard Manure/Compost, addition of N, soil amendments, mulches etc. are recommended.
- Mechanical and biological measures should have scientific input.
- The lease conditions must have a clause that mining company should remove the top soil first and keep it dumped separately. The company should level up the area and spread soil on the top before taking afforestation programme.
- Initially for first 3-5 years there should not be any biotic interference. Severely affected sites should be fenced.
- Technologies developed/available for stabilization of minespoils and landslides should be immediately passed on to the user agencies.
- People's involvement is recommended at all stages.

Future Line of Work

Several long term studies have been conducted for the reclamation of mined lands in developed countries like USA, UK, France etc., but in India very meagre information for the rehabilitation of minespoils of various mines on scientific lines is available. Since the reclamation measures vary with the type of minespoil, site and physiographic conditions, the following future line of work is suggested for minespoils and landslides rehabilitation:

- Characterization of various minespoils and landslides for important physical, chemical and biological properties should be done e.g. compaction, bulk density, available moisture, organic matter content, infiltration, pH, texture, structure, fertility parameters, toxic elements, micro-organism etc. These will be helpful in defining limitation for plant growth.
- To develop soil working and planting techniques e.g. wattling, trenching, holes, pits, brick planting, direct seeding, spraying of seeds in inaccessible areas by treating with clay, mud or hay etc. for minespoils and landslides.
- To have studies on the top soiling and mixing, use of manures and fertilizers, mulches, hay compost, sewage sludge, flyash, papermill sludge etc. and to evaluate their effect for different minespoils and sites.
- In steep slopes use of geo-textile measures like geo-jute, coir net *(Bhoovastra)* etc. are required to be studied.
- Selection and evaluation of such plant materials which add nitrogen, have aesthetics, have maximum soil binding characteristics and are also ecologically and economically sound.
- Scientific mining methods causing least damage to environment should be evolved.
- Elaborate and systematic long term monitoring of rehabilitation works be made to focus on important ecological and economic dimensions.
- Local level institutions be established to ensure effective people's participation.

ACKNOWLEDGEMENTS

The authors are grateful to Director, Central Soil and Water Conservation Research and Training Institute,

Dehradun, for necessary guidance and encouragement provided for writing this manuscript. Help rendered for typing this paper by Smt. Mamta Negi is also acknowledged.

REFERENCES

Anonymous (1992). Status of Mining Leases in India. Indian Bureau of Mines, Nagpur.

Apel, L.M. (1983). Leachability and revegetation of solid waste from mining. EPA-600/S_2-82-093. U.S. Environ. Prot. Agency, Cincinnati, Ohio.

Bansal, R.C. and Mathur, H.N. (1976). Landslides—the Nightmares of hill roads. *Soil Conserv. Digest,* 4(1): 36-37.

Barfied, B.J., Warner, R.C. and Haan, C.T. (1981). Applied hydrology and sedimentology for disturbed areas. Chapter I. pp. 1-5.

Berry, C.R. (1983). Growth response to four hardwood species to spot fertilization by nutrient tablets in the Tennessee Copper Basin. *Reclamation and Revegetation Research,* 2(3): 167-175.

Byrnes, W.R., McFee, W.W. and Steinhardt, G.C. (1982). Soil compaction related to agricultural and construction operations. Purdue Univ. Agric. Exp. Stn. Bull. 397, West Lafayettee, Indiana.

Chaturvedi, J.K. (1983). Afforestation of Bauxite minded area in Central India. *Indian Forester,* 109(7): 458-465.

Chichester, F.W. (1983). Premining evaluation of forage grass growth on minespoil materials from an East-Central Texas. Lignite Site: 2. Soil profile horizons. *Soil Sci.,* 135(4): 236-244.

Chinnamani, S., Venkataraman, C. and Tejwani, K.G. (1980). Landslides and landslips in Penisular India. Bull No. 4, CSWCRTI, Dehradun, India p. 30.

Dadhwal, K.S. and Katiyar, V.S. (1989). Effect of Farmyard manure on growth, yield and chemical composition of bhabar in a minespoil. *J. Indian Soc. Soil Sci.,* 37(2): 412-415.

Dadhwal, K.S. and Katiyar, V.S. (1990). Measures for the reclamation of abandoned mine lands. *Himalaya (man and nature),* 13(9): 14-18

Dadhwal, K.S., singh, B. and Narain, P. (1991a). Effect of placing and mixing of normal soil on the performance of two grasses in limestone minespoil *J. Indian Soc. Soil Sci.,* 39(4): 751-756.

Dadhwal, K.S., Singh, B. and Narain, P. (1991b). Effects of nitrogen application on *Eulaliopsis binata* grass in a limestone minespoil from outer Himalayas of Uttar Pradesh. *J. Indian Soc. Soil Sci.*, 39(4): 786-788.

Dadhwal, K.S., Singh, B., Narain, P. and Katiyar, V.S. (1998). Techniques for the rehabilitation of degraded mined lands. *Indian J. Soil Conserv.*, 16(2): 51-62.

Dadhwal, K.S. and Singh, B. (1993). Rooting behaviour of different plant species in limestone mine area. *Indian Forester,* 119(1): 71-74.

Dadhwal, K.S., Juyal, G.P., Katiyar, V.S and Joshie, P. (1998). Characterization and staiblization of minespoils. In : Soil and Water Conservation—Challenges and Opportunities (L.S. Bhushan, I.P. Abrol and M.S.R.M. Rao eds.), Vol. 2, 8th ISCO Conf., Oxford & IBH Pub. Pvt. Ltd., New Delhi, pp. 1345-1354.

Deb A.K., Joshie, S.S. and Gholap, S.D. (1969). Landslides-causes and remedial measures, Paper presented at Nat. Semi , Hill Roads slope stablization, Boarder Roads Organization, New Delhi.

Deely, D.J. and Borden, F.Y. (1973). In "Ecology and reclamation of devastated land" (R.J. Hutnik and G. Davis, eds.) Vol. 1 pp. 69-79. Gordon and Breach, New York.

Dhruvanarayana, V.V., Katiyar, V.S and Dadhwal, K.S. (1987). Mined area reclamation in Mussoorie Hills. *Indian J. Soil Conserv.*, 15(2): 82-90.

Dhyani, S.K., Dadhwal, K.S. and Katiyar, V.S. (1988). Biological measures for rehabilitation of abandoned minespoil areas. *Indian Farming,* 38(7): 15-18.

Haruyama, M. (1980). Mass movements in the slopes with pyroclastic deposits. Proc. Int. Symp. on landslides, New Delhi.

Johnson, M.S. (1976). Plant growth on flourspar mine tailings. *J. Soil Wat. Conserv.* 31(1): 17-20.

Juyal, G.P., Katiyar, V.S., Sastry, G., Singh, G., Joshie, P. and Arya, R.K. (1991). Geojute for rehabilitation of steep mine spoil areas. Bulletin No. T-26/D-19, CSWCRTI, Derhadun.

Juyal, G.P., Dadhwal, K.S. and Joshie, P. (1995). USDA method for minespoil classification and its application at Sahastradhra mined watershed. *Indian J. Soil Conserv.*, 23(3): 196-199.

Juyal, G.P. and Dadhwal, K.S. (1996). Geojute for erosion control with special reference to minespoil rehabilitation. *Indian J. Soil Conserv.*, 24(3): 179-186.

Juyal, G.P., Katiyar, V.S., Dadhwal, K.S. and Joshie, P. (1998). Reclamation of mine spoil in outer Himalayas. In: Soil and water conservation-challenges and opportunities (Eds. Bhushan, L.S., Abrol, I.P. and Rao, M.S.R.M.), Vol. 2, IASWC, Dehradun, India, pp. 1384-1392.

Katiyar, V.S., Dadhwal, K.S. and Dhyani, S.K. (1988). Soil Conservation measures in minespoil stabilization (K.A. Shankarnarayana ed.), Wastelands Development and their utilization, Scientific Publishers, Jodhpur, pp. 226-236.

Khoshoo, T.N. (1984). Land reclamation in open cast mines. 27th Holland Memorial Lecture, Mining, Geological and Metallurgical Instt. of India, Calcutta.

Krishnaswami, V.S. and Jain, M.S. (1975). A review of some of the major landslides in northern and north-western Himalyas. Proceedings of Seminar on Landslides and toe Erosion Problems with Special Reference to Himalayan Region, Indian Society of Engineering Geology, Gangtok, Sikkim.

Lindemann, W.C., Lindsey, D.L. and Fresquez, P.R. (1984). Amendment of minespoil to increase the number and activity of micro-organisms. *Soil Sci. Soc. Am. J.,* 48(3): 574-578.

McNab, W.H. and Berry, C.R. (1985). Distribution of above ground biomass in three pine species planted on a devastated site amended with sewage sludge or inorganic fertilizer. *Forest Sci.,* 31(2): 373-382.

McSweeney, K., Jansen, I.J. and Dancer, W.S. (1981). Sub-surface horizon blending—an alternative strategy to B-horizon replacement for the construction of post mine soil. *Soil Sci. Soc. Am. J.,* 45: 794-799.

Mielke, L.N. and Schepers, J.S. (1986). Plant response to top soil thickness on an eroded loess soil. *J. Soil Wat. Conserv.,* 41(1): 59-63.

Narayana, M.P. (1986). Mined area reclamation—The Nyveli experience. Paper presented in Indo-British workshop on mined area reclamation, New Delhi.

Nimah, M.N., Ryan, J. and Chaudhary, M.A. (1983). Effect of synthetic conditions on soil water retention, hydraulic conductivity, porosity and aggregation. *Soil Sci. Soc. Am. J.,* 47(4): 742-745.

Plass, W.T. and Capp, J.P. (1974). Physical and chemical characteristics of surface minespoil treated with flyash. *J. Soil Wat. Cons.,* 29(3): 119-121.

Power, J.F., Ries, R.E. and Sandoval, F.M. (1978). Reclamation of coal mined land in the Northern Great Plains. *J. Soil Wat. Cons.*, 33(2) : 69-74.

Ramprasad and Shukla, P.K. (1985). Reclamation and revegetation of coal mine overburdens in Madhya Pradesh. *J. Trop. For.*, 1(1): 79-84.

Sastry, G., Mathur, H.N. and Tejwani, K.G. (1981). Landslide control in north-western outer Himalayas. Bull No.R-8/D-6, CSWCRTI, Dehradun.

Schuman, G.E., Taylor, E.M., Rauzi, F.Jr. and Pinchak, B.A. (1985). Revegetation of mined land influence of top soil depth and mulching method. *J. Soil Wat. Cons.*, 40(2) : 249-252.

Selby, M.J. (1982). Hill slope materials and processes (chap. 5), Oxford Univ. Press, Oxford, p. 260.

Soni, P., Mathur, H.H., Naithani, S., Kumar, O. and Vasistha, H.B. (1986). Rehabilitating mining land in Doon Valley. *Himalaya (Man and Nature)* 9 (10): 1-2.

U.S. Environmental Protection Agency (1976). "Erosion and sediment control, surface mining in the eastern, U.S., Vol. I," EPA-625/3-76-006, U.S. Govt. Print. Office, Washington, D.C.

Uresk, E.W. and Yamomoto, T. (1986). Growth of forbs, shrubs, and trees on bentonite minespoil under green house conditions. *J. Range Mmgmt.*, 39(2): 113-117.

Valdiya, K.S. (1988). Environmental impacts of mining activities. Mining and environment in India. HR Publ. Tallital-Nainital (Uttaranchal): 29-42.

Yamanka, K. and Holl, F.B. (1984). Effects of N and seeding rate on grass-legume mixtures on coal mine spoils. *Agron. J.*, 76(6): 895-900.

14

The Fourth Law of Thermodynamics in Ecologic Research

Swapan C Deb

The *science of thermodynamics* deals with relations between different forms of energy. Classical thermodynamics focuses on transformations of energy from one form to other, relatively close to the thermodynamic equilibrium. Onsager[1] and Prigogine[2] have initiated a development of classical thermodynamics. A wider theoretical application of thermodynamics was not initiated until the late seventies[3]. Thermodynamics are based on general laws of nature, namely *the first, second* and *third law*. It has proposed to introduce another tentative, fourth law thermodynamics. Logically from these laws, it is possible to correlate many of the observable properties of ecosystems (eg change of ecological structure and the species composition). The thermodynamic approach is particularly well fitted as a tool to describe ecosystems from a holistic point of view because it is based on the macroscopic flows of energy and mass[4,5].

Ecosystems are non-isolated (open to energy), steadily changing and are maintained far from thermodynamic equilibrium. The ecosystems have a tendency to move away from thermodynamic equilibrium losing entropy or gaining negentrophy and information, due to the through-flow of energy. The maintenance of the ecosystems far from equilibrium requires a steady flow of energy to cover the energy needed to maintain a system far from equilibrium and hence prevent it form developing towa ds thermodynamic equilibrium. The available work of the system (i.e. *exergy*) is a measure of the distance from thermodynamic equilibrium. Therefore, it is possible to formulate that *ecosystems attempt to develop toward a higher level of exergy.*[4]

Ecosystems are soft and adaptable systems and are able to meet changes in external factors or impacts with many varying regulation processes on different trophic levels. Despite the relatively major changes in environmental conditions, the results are often only minor changes in the function of the ecosystem. In spite of changes in external factors, the state variables (not necessarily species) are maintained almost unchanged.

The modern *Neo-Darwinian theory* fanned out to include all the recent developments in genetics and in evolutionary ecology, which is able to describe the very complex competition among species. Darwin's theory states that the species that are best fitted to the prevailing conditions in the ecosystem will survive. Hence, it is preferable that the *life is a matter of survival* and *growth.* The species, or rather combination of species, may be connoted the fittest and will be selected due to their properties. Darwin's theory may, in other words, be used to describe the changes in ecological structure and species composition, but cannot be directly applied quantitatively with the present formulation, for instance in ecological modelling. Thermodynamics may, however, offer a concept, which can be used to give a *quantitative description of ecosystem developmental tendencies.*

The present paper reviews the tentative fourth law of thermodynamics, its importance, developmental facets and possibilities in the context of ecologic research.

EXERGY AND ECOSYSTEM THEORY

Exergy is an important tool to be applied as a quantitative description of ecosystem development. It is the distance from thermodynamic equilibrium and covers both the size of the organised structure and its content of thermodynamic information.[6] *Growth* may be defined as a quantitative expression of the increase of organised structure. Thermodynamically, *growth means that the system is moving away from thermodynamic equilibrium by increasing its organised structure.* At equilibrium, the system cannot do any work, the components are inorganic and hence have the lowest possible free energy, and gradients. Biological systems in particular have many possibilities to grow or to move away from thermodynamic equilibrium. Thus, it is crucial in ecology to know which pathways among the possible ones an ecosystem will select for growth.

The ecosystems must be open, or at least non-isolated. This is absolutely necessary for their existence. A flow of energy through the system is sufficient to form an ordered structure (dissipative structure)[7]. Later, Morowitz[8] calls this as the fourth law of thermodynamics. This law could encompass a statement about ordered structure of the system, or factors determine how an ecosystem will develop. This expanded version was formulated as a tentative fourth law of thermodynamics[9], but was already expressed with a slightly different formulation in Jorgensen and Mejer,[10] Mejer and Jorgensen[11] and in Jorgensen[12].

EXERGY AND DARWIN'S THEORY

Darwin and neo-Darwinism postulated that the best-fitted species (best coordinated to the prevailing conditions), would survive. Survival means that the biomass of the species will be maintained or maybe even increased. An organism or a population is exposed to many constraints, determined by the forcing functions on the ecosystems and the other organisms living in it. Thermodynamically, the organisms that have their properties better coordinated to the prevailing conditions will be able to contribute most to the work content or exergy of the system due to their biomass with embodied information[9,13,14,15]. *The exergy or chemical energy, which can be used to do work in mineral oil, is about 42 kJ/g.* The chemical energy (free energy, exergy) of biomass with an average composition of proteins, carbohydrates and fat can be calculated to be approximately 18.7 kJ/g[16].

Brown[3], defines fitness as the rate at which resources in excess of those required for maintenance can be utilised for reproduction. He uses dW/dt, called *reproductive power* to find the optimal *body mass, W*. The size yielding the biggest increase of the biomass corresponding to the biggest increase of the exergy in the system.

An ecosystem encompasses many species. They cannot all obtain the biggest biomass independently of the other species—the species are interdependent. Darwin considered this complication and expressed it as 'prevailing conditions', and is anticipated to include all the abiological and biological constraints imposed on the species, i.e., including constraints originating from other species. The evolution and co-evolution over a very long period have implied that the species have adapted to each other. They were able to find ways to move further away from thermodynamic equilibrium (get more growth) if they co-operated by adjusting their properties to each other and the prevailing external factors[17]. The effect of this co-operation is consistent with Patten.[18] He shows that the indirect effect often exceeds the direct one. For instance, a predator—prey relationship may also be beneficial for the prey due to a number of factors including faster circulation of the nutrients.

Darwin's Theory presumes that populations consist of individuals, which:

1. on an average produce more offspring than needed to replace them upon death—these are the properties of high reproduction. Translated into thermodynamics: more possible pathways for utilisation of the energy flow are developed than the system and its energy flow can sustain. It implies that a competition even among pathways, that are only slightly different, will be established.
2. have offspring resemble their parents more than they resemble randomly chosen individuals in the same population—this is the property of inheritance. Thermodynamically, it means that the properties, which showed a better ability to utilise the energy flow to move as far away from thermodynamic equilibrium as possible, by construction of more

biomass, to a large extent will be preserved. Genetics can explain how this is possible.

3. vary in heritable traits influencing reproduction and survival due to differences in fitness to the prevailing conditions—this is the property of variation. Modernised neo-Darwinism is able to give a long list of mechanisms that can create new pathways. It implies that new possibilities are steadily created to meet the challenge of utilizing the energy flow to move away from thermodynamic equilibrium. The possibilities are examined under the prevailing conditions, and those, which are successful, are preserved according to the properties of inheritance.

Evolution can continue on the shoulders of the already found successful solutions and therefore steadily find new and better solutions, i.e. select the best genes, among all the present genes including those that are continuously emerging by mutations and sexual recombination.

It implies that selection processes to facilitate 'best possible survival' under the prevailing conditions are changing the properties. For plants this includes grazing and for the grazer, the availability of food. The species cannot change the properties of other species directly, but all species must consider other species in their effort to find a feasible combination of properties that is able to offer a higher probability of survival. This explains how species adapt to each other (coevolve), and how they can co-operate on the joint goal to move, as much as possible, away from thermodynamic equilibrium. In principle, each of the species is striving to get the highest possible growth for its species. As these goals cannot be reached if the species don't adapt to other species (as they are also part of the life conditions), the result will be that the species together move away from thermodynamic equilibrium, i.e. give the system the highest possible exergy. It will often coincide with the highest, or close to the highest, biomass for most species, at least on a long-term basis.

The exergy, as it measures the distance from thermodynamic equilibrium of the entire ecosystem, seems to be a good tool to quantify survival and growth in the Darwinian sense for the entire ecosystem. Calculations of the exergy of ecosystems make it possible to unite the chemical energy of the organic matter and the information, in the sense of Boltzmann,[13] embodied in the species.

ESTIMATION OF THE EXERGY OF AN ECOSYSTEM

Exergy of an ecosystem can be estimated using the principle of thermodynamics. Jorgensen[4] considered a reference environment that represents the system (ecosystem) at thermodynamic equilibrium, where all the components are inorganic at the highest possible oxidation state (maximum free energy is utilised to do work) and homogeneously distributed in the system (no gradients) the situation (Figure 1). As the chemical energy embodied in the organic components and the biological structure contributes by far the most to the exergy content of the system, there seems to be no reason to assume a (minor) temperature and pressure difference between the system and the reference environment. Under that circumstances, the exergy content of the system as coming entirely from the chemical energy: $_{-c}(\mu_c - (\mu_{co})$ Ni. These calculations that exergy of the system compared with the same system at the same temperature and pressure but in the form of an inorganic soup without any life, biological structure, information or organic molecules. As $_c(\mu_c - (\mu_{co})$ can be found from the definition of the chemical potential replacing activities by concentrations, we get the following expressions from exergy.

$$Ex = \underset{i=0}{\overset{i=n}{RT}} - C_i \ln C_i / C_{i,o} \quad (1)$$

Where R is the gas constant, T is the temperature of the environment (Figure 1), while C_i is the concentration of the i'th component expressed in a suitable unit, e.g. for phytoplankton in a lake C_i could be expressed as Mg/l of a focal nutrient. $C_{i,o}$ is the concentration of the i'th component at thermodynamic equilibrium and n is the number of components. $C_{i,o}$ is of course a very small concentration (except for i=o, which is considered to cover all the inorganic

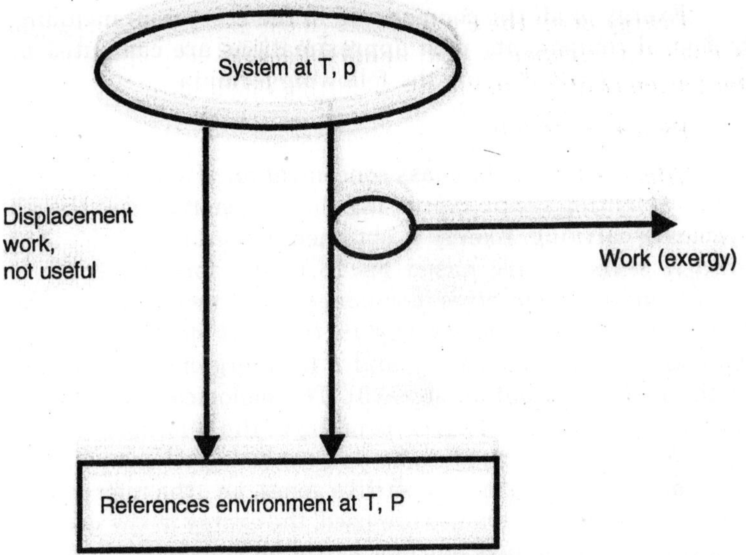

Fig. 1: The exergy of the system is calculated in the text for the system relatively to a reference environment of the same system at the same temperature and pressure, but as an organic soup with no life, biological structure, information or organic molecules[5].

compounds) or even biological components spontaneously in an inorganic soup at thermodynamic equilibrium. Notice the accordance between equations 1 (above). Exergy can be interpreted as the sum of the chemical exergy of the components and Kullbach's measure of information[11,17-19].

The problem by application of equation (1) is related to the size of C_{io}[16]. For dead organic matter, detritus, which is given the index 1, it can be found from classical thermodynamics:

$$\mu_1 = \mu_{10} + RT \text{ In } C_1/C_{10} \tag{2}$$

Where μ indicates the chemical potential. The difference $\mu_1 = \mu_{10}$ is known for organic matter, e.g. detritus is a mixture of carbohydrates, fats and proteins. If these figures are considered, we get approximately 18.7 kJ/g detritus.

Exergy of all the components in the ecosystem including biological components with approximations are computed by Jorgensen *et al*[15,16] using the following formula:

$$Ex = \bullet_i C_i (\bullet = \beta) \tag{3}$$

Where Ci is the biomass concentration of species i and $\bullet_i$ is the weighting factor expressing the information that the ith species is carrying. Exergy is obtained in kJ/unit of volume or area if •-values are based on 18.7 kJ/g for detritus and correspondingly for other components. It may, however, be convenient to express exergy in the unit detritus exergy equivalent (• for detritus=1 and • for inorganic components at the highest oxidation state=0). The biological components then have •-values > 1 corresponding to the information they carry. Table 1 gives •-values for various biological components, calculated from their non-repetitive genes, i.e., the information they carry. If these •-values are used, the exergy in the detritus equivalent is obtained but can easily be converted to kJ by multiplication by 18.7[11,14,15].

The equation can be applied to calculate exergy for important ecosystem components. The total exergy of an ecosystem cannot be calculated exactly. To measure the exergy, concentrations of all components of an ecosystem should be examined carefully. However, it can be calculated the contributions from the dominant components, which are of interest for a particular problem.

The calculation of exergy accounts for the chemical energy in the organic matter and for the information that originates in the extremely small probability to form living inorganic matter. The weighting factors could also be considered quality factors reflecting how developed the various groups are due to their embodied information. In principle, it should include all the components in the calculations. All inorganic compounds should also be included but their contribution to the exergy of the system is minor and may therefore be excluded from these approximate calculations.

The calculations are also consistent with the classical application of thermodynamics on chemical equilibria. If we consider the chemical reaction zooplankton + oxygen <—>

carbon dioxide + water + nutrients, the mass constant, K, can in principle be defined in the usual way. The very low zooplankton concentration, due to the low probability of the presence of zooplankton at thermodynamic equilibrium, will be reflected in a huge K-value, which can be translated to a high free energy equal to exergy in this case.

Table 1. Approximate number of non repetitive genes

Organisms	*Number of Information Genes*	*Conversion factor*
Detritus	0	1
Minimal Cell[8]	470	2.7
Bacteria	600	3.0
Algae	850	3.9

Sources: T.Cavalier-Smith[20], Li and Grauer[21] and Lewin.[22]

Based on number of information genes and the exergy content of the organic matter in the various organisms, compared with the exergy contained in detritus 1g detritus has about 18.7 kJ exergy (=energy which can do work).

The present calculations do not include the information embodied in the structure of the ecosystem network, i.e. the relationships between the various components, which are represented by the network. The information of the network encompasses the information of the components. Ulanowicz,[23] calculated this by as a contribution to ascendancy.

All computations of exergy will have the following shortcomings:

1. The Computations will be based upon either a model or a limited number of measurements. The results of the computations are, therefore, more appropriate for finding a relative difference in exergy by a comparison of an ecosystem under different conditions.
2. The calculations—as with all calculations in thermodynamics—are based upon approximations and assumptions. But as it draws conclusions on the basis

of the differences in exergy rather than on the basis of absolute values, the results may still be applicable in ecosystem theoretical contexts.

In addition, it implies that the computed exergy will be entirely related to the chemical composition and the biological structure with its information. This is the major contribution under all circumstances, as only minor temperature and pressure differences are realistic between an ecosystem and its environment. Due to the above-mentioned shortcomings and approximations, the exergy calculated in accordance with equation (3) should be denoted an exergy index.

THE TENTATIVE FOURTH LAW OF THERMODYNAMICS

Applying Darwin's theory simultaneously on all components of an ecosystem by translating survival and growth into the concept of exergy, makes it possible to formulate a tentative fourth law of thermodynamics. It a system receives a through-flow of exergy, the system will utilise this exergy to move away from thermodynamic equilibrium. If the system is offered more than one pathway to move away from thermodynamic equilibrium, the one yielding most stored exergy, i.e. with the most ordered structure or the longest distance to thermodynamic equilibrium by the prevailing conditions will have a propensity to be selected.

As it is not possible to prove the first, second and third laws of thermodynamics by *deductive methods*, the tentative fourth law can only at the best be proved by *inductive methods*. It implies that fourth law should be investigated in as many cases as possible. Several modelling cases have been examined and they have all approved the tentative law[9,24,25]. The law has particularly been used successfully to develop models with dynamic structure[9,24,26-30]. The term *'have a propensity to be selected'* is used in the formulation to indicate that it is hardly possible to make deterministic statements about the development of an ecosystem,[31] due to the random character of many of the factors and to the above-mentioned shortcomings in calculations. Ulanowicz uses the term that an ecosystem is ontic open.

It may be considered strong support that the order of microbial processes at pH = 7.0 and 25°C is oxidation by oxygen (aerobic respiration), by nitrate, by Mn(IV), by iron (III), by sulphate and by carbon dioxide in accordance with a declining number of produced ATPs which measure the exergy stored in the cells.[32] Table 2 gives the exergy available for storage in ATPs and released by these oxidation processes. The potential energy in aerobic heterotrophs is only slightly greater than that for denitrifiers, which often coexist in upland soils.[33] Several observations have supported the tentative fourth law of thermodynamics.

Table 2. kJ/equiv. available to build ATP for various oxidation processes of organic matter at pH = 7.0 and 25ºC

Reaction	*Available Kj/equiv*
$CH_2O + O_2 \rightarrow CO_2 + H_2O$	125
$CH_2O + 0.8\ NO_3 \rightarrow 0.8H^+ \rightarrow CO_2 + 0.4\ N_2 + 1.4\ H_2O$	119
$CH_2O + 2\ MnO_2 + H^+ \rightarrow CO_2 + 2Mn^2 + 3\ H_2O$	85
$CH_2O + 4FeOOh + 8H^+ \rightarrow CO_2 + 7H_2O + Fe^2$	27
$CH_2O + 0.5SO_4^{2-} + 0.5H^+ \rightarrow CO_2 + 0.5\ HS^-\ H_2O$	26
$CH_2O + 0.5CO_2 \rightarrow CO_2 + 0.5CH_4$	23

Numerous experiments have been performed to imitate the formation of organic matter in the primeval atmosphere on earth $4*10^9$ years ago.[14] Various sources of energy have been sent through a gas mixture of carbon dioxide, ammonia and methane. Analyses have shown that a wide spectrum of various compounds, including several amino acids, is formed under these circumstances, but generally only compounds with rather large free energy (i.e. high exergy storage) will form an appreciable part of the mixture.[34]

The entire process of evolution has worked towards organisms with an increasing number of information genes (genes actually utilised) and more types of cells, i.e., towards storage of more exergy due to the increased information content.

At the biochemical level, different plants operate three different biochemical pathways for the process of photosynthesis: (a) *the C3 or Calvin Benson cycle,* (b) *the C4 pathway* and (c) the *crassulacean acid metabolism (CAM) pathway*. The latter pathway is less efficient than the two other possible pathways, measured as grams (g) of plant biomass formed per unit of energy received. Plants using the CAM pathway can survive in harsh, arid environments, however, plants following C3 and C4 pathways cannot. Photosynthesis will, however, switch to C3 as soon as the availability of water is sufficient.[35] The CAM pathways give, in other words, the highest exergy storage under harsh, arid conditions, while the two other pathways give the highest exergy storage under other conditions. These observations are completely in accordance with the tentative fourth law of thermodynamics. Extensive investigations are needed to offer a full acceptance of the hypothesis of tentative of fourth law. At this stage, it could, however, be considered a useful working hypothesis.

The tentative fourth law of thermodynamics may also be considered as an extended version of 'Le Chatelier's Principle'. Formation of biomass may be described as:

Energy + nutrients = molecules with
more free energy (exergy) and organisation (4)

If energy is pumped into a system, the equilibrium will, according to Le Chatelier's principle, shift towards a utilization of the energy. It means that molecules with more free energy and organisation are formed. If more pathways are offered, the pathways, that give most relief, i.e. use most energy and thereby form molecules with most embodied free energy (exergy) will win according to the proposed tentative law of thermodynamics.

It is noticeable that the change in exergy is not necessarily > 0, as it depends on the changes in the resources of the ecosystem. However, the ecosystem attempts to reach the highest possible exergy level under the prevailing condition and with the available gene pool ready to challenge the currently changed conditions imposed by the currently changed forcing functions. If it compares with Figure 2, where the reactions of exergy to an increase and a decrease of the nutrient concentration in a lake are shown.

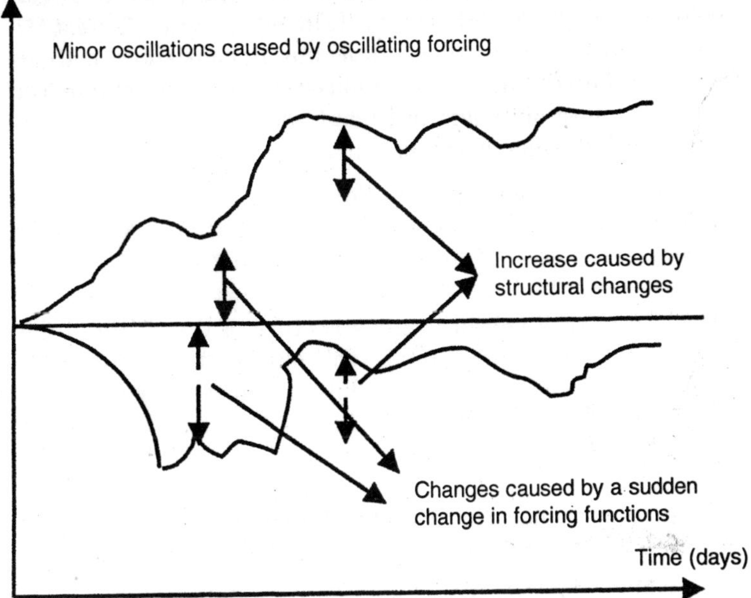

Fig. 2: Exergy response to increased and decreased nutrient concentration[4]

The different pathways available in an ecosystem compete in a very complex way, because the processes are many, complex and dependent on many factors: (1) at least somewhere in the order of 20 elements, of which some may be limiting, (2) competition from the other possible pathways, (3) temperature, (4) light (for photosynthetic pathways only), (5) ability to utilise the combined resources, (6) initial conditions, and (7) variations of conditions in time and space.

The highest exergy depends on the selection of the organization. Two systems with the same prevailing conditions will consequently not necessarily select the same species and food web, because the two systems most probably have a different history and therefore, different initial conditions.

Decomposition rates are of great importance, as they determine not only the ability of the different products (organisms) to maintain their concentrations, but also at which rate the inorganic compounds are recycled and can be reused.

Measurements in ecosystem show that the most abundant resource (nutrient), relative to its need, recycles fastest.[36,37] Model studies[1] have shown that this is consistent with the tentative fourth law of thermodynamics. Ecosystems that recycle the most abundant nutrient (resource) fastest will obtain the highest storage of exergy.

The two hypotheses, *'to optimise exergy dissipation or exergy captured'* and *'to optimise the exergy of the system under the prevailing conditions'* are two aspects for the analysis of an ecosystem starting from the early stage to mature stage. Increased exergy stored in the structure of the system under development will also enable the system to capture more exergy, as already mentioned, but the discrepancy between the two theories occurs when the system has attained the maximum rate of capturing exergy (85-90% of the incoming solar radiation). The tentative fourth law of thermodynamics can be used to describe the development both for a exergy captured and the exergy dissipation in a mature system.

GROWTH AND DEVELOPMENT OF ECOSYSTEMS

The results of a flow of energy through an ecosystem by the following equation:

Ex + Ex(t) → Ex(t+1) + energy utilised (lost by dissipation) for maintenance + reflected exergy. (5)

The two phenomena, *energy storage* and *exergy utilisation* are parallel during the successional development of ecosystem from an early to a mature stage.[38] An ecosystem at an early stage of development, for instance an agricultural field has only a small exergy storage and exergy utilisation. The biomass per square meter is small (compared with the mature system), i.e. the exergy storage is small. The structure is simple and only little energy (exergy) is needed for respiration or growth as they are both to some degree proportional to the biomass. The total surface area of the plants is furthermore small, which implies that they catch and utilise only a fraction of the incoming solar radiation.

As the system develops, the structure becomes more complicated, animals with more information per unit of biomass, i.e. with more genes, populate the ecosystem and the total biomass per square metre increases. It implies that *exergy storage* as well as the *exergy needed for maintenance both increase*. A very mature ecosystem for instance a natural forest has a very complex, organised structure and very well organised food webs. It contains a very high concentration of biomass per square metre and contains a lot of information in a wide variety of organisms. The entire structure tries to utilise solar radiation either directly or indirectly, resulting in a high utilisation of the solar exergy flux. The catabolic energy demand is related to the total biomass and the overall organisation. It represents the exergy needed for maintaining the ecosystem far from thermodynamic equilibrium, in spite of the second-law tendency of the system to develop towards thermodynamic equilibrium.

The development of ecosystems may also be described as a steady growth of a gradient between the ecosystem and thermodynamic equilibrium.[39-42] The force to break down the gradient is represented by *respiration* and *evapotranspiration* that *spend exergy* and *produce entropy*. As long as the exergy received from solar radiation can compensate for this need of exergy to maintain the gradient, it is possible for the system to stay far from thermodynamic equilibrium.[42] If even more exergy can be captured than needed for maintenance of the gradient, the surplus exergy increases the stored exergy, which means that the system moves further away from thermodynamic equillibrium and thereby increases the gradient even more.[1]

When an ecosystem has reached the state of maturity where the system captured almost all the solar energy that is physically possible, it is no longer possible to increase the amount of energy captured. The system can therefore in this situation not increase further dissipation and it cannot exceed the energy captured on a long-term basis. Moreover, the amount of inorganic matter available for further growth must in this stage be very limited, as most inorganic matter has been utilised to build up the very developed structure. The amount of information stored in biomass may, still increase in a mature ecosystem due to: (1) immigration of (slightly) better fitted species, and (2) emergence of new genes of genetic combinations. This latter possibility is covered by the concept of 'evolution'.

The system stops the growth of biomass when the most limiting inorganic component has been fully utilised for biomass construction. Nutrients and water are often the limiting factors in growth of plants. These resources cycle, which gives possibilities for formation of new biomass with perhaps more information, but the total biomass, cannot be changed by this reallocation of resources. This constraint by the laws of conservation is essential for the development of more and more complex living structures. As living organisms compete for limited food supplies, they invent and develop thousands of new and ingenious strategies.[39] Some species invest in both for capturing prey (e.g. movement) and for avoiding predators (e.g. speed for defence). Others use protective armour or chemical poisons. Each species defines the terms under which it engages in the harsh business of life. Better feedbacks to assure maintenance of a high biomass level to changed circumstances, better buffer capacities, better specialisation to populate all possible ecological niches and better adapted organisms to meet the variability in forcing functions are all developed. Therefore, biomass is maintained at the highest level over a longer time and the information will increase due to the steady development of better feedbacks and more self-organisation. Both contributions are reflected in a higher exergy. The exergy of the mature system can therefore still grow further, namely, by increase of the information. In order words, the system better utilises its resources and becomes more fitted to the prevailing conditions. Adaptation and specialisation require information, which implies that a better fitness to prevailing conditions is more probable by a system with more stored information in the genes.

Neither exergy nor information is conserved. All transfers of energy lose exergy. Exergy and information are also lost by the death of an organism as (•= β) in equation[19] decreases from a *value >>1 to 1* (detritus energy equivalent is applied as unit). Exergy and information may, however, be gained when phytoplankton is converted to zooplankton by grazing. Energy and information are, therefore, not cycling in the same manner, as we know for mass and energy. Their distribution in ecosystems will as, energy and matter, anyhow follow a complicated pattern, determined by the life processes.

The two hypotheses of *maximisation of exergy storage* and of exergy capture are, as already pointed out, not consistent when we have to describe the further development of an ecosystem which has reached maturity. In this domain Mauersberger's minimum principle may be valid.[40,41] *Ecosystems locally decrease entropy production by tansporting energy-matter from more probable to less probable spatial locations.* Thereby less exergy is lost and more exergy can be stored.

A possible hypothetical formulation of the tentative fourth law of thermodynamics trying to unite with hypotheses of Kay and Schneider[42,43] valid for a system under development and with the hypotheses of Mauersberger for mature ecosystems may be: *If a system is moved away from thermodynamic equilibrium by application of a flow of exergy, it will utilise all avenues available, to build up as much dissipative structure (store as much exergy) as possible.* An ecosystem at an early stage will try to store more exergy by increasing the captured—decrease the reflection, while a mature ecosystem will gain more exergy by decreasing the exergy lost by the maintenance processes—mainly achieved by increase of the information content of the system.

The overall description of the development of ecosystems is illustrated in Figure 3. In the first phase, the structure and its biomass increases rapidly mainly due to the rapid growth of *r-strategists.* The gradients, the amount of exergy captured by the system and the exergy required for maintenance are all increasing. A transition phase is shown between the first and third phase. In the third phase, the limiting elements are practically used up, which implies that a further increase in the physical structure measured by the biomass is not possible. A better use of the resources and more storage of information, however, can continue and will work hand in hand. More exergy cannot be captured or dissipated but the resources can be reallocated to give more exergy stored and a development towards relatively less dissipation of exergy (Mauersberger's principle).

A development towards *k-strategists* is, thus, favoured as they are bigger in size and thereby have less specific exergy needs. Microorganisms also shift from *r-strategists* with quick exploitation of the resources to *k-strategists* with a slow exploitation of resources.[44] Late successional plants with k-strategy produce litter that is poor in nutrients and simple sugars but high in lignin,[45] which change the physico-chemical environment of the topsoils. These processes cause modifications in the composition and activity of microbial communities, favouring the k-selected organisms. The shift from r-strategist to k-strategist may be considered a process with synergistic effects.

A forest captures a lot more exergy than a desert or a grassland, but a 50-year-old forest or a 200-years-old forest or even an old rain forest capture approximately the same amount, almost 90% of the incoming solar radiation[43]. It can however been shown[14], that the stored exergy increase when a forest gets older. Usually a rain forest has more exergy stored than a temperate forest.

The exergy utilisation for different types of systems[43] is shown in Table 3. In the same table the exergy storage is shown for some typical 'average' systems. The results of Table 3 are completely consistent with the illustration of the development of ecosystem shown in Figure 3. As seen, there is a steep linear relationship between exergy storage and exergy capture, when the system is under development from the early to the mature stage, but a mature system may still develop its exergy storage, although the exergy captured has attained the practical maximum of about 70-80% of the total solar energy received by radiation. This points towards Ex_{bio} (exergy storage) as a general optimiser during the entire development from the early stage of the nature stage.

Johnson[48] concludes that 'ecosystem structure is a function of two antagonistic trends: *one towards a symmetrical state resulting in least dissipation and the other toward a state of maximum attainable dissipation'*. The latter is obtained by

Table 3. Energy utilisation and exergy storage[43]

Ecosystem	*% Exergy utilisation*	*Exergy storage KJ/m²*
Quarry	0	6
Desert	73	2
Clear Cut	594	49
Grassland	940	59
Fir plantation	12700	70
Natural forest	26000	71
Old deciduosu forest	38000	72
Tropical rainforest	64000	70

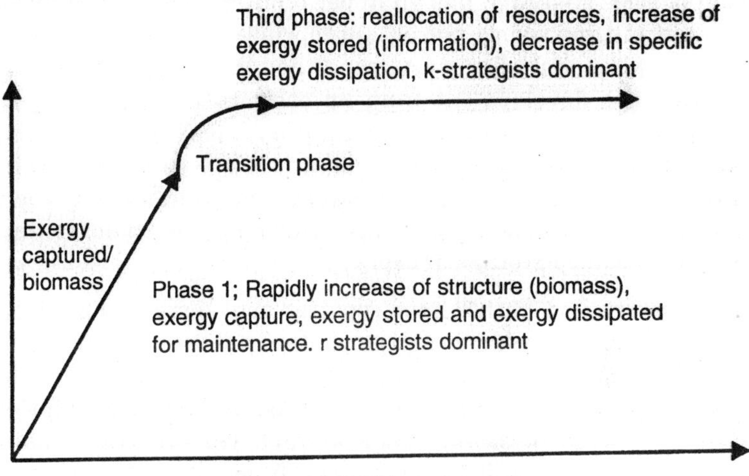

Fig. 3: The graph shows the exergy captured (or biomass) versus the exergy stored as biomass and information in the ecological structure. When the captured exergy has reached the physically possible level, further development of exergy storage is still possible by more cycling and by increase of the information embodied in the biomass.

a rapid growth of biomass and structure, implying a more effective capture of the exergy contained in the solar radiation. The first trend is obtained by a reallocation of biochemical elements, which due to mass conservation limit the amount of biomass and structure. A development towards more effective organisms (less dissipation relative to the biomass which for instance is the case for bigger organisms),[47] requiring less exergy for maintenance, will imply that although (almost) the same amount of exergy is captured, the stored exergy can still increase. Inevitably, the configurations of the interactions become more mutualistic, self-reinforcing and self-entailing.

It is also of important to notice the basic difference between the two descriptors, exergy storage and exergy degradation. The latter is a differential quantity and is measured in a unit which includes time, for instance kJ, $(m^2 24h)^1$. As the available exergy per unit of time in the form of solar radiation has an upper limit, exergy degradation per unit of time seems not an appropriate descriptor of long-term development of ecosystems, although it may be used to describe the development from an early to a mature stage. Exergy storage is an integrated quantity, measured in a unit not including time, for instance kJ/m^2. It has in principle no upper limit, as the storage of exergy can steadily increase. Each new day brings more exergy, which may be added to the already obtained storage of exergy.

Pattern and Fath[48] have reported that increased cycling implies increased exergy storage at steady stage conditions. It was shown as a general mathematical consequence of steady state network theory. Ecosystems are dynamic. It is therefore a simplification to assume steady state. On the other side, steady state may be interpreted as a freezing of the system, *which* gives a realistic comparison of two situations: *with* and without cycling. It seems therefore appropriate[45] to set up the following pertinent hypothesis: *Increased cycling implies that the exergy storage and the exergy through flow increase simultaneously with a decrease in specific exergy dissipation. It is in accordance with the interpretation of the tentative fourth law of thermodynamics and Mauersberger's principle applied for mature ecosystem.*

When a forest is burned, complex organic compounds are converted into inorganic compounds. Some of the inorganic compounds from the wood are lost as particles of ash that are blown away or as vapours that escape into the atmosphere and are distributed widely. Other compounds are deposited on the soil surface. These are highly soluble in water and readily available for vegetation uptake. Therefore, immediately after a fire there is an increase in the availability of chemical elements, which are taken up rapidly, especially if there is a moderate amount of rainfall.[49] The inorganic nutrients can then lead to a pulse in growth of vegetation. This in turn provides an increase in nutritious food for herbivores. Challenges to find new opportunities to move from thermodynamic equilibrium therefore may have a long-term positive effect on the growth of ecosystems in the broadest sense of this concept (*Holling's cycle*)[50]

A modified Holling's cycle (Figure 4) is presented by Ulanowicz[31]. Holling cycle is modified in almost the same way as Figure 4. Only the x-axis is not specific exergy = exergy/ total biomass, but instead mutual information of flow structure. The basic idea is, however, the same. The renewal phase corresponds to rapidly increased biomass exploitation and a rapid increase in the level of information where conservation has very slow increases in both biomass and information. The destruction phase will, due to an external impact (forcing function), reduce both the amount of biomass and the information stored in this biomass. Therefore, new possibilities are created for the utilisation of emergent mutations and sexual recombination. After each round in Holling's cycle the biomass can probably hardly be higher as it is limited by the presence of essential elements, but due to the current test of new mutations and sexual recombination, new and perhaps better combinations of properties will emerge. Consequently, there is a propensity that the exergy and the specific exergy may have increased.

It is observed that the tentative fourth law of thermodynamics can explain in all phases of the ecosystems development. Due to its general character, it is necessary to test the tentative law with many more case studies before it can be recommended to use generally.

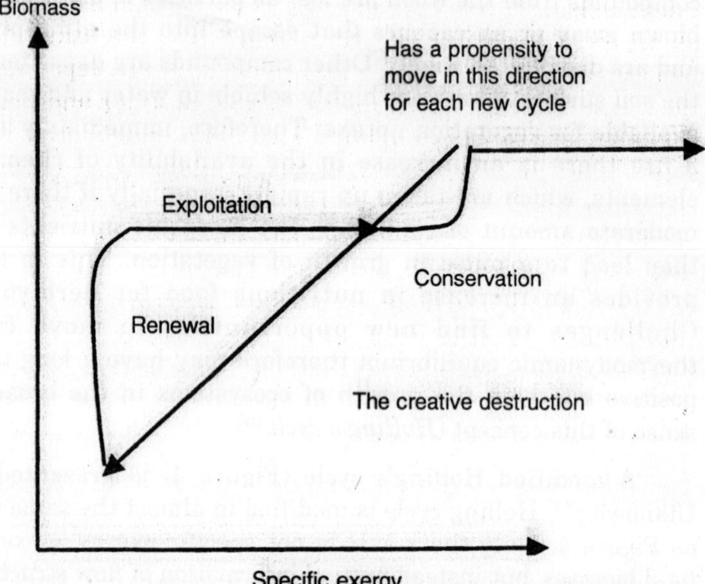

Fig. 4: The Holling's cycle. A mature system has only very scarce available inorganic matter to build new biomass and therefore to test new better possible solution to moving away from thermodynamic equilibrium. By introduction of a destruction phase is created possibilities in the long run to move further away from thermodynamic equilibrium.[50]

REFERENCES

Boltzmann, L. (1905). The Second Law of Thermodynamics. (Populare Schriften, Essay no. 3 (address to Imperial Academy of Science in 1886). Reprinted in English in: Theoretical Physics and Philosophical Problems, *Selected Writings of L. Boltzmann.* Dreidel, Dordrecht.

Brown, J.H. (1995). Macroecology. *The University of Chicago Press,* Chicago, IL.

Botkin and Keller. (1995). Environmental Science, Earth as a living Planet. John Wiley and Sons, NY, p. 630.

Carter, J.P., Hsiao, Y.H., Spiro, S. and Richardson, D.J. (1995). Soil and Sediment Bacteria capable of aerobic nitrate respiration, Appl. Environ. Microbiol, 61: 2852-2858.

Cavalier-Smith, T. (1985). The Evolution of Genome Size. Wiley Publication, Chichester. p. 48.

Coffaro, G. and Bocci, M. (1997a). How *Ulva rigida* and *Zostera marina* compete for resources: a quantitative approach with application to the Lagoon of Venice, Ecol. Modelling 102:81-96.

Evans, R.B. (1966). A proof that Exergy is the only consistent measure of Potential Work, Ph.D. Thesis, Dartmouth College, Hanover, NH.

Gerson, U. and Chet, I. (1981). *Are allochthonous and autochtonous soil microorganisms r- and k-selected?* Rev. Ecol. Biol. Sol. 18:285-289.

Heal, O.W. and Dighton, J. (1986), Nutrient cycling and decomposition in natural terrestrial ecosystems, In: M.J. Mitchell, J.P. Nakas (eds.): Microfloral and faunal interactions in natural and ago ecosystems, Nijhoff & Junk Dordrecht, pp. 14-73.

Holling, C.S. (1986). The resilience of terrestrial ecosystems: Local surprise and global change. In: W.C. Clark and R.E. Munn (eds.) Sustainable Development of the Biosphere. Cambridge University Press, Cambridge, 292-317.

Johnson, L. (1990). The thermodynamics of ecosystem. In: O. Hutzinger, (ed.). The Handbook of Environmental Chemistry, Vol. 1. The Natural Environmental and the Biogeochemical Cycles, Springer-Verlag, Heidelberg. pp. 2-26.

Jorgensen, S.E. (1992a). Development of models able to account for changes in species composition. Ecol. Modelling. 62:195-208.

Jorgensen, S.E. (2000). The Tentative Fourth Law of Thermodynamics, In: Jorgensen, S.E. and Muller, F. (eds.) Handbook of Ecosystem Theories and Management, Lewis Publisher, London, II.13:(161-173).

Jorgensen, S.E. and Mejer, H.F. (1979). A holistic approach to ecological modelling. *Ecol. Modelling* 7: 169-189.

Jorgensen, S.E. and Mejer, H.F. (1977). Ecological buffer capacity. Ecol. Modelling 3: 39-61.

Jorgensen, S.E. and Mejer, H.F. (1977). Ecological buffer capacity. *Ecol. Modelling* 3:39-61.

Jorgensen, S.E. (1995). Modelling in Ecotoxicology, *Elsevier Publication,* Amsterdam.

Jorgensen, S.E. (1997). Integration of Ecosystem Theories: A Pattern. 2. Edition. Kluwer Academic Publication Dordrecht. Boston. London. p. 400.

Jorgensen, S.E., Patten, B.C. and Straskraba, M. (1998). Ecosystem emerging: 3 Openness. *Ecol. Modelling* (in press).

Jorgensen, S.E. Nielsen, S.N. and Mejer, H. (1995). Energy, environ, exergy and ecological modelling. *Ecol. Modelling,* 77:99-109.

Jorgensen, S.E. (1982). A holistic approach to ecological modelling by application of thermodynamics. In: W. Mitsch *et al.*, (eds.). Systems and Energy. *Ann Arbor Press, Ann Arbor, Ml.*

Jorgensen, S.E. (1986). Structural dynamic model. Ecol. Modelling, 31:1-9.

Jorgensen, S.E. (1988). Use of models as an experimental tool to show that structural changes are accompanie by increased exergy, Ecol. Modelling, 41:117-126.

Jorgensen, S.E. (1990). Modelling in Ecotoxicology. Elsevier Publishers, Amsterdam.

Jorgensen, S.E. and Padisak, J. (1997). Does the intermediate disturbance hypothesis comply with thermodynamics? Hydrobiologia 323:9-21.

Jorgensen, S.E. and De Bernardi, R. (1997). The application of a model with dynamic structure to simulate the effect of mass fish mortality on zooplankton structure in Lago di Annone, Hydrobiologia 356:87-96.

Kay, J. and Schneider, E.D. (1990). On the applicability of non-equilibrium thermodynamics to living systems (internal paper). Waterloo University, Ontario, Canada.

Kay, J. and Schneider, E.D. (1992). Thermodynamics and measures of ecological integrity. In: Proc. Ecological Indicators, Elsevier Publication, Amsterdam. pp. 159-182.

Kooijman, S.A.L.M. (1993). Dynamic Energy Budgets in Biological Systems. Theory and Applications in Ecotoxicology, Wiley Publication, New York.

Lewin, B. (1994). Genes V. Oxford University Press, Oxford, p. 620.

Li, W.H. and Grauer, D. (1991). Fundamentals of Molecular Evolution, Sinauer, Sunderland, Massachusetts. p. 430.

Mauersberger, P. (1983). General Principles in Deterministic Water Quality Modeling. In: G.T. Orlob (ed.). Mathematical Modeling of Water Quality: Streams, Lakes and Reservoirs (International Series on Applied Systems Analysis, 12). Wiley Publication, New York, 42-115.

Mauersberger, P. (1995). Entropy control of complex ecological processes. In: B.C. Pattern and S.E. Jorgensen (editors). Complex Ecology: The Part-Whole Relation in Ecosystems. *Prentice-Hall Englewood Cliffs,* NJ, 130-165.

Mejer, H.F. and Jorgensen, S.E. (1979). Energy and Ecological buffer capacity. In: S.E. Jorgensen (ed.). State-of-the-Art of Ecological Modelling. (Environmental Sciences and Applications, 7.). Proceedings of a Conference on Ecological Modelling, 28. August-2. September 1978, Copenhagen. International Society for Ecological Modelling, Copenhagen, 829-836.

Morowitz, H.J. (1992). Beginnings of Cellular Life. *Yale University Press,* London.

Morowitz, H.J. (1968). Energy Flow in Biology. Academic Press, New York.

Nicolis, G. and Prigogine, I. (1989). Exploring Complexity: An Introduction. *Freeman Publication,* New York.

Nielsen, S.N. (1992). Application of maximum exergy in structural dynamic models. Ph.D. Thesis, DFH. Institute A, Section of Environmental Chemistry, Compenhagen, Denmark, p. 51.

Odum, E.P. (1969). The strategy of ecosystem development. *Science* 164:262-270.

Onsager, L. (1931). Reciprocal relations in irreversible process, I Phys. Rev. 37:405-426.

Patten, B.C. and Fath, B.C. (1999). Environ Theory and Analysis (in Press).

Patten, B.C. and Faith, B.C. (1999). Environ Theory and Analysis, Prentice-Hall Publisher (in press).

Prigogine, I. (1947). Etude Thermodynamique des process irreversibles *Desoer. Liese.*

Reeves, H. (1991). The Hour of our Delight. Cosmic, Evolution, Order and Complexity. Freeman Publication, New York. p. 246.

Schlesinger, W.H. (1997). Biogeochemistry. An analysis of Global Change (2nd Ed), Academic Press, Tokyo p. 680.

Shugart, H.H. (1998). Terrestrial ecosystems in changing environments. Cambridge University Press, Cambridge 10-32.

Straskraba, M. Jorgensen, S.E. and Patten, B.C. (1999). Ecosystem Emerging: Dissipation, Ecol. Modelling (In press, personal communication).

Svirezhev, Y. (1998). Thermodynamic Orientors: How to use Thermodynamic Concepts in Ecology, In: Eco Targets, Goal Functions, and Orientors, Springer Verleg Publications, pp. 102-122.

Ulanowicz, R.E. (1997). Ecology, The Ascendent Perspectives. Columbia University Press. New York, pp. 200-202.

Ulanowicz, R.E. (1991). Formal Agency in Ecosystem Development, Theoretical Studies of Ecosystems: The Network Perspective, Cambridge University Press, Cambridge, p. 201.

Vollenweider, R.A. (1975). Input-output models with special reference to the phosphorus loading concept in Limnology. Schweiz Z Hydrol. 37:53-84.

15

Cytochrome P450 and Other Biotransformation Activity in Aquatic Organisms: Potential Biomarkers to Environmental Pollution

S. Arun and *P. Subramanian*

Increased world industrialization since the end of the second World War and the concomitant increased demand for chemicals have caused both developed and developing nations to face increasing ecological problems due to the release of toxic contaminants into the environment. Some of these hazardous contaminants are Polynuclear Aromatic Hydrocarbons (PAH's) Poly Chloro Biphenyls (PCB's), Poly Chlorinated dibenzo-p-dioxins, Organophosphates and organochlorines. Uptake and accumulation of these hazardous substances in the tissues of aquatic organisms occur from the sediment, contaminated water column and food chain (Livingstone, 1993) which causes biological damages to these organisms. This alarming situation has led to a growing interest among scientists to study the metabolism of xenobiotics by aquatic organisms. They are not defenceless to the adverse effect of these xenobiotic compounds and enzymatically they have some intrinsic detoxifying mechanisms to eliminate the accumulated xenobiotic compound from their body. The enzymes which are involved in metabolism/removal of xenobiotic compounds are known as biotransformation enzymes. The induction response of these biotransformation enzymes in aquatic organisms to organic contaminants would also be used as biomarkers in monitoring environmental pollution (Payne, 1976). Generally these protective, biotransformation enzymes present in all living organisms convert the lipophilic (fat-soluble) chemicals into more water-soluble (hydrophilic) forms, which could be consequently eliminated from the organisms by excretion. Thus, the process involved in the conversion of toxic chemicals into less toxic by-products is referred to as biotranformation. This biotransformation process is normally divided into two phases, known as phase I functionalization reaction and phase II conjugative reactions (Livingstone 1994)). In addition antioxidant enzymes also may be of importance in scavenging reactive oxygen species (Fig.1).

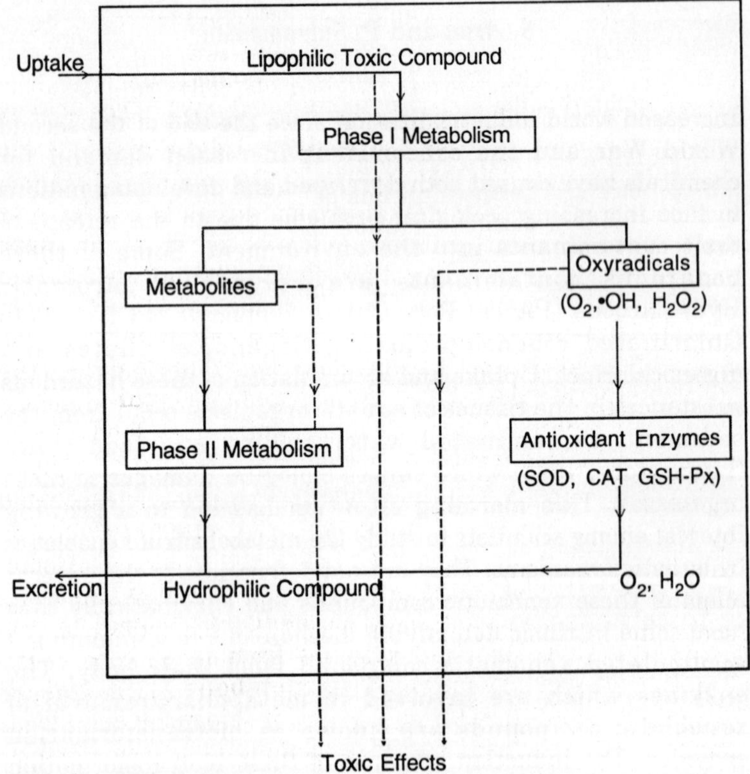

Fig. 1: Schematic representation of detoxificaiton (phase I, II and antioxidant mechanism) process.

BIOTRANSFORMATION ENZYMES

Phase I Metabolism

In Phase I metabolism, the oxidative reaction plays a main role in the removal of xenobiotic compounds. In oxidative reaction particularly, mixed function oxygenase (MFO) enzymes insert one atom of oxygen into the xenobiotic compounds, such as drugs, pesticides, polycyclic aromatic hydrocarbons and convert these lipophilic compounds to hydrophilic compounds. The cytochrome P450 serves as both the oxygen and substrate-binding focus for the MFO reaction and undergoes cyclic oxidation/reduction of the haem iron during catalysis (Livingstone, 1991). It consists of single polypeptide chain with iron-protoporphyrin IX loosely bound by hydrophobic forces, electrostatic and covalent bonds. The cytochrome P450 monoxygenase functions in association with several microsomal proteins, located in the endoplasmic reticulum together with P450 isoenzymes, that transfer electrons to reduce cytochrome P450 during its catalytic cycle. These components include NADPH cytochrome c reductase, cytochrome b_5, and NADPH cytochrome c reductase. Electrons from NADPH or NADH are transferred through these compounds and cytochrome P450 (CYP450) inserts one atom of oxygen into the substrate and reduces the second oxygen atom to form water. There have been two major views on the function of CYP 450 enzymes i.e., it is involved in the metabolism of endogenous compounds (steroid hormones, vitamins and bile acids) and exogenous compounds such as xenobiotics (Jakoby, 1980). The mechanism of xenobiotic metabolism by heme protein (CYP 450) was clearly explicated by Hollebone (1986) and Hollebone *et al.*, (1995) (Fig. 2). In this model, the lipid bound protein is an oligomeric globin, capable of both rotation and translation within the lipid. As some sites of the globin protein would face the cytosol, oxygen molecules can be captured from the aqueous medium of the cell. By applying antibonding electron density to this trapped oxygent at the heme site, the protein is capable of heterolysis such as an oxygen molecule to a hydroxide ion, which is released back to the cytosol, and a positive oxygen atomic ion, which remains bound to the active site by back bonding to the

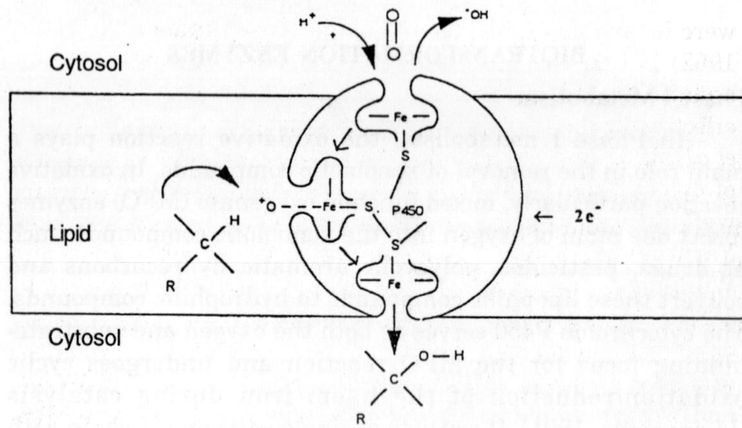

Fig. 2: Metabolism of lipophlic xenobiotic compound into hydrophilic compound by cytochrome P 450.

iron (II) state of the heme. This activated site is protected by a tyrosine cover. After rotation of globin and opening of the lipophilic tyrosine, this hydrophobic, activated site can encounter and adsorb hydrophobic xenobiotic compounds which dissolved in the lipid bilayer. Then, the oxygen positive ion oxidizes the xenobiotic compound. After oxidation, the hydrophilic hydroxylated metabolite could force rotation of the protein and permit release of the less toxic metabolites into lumen for conjugation.

Phase I Metabolism in Aquatic Organisms

Cytochrome was first discovered in liver by Carfinkel (1957) and Klingenberg (1958) as a carbon monoxide binding pigment and named P450 from the position at 450 nm of the absorption peak of the carbon monoxide complex of the reduced pigment (Omura and Sato, 1964). It is an oxygen activating oxidase (Omura *et al.*, 1965) that acts as the locus of substrate activation in MFO reactions (Schenkman *et al.*, 1967). The metabolism of xenobiotic compounds to various oxidative derivatives was demonstrated in mammals as early as the 1940s. Many of these reactions were shown to be carried out by MFO, which was subsequently determined to be cytochrome P450 dependent (Conney, 1982). Early studies of detoxification systems in aquatic organisms seemed to indicate that they were incapable of microsomal oxidations (Brodie and Maickel, 1962) but in the late 1960s, few studies demonstrated that aquatic organisms possessed oxidative enzymes which were effective in the transformation of xenobiotic compounds (Chan *et al.*, 1967 and Anderson, 1968). Consequently, the existence of cytochrome P450 dependent monoxygenase activities was well documented in fish (Dewaide 1971; Chambers and Yarbrough, 1976; Ahokas, 1977; Elcombe and Lech, 1979; Williams *et al.,* 1989; Forlin *et al.,* 1995; Durand and Cravedi, 1989). Complementary function of these enzymes in metabolism or conversion of toxic xenobiotic compounds such as polycyclic aromatic hydrocarbons, polychlorinated biphenyles, pesticides and other toxic drugs was also studied in fishes (Stegeman 1978; Chamber, 1979; James and Bend, 1980; Payne and Fancy, 1982; Maemura and Omura, 1983; Payne *et al.,* 1984; Forlin *et al.,* 1985; Tan and Melius 1986; Andersson *et al.,* 1987; Lindstrom-Seppa and Oikari, 1988, 1990; Elskus and Stegeman, 1989; Boon *et al.,* 1992; Otto *et al.,* 1994; Slederink *et al.,* 1995 and Woodin *et al.,* 1997).

Despite a considerable number of works carried out in Phase I metabolism of aquatic organisms, fewer studies were conducted on invertebrate species relative to vertebrate species. An attempt was made on sponges for detecting biotransformation enzymes. But the activity of monooxygenase enzymes was not detected in *Tethya limski, T. lyncurium,* and *Suberites domuncula* (Zahn *et al.,* 1981, 1982, 1983). Similarly, metabolism of PAH was reported as being absent in coelenterate species (Metridium sp.) (Lee, 1981). In addition, cytochrome P450 and cytochrome b_5 (cyt.b_5), activity were not detected in the parasitic nematode *Haemonchus contortus* (Kawalek *et al.,* 1984). In contrast, Free-living nematodes such as *Panagrellus rediviv*, *Turbatria ceti* and *Caenorhabditis etegans*, were presumed to have the MFO system (Chitwood *et al.,* 1987). Studies on Annelida reveal the existence of cytochrome P450 and MFO activity and its induction towards xenobiotics (Fries and Lee, 1984; Lee *et al.,* 1981). Elmamlouk *et al.,* (1974) observed the presence of cytochrome P450 in hepatopancrease of the lobster, *Homarus americanus* and deduced that the level of CYP 450 was lower when compared to the mammalian liver CYP 450 and also observed that this enzyme was not involved in the process of xenobiotic metabolism in the lobster. In contrast, Foureman *et al.*, (Foureman, 1978) observed that *Homarus americanus* has the ability to metabolise the xenobiotics benzo-a-pyrene (Bap) slowly when compared to mammals. James *et al.*, (James *et al.*, (1979) had shown that hepatopancreatic microsomes were unable to catalyse substrate oxidation because these microsomes contained inhibitors of NADPH and CYP 450 dependent monooxygenases. After rectifying these difficulties by removing this monooxygenase inhibitors from the microsomes, James *et al.*, (1982) found that the lobster could catalyse the oxidation of several substrates. Singer and Lee (1977) Pohl *et al.*, (1974) and Bihari *et al.*, (1984) reported that MFO activity in hepatopancreas of crustaceans was either very low or undetectable. James (1979) detected the MFO enzymes in several marine species and compared them with each other. The Spiny lobster, *Panuliros argus*, (hepatic microsomes) contained high level of cytochrome P450 when compared to other teleosts and elasmobranches. However, the activity of NADPH cytochrome c.reductase was low in crustacean species (Spiny lobster and Blue crab) when compared to other species of teleosts and elasmobranches. Stegeman and Kaplan (Stegeman and Kaplan 1981) noted the presence of cytochrome P405 and its MFO enzymes, such as NADPH and NADH cyt.c. reductase and Cyt.b_5 in the subcellular fraction of *Balanus eburneus* crustaceans. James and Little (1984) inferred that 3 methylcholanthrene (3-MC) did not induce MFO activity (ethoxyresorufin O deethylase; EROD) and CYP 450 activity in the spiny lobster. On the other hand, Lee *et al.*, (1981) reported an increased activity in CYP 450 and monooxygenase enzymes in crab hepatopancreas after exposure to oil or polychlorinated biphenyls. Batel *et al.*, (1983) also detected the increased monooxygenase activity in the crab Pilumnus spinifer after exposure to diesel 2 oil. Jewell and Winston (1989) compared the hepatic and green gland microsomal monooxygenase enzymes (CYP-450, cytochrome b_5, NADH and NADPH cytochrome c.reductase) with rat liver. CYP 450 content of hepatic microsomes was 10 fold higher than that of the green gland and comparable to that of rat's liver. NADPH cytochrome c.reductase activity in hepatic microsomes was approximately 2% of that found in rat liver microsomes. In molluscs, an increased activity of CYP 450 and its MFO were observed after exposure to mytilids (*Mytilus edulis* and *M. galloprovincialis*) and gastropods (*Littorea littorea* and *T. haemastoma*) to PAH and other hydrocarbons. (Livingstone *et al.,* 1985; Gilewicz *et al.,* 1984; Livingstone *et al.,* 1986). Early studies failed to detect the MFO activity in echinoderms but in 1989 Den Besten and Livingstone detected CYP 450 and its MFO activity in several echinodern species such as *Asterias rubens, Marthasterias glacialis, Holothuria forskali* and *Echinus esculentus.*

Phase II Metabolism

After oxidative function on xenobiotics (phase I metabolism) to form polar metabolites, secondary processes come into play to convert the metabolites into water soluble compounds for excretion through conjugation reactions.

A group of enzymes which catalyse the conjugation of electrophilic substances with endogenous reduced glutathione have been referred to as glutathione-s-transferases (GST). The function of these enzyme is to conjugate potentially harmful electrophils with endogenous glutathione (GSH) and thus protect other nucleophilic centres such as proteins and nucleic acids, as well as means of excretion after conjugation with glutathione. As was classified by Boyland and Chaseaud (1969) the reaction of GST with electrophilic compound is given below.

$$\text{Benzyl chloride + Glutathione} \xrightarrow{\text{GST}} \text{Benzyl glutathione + HCl}$$

In this reaction, with the help of GST enzymes the lipophilic benzyl chloride would convert into hydrophilic benzyl glutathione. Here the GST enzymes enhances the attachment of glutathione (water soluble mioety) to the lipophilic compound and convert into water soluble compound and eliminate it through excretion.

Phase II metabolizing enzymes in aquatic organisms

Conjugating enzyme (Glutathione s-transferases)

Glutathione S-transferases play a main role in conjugation in the invertebrate and vertebrate species. Early studies of detoxification system in aquatic organism seemed to indicate that they were incapable of conjugations. However, it was later demonstrated that some fish species could detoxify xenobiotics and that the toxicity of several xenobiotics which could be related to hepatic metabolic activity (Buchler, 1966; Lech, 1974). Thus, failure to detect these enzyme in aquatic organism, particularly in invertebrates, might be due to the naturally occurring inhibitors. It was suggested then that endogenous quinones or polyphenols might be involved in inhibition of the GST enzyme in crustaceans and molluscs (Dierickx, 1984). Clark *et al.*, (1984) succeeded in detecting the glutathione s-transferase enzymes in New Zealand grass crab (*Costelytra zealundica*) by rapid fractionation of the homogenate preparation or by inclusion of GSH, cystine or other thiols in the homogenization medium.

The presence of GST enzyme in various tissues of fishes was documented by many workers. Bauermeister *et al.*, (1983) identified the activity of GST enzymes in different tissues of rainbow trout and Ramage and Nimmo (1984) scrutinized the substrate specificities and subunit composition of GST in rainbow trout (*Salmo gairdneri*). Andersson *et al.*, (1985) analysed the activity of these enzymes in rainbow trout during starvation. Del Boccio *et al.*, (1987) detected the tissue distribution of GST in female adult *Bufo bufo* while Aceto *et al.*, (1994) also observed the tissue distribution of the enzyme in the trout, *Salmo iridaeus*. Later in 1994, Di Ilio *et al.*, compared the activity of GST enzyme in the liver of *Bufo bufo* with *Salmo iridaeus*. In addition, many studies were carried out on the induction of GST enzyme in response to various xenobiotic compounds. Induction of GST enzyme was observed in white fish *Coregoneus muksun* when exposed to pulp and paper mill effluent (Lindstrom-Seppa and Oikari 1989). In contrast decreased activity of GST was reported in lake trout exposed to pulpmill effluent by Oikari *et al.*, (1988). Boon *et al.*, (1992) observed increased GST activity in plaice (*Pleuronectes platessa*) 16 days after exposure to polychlorinated biphenyls. Similarly, Andersson *et al.*, (1985); Goksoyr *et al.*, (1987); Zhang *et al.*, (1990) Celander *et al.*, (1993) were noted induction of GST enzymes in fishes when exposed to β-Napthoflavone (BNF). Celander *et al.*, (1994) compared the effect of crude oil and β-napthoflavin on hepatic GST enzyme of viviparous blenny (*Zoarces viviparous*). They found that induction of GST enzymes in response to β-napthoflavin but not to crude oil. On the other hand, Huuskonen and Lindstrom-Seppa (1995) reported that there was no induction in perch (*perca fluviatilis*) GST activity in response to pulp mill effluents and deduced that, in this test animal, the conjugative enzyme could not serve as a biomarker for unbleached pulp mill effluent. Lemaire *et al.*, (1996) also examined the induction of GST enzyme in rainbow trout exposed to BNF and compared with the Sea bass and dab (*Dicentrachus labrax* and *Limanda limanda*) which were exposed to 3-methylcholanthrene (3MC). Induced state of GST was observed only in rainbow trout (*Oncorhynchus mykiss*) but not in Sea bass and dab.

Information on GST enzyme was also limited in invertebrates when compared to vertebrate species. In porifera activity of GST was not detected in *Geodia barretti* (Stenerson *et al.*, 1987). In contrast, activity of GST was detected in the Coelenterates, hydra (*Hydra attenvata*), Seapen (Virgularin sp.) and Sea anemone (Bolocera sp.) (Stenerson *et al.*, 1987). Douch and Buchanan (1978) assessed the activity of GST in the cestode *Moniezia expansa* and the nematode *Ascaris suum*. A number of studies were conducted in molluscs viz, *M. arenaria* and *M. edulis* (James et al 1979), *Cerithidea stelleri* (Schlenk and buhler 1988), *C.obtusa* (Balabaskaran *et al.*, 1986), *Littorea littorea* (Stenerson et al 1987). In echinodermates the activity of GST was observed in a few species of star fish (*Asterias rubens*), sea urchin (*Echinus esculentus*) and sea cucumber (*Parastichopus tremulus*) (Stenerson *et al.*, 1987). James *et al.*, (1979) observed the phase II metabolizing enzymes in several marine crustacean species and Stenersen et al., (1987) detected the enzyme's activity in brine shrimp (*Artemia salina*). Arun and Subramanian (1997 & 1998) also measured the GST enzyme activity in freshwater prawn *Macrobrachium malcomsonii* and *M. lamarrei lamarrei*. Lindstrom Seppa et al., (1983) quantified the activity of GST enzymes in the subcellular fraction of *Astacus astacus*, apart from the detection of GST enzymes in its tissues. Induction of these enzymes in response to various pollutants may serve as biomarker. Lee *et al.*, (1988) observed increased activity of GST enzyme in the crustaceans *Callinectes sapidus* and *Penaeus aztecus* when exposed to 2,6-ditertiary-butyl-4-hydroxy toluene. On the other hand, Almar *et al.*, (1987) detected a decreased activity of GST enzymes in *P.clarki* when exposed to cadmium. In *Astacus astacus* induction of GST was detected after exposure to Bap and Aroclor 1254. In contrast, no activity was observed in the same animal when exposed to 7,8-napthoflavone (Lindstrome Seppa and Hanninen 1986). These results suggest that the conjugative enzyme (GST) in *A. astacus* may serve as a biomarker to Aroclor 1254 but not to 7,8-napthoflavone. So, without determining the induction of GST enzyme to certain pollutants, we cannot consider the GST enzyme as a biomarker for certain pollutants.

OXYRADICALS AND ANTIOXIDANTS

Sources of Oxyradicals

The paradox of aerobic life is that higher eukaryotic aerobic organisms cannot exist without oxygen, yet oxygen is inherently dangerous to their existence (Davis 1995). Concerted tetravalent reduction of oxygen by the mitochondrial electron-transport chain, to produce water, is considered to be a safe process.

$$O_2 + 4H^+ + 4e^- \longrightarrow 2H_2O$$

However, the molecular oxygen may undergo univalent reduction to produce harmful intermediates, termed reactive oxygen species (ROS) (so called oxyradicals) and these appear to be largely responsible for oxygen toxicity.

The product of one electron reduction of molecular oxygen is the superoxide anion radical (O_2^-),

$$O_2 + 4e^- \longrightarrow O_2^-$$

This anion radical may undergo a dismutation reaction to produce hydrogen peroxide (H_2O_2) which is the product of bivalent reduction of oxygen,

$$2O_2 + 4H^+ \longrightarrow H_2O_2 + O_2$$

The trivalent reduction of oxygen produces the highly reactive hydroxy radical (^oOH) which may be generated by the interaction of O_2^- and H_2O_2 via the Haber-Weiss reaction

$$H_2O_2 + O_2^- \longrightarrow HO^o + HO^- + O_2$$

Although, these reactions are kinetically slow (Halliwell, 1978), transition metals such as iron and copper can also serve as catalysts. The iron catalyzed in the Haber-Weiss reaction is depicted in following reactions:

$$O_2^- + Fe^{3+} \longrightarrow O_2 + Fe^{2+}$$

$$Fe^{2+} + H_2O_2 \longrightarrow HO^{o} + OH^- + Fe^{3+}$$

In addition, during phase I metabolism, CYP 450 and microsomal MFO system are also involved in the generation of oxyradicals (Livingstone 1991). In mixed function oxygenase reaction, if substrate oxidation is not tightly coupled to electron flow, there may be a possibility for the formation of $H_2O_2^-$.

$$O_2 + NADPH + H^+ \longrightarrow H_2O_2 + NADP$$

Antixoidant defences (removal of oxyradicals)

Antioxidant enzymes

Oxyradicals are capable of inducing oxidative tissue damage, lipid peroxidation, nucleic acid damage, enzyme inactivation and protein degradation (Borg and Schaich 1984). In order to prevent damage to cellular components, there are numerous enzymatic antioxidant defences designed to scavenge oxyradicals in the cell.

Superoxide dismutase (SOD) scavenge O_2^- by a rapid dismutation reaction.

$$2O_2^- + 2H^+ \xrightarrow{SOD} H_2O_2 + O_2$$

<

As mentioned previously, the production of H_2O_2 within the cell may lead to the production of HO^{o} and subsequent cellular damage via the metal catalysed Haber-Weiss reaction. So, it is necessary to remove H_0O_2 Catalase (CAT) rapidly dismutases H_2O_2 to water and oxygen.

$$2H_2O_2 \xrightarrow{CAT} 2H_2O + O_2$$

Glutathione peroxidase (GSH-Px) catalyses the reduction of organic peroxides or H_2O_2. Glutathione, a non-enzymatic antioxidant, provides the reducing equivalents needed by GSH-Px to carry out the reduction of organic peroxides to the corresponding alcohols and water.

$$2GSH + R\text{-}OO.H \xrightarrow{GSH\text{ - }Px} GSSG + ROH + H_2O$$

Antioxidant enzymes in Aquatic Organisms

Most of the work, on antioxidant enzymes, so far carried out in vertebrates has been restricted to higher groups, particularly nammals. In 1972 Rabie *et al.*, reported the baseline activities of catalase in fishes. They compare the activities of catalase in erythrocytes of modern teleost and a primitive species (*Polypterus senegalus*). They found positive activity of CAT in modern teleosts but not in Polypterus species. The complete absence of catalase from the erythrocytes of the Polypterus, (a species which is taxonomically regarded as a living fossil) supports the contention that catalase is not a fossil enzyme. They presumed the lack of catalase was due to the absence of the appropriate stimulus at a time, when the aquatic life in swamps was anaerobic. Smith (1976) examined the catalase activity in erythrocytes of modern teleosts and a milk fish (*Chanos chanos*, which is intermediate between teleosts and polypterus) and reported its absence (redundant) as in Polypterus.

Hoffert *et al.*, (1975) scrutinized the toxicity of oxygen in invertebrates and vertebrates when exposed to hyperoxygen but did not assess the antioxidant defences against the hyperoxic toxicity. In 1983, Desrochers and Hoffert reported that superoxide dismutase provides protection against the hyperoxia in the retina of the rainbow trout (*Salmo gairdueri*). Simon *et al.*, (1977) compared the activity of SOD in different tissues of turtle, rabbit and mouse. The specific activity of SOD in the liver of turtle was more or less similar to that in the mouse or rabbit. Matkovics *et al.*, (1977) reported the activities of SOD, CAT and peroxidase in various tissues of snail, frog, pigeon, rat and pig. In addition, they compared carp's enzyme activities with these animals activity. Relative to terrestrial animals, SOD activity was low in carp and high for peroxidase. Mazeaud *et al.*, (1979) observed that, relative to human erythrocytes, carp cells displayed slightly lower SOD activities, very low catalase activities, and markedly higher GSH-Px activities. Smith and Shrift (1979) analysed the GSH-Px enzymes in vertebrate and invertebrate species and suggested that the presence of the enzymes activity in so many animals groups reveals the widespread occurrence of genetic information for the specific assimilation of the selenium atom. Aksnes and Njaa (1981) compared the SOD, CAT, GSH-Px activities in nine fish species and analysed these activities in various tissues of mackerel and saithe. They found an absence of GSP-Px in Capelin and in the white muscle and the intestinal tract of mackerel. Gabryelak *et al.*, (1983) reported the seasonal variation in the activity of these antioxidant enzymes in freshwater fish species. In 1984, Witas *et al.*, compared the SOD and CAT activities of fish (Red-blood fish, White-blood fish) and other Antarctic vertebrates (mammals, birds). Barddon-Galloway and Balthrop (1985) characterized the Se-GSH Px enzymes in Black Sea bass (*Centropristis striata*) and reported that selenium was found in GSH-peroxidase fraction which plays an important role in marine fish. Apart from this information Di Ilio *et al.*, (1986) analysed the GSH-Px enzyme activities in *Bufo bufo* during embryonic and larval development. Aceto *et al.*, (1994) also reported the antioxidant enzyme's activity in *Salmo iridaeus* during embryonic and larval development. They observed a gradual increase of superoxide dismutase and catalase activity throughout embryo development. Peters and Livingstone (1996) also detected these enzymes activities in embryonic and early larval stages of turbot and found how these enzymes aid the freshly-hatched larvae to survive even in unfavourable environmental conditions.

Production of oxyradicals during the Phase I metabolism of xenobiotics cannot be ruled out. Increased activity of antioxidant enzymes with exposure to organic xenobiotics have been observed in *Limanda limanda* (1993a). In contrast, Otto *et al.*, (1994) noted that the antioxidant enzymes did not appear to respond to exposure to contaminated sediments. Xenobiotic-stimulated, NAD(P)H-dependent reactive oxygen species production was investigated in 11,600 g supernatants of 4-day old larvae of turbot (Scophthalmus maximus) by Peters *et al.*,

(1996). Some field studies revealed the elevation of hepatic catalase and SOD activities with exposure to PAHs and PCBs in Goby and Mussel (Livingstone *et al.*, 1995). In contrast to this field study, no elevation of hepatic Se-GPx activity with PAH-exposure in *L.limanda* (Livigstone *et al.*, 1993a) was observed but increased activity was seen after 5 or 11 days in 3 MC-treated *L.limanda* (Lemaire *et al.*, 1996). In aquatic organisms a limited number of works have been made on antioxidant enzymes in invertebrates when compared to vertebrates including mammals. Mc Cord and Fridorich (1969) and Shick and Dykens (1985) observed the activity of SOD and CAT in four species of sponges (*Carteriospongia foliascens, Ircina ramosa, Neofibularia irata* and Phyllospongia sp.) but not the presence of GSH-Px.

In Coelenterates, SOD and CAT activities were detected in a Sea anemone, *Anthopleura clegantissima* (Dykens and stick 1985). In trematodes and cestodes, activity of catalase was not observed (Barrett and Beis 1982). Absence of these enzyme in these species may be due to parasitic mode of life or absence/ low levels of oxygen in their environment. Similarly, Blum and Fridovich (1984) found no evidence of catalase activity in giant tube worm *Riftia pachyptila* but SOD and GSH-Px activities were detected. Both SOD and CAT activities were detected in different species of planarian. In the same way they found these enzyme's activity in Annelida (Nereidae sp and Paraonidae sp.). In molluscs, a correlation between SOD and CAT activities was observed in the siphon of *T.maxima* (Stick and Dykens 1985). The activity of catalase was absent in deep-sea hydrothermal vent clam *Calyptogena magnifica* but the GSH-Px activity was observed. Se-dependent GSH-Px was observed in *M.edulis* (H_2O_2 was used as a substrate) (Winston et al 1990). In 1992, Livingstone *et al.,* reported the properties of antioxidant enzymes (SOD, CAT, Se-GSH-Px, and Total GSH-Px) in the digestive gland of *M.edulis*. Studies of Gamble *et al.*, (1995) reveals the high activity of antioxidant enzymes in major digestive detoxification tissues of *M.edulis* and NADPH, NADH dependent O_2^- and OH^o production were observed in this species by Livingstone *et al.,* (1989) and Winstone *et al.*, (1990). Wenning *et al.,* (1988) observed an elevated level of

CAT and SOD in the digestive gland of *G.demissa* when exposed to paraquat. But in the field studies, no change or decreased level of these enzymes were observed in whole body of *M.edulis* exposed to diesel oil, PAH and PCB (Suteau *et al.*, 1988). Wenning and Guilio (1988) revealed that paraquat can produce reactive oxygen species in ribbed mussels and wedge clams and antioxidant enzymes would give protection to this organisms from oxidative tissue damages. Livingstone *et al.*, (1995) demonstrated the usefulness of SOD and CAT enzymes as biomarkers for organic pollution in fish and molluscs. Garcia Martinez *et al.*, (1995) and Garzia Martinez and Livingstone (1995) investigated the potential of nitrofurantion and benzo(a) pyrene diones to stimulate the reactive oxygen production in *M.edulis*. In Crustacea, hepatic microsomes of *P.clarkii* were involved in the NADH, NADPH-dependent production of OH° (Jewell and Winston 1989). Addition of catalase inhibited the production of OH° and indicated that H_2O_2 serves as the precursor of OH via a Fentan reaction i.e. the antioxidant enzyme, catalase, destroy the H_2O_2 and prevents the production of OH° by H_2O_2. Yokota (1970) found the activity of catalase in some crustacean species (*Leucosia longiforns, Hemigrapsus sanguineus, Portunus trituberculatus and Eriocheir japonicus).* In 1994, Bell and Smith reported the occurrence and distribution of catalase, SOD, GSH-Px enzymes in the haemolymph of a crab (Carcinus maenas). In echinoderms, information on the activity of these antioxidant enzymes was scanty.

CONCLUSIONS

Biotranformation enzymes in aquatic organisms can provide new information such as detoxifying activity against specific xenobiotic compounds, its role as a biomarker for the measurement of environmental pollutants, etc. Several works reveal the occurrence of biotransformation (Phase I and Phase II metabolizing enzymes) and antioxidant enzymes in aquatic vertebrates rather than invertebrates. Invertebrate species particularly molluscs and crustaceans exhibit bioindicator characteristics relevant to the field of environmental management (Table 1). Within these two phyla, crustacean species were less studied and represents a focal area for future

Table 1. Responses of biotransformation and antioxidant enzymes as a biomarker to various pollutant

Animal	*Biotransformation enzymes*	*Antioxidant enzymes*	*Pollutants / Pro-oxidant*	*Response*	*Reference*
1	2	3	4	5	6
Vertebrates					
Read Sea Bream	CYP 4501A	–	3-methyl cholanthrene	Increase	Mizukami *et al.*, (1995)
Ranbow trout	EROD	–	3-methyl cholanthrene	Increase	Stegeman *et al.*, (1981)
Rainbow trout	EROD	–	Pulpmill effluent	Increase	Lindstrom-Seppa and Oikari (1990)
Common Carp	-	GSH-Px	Paraquat	Increase	Gabryelak and Klekot (1985)
Feral spot	-	SOD	PAH	Increase	Roberts *et al.*, (1987)
Flounder	EROD	-	PAH	Increase	Payne *et al.*, (1987)
Cat fish	-	CAT	PAH	Increase	Di Giulio *et al.*, (1989)
Flounder	EROD	-	Petrochemical compound	Increase	Sulaiman *et al.*, (1991)
Dab	EROD	-	PAH	Increase	Lange *et al.*, (1992)
Viviparous blenny	CYP 450	-	BNF	Increase	Celander *et al.*, (1994)
Goby	CYP 450	-	PAH	Increase	Livingstone *et al.*, (!995)
Rainbow trout	EROD	-	BNF	Increase	Lemaire *et al.*, (1996)

(Table Contd...)

Table 1. (Contd...)

1	2	3	4	5	6
Anopalrchus purpurescens	CYP 450	–	BNF	Increase	Woodin *et al.*, (1997)
Invertebrates					
Coelenterates					
A-elangantissima		SOD&CAT	Hyper Oxygen	Increase	Dykens and Stick (1984)
Plantyhilminthus					
Phagocata velatn	*NADH-Cyt.red*	–	*DDT*	*Increase*	*Baldwin and Wells (1978)*
Polychaetes					
N. sccinea	*CYP 450*	–	*Crude oil*	*Increase*	*Lee et al., (1981)*
Molluscs					
Mytilus galloprovincials	CYP-450	–	Hydrocarbon	Increase	Gilewicz *et al.*, (1984)
M..edulis	*CYP-450*	–	*Diesel oil*	*Increase*	*Livingstone et al., (1985)*
Thais haemastoma	*CYP-450 reductase*	–	*Crude oil*	*Increase*	*Livingstone et al., (1986)*
G.demissa	–	*SOD & CAT*	*Paraquat*	*Increase*	*Wenning et al., (1988)*
Littorina littorea	CYP 450	–	Diesel oil	Increase	Livingstone *et al.*, (1985)
Crustacea					
Fiddler crab					
blue crab	CYP 450	–	Hydrocarbon	Increase	Lee and Conner (1981)
C.sapidus	*CYP 450*	–	*Arcolor 1254*	*Increase*	*Lee et al (1982)*
M.crispta	CYP 450	–	3-MC & Benzo a-pyrene	Increase	Batel et al (1988)
M.malcolmsonii	*GST*	*SOD, CAT GSH-Px*	*Hyper oxygen*	*Increase*	*Arun and Subramanian (1998)*
M.lamarrei lamarei	*GST*	–	*Hyper oxygen*	*Increase*	*Arun and Subramanian (1997)*

research. Similarly detailed research is also meager about toxicant-mediated oxyradical production and detoxifying mechanism in both vertebrate and invertebrate species. Generally, invertebrate species are more sensitive, even to low level of pollutants when compared to vertebrates. Thus, analysis of these biomarker enzymes in invertebrates would be useful to environmental managers to assess water quality at early stages of contamination. On the other hand, larvae of the both vertebrate and invertebrate species are more sensitive and susceptible to environmental changes. To our knowledge, there is little or no quantitative data on xenobiotic metabolising enzymes in the embryonic development of aquatic organisms. Further research on this aspect may clarify whether these aquatic organisms acquire/develop these defence mechanism from the embryonic or in later stages. In future, much remains to be learned about the use of these biotransformation enzymes in the field, as a biomarker for environmental pollutants.

ACKNOWLEDGEMENTS

We thank Dr. D.R. Livingstone (Plymouth, England) for providing enough materials for this work and also thank Dr. A.S. Rao (Dept. of Biotechnology, Bharathidasan University) for making critical comments in this manuscript. The senior author wishes to thank CSIR-India for providing financial support in the form of Research Associateship.

REFERENCES

Aceto. A., Amicarelli. F., Sacchetta. P., Dragani. B., Bucciarelli. T. Masciocco. L., Miranda. M., Di Ilio. C., (1994). Developmental aspects of detoxifying enzymes in fish (*Salmo iridaeus*) Free Rad. Res. 21: 285-294.

Ahokas. J.T. (1977). Cytochrome P450 and mixed function oxidase of trout. Salmo trutta lacustris. Thesis, Acta. Univ. Ovl. D20. Pharmacol. Physiol. 2.

Aksnes. A. and Najaa. L.R. (1981). Catalase, glutathione peroxidase and superoxide dismutase in different fish species. Comp. Biochem. Physiol. 69B: 893-896.

Almar. M.A., Diaz-Mayans. J., and Romero. F.J. (1987). Glutathione content and Glutathione S-transferase activity in midgut gland of *Procambarus clarkii* sex differences, the effect of fasting and their implications in cadmium toxicity. Comp. Biochem. Physiol. 87C: 433-435.

Anderson. R.H. (1968). Drug metabolism in marine vertebrates. Fed. Proc. 26: 1047-1055.

Andersson. T., Bengtsson. B.E., Forlin. L., Hardig. J. and Larsson. A. (1987). Long term effects of bleached kraft mill effluents on carbohydrate metabolism and hepatic exnobiotic biotransformation enzymes in fish. Ecotoxic. Envir. Saf. 13: 53-60.

Andersson. T., Kovisaari. U. and Forlin. L. (1985). Xenobiotic biotransformation in the rainbow trout liver and kidney during starvation. Comp. Biochem. Physiol. 82C: 221-225.

Arun. S. and Subramanian. P. (1998). Antioxidant enzymes in freshwater prawn *Macrobrachium malcolmsonii* during embryonic and larval development. Comp. Biochem. Physiol (B). 121: 273-277.

Arun. S., and Subramanian. P. (1997). Glutathione-S-transferases activity during embryonic and larval development in freshwater prawn *M.lamarrei lamarrei* Current Science. 73: 107-109.

Balabaskaran. S., Chew. S., Segaran. M. (1986). Studies on glutathione S-transferase in molluscs. Comp. Biochem. Physiol. 85B: 183-192.

Baldwin. R.L., Wells M.R. (1978). Effect of DDT on NADH-cytochrome b_5 reductase activity in the freshwater planarian *Phagocata velata*. Bull Environ. Contam. Toxicol. 19: 428-430.

Barddon-Galloway. S. and Balthrop. J.E. (1985). Se-dependent GSH-Peroxidase isolated from black sea bass (*Centropristis striata*). Comp. Biochem. Physiol. 82C: 297-300.

Barrett. J., Beis. I. (1982). Catalase in free-living and parasite platyhelmiths. Experientia. 38: 536.

Batel. R., Bihari. N. and Zahu. R.K. (1988). 3-methylcholanthrene does induce mixed function oxidase activity in hepatopancrease of spiny crab *Maja crispata*. Comp. Biochem. Physiol. 90C: 435-438.

Batel. R., Bihari. N., Kurelec. B., Muller. W.E.G. and Zahu. R.K. (1983). Presence of mixed function oxygenase in some marine invertebrates. Thalassia Jugosl. 19: 23-29.

Bavermeister. A., Lewendon. A., Ramange. P.I.N., Nimmo. I.A. (1983). Distribution and some properties of the glutathione S-transferase and y-glutamyl transpeptidase activities of rainbow trout. Comp. Biochem. Physiol. 74C: 89-93.

Bell. K.L., Smith. V.J. (1995). Occurrence and distribution of antioxidant enzymes activity in the blood of *Carcinus maenas*. Mar. Biol. 123: 829-836.

Bihari. N., Batel. R., Kurelec. B., and Zahn. R.K. (1984). Tissue distribution seasonal variation and induction of benzo(a)pyrene monooxygenase activity in the crab *Maja Crispata*. Sci. Total. Environ. 35: 41-51.

Blum. J. Fridovich. I. (1984). Enzymatic defense against oxygen toxicity in the hydro thermal vent animals, *Riftia pachyptila* and *Calyptogena magnifica*. Arch. Biochem. Biophys. 228: 617-620.

Boon. J.P., Everaarts. J.M., Hillebrand. M.T.J., Eggens. M.L., Pijnenburg. J. and Goksoyr. A. (1992). Changes in levels of hepatic biotransformation enzymes and haemoglobinn levels in female plaice (*Pleuronectes platessa*) after oral administration of a technical polychlorinated biphenyl mixture (Clophen A40). The Science of the Total Environment. 114: 113-133.

Borg. D.C. and Schaich. K.M. (1984). Cytotoxicity from coupled redox cycling of autooxidizing xenobiotics and metals. Israel. J. Chem. 24: 38-53.

Boyland. E. and Chasseaud. L.F. (1969). The role of glutathione and glutathione-S-transferases in mercapturic acid biosynthesis. Adv. Enzymol. 32: 173-219.

Brodie. B.B. and Maickel. R.P. (1962). Comparative biochemistry of drug metabolism. In Proc. Ist. Int.Pharmac. Meetg. (Eds, Brodie. B.B. and Erdos. E.G.) Mac Millan. New York. 6-299.

Buchler. D.R. (1966). Hepatic drug metabolism in fishes. Fed. Proc. Fedn. Am. Socs. Exp. Biol. 25: 343.

Carfinkel, D. (1957). Isolation and properties of cytochrome b_5 from pig liver. Archs. Biochem. Biophys. 71: 111-120.

Celander. M., Leaver. M.J., George. S.G. and Forlin. L. (1993). Induction of cytochrome P450 IAI and conjugating enzymes in rainbow trout: a time course study. Comp. Biochem. Physiol. 106C: 343-349.

Celander. M., Naf. C., Broman. D., Forlin. L., (1994). Temporal aspects of induction of hepatic cytochrome P450 IA and conjugating enzymes in a viviparous blenny (*Zoarces viviparus*) treated with petroleum hydrocarbons. Aquat. Toxiçol., 29: 183-196.

Chamber. J.E. (1979). Induction of microsomal mixed function oxidase system components in Striped mullet by short-term exposure to crude oil. Toxicol. Lett. 4: 227-230.

Chambers. J.E. and Yarbrough. J.D. (1976). Xenobiotic biotransformation systems in fishes. Comp. Biochem. Physiol. 55C: 77-84.

Chan. J.M. Gillett. J.W. and Terriere. L.C. (1967). Interaction between microsomal electron transport systems of trout and male rat in cyclodiene epoxidation. Comp. Biochem. Physiol. 20: 731-742.

Chitwood. D.J., Lusby. W.R., Salt. T.A., (1987). Sterol metabolism in the nematodes *Panagrellus re*, *Turbatrix aceti* and *Caenorhabditis elegans*. Comp. Biochem. Physiol. 86B: 103-107.

Clark. A.G. Dick. G.L. and Smith. J.N. (1984). Kinetic studies on the glutathione S-transferases from the Larvae of *Costelyta zealandicu* Biochem. J. 217: 51-58.

Conney. A.H. (1982). Induction of microsomal enzymes by foreign chemicals and carcenogenesis by polycyclic aromatic hydrocarbons. G.H.A. Clowes memorial lecture. Cancer. Res. 42: 4875-4917.

Davies, K.J.A. (1995). Oxidative stress; the paradox of aerobic life. In free radicals and oxidative stress: environment, drugs and food additives (Rice-Evans. C. Halliwell. B& Lunt. G.C. eds) pp. 1-31. London. Portland press.

Del-Boccio. G., Di-Ilio. C., Miranda. M., Manilla. A. Zarivi. O., Bonfigu. A. and Federici. G. (1987). Glutathione transferase activity during *Bufo bufo* development Comp. Biochem. Physiol. 86B: 749-753.

Desrochers. P.E. and Hoffert. J.R. (1983). Superoxide dismutase provides protection against the hyperoxia in the retina of the rainbow trout (*Salmo gairdneri*) Comp. Biochem. Physiol. 76B: 241-247.

Dewaide. J.H. (1971). Metabolism of xenobiotics. Comparative and Kinetic studies as a basis for environmental pharmacology. Thesis. Leijn, Nijmegen.

Di Guilio, R.T., Habig. C. and Wolf. T. (1989). Phase I and Phase II biotransformation enzyme activities in channel cat fish exposed to contaminated sediment. The toxicologist 9, 43 (abs).

Di Ilio. C. Del Boccio. G. Miranda. M. Manilla. A. Zarivi. O. Federici. G. (1986). Glutathione peroxidase and glutathione reductase activities during *Bufo bufo* development. Comp. Biochem. Physiol. 83C: 9-12.

Di Ilio. C., Aceto. A., Dragani. B., Sacchetta. P., Bucciarelli. T., Augemcci. S., Miranda. M., Poma. A., Amicarelli. F. and

Federici. G. (1994). Developmental aspects of Amphibian (*Bufo bufo*) and fish (*Salmo iridaeus*) embryo gluthione transferase (1994) Life chemistry reports. 12:61-64.

Dierckx. P.J. (1984). Glutathione S-transferases in aquatic macro-invertebrates and its interaction with different micro-pollutatns. Sci. Tot. Envir. 40: 93-102.

Douch. P.G.C., Buchanan. L.L. (1978). Glutathione conjugation of some xenobiotics by *Ascaris summ* and *Moniezia expansa*. Xenobiotica. 8: 171-176.

Durand. E.F.P. and Cravedi. J.P. (1989). Charecterisation of xenobiotic metabolizing enzymes in sturgeon (*Acipenser baeri*) Comp. Biochem. Physiol. 93(B): 921-928.

Dykens. J.A. and Stick. J.M. (1985). Photobiology of the symbiotic Sea anemone, *Anthopleura elegantissima*: defences against photodynamic effects and seasonaly photoaccimatization. Biol. Bull. 167: 683-697.

Elcombe. C.R. and Lech. J.J. (1979). Induction and characterization of heam proteins P450 and monooxygenation in rainbow trout (*Salmo gairdneri*). Toxicol. Appl. Pharmocol. 49: 437-450.

Elmamlouk. T.H., Gessner. T., Brownie. A.C. (1974). Occurrence of cytochrome P450 in hepatopancreas of *Homarus americanus*. Comp. Biochem. Physiol. 48B: 419-425.

Elskus. A.A. and Stegeman. J.J. (1989). Further consideration of phenobarbital effects on cytochrome P450 activity in the kill fish, *Fundulus heteroclitus* Comp. Biochem. Physiol. 92(C): 223-230.

Forlin, L., Andersson. T., Bengtsson. B.E., Hardig. J., and Larsson. A., (1985). Effects of pulp bleached plant effluents on hepatic xenobiotic biotransformation enzymes in fish: Laboratory and field sutdies. Mar. Envir. Res. 17: 109-112.

Forlin. L., Lemaire. P., Livingstone. D.R. (1995). Comparative studies of hepatic xenobiotic metabolizing and antioxidant enzymes in different fish species. Mar. Environ. Res. 24: 27-30.

Foureman. G.L., Ben-Zvi. Z., Dostal. L., Fouts. J.R., Bend. J.R. (1978). Distribution of ^{14}C-benzo(a)pyrene in the lobster, *Homarus americanus*, at various times after a single injection into the pericandial sinus. Bull. Mt. Desert. Inst. Biol. Lab. 18: 93-95.

Fries. C.R., Lee. F.R. (1984). Pollutant effect on the mixed function oxygenase (MFO) and reproductive systems of the marine polychalt *Nereis virens*. Mar. Biol. 79: 187-193.

Gabryelak. T. and Klekot. J, (1985). The effect of paraquat on the peroxide metabolism enzymes in erythrocytes of freshwater fish species. Comp. Biochem. Physiol., 81C: 415-418.

Gabryelak. T., Piatkowska. M., Leyko. W., Peres. G. (1983). Seasonal variations in the activities of peroxide Metabolism enzymes in erythrocytes of freshwater fish species. Comp. Bioche. Physiol. 75C: 383-385.

Gamble. S.C., Goldfrab. P.S., Porte. C. and Livingstone. D.R. (1995). GSH-Px and other antioxidant enzyme function in marine invertebrates (*Mytilus edulis, Pecten maximus, Carcinus maenas* and *Asterias rubens*) Mar. Envr. Res. 39: 191-195.

Garcia Martinez. P. and Livingstone. D.R. (1995). Benzo(a)pyrene-Stimulated oxyradical production by microsomes of digestive gland of the common mussel, *Mytilus edulis* L. Mar. Environ. Res. 39: 185-189.

Garcia Martinez. P., Winstone. G.W. Metash-Dickey. C., O'Hara SCM., Livingstone. D.R. (1995). Nitrofurantion-stimulated reactive oxygen species production and Genotoxicity in digestive gland microsomes and cytosol of the common Mussel (*Mytilus edulis* L.). Toxicol. Appl. Pharmacol. 131: 332-341.

Gilewicz. M., Guillqume. J.R., Carles. D., Leveau. M., Bertrand. J.C. (1984). Effects of petroleum hydrocarbon on the cytochrome P450 content of the mollusc bivalve *Mytilus galloprovincialis*. Mar. Biol. 80: 155-159.

Goksoyr. A., Andersson. T., Hansson. T., Klug Soyr. J., Zhang. Y., Forlin. L. (1987). Species characteristics of the hepatic xenobiotic and steroid biotransformation system of two teleost fish, Atlantic Cod (*Gadus morhua*) and rainbow trout (*Salmo gairdneri*) Toxicology and Applied Pharmacology. 89: 347-360.

Halliwell. B. (1978). Biochemical mechanisms accounting for the toxic action of oxygen on living organisms: the key role of superoxide dismutase. Cell. Biol. Int. Rept. 2: 113-128.

Hellebone (1986). Categorization of lipophilic xenobiotic by enthalnic structure–function responses of hepatic mixed function oxidase. Drug Metab. Rev.17: 93-143.

Hoffert. J.R. Baeyens. D.A. and Fromm. P.O. (1975). In vivo oxygen toxicity in invertebrates and vertebrates. Comp. Biochem. Physiol. 52A: 351-353.

Hollebone. B.R., Brown Lee. L.J., Davis. D., Michelin. N., Purdy. D., (1995). A mechanistic structure – activity relationship for hepatic polysubstrate monooxygenase. Envt. Tox col. Chem. 14: 29-41.

Huuskonen. S. and Lindstrom-Seppa. P. (1995). Hepatic cytochrome P450 IA and other biotransformation activities in perch (*Perca fluviatilis*): the effects of unbleached pulp mill effluents. Aquatic Toxicology. 31: 27-41.

Jakoby, W.B. (1980). in Enzymatic Basis of Detoxification (Jakoby, W.B. Eds) Vol. 1: pp 1-6, Academic Press, New York.

James. M.O. and Bend. J.R. (1980). Polycyclic aromatic hydrocarbon induction of cytochrome P450 dependent mixed-function oxidases in marine fish. Toxicol. Appl. Pharmacol. 54: 117-133.

James. M.O. and Little. P.J. (1984). 3-methylcholanthrene does not induce invitro xenobiotic metabolism in spiny lobster hepatopancreas of affect in vivo disposition of benzo(a)pyrene. Comp. Biochem. Physiol. 78C: 241-245.

James. M.O., Bowen. E.R., Dansette. P.M. Bend. J.R. (1979a). Epoxide hydrase and glutathione S-transferase activites with selected alkene and arene oxides in several marine species Chem. Biol. Interact. 25: 321-344.

James. M.O., Khan. M.A.Q., Bend. J.R. (1979). Hepatic microsomal mixed-function oxidase activities in several marine species common to coastal Florida. Comp. Biochem. Physiol. 62C: 155-164.

James. M.O., Sherman. B., Fisher. S.A., Bend. J.R. (1982). Benzo(a) pyrene metabolism in reconstituted monooxygenase systems containing cytochrome P450 from lobster (*Homarus americanus*) hepatopancreas fraction and NADPH-cytochrome P450 reductase from pig liver. Bull. Mt. Desert. Isl. Biol. Lab. 22: 37-39.

Jewel. S.E.C. and Winston. G.W. (1989). Characterization of the microsomal mixed-function oxygenase system of the hepatopancreas and green gland of the red swamp cray fish, *Procambarus clarkii*. Comp. Biochem. Physiol. 92B: 329-339.

Jewell. C.S.E. and Winston. G.W. (1989). Oxyradical production by hepatopancreas microsomes from the red swamp cray fish, *Procambrus clarki* Aquat. Toxicol. 14: 27-46.

Kawalek. J.C., Rew. R.S., Heauner. J. (1984). Glutathione-S-transferase, a possible drug-metabolizing enzyme in *Haemonchus contortus*. Comparative activity of a cambendazole-resistant and susceptible strain. Int. J.Parasitol. 14: 173-175.

Klingenberg, M. (1958). Pigments of rat liver microsomes. Archs. Biochem. Biophys. 75: 376-386.

Lange. U., Jedamski-Grymlas. J., Siebers. D., & Karbe. L., (1992). Ethoxyresorufin O-deethylase and cytochrome P450 in the liver of dab [*Limanda limanda* (L)] from the central and southern North Sea. Mar. Pollut. Bull. 24: 446-451.

Lech. J.J. (1974). Glucuronide formation in rainbow trout effect of salicylamide on the acute toxicity, conjugation and excretion of 3-trifluoromethyl-4-nitrophenol. Biochem. Pharmacol., 23: 2403-2410.

Lee. R.F. (1981). Mixed function oxygenase (MFO) in marine invertebrates. Mar. Biol. Lett. 2: 87-105.

Lee. R.F. and Conner. J.W. (1981). Responses of detoxification/toxication system in marine invertebrates to organic pollution. PRIMA Newslett. 2:4-5.

Lee. R.F. Conner. J.W., Page. D., Ray. L.E. and Giam. C.S. (1982). Cytochrome P450 dependent mixed function oxygenase system in marsh carbs. In: Vernberg W.B, Calabrese A. Thurberg F.P, Vernberg F.J. (eds). Physiological mechanisms of marine pollutant toxicity. Academic press. New York. pp. 147-159.

Lee. R.F., Keeran. W.S., Pickwell. G.V. (1988). Marine invertebrate glutathione S-transferases, Purification charecterization and induction Mar. Env. Res. 24: 97-100.

Lee. R.F., Singer. S.C., Page. D.S. (1981). Responses of cytochrome P450 systems in marine crab and polychaetes to organic pollutants. Aquat. Toxicol. 1: 355-365.

Lemaire. P., Forlin. L., Livingstone. D.R. (1996). Responses of hepatic biotransformation and antioxidant enzymes to CYPIA-inducers (3-methyl cholanthrene, β-napthoflavone) in Sea bass (*Dicentrarchus labrax*), dab (*Limanda limanda*) and rainbow trout (*Oncorhynchus mykiss*). Aquatic Toxicology. 36: 141-160.

Lindstrom-Seppa. P. and Oikari. A. (1988). Hepatic xenobiotic biotransformation in fishes exposed to pulpmill effluents. Wat. Sci. Tech. 20: 167-170.

Lindstrom-Seppa. P. and Oikari. A. (1990). Biotransformation activities of feral fish in water receiving bleached pulpmill effluents. Envir. Toxicol. Chem. 9: 1415-1424.

Lindstrom-Seppa. P., Hanninen. O. (1986). Induction of cytochrome P450 mediated monooxygenase reactions and conjugation activities in freshwater cray fish (*Astacus astacus*) Arch. Toxicol. Suppl. 9: 374-377.

Lindstrom-Seppa. P., Kovivusaasri. U., Hanninen. O. (1983). Metabolism of foreign compounds in freshwater cray fish (*Astacus astacus* L.) Aquat. Toxicol. 3: 35-46.

Lindstrom-Seppa. P., Oikari. A. (1989). Biotransmation and other physiological responses in white fish caged in a lake receiving pulp and paper mill effluents. Ecotoxicology and Environmental Safety. 18: 191-203.

Livingstone, D.R. (1991). in Advances in Comparative and Environmental Physiology (Gilles, R.Eds.) Springer-Verlag Berlin. Heidelberg. Vol. 7; pp 45-185.

Livingstone, D.R. (1993). Biotechnology and pollution monitoring use of molecular Biomarkers in Aquatic environment 57, 195-211.

Livingstone, D.R. (1994). Recent developments in marine invertebrate organic xenobiotic metabolism. Toxicology & Ecotoxicology News. 1:88-95.

Livingstone. D.R. Lips. F., Garcia Martinez. P. and Pipe. R.K. (1992). antioxidant enzymes in the digestive gland of the common mussel, *Mytilus edulis* Mar. Biol. 112: 265-276.

Livingstone. D.R. Martinez. G.P., Winston. G.W. (1989). Menadione-stimulated oxyradical production in digestive gland microsomes of the common mussel, *Mytilus edulis* L. Aquat. Toxicol. 15: 213-236.

Livingstone. D.R. Stickle. W.B., Kapper. M., Wang. S. (1986). Microsomal detoxification enzyme responses of the marine snail, *Thais haemastoma*, to laboratory oil exposure. Bull. Environ. Contam. Toxicol. 36: 843-850.

Livingstone. D.R., Lemaire. P., Matthews. A., Peters. L., Bucke. D., Law. R.J. (1993a). Pro-oxidant, antioxidant and 7-ethooxyresorufin O-deethylase (EROD) activity responses in liver of dab (*Limanda limanda*) exposed to sediment contaminated with hydrocarbons and other chemicals. Mar. Poll. Bull. 26: 602-606.

Livingstone. D.R., Lemaire. P., Matthews. A., Peters. L.D., Porte. C., Fitzpatrick. P.J., Forlin. L., Nasci. C., Fossato. V., Wootton. N. and Goldfarb. P. (1995). Assessment of the impact of organic pollutants of Goby (*Zozterisessor ophiocephalus*) and mussel (*Mytilus galloprovincialis*) from the veniece Lagoon; Italy: Biochemical studies. Mar. Envr. Res. 39: 235-240.

Livingstone. D.R., Moore. M.N., Lowe. D.M., Nasci. C., Farrar. S.V. (1985). Responses of the cytochrome P450 monooxygenase system to diesel oil in the common mussel, *Mytilus edulis*. L., and the periwinkle *Littorina littorea* L. Aquat. Toxicol. 7: 79-91.

Maemura. S., and Omura. T., (1983). Drug oxidizing monooxygenase system in Liver microsomes of gold fish (*Carassus aurtus*) Comp. Bioche. Physiol. 76(C): 45-51.

Matkovics. B., Novak. R., Hanh. H.D., Szabo. L., Varga. S.I. and Zalesna. G. (1977). A comparative study of some more important experimental animal peroxide metabolism enzymes. Comp. Biochem. Physiol. 56B: 31-34.

Mazeaud. F., Maral. J., Michelson. A.M. (1979). Distribution of superoxide dismutase and glutathione peroxidase in the carp: erythrocyte manganese SOD. Biochem. Biophys. Res. commun. 86: 1161-1168.

Mc Cord. J.M. & Fridovich. I. (1969). Superoxide dismutase: an enzymatic function for erythrocuprein (hemocuprein). J. Biol. Chem. 244: 6049-6055.

Mizukami. Y, Okauchi. M., Arizono. K., and Ariyoshi. T. (1995). Induction of cytochrome P4501A by 3-methylcholanthrene in Red Sea Bream *Pagrus major*. Fishereis science. 61: 599-603.

Oikari. A. Lindstrom-Seppa. P., Kukkonen. J. (1988). Subchronic metabolic effects and toxicity of a stimulated pulpmill effluent on juvenile lake trout, *Salmo trutta*, *M. Lacustris*. Ecotoxicol. Environ. Safety. 16: 202-218.

Omura, T. and Sato, R. (1964). The carbon monoxide binding pigment of liver microsomes. J. Biol. Chem. 239: 2370-2378.

Omura. T., Sato. R., Cooper. D.Y. Rosenthal. O. and Estabrook. R.W. (1965). Function of cytochrome P450 of microsomes Fedn. Proc. Fedn. Am. Soc Exp. Biol. 241: 1181-1189.

Otto. D.M.E. Lindstrom-Seppa. P. and Sen. C.K. (1994) Cytochrome P450 dependant enzymes and oxidant-mediated responses in rainbow trout exposed to contaminated sediments. Ecotoxicol. Envir. Safe. 27: 265-280.

Payne, J.F. (1976). Field evaluation of benzopyrene hydroxylase induction as a monitor for marine pollution. Science. 191: 940-946.

Payne, J.F., Bauld. C., Dey. A.C., Kiceniuk. J.W. and Williams. U (1984). Selectivity of mixed-function oxygenase enzyme induction in flounder (*Pseudopleuronectes americanus*) collected at the site of the Baie verte, Newfoundland oil spill. Comp. Biochem. Physiol., 79(C): 15-19.

Payne. J.F. and Fancey. L.L. (1982). Effect of Long-term exposure to petroleum on mixed function oxygenases in fish. Further support for use of the enzyme system in biological monitoring. Chemosphere. 11: 207-214.

Payne. J.F., Fancy. L., Rhimutula. A.D. and Porter. E.L. (1987). Review and perspective on the use of mixed function oxygenase enzymes in Biological monitoring. Comp. Biochem. Physiol., 86C: 233-245.

Peters. L.D. and Livingstone. D.R. (1996). Antioxidant enzyme activities in embryologic and early larval stages of turbot. J. Fish. Biol. 49: 96-97.

Peters. L.D. O'Hara. S.C.M. and Livingstone. D.R. (1996). Benzo(a) prene metabolism and xenobiotic-stimulated reactive oxygen species generation by subcellular fraction of Larvae of turbot (*Scophthalmus maximus*. L) Comp. Biochem. Physiol. 114C: 221-227.

Pohl. R.J., Bend. J.R., Guarino. A.M. and Fouts. J.R. (1974). Hepatic microsomal mixed function oxidase activity of several marine species from coastal marine. Drug. Metab. Dispos. 2: 545-555.

Rabie. F. Magid. A.M.A. Gummaa. K.A. and Karrar. O. (1972). Evolution of catalase in fish. Comp. Biochem. Physiol. 43A: 1053-1055.

Ramage. P.I.N. and Nimmo. I.A. (1984). The subtrate specificities and subunit compositions of the hepatic glutathinone S-transferases of rainbow trout (*Salmo gairdneri*) Comp. Biochem. Physiol. 78B: 189-194.

Roberts. M.H., Sved. D.W., Fetton. S.P. (1987). Temporal changes in AHH and SOD activities in feral spot from the Elizabeth River, a polluted sub-estuary. Mar. Environ. Res. 23: 89-101.

Schenkman. J.B., Remner. H. and Estabrook. R.W. (1967). Spectral studies of drug interactions with hepatic microsomal cytochrome. Molec. Pharmacol. 3: 113-123.

Schlenk. D., Buhler. D.R. (1988). Cytochrome P450 and Phase II activities in the gum boot *Cryptochiton stelleri* Aquat. Toxicol. 13: 167-182.

Simon. L.M. Theodore. J.L.V.J. and Robin. I.D. (1977). Some comparative aspects of the organ distribution of superoxide dismutase activity in the freshwater turtle, *Pseudemys scripta elegans*. Comp. Biochem. Physiol. 58B: 377-379.

Singer. S.C. and Lee. R.F. (1977). Mixed function oxygenase activity in blue crab *Callinectes sapidus*: tissue distribution and correlation with changes during molting and development. Biol. Bull. 153: 377-386.

Sleiderink. H.M. Everaarts. J.M., Goksoyr. A. and Boon. J.P. (1995). Hepatic cytochrome P450 IA induction in Dab (*Limanda Limanda*) after oral dosing with the polychlorinated biphenyl mixture Clophen A40. Environ. Toxicol. Chem. 14: 679-687.

Smith. A.C. (1976) Catalase in fish red blood cells. Comp. Biochem. Physiol. 54B: 331-332.

Smith. J. and Shrift (1979). Phylogenetic distribution of glutathione peroxidase. Comp. Biochem. Physiol. 63B: 39-44.

Stegeman. J.J. (1978). Influence of environmental contamination of cytochrome P450 mixed function oxygenases in fish: Implication for recovery in the Wild Harbour Marsh. J. Fish. Res. Bd. Can. 35: 658-674.

Stegeman. J.J. and Kaplan. H.B. (1981). Mixed function oxygenase activity and Benzo(a)pyrene metabolism in the Barnacle *Balanus eburneus* (Crustacea: CIRRIPEDIA). Comp. Biochem. Physiol. 68C: 55-61.

Stegeman. J.J., Koltz. A.V., Woodin. B.R., and Pajor. A.M. (1981). Induction of hepatic cytochrome P-450 in fish and the indication of environmental induction in scup (*Stenotomus chrysops*). Aquatic Toxicol., 1: 197-212.

Stenerson. J., Kobro. S., Bjerke. M., Arend. U. (1987). Glutathione transferases in aquatic and terrestrial animals from nine phyla. Comp. Biochem. Physiol. 86C: 73-82.

Stick. J.M. and Dykens. J.A. (1985). Oxygen detoxification in algalinvertebrate Symbioses from the Great Barrier Reef. Ocecologia. 66:33-41.

Sulaiman. N., Burke. M.D. and George. S.G. (1991). Sublethal assessment of pollutant impact on flounders in an industrial estuary using hepatic biochemical indices. Marine Ecology Progress Series. 68: 207-212.

Suteau. P. Daubeze. M. Migaud, M.L. Narbonne. J.F. (1988). PAH-metabolizing enzymes in whole mussels as biochemical tests for chemical pollution monitoring. Mar. Ecol. Prog. Ser. 46: 45-49.

Tan. B. and Melius. P. (1986). Polynuclear aromatic hydrocarbon metabolism in fishes. Comp. Biochem. Physiol. 83(C): 217-224.

Wenning. R.J. and DiGiulio. R.T. (1988). Microsomal enzyme activities superoxide production and antioxidant defences in ribbed mussels (*Geukensia demissa*) and wedge clams (*Rangia cuneata*). Comp. Biochem. Physiol. 90C: 21-28.

Wenning. R.J., Di-Giulio. R.T. and Gallagher. E.P. (1988). Oxidant-mediated biochemical effects of paraquat in the ribbed mussel, *Gueskensia demissa*. Aquat. Toxicol. 12: 157-170.

Williams. D.E., Masters. B.S.S., Lech. J.J., Buhler. D.R. (1989). Sex differences in cytochrome P450 isozyme composition and activity in kidney microsomes of mature Rainbow trout. Biochemical Pharmacology. 35: 2017-2023.

Winston. G.W., Livingstone. D.R. and Lips. F. (1990). Oxygen reduction metabolism by the digestive gland of the common marine mussle *Mytilus edulis* L. J. Exp. Zool. 255: 296-308.

Witas. H.T. Gabryelak and Matkovics (1984). Comparative studies on superoxide dismutase and catalase in livers of fish and other Antarctic vertebrates. Comp. Biochem. Physiol. 77C: 409-411.

Woodin. B.R. Smolowitz. R.M., Stegeman. J.J. (1997). Induction of cytochrome P450 IA in the intertidal fish *Anoplarchus purpurescens* by prudhoe Bay crude oil and environmental induction in fish from prince William sound. Environ. Sci. Technol. 31: 1198-1205.

Yokota. S. (1970). Comparative studies on the ultrastructure of hepatic microbodies. I. Hepatopancreas microbodies of the marine mollusca and crustacea. Zool. Mag. (Tokyo) 79: 296-301.

Zahn. R.K., Kurelec. B., Zahn-Daimler. G., Muller W.E.G., Rijavec. M.,Batel. R., Given. R., Pondeljak. V.R. (1982). The effect of benzo(a) pyrene on sponges as model organisms in marine pollution. Chem. Interact. 39: 205-220.

Zahn. R.K., Zahn. G., Muller. W.E.G., Kurelec. B., Rijavec. M., Batel. R., Given.R. (1981). Assessing sequences of a marine pollution by hydrocarbon using sponges as model organisms. Sci. Environ. 20: 147-169.

Zahn. R.K., Zahn-Daimler. G., Muller. W.E.G., Michaelis. M.L., Kurelec. B., Rijavec. M., Batel. R. (1983). DNA damage by PAH and repair in a marine sponge. Sci.Total Environ. 26:137-156.

Zhang. Y.S. Andersson. T. and Forlin. L. (1990). Induction of hepatic xenobiotic biotransformation enzymes in rainbow trout by β-napthcflavone. Time course studies Comp. Biochem. Physiol. 95(B): 247-253.

16

Multidimensional Problem of Environmental Refugees and National Obligation

Datta, Abhijit; Dutta, Sunita; Pandey, P.N., Kaviraj, Shanta

In recent years, millions of people are forced from their abode and homelands due to sudden catastrophes or from the combined effects of land abuse, overcrowding, industrialization, urbanization, pollution, sudden alteration in climate and other assaults on ecosystems throughout the world. These effects are further magnified by many factors including chronic poverty, economic failure, unsustainable development, faulty technology and depleted natural resources (Leiderman, 1995). Environmental Refugees here refers to those displaced by environmental degradation. Such degradation may be the result of economic activity, such as the displacement of fishermen due to pollution of the upstream, or displacement due to construction of a large dam. Such displacement is considered to be direct displacement. But if the displacement takes place as a result at a more interactive pattern of development, the resulting environmental refugees represent indirect displacement. A environmental refugee is thus a form of development-induced displacement that cuts across the distinction between direct and indirect displacement. Environmental refugees are people fleeing the environment, as they fear for there lives. They suffer a bleak future as their homeland may be totally destroyed, contaminated or transformed. In the end they have little or no hope of return.

One of the greatest environmental disasters was the Chernobyl nuclear accident in April 1986, on the Ukraine–Byelorussia borders (Anspaugh, 1988, Wilke, 1986). The bursting of the reactor exposed 400 million persons to dangerous radioactive substances hundreds of thousands fled the blast area (Lofstedt and White, 1990). In 1988 approximately 140,000 people evacuated were tested for radiation effects (Milne, 1988). More than 11,600 children and more than 2000 mothers were sent for medical treatment to Cuba from 1990 to 1993 (Martin, 1993).

The Chernobyl caused only a fraction of the environmental refugees in the world. But such disasters force us to realise the impacts such episodes leave behind. There are already may be at least 25 million environmental refugees worldwide, and millions more "proto-refugees" suffering from near-disaster condition (Leiderman, 1995)

The scientific community have at last come to realise, from growing number of articles, that environmental degradation is a root cause of refugee flow and that the number of displaced persons is much larger than the documented statistics. Jacobson (1988) noticed that "environmental refugees have become the single largest class of displaced persons in the world" Homer-Dixon (1991) stated that environmental degradation is likely to produce waves of environmental refugees that spill across borders with destabilizing effects both on the domestic environment and international relations. The Executive Director of UNEP, Mustafa Tolba (1985) stated that "these people are the millions fleeing the drought of Northern Africa. The victims of Bhopal and the thousands made homeless by Mexico earthquake. They are environmental refugees" (Leiderman1995). Official figure estimates the number of environmental refugees to be 17 million more. About half of these are most probably, in sub-Saharan Africa (Jacobson, 1988). Trolldalen *et al.*, 1992; Westing, 1992). Myers (1993) estimated it to be 25 million and the number are increasing rapidly. The Intergovernmental panel on climate change (IPCC,1990) was of the view that the gravest affect of climate Chang and green house effect may result into migration of millions of people who will be displaced due to shore line erosion, coastal flooding and agriculture disruption. Myers (1993) projected environmental refugees in a green house affected world (in year 2050) at 150 million persons.

People are displaced due to environmental degradation. They flee because they are forced to leave against their will as their options to stay there have exhausted. For example, those in the shadow of Bhopal disaster or Chernobyl flight was the only option left to survive.

In recent days the environmental refugees are considered a "threat to security". Belterton (1992) stated that U.S. military may be required to guard the borders with Mexico to prevent the immigration of fleeing environmental refugees from Third World (Linderman, 1995). It is not unethical to say that too many people pose a problem as people pollute, bring about congestion, cause irreversible degradation of the area and deplete the local resources. It will leave our children a more poorer, more crowded and more divided country (Lamm and Imhoff, 1985).

The purpose of this paper is to clear the concept of linkage between environmental degradation and population displacement and identify the myriad of issues related to this national and international problem. The purpose is also to identify the national obligations and what types of policy recommendations can be adopted to solve this deep-rooted human suffering.

Identifying Environmental Refugees

It will be unethical to restrict the concept of refugees to only those who run away from their countries but should also include those fleeing villages, cities, districts or bio-regions within countries, as environmental disasters are place specific. Nations have now started recognizing these "internal refugees" and their numbers are as high as those crossing the borders. This will enable us to realize the actual magnitude of humanity's uprooting.

There is a distinct difference between "refugees" and "migrants". While refugees are "pushed" from danger, migrants are "pulled" or attracted toward a brighter destination (Myers, 1995). The refuges not only flee from danger but they do not expect any support or protection from their government. Instead, they "experience abandonment, ambivalence, indifference, incompetence, or outright antagonism

(Leiderman,1995) . Sometime the government refuses to accept that such a problem exists or take the benefit of the episode to drive away unwanted people out of the state or country.

For example, Somalia's subsistence crisis was to mass-transfer thousands of refuges to Sudan (Newland 1995). Another problem that such refugee face is that they are not welcome everywhere and thus face trouble settling anywhere. Take the example of Chernobyl refugees. People resisted allowing irradiated refugees for fear of intermarriage consequences. On April 26, 1993 Byelorussian media announced that the rate of defects in newborns had increased by 18% and the number of oncological diseases had trebled (Goudie, 1994). In the process they become vulnerable, impoverished, desperate for signs of hope, and driven away by forces beyond their control (Leiderman, 1995).

The Role of Environment in Population Movement

(a) *The advocates*

Human beings have degraded the environment more than any other species in the world. The degradation reached beyond repair at places. They have endangered their own existence by their own actions and those of others. These adverse activities include land denudation, generating and dumping toxic wastes, indiscriminate mining, short cutting natural processes, disrupting hydrological cycles, and driving species to extinction. The disaster of such activities is magnified when undertaken in sensitive areas. But the greatest threat is from the massive scale at which such developmental activities are undertaken worldwide and the degradation surpass the pollution control prevention measures, ecological restoration and self recovery process. It has became a reality that environment fails to provide the shelter that we require. With the population crossing the ten billion marks by 2050, million will face trouble staying at the place where they are as the safety and resources diminish.

Myers (1995) estimated that there is approximately 25 million environmental refugees i.e., one per 200 persons. It might be doubled by 2010 and one hundred million more by 2050 if environmental degradation remains unabated, if global warming causes higher sea levels, stronger flood and storms and unfavorable weather patterns. Thus by 2050 there will one per 100 inhabitants.

The scientific community has at last started believing the causes that produce such massive scale of human migration. These include –landlessness, deforestation, desertification, soil erosion, salinization and water logging of irrigated lands, water scarcity and drought, agricultural stress, depletion of bio-diversity, extremes of climates, governmental shortcomings, war, toxic chemical, waste disposal, radiation and other pollution problems that scar the ecosystems.

Although there is growing awareness of the relation between environmental change and human movement, the traditional literature on migration has ignored the connection. In their report to the Trilateral Commission (International Migration Challenges in a New Era), Meissner *et at.* (1993) never mentioned environment or natural resources. Rogers (1992) proposed four key indicators for population movement but again environment was not mentioned (Leiderman, 1955)

- Population growth;
- Economic restructuring;
- Increasing economic disparities; and
- Increasing refugee flow.

Other recent reviews on the causes of irrigation fail to include environmental degradation or resource depletion as potential factors. This is in sharp contrast to the statements in The State of The World Refugees (UNHCR, 1993), which clearly recognizes environmental degradation as the root cause of population displacement.

Contrary to this is the growing number of literature advocating that the interrelationship between migratory responses and environmental degradation (Hall and Hanson, 1992; Kavanagh and Lonergar, 1992; Fornos 1993; Stoett, 1993; Lee, 1996; Suhrke, 1994). Emphasis has been given to the plight of "ecological refugees" or "environmental refugees" (El-Hinnawi, 1985; UNHCR, 1993). Unfortunately the World Commission on Environment and Development (WCED) considers the environmentally induced population displacement as a recent phenomenon (WCED, 1987). But there is little doubt that people have fled because their land has been degraded due to natural disasters, warfare, over-exploitation, recurrent droughts or extreme floods.

From the writings of EL-Hinnawi (1985) Jacobson (1988) and Myers (1993,1995) it becomes clear that environmental degradation will produce "waves of refugees". Suhrke (1994) labels this group the "maximlists". Supporting their agreements is the fact that environmental catastrophes such as floods, droughts and earthquakes are displacing huge number of people because population density in regions prone to such disaster, is increasing rapidly.

After its first inception in 1985 by El-Hinnawi in his United Nations Environment Programme (UNEP), the term "environmental refugees" is frequently being used in the contemporary literature on environment and development. It has been defined by El-Hinnawi (1985) as:

"........ Those people who have been forced to leave their traditional habitat, temporarily or permanently, because of marked environmental disruption (natural and/or triggered by people) that jeopardized their existence and/or seriously affected the quality of their life."

(b) The Contrarians

Despite all the claims, there has been little research to find out the actual role of environment in population movement. Is it possible to isolate environmental causes from the complex set of variables that affect human migration. The contrarians consider the argument presented by the "maximalists" as incorrect and based on superficial information. They consider the statistic presented by them are nothing but educated guesswork and have little empirical evidence to authenticate these data (Mougeot, 1992).

Secondly can the environmental issues be separated from socio-economic and political causes? It should be realized that destruction of the ecosystem is socially and spatially related. Except for a few, most industrial accidents have resulted in short-term displacement and only a few have been displaced permanently from their homes.

Development projects, although displaces large population, should be treated separately. These projects also involve resettlement programmes and are unlikely to produce "waves of environmental refugees", although the environmentalists and NGOs agree that such resettlement programmes are only a piece-meal approach.

It is also argued that the largest chunk of such refugees come from poorer countries of the world, like Bangladesh, Central America, Haiti, South Korea where the population growth and poverty have together magnified the problem and have made them susceptible to environmental disasters. (Ogata, the UN High Commissioner on Refugees, 1992). Some also claim that (Lee, 1996) disasters result in increased death rather than increased displacement.

Thus removing environmental process from the social, economic and political processes is next to impossible.

Link Between Environment and Displacement

Numerous examples are cited to link environmental change and human displacement. Common examples are Sahel in Africa, El Salvador, Haiti, Bangladesh (El Hinnawi, 1985: Hall and Hanson, 1992: Suhrke, 1992; Myers, 1995). These countries have no doubt experienced severe environmental stress but there are several other factors like rapid population growth, civil war, inequalities in land distribution, poverty etc. that was added to the process. The Kissinger Report of 1994 (Lonergan 2000) has linked the conflict of El Salvador to poverty and inequality, which has forced millions to migrate. It cannot be denied that sufficient environment damage has also taken place in the country, which might have triggered political and social instability resulting into economic stagnation and poverty.

Sahel is also severely affected by droughts and famine. But poverty, marginal agricultural land, war, inflation and landlessness have worsened the situation. The same is true to all cases, which are used as "evidence" of environmental refugees. Thus environmental degradation can be an important cause but not the sole cause in generating environmental refugees.

Environmental Refugees from Third World Countries

Although environmental refugees are worldwide there are some large pockets: 4 million in the Horn of Africa, 2 million in Sahel, about 3 million in rural areas of Central China and about half a million in Mexico and Central America (Myers,1995). The principal reason being landabuse, agriculture collapse, drought, starvation and civil war.

These places have some common characteristics. They are mostly places with high population densities, predominantly rural agricultural, poor soils, people are extremely poor and suffer from malnutrition, food and water shortage. They appear in cities, which are already overcrowded and have millions of jobless people, congested, filthy and are forced to live on bare subsistence. They are also forced from areas, which are targeted for rapid industrialization under national programmes. These projects are usually opposed but proceed anyway with little or no care for land destruction and/or refugees caused (Shiva, 1991). In India, statistics show that about 20 million have been displaced by official projects including dams, mines, forestpreserves and irrigation (International Organization for Migration and Refugee Policy Group, 1992).

In Egypt, more than a million hectares of soil were stripped from productive farms to make bricks for the Aswan Dam. Saddam Hussain and his activities have destroyed thousands of sq. km. forcing thousands to be trapped there or fleeing to Iran. In the vast Asian Desert, the former Soviet Union's thirst for water to irrigate and expand cotton farming has intercepted the headwaters of the Aral Sea. The sea has shrunk from 60,000 sq. km. to 50,000 sq. km. With the consequence that the salinity has risen tremendously destroying the fisheries that once has a capture capacity of 50,000 tons per annum. As a result the age-old fishermen's villages have been wiped out (Eustis and Micklin, 1993: Glanz and Zonn, 1991).

During Vietnam War in 1982, according to a UN study, 15000 sq. km. of mangrove forest had been destroyed.

Numerous bomb craters halted farming, forcing millions to become environmental refugees.

In Eastern Europe, during soviet rule, more than 100 villages and Town in Northern Bohemia (now in Czech Republic) were devastated and more than two hundred thousand people suddenly became environmental refugees. (Pinos, 1993).

The much take about Narmada Project (India), which envisages the construction of 30 big, 135 medium and over 3000 minor irrigation dams is expected to displace over two lakh tribals, devastate 40,000 hectares of prime teak forests and 84,000 hectares of cultivable land in the states of Gujrat and Madhya Pradesh. The Singrauli Super Thermal Project in Madhya Pradesh has already submerged or cleared several hundred hectares of land displacing a large number of tribals due to bar planning.

In the case of the Tehri Dam it is expected to displace 85,600 persons by submerging the town of Tehri and 95 villages, 72 of them partially (Saad Ali, 1991). An international conference in San Francisco in 1988 termed it the "second most dangerous irrigation project in the world", apart from the fact that it is located in a seismic zone.

Environmental Refugees from Developed Countries

Fewer environmental refugees are found in industrial countries than in the Third World. Moreover unlike huge "point" sources, there are smaller "non-point" sources. The refugees are generated mainly due to exposure to toxic chemical, air pollution, and contamination of water supply. Exposure to such contaminants can cause pathological symptoms and force victims to flee temporarily or permanently. An emerging problem is the "sick building syndrome", a complex condition where various factors like chemicals, pathogenic microbes and poor ventilation interact to make people so ill that they flee.

Apart from this, coming up of big cities in the coastlines, over earthquake zones, dense forests, mountain foothills may also contribute to refugee problem. There have been cases of refugee generation as a result of earthquake, landslides hurricane in Japan, Australia, Europe and U.S. The Hurricane Andrew, that struck the coast of Florida in 1992, created half a million environmental refugees.

Environmental Problems in Future

In 1990, the Intergovernmental Panel on Climate Change (IPCC) predicted that the greatest effect of climate change on population displacement will be sea-level rise, severe storms, coastline erosion, coastal hooding and agricultural disruption. The IPCC estimated that upto 360,000 km. of coastline might be affected in future. Based on this, Myers (1992) predicted "environment refugees" in a green house affected world to be 150 million by the year 2050. Scientists have predicted 13 million climate change environment refugees in Bangladesh, 16 million in Egypt, up to 73 million in China, 20 million in India and a million more in small countries and 50 million more for agriculturally dislocated areas by the year 2050 (Myers, 1995).

While predicting the future one should not forget the "adaptation mechanism". It is true that climatic changes will have significant influence in some regions, but these changes will take place slowly giving enough time to regions and communities to adapt. Although, most affected will be the poor with few options while the rich will have resources, means and government support to adapt faster in the face of environmental change. There is thus an urgent need for in-depth studies on the adaptation mechanism of different communities. How different regions cope with environmental stresses? Why resource scarcity in some regions do not cause major population immigration, etc?

What conclusion can be drawn from the above discussion?

Four general conclusions can be drawn for the above:

1. A generalized statement that environmental degradation results in population displacement overshadows a number of complexities making the prediction difficult.
2. It is very difficult to separate specific role of environmental degradation and human displacement, especially those that are "voluntary" in nature.

One should remember that displacement takes place due to combination of environmental, economic, social and political stigmata. Thus segregating environmental processes from other factors is not only difficult but also distortion of reality.

3. Various literatures assure that human movement from adverse situations is a certain means of relief from environmental pressure.

Although there is an old Chinese proverb which states that "of thirty ways to escape danger, running away is the best" (El Hainnawi, 1985) but it is not always true and movement may not always reduce the environmental stress. Rather one type of environmental stress may be replaced by another (economic, social, political or other environmental stresses).

4. One important point, which is usually overlooked is that whether such movements are of temporary or permanent nature.

Question thus arises whether migrants desire to return back to their homing ground, if that option is open? There are thus three important stages in the displacement process- (Lonergan, 2000)

- **Survival** – utilizing displacement as a need for acquiring relief from environmental stresses;
- **Recovery** – When migrants use their movement to recover from the immediate distress and consolidate their position; and
- **Improvement** - a migrant may use the opportunity for improving their position and future prospects, in which case a return to their home is not likely to occur.

A statistics of all the above attitudes will decide the magnitude of the problem and severity of the environmental crises.

The future National and International Politics to curb the severity of environmental refugees crises.

The four conclusions mentioned above pose difficulties in developing a scientific policy to deal with issues of environmental crises and population displacement. Taking into account these difficulties, two sets of recommendations can be formulated:

- Policy recommendation for assisting communities and areas under stress that may cause human migration;
- Specific policy recommendation for agencies involved in solving crisis.

Thus, despite of the complex nature of displacement, greater emphasis should be given in national developmental programmes. Major emphasis should be given or promoting sustainable development not only in the field of ecology but also in nation's economic and social policies for ensuring human security. Under this broad framework micro-policies should include:

- Development of a system that will help to anticipate displacement in case of environmental crisis,
- Identification of adaptation mechanisms, and how such mechanisms can be imbibed in vulnerable communities and regions.
- Development of case studies to understand how environmental degradation influences population movement and what can be done to assist those affected;
- Development of better working relationships among human rights, environment, population and NGO's related to such organizations;
- Development of politics to include refugees in the development programmes to assist the affected;
- Recognition of causes of environmental degradation and formation of policy framework to ensure minimum environmental impacts;
- Development of an international policy to assist those countries most vulnerable to future environmental change; and the environment.
- Recognizing the link between human rights and the environment.

It is thus clear that environmental degradation, resource depletion, population growth and inequitable distribution of income and/or resources lead to insecurity and population movement. According to the World Commission on Environment and Development (WCED, 1987)-

"Poverty is a major cause of global environmental problems. It is therefore futile to attempt to deal with environmental problems without a broader perspective that encompasses the factors underlying world poverty and international equality".

In conclusion it can be said that displacement is not only loss of land or homing ground. It involves a total loss of identity; an identity that had developed through generation amidst the surrounding ecosystem including the land, the forests, the tree and other biotic and a biotic components. In other words displacement degrades human dignity. The actual trauma of displacement can only be appreciated when human dignity receives its true respect. Steps should be taken to remedy the damage and prevent its recurrence.

REFERENCES

Anspaugh, Lynn R, *et al.*, (1988). The global impact of the Chernobyl reactor accident; Science, 24: Dec. 18.

El-Hinnawi, Essam (1985). Environ. Refugees. U.N. Environ. Programme, Nairobi, Kenya.

Eustis, Mark and Philip Micklin (1993). Anatomy of a disaster: A bird's eye view of a man-sized problem. Earthwatch, 12:4 (May/ June).

Fornos, Werner, (1992). Desperate Departures: The flight of Environ. Refugees. Washington DC. The Population Institute.

Glanz, M.H. and Igor Zonn (1991). [Aral sea] A quiet Chernobyl. The World & I Magazine, 6:9 (Sept.).

Goudie, Dong (1994). Private communication source: FESCHBACH@ guvax.acc. Georgetown. eda. March 29.

Hall. J. and A.Hanson, (1992). A New kind of sharing. Ottawa: IDRC Books.

Homer-Dixon, T. (1991). On the threshold: environmental changes as causes of acute conflicts. International security, 16(2): 76-116.

Jacobson, Jodi (1988). Environ. Refugees: A Yarstick of Habitability, Worldwatch Paper 86,

Jacobson, Jodi, (1988). Environ. Refugees: A yardstick of habitability Worldwatch Paper 86, Washigton DC. Worldwatch Institute.

Kavanagh, B. and S. Lonergan (1992). Environmental Degradation. Population Displacement and Global Security. Canadian Global Change Programme, Tech. Report.

Lamm R.and G. Imhoff. (1985). The Immigration time bomb: the fragmenting of America. N.Y. Refugees, 84: 11-12.

Lec, Shin-Wha (1996): In Limbo:Environ. Refugees in the Third world. Paper presented at the NATO Advanced Research Workshop on conflict and the environ, Norway.

Lofstedt R,E, and White, A.L. (1990). Cernobyl: four years later, The repercussions continue, Environment, 32: 3.

Lonergan, Steve (2000). The Role of Environ. Degradation in Population Displacement. Open Document http/://escp.sc-edu/ ecsplib. usf./

Martin, Luis and Linda Martin (1993). "Children of Chernobyl, Rich Winkel, Hawana Cuba. Apr. 23.

Meissner, Dois M, *et al.*, (1993). International Migration: Challenges in a New Era. 44, N.York. Trilateral Commission.

Meyers, N (1995). Environ. Exodus: An emergent crisis in the global arena, Wasluigton,D.C, Climate Inst.

Milne, Roger, (1988). Chernobyl disaster to become national park, New Scientist, Sept. 15.

Mongeout, Lue. (1992). Outmigration induced by environ. Degradation Environ, Dept. World Bank, Washington. D.C.

Myers, Norman (1993). Environ. Refugees in a globally warmed world, Bioscience, 43:11 Dec. 95., 52-61.

Newland, Kathleen, (1995). Environ. Degradation and refugee crisis. Source: Stuart M. Leiderman, 1995.

Pinos, Jan. (1993). Plight of Libkovice. Acid News, Oct.

Saad Ali (1991). Environ. Refugees—Multi Dimensional Problem. Hindu Survey of Environ; Hindu. pp. 34-35.

Shiva, Vandana (1991). Ecology and the Politics of Survival Conflicts Over Natural Resources in India. U.N. University Press, Tokyo, & Sage pub.

Stoett, P.J. (1993). Environ. Refugees: Conceptual problems and international mitigation. For Lonergan, 2000.

Stuart M, Leiderman, (1995). Environ Refugees and Ecological Restoration Environ. Response, 4th World Project Durham, 1995. web.http:// pab pages, unh.edu/v Leiderman.

Suhrke, Astri (1994). Environ. Degradation and Population Flows. Journal of International Affairs, 47(2).

Trolldalen *et. al,* (1992). Environ. Refugees- a discussion paper. World Foundation for Environ. and Develop and the Norwegion Refugee Concil. Oslo.

United Nations High Commission for Refugees (1993). The State of the World's Refugees: The Challenge of Protection, Geneva, Switzerland.

Westing. A.H., (1992). Environ. refugees: a growing category of displaced persons. Environmental Conservation, 19(3): 201-207.

Wilke, Tom and Roger Milne, (1986). The world's worst nuclear accident, New Scientist, May 1. 1986.

World Commission on Environ. and Develop, (WCED) (1987). Our common future, Oxford Univ. Press.

17

Potentials of Earthworm Resources in Environmental Amelioration

G. Tripathi and *P. Bhardwaj*

The increase in human population is increasing undesirable wastes in our environment. These are imposing a great waste disposal problem and creating health hazards. Recently, advancements have been made in order to properly recycle the hitherto untapped and so-called 'waste materials' through the earthworms by vermicomposting so as to make available cheap and high quality protein and nutrient rich source by ecologically sound means, yet keeping simultaneously check on pollution. Uses of earthworm are many and their advantages are manifold (Fig. 1). The integrated waste management-cum-vermiculture technology may be useful in improving the rural economy at the same time (Satchell, 1983; Dash and Senapati, 1986). The increasing awareness of the role of earthworms as bio-fertilizers in agricultural production, and cheap protein sources for fish, poultry and human feed has lead to the

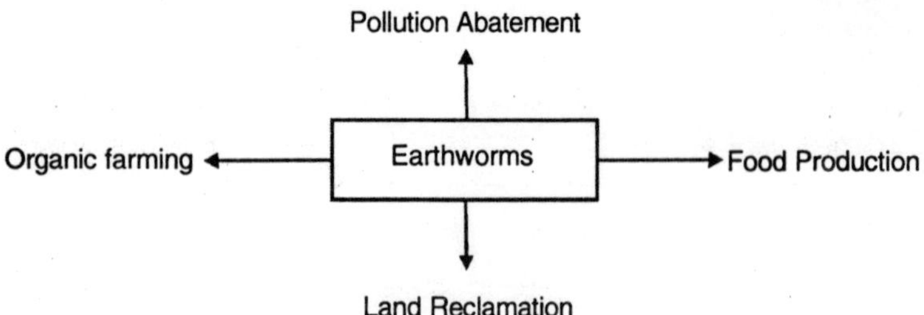

Fig. 1: Use of earthworms

development of suitable technologies in certain advanced countries for their mass culture on a commercial scale (Sabine, 1978). Vermicomposting in recent years has gained importance because of its more economic value over traditional methods of composting.

Vermiculture biotechnology provides a high grade nutrient rich compost of significant value as an organic biofertilizer, soil conditioner and tonics for plantation which is indeed key to overcome energy crisis, maintain the environment and improve the economy. One estimate is that 4 tons of vermicompost may replace the use of 40 tons cow manure by giving 36% high yield. Earthworms have the potential to replace chemical fertilizer to some extent and create a better environment for development and growth of useful organisms to fight pests and pathogens of plants. Developed countries are earning millions of dollars from vermicomposting programme. Earthworms have 70-80% protein on dry mass basis. This protein is of high quality and has a good balance of essential amino acids and is especially rich in lysine. Earthworms are used as a protein rich food sources for fishery, pigery and poultry industries. They are used as bait for fish throughout the world. The earthworms are highly resistant to many pesticides and heavy metals. They accumulate them in their tissues and may overcome the effects of such toxic chemicals by increasing mucus secretion, restricting the movements and increasing the reproductive potential upto certain concentration levels. Since the earthworms have the ability to take up in their tissues a number of unwanted chemicals, they may be used as bioindicators of pollution. These invertebrates that are designated as secondary decomposers in soil by their activity (Lavelle, *et al.,* 1997), also play a significant role in management and reclamation of degraded pedoecosystem. Earthworms contribute to distribution of surface litter, spatial heterogeneity, microbial activity (Bohlen, *et al,* 1997). The introduction of earthworms in degraded land enriches the exhausted soil with nutrients. The success of soil or land reclamation depends upon identification of proper species suitable for different agroclimatic zones of a country and occurrence of threshold number of earthworms in a particular zone. Such practices are already on the way to improve the quality of exhausted or degraded soil in some countries.

Conversion of Negative Assets Into Profit

Organic wastes, which are produced all over the globe, create major environmental and disposal problems. These wastes are mainly found from waste water, sewage, animal, urban and commercial sources. These organic wastes pose environmental problems like occupying vast areas, dispersing foul odours and forming breeding grounds for most of the pathogenic microorganisms and mosquito vectors. Under such a situation, recycling of wastes is the only alternative. Earthworms probably form the major soil invertebrates that contribute to recycling of organic matter or we can say that because of their food and feeding habits, earthworms are nature's most useful converters of wastes especially those of agricultural origin (Kale, 2001). The most efficient methods of waste management known to man was vermicomposting. This is the latest biotechnology wherein the earthworms are used in composting waste biomass. It helps to process waste simultaneously giving biofertilizers and protein for both livestock and aquaculture (Madan and Sharma, 1986). Various major research programs on vermitechnology were conducted all over the world with two main aims: (i) to convert various organic wastes into useful materials that could be used in agricultural lands to improve soil structure and fertility and which also would have considerable potential in horticulture as a plant growth medium or component of commercial potting compost; (ii) biomass production of worms from the worm-worked wastes and process them into a highly nutritious protein feed supplement for fish, poultry and pigs (Edwards, 1988) (Fig. 2). Earthworm casting or vermicompost are the pool of concentrated nutrients. Total nitrogen, organic matter, nitrate, nitrogen, phosphorus, potassium, sodium and magnesium were at a higher level in casting than in the surrounding soil (Dash and Patra, 1979, Reddy, 1983, Bhadauria and Ramakrishnan, 1989; 1996). The castings of *Pontoscolex corethrurus* were rich in soluble calcium and carbonates. Soluble carbonates contribute to the

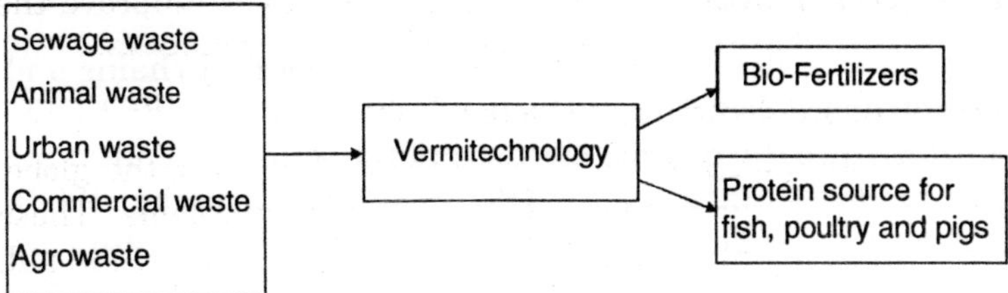

Fig. 2. Importance of vermitechnology

exchangeable base content of castings (Kale and Krishnamoorthy, 1980). Several recent works has been done on earthworm ecology of vermiculture, and the relationship among earthworms, waste recycle and soil fertility (Tomati and Grappelli, 1983; Tomati *et al.,* 1986; Edwards, 1988; Haimi and Huhta, 1995). Different laboratories have tried the possibility of utilizing earthworms to break down organic wastes, which have been the causative agents of much organic pollution. By regulating the moisture levels and mining the ingredients in proper ratios that can be accepted by the earthworms, coffee pulp (Arellano *et al.,* 1995) and pig solids (Domenquez and Edwards, 1991) could all be converted into good quality soil additives, along with the biomass production of earthworms. Vermicomposting facilities are reported to be already in commercial operation in Japan, Canada and USA. In Asia, vermicomposting is being efficiently practiced in the Philippines.

Earthworms such as *Eisenia fetida, Eudrilus eugeniae* and *Pcrionyx excavatus* are easily adaptable to agricultural wastes like sugarcane thrash, paper pulp, activated sludge, faceal of cow, sheep, horse and biogas sludges, of poultry dropping. These days sewage sludge and animal slurries are useful as soil conditioners and fertilizers due to their organic constituents and high levels of inorganic nutrients. The presence of earthworm in a sludge produces increased O_2 consumption, decreased anaerobic decomposition and increased mineralization (Mitchel and Horner, 1980). Many of these effects stimulate microbial activity by tunnelling and casting of the earthworms which increase aeration and surface area of the sludge.

Biological Indicator

Many toxic materials accumulate along food chains and the detritvore-decomposer levels are frequently the first to be affected since the organic matter and the soil are the ultimate sink for most contaminants (Reinecke and Reinecke, 1998). Conventional chemicals methods of analysis of pollutants are costly and give no indication of the bio-availability of material present in the environment. Alternative methods for this purpose have been proposed by the European Economics Community (EEC), OECD and also by FAO. Results indicate that since earthworm form a major component of the soil fauna as well as their sensitiveness to these toxic materials, they can serve as convenient bioindicators of pollution soils. Because earthworms are common in many soils they are vulnerable to the impacts on soil. They are selected for assessment of environmental risk for various reasons. They improve soil fertility and form food for a variety of animals. Not only are they general representative of soil fauna but they are also practical to use in both laboratory and field tests. Because of their relatively large size they are convenient to handle and can reach a relatively high biomass in some soils. Their behaviour brings them in close contact with the soil and they have short life cycles, allowing for long-term studies for successive generations.

Earthworms have been described as monitors of heavy metal pollution (Lee, 1985) and of organic pollutants. (Tarradellas *et al.,* 1982; Matrinucci *et al.,* 1983; Ebing *et al.,* 1984). They have been described as bioindicator of forest site quality (Muys and Granval, 1997). Some workers in their studies showed that pollution of water can be controlled economically by earthworm culture.

Removal of Toxic Metals

The industries and the power plants have been bleaching out volumes of toxic chemicals and effluents including heavy metals in the environment. Many groups of pesticides including organophosphate, carbonates and fungicides often based on heavy metal ions, contribute significantly to environmental pollution and thus adversely effect other organisms including human health. The earthworms are highly resistant to many pesticides and heavy metals. Their body can detect the adversities in the environment (Kale, 2001). They have the capacity to accumulate pesticides and heavy metals in their tissue (Fig. 3). The studies have been shown that the earthworm concentrate cadmium, cobalt, mercury, zinc and lead. The accumulation of toxic chemicals in earthworm tissue is

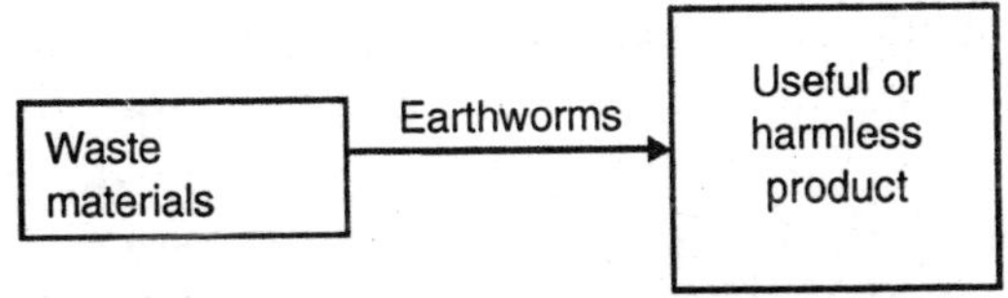

Fig. 3. Earthworm as bioindicator of pollution

ecologically important because they are the important component in the food chain of several species of birds and mammals. The bioaccumulation of heavy metals or other toxicants in the tissue of earthworms depends on the soil properties, pH, calcium concentration and so on. Peramaki *et al.*, (1992) have clearly demonstrated the importance of pH and soil calcium. Their results showed an increasing accumulation of cadmium in the earthworm *Aporrectodea caliginosa* at lower pH values and confirm the finding of Ma *et al.*, (1982) on the importance of pH. Therefore, the use of earthworms as biological monitors to determine the concentration of heavy metals and other pollutants in their tissue and to estimate the degree of pollution is of great significant value in controlling environmental pollution.

Impact Assessment

Legislation in many countries has recently focused the attention of workers on the need for sensitive organisms from the soil for the use in research and environmental monitoring (Reinecke and Reinecke, 1998). No doubt earthworms may play an important role in ecotoxicological evaluation. The effect of chemicals on earthworm and methods adopted for investigating these effects have been reviewed by Satchell (1955), Dowey (1963), Edwards and Thompson (1973) and Edwards and Bohlen (1992). An important aim in earthworm ecotoxicology is to be able to predict effects of harmful chemicals in the field on the basis of laboratory experiments. The effects of chemicals on earthworms have been assessed in the laboratory by using pots, boxes or other containers. These tests produce more consistent and reproducible results because here standard number of test earthworms of a single species are used and they are forced to have intimate contact with the toxic compounds well documented laboratory protocols for earthworm reproduction tests have been developed but little is known of effects on earthworm reproduction in the field. In contrast to laboratory tests, it is extremely difficult to assess the relative toxicity of different pesticides in the field. This may be due to the considerable variability between test sites, soils, size of plots, formulation dose methods of application and type of crop grown (Edwards and Bater, 1992). Although field tests have been adequately conducted to identify those chemicals which are extremely toxic to earthworm, it is much more difficult to identify accurately, compounds which are only moderately or slightly toxic to earthworms. Much effect has been put into earthworms test development (Eijsackers, 1998). The Organization for Economic Coordination and Development (OECD) and the International Standards Organization (ISO) have carried out a number of tests and standardization studies. In a recent study for OECD on the selection of a set of laboratory ecotoxicity test, the earthworm test secured 6th position for the OECD tests 13 for the test of environmental protection Agency (EPA), and 18 for the ISO tests. Standardization of earthworms for toxicity tests has also been done by the National Institute for Human Health and the Environment (RIVM-NL) and the International Organization for Biological Control (IOBC). If more ecologically relevant earthworm tests should be accepted, a lot of work has to be done. Only in this way an ecological sound risk assessment of environmental contaminants and potential contaminating activities and situations be achieved (Eijsackers, 1998).

Improvement of Degraded Land

Modern agricultural practices are resulting in an increasing impact on environment, causing a various decline in the natural resources of arable lands. Management of natural resources on a sound ecological basis confronts us with problems of a biological origin (Christensen, 1988). The land bioreclamation deals with processes, which determine the levels of organisms and is the subject of the action of living things. Therefore, it is inevitable that the biological methods in land reclamation provide food possibilities for upgradation of conservation of soil fertility on a sustainable basis. Due to close coupling of earthworms to processes and functions of ecosystems. Land reclamation by earthworm is an exciting approach to improve the degraded lands in various parts of the globe. Barrett (1949) was the first who reported that the soil structure and fertility can be increased by manipulating earthworm populations in the field or by adding earthworms to soil with no earthworm present. The ways in which earthworms contributed to soil turnover, structure and fertility have been summarized by Edwards and Lofty (1977), Lee (1985) and Edwards and Neuhauser (1988).

They form extensive burrows by literally eating their way through the soil and deposit the ingested material as surface casts. In this way, they mix large quantities of soil from deeper layers to the surface and take down decaying organic matter into the soil. Earthworms also increase the aeration and water holding capacity of the soil. The earthworms are beneficial in terms of plant growth and crop yields in lacking soils (Fig. 4).

The introduction of earthworms of new habitats has been reported by several workers in New Zealand and Europe that lacked native earthworms (Stockdill, 1959, 1982; Stockdill and Cossens, 1966; Marinissen, 1991; Stein *et al.*, 1992). Ghilarov and Mamajev, (1966) inoculated earthworms into reclaimed

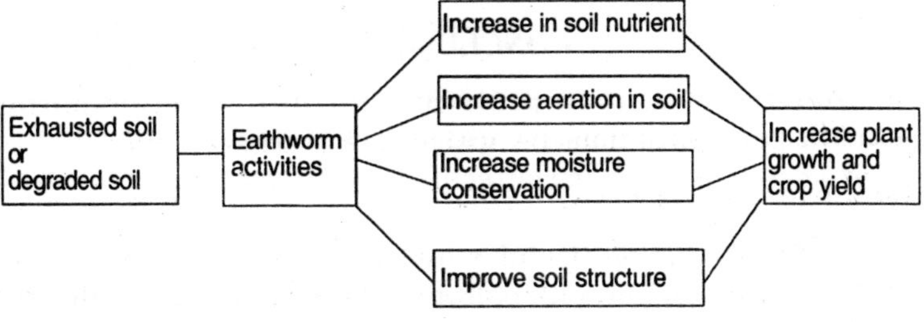

Fig. 4. Role of earthworm in improving degraded land

irrigated land in Uzbekistan and reported considerable improvements in soil structure and fertility. The introduction of mixture of *Aporrectodae caliginosa, A. longa, Lumbricus rubellus* and *L. terrestris* in polder grasslands has increased 10% productivity in the Netherlands (Hoogerkamp *et al;* 1983). Edwards and Lofty (1980) inoculated *L. terrestris* and *A. longa* into sterilized soil in England and recorded increased root growth and grain yield in direct drilled cereals. They demonstrated clearly that the increased root growth after earthworm inoculation was due to the formation of channels that were lined with more available nutriends than the surrounding soils. This was due to casting materials rich in plant nutrients by earthworms around the circumference of their burrows. Julka and Mukherjee (1986) demonstrated the potentiality of earthworms in decomposition of organic matter of sterilized soil. The success of the land or soil reclamation depends on identification of proper species suitable for different agro-climatic zones of a country and then threshold numbers. In Indian conditions *Polypheretima elongata* is suitable for soil reclamation. This species produces more mucus which binds available nutriends. It helps in nutrient restoration and moisture conservation (Singh and Rai, 1998). When the earthworms die during unfavourable conditions, their microbial decomposition releases considerable amount of nitrogen and other nutrients (Julka, 1988). It has been estimated that the decomposed worm tissue release nitrogen as 25% in the form of nitrate, 45% as ammonia, about 3% as soluble organic compounds, and rest 27% is unaccounted (Satchell, 1967). Thus introduction of earthworms in agricultural land improves the yield of most of the crop.

CONCLUSIONS

Agricultural, food processing or any cellulose based industry generates massive quantities of wastes. Recycling of these wastes through vermitechnology reduces the problem of non-utilization of agrowastes and turn negative assets into profits, by producing useful materials and at the same time minimizing environmental pollution vermitech i.e., introducing earthworms into soil or to form vermicompost is the most natural and best among all sustainable agricultural practices.

Nutrients in earthworm castings are predigested and hence are readily soluble in water for uptake by plants. Uncultivated land may be reclaimed for agricultural use by the inoculation of earthworms. They make exhausted soils more compact and its poor structure changes into deep friable top-soil, which in turn is beneficial for plant growth and crop yield. Vermicomposting biotechnology is very useful in order to minimize environmental pollution and obtain good organic fertilizers. It also gives high quality protein for both livestock and aquaculture. Since earthworms form a major component of the soil fauna as well as their sensitiveness to toxic heavy metals, pesticides and other lethal materials, they can serve as convenient bioindicators of polluted soils. In short we can say that earthworms are known as compost manufacturer, protein producer, land reclaimer and environmental bioindicators.

ACKNOWLEDGMENTS

The authors are grateful to the Department of Biotechnology (DBT), New Delhi, for providing financial support.

REFERENCES

Arellano, R.P., Barois, I. and Arand, E. (1995). Earthworm carrying capacity for coffee pulp using *Eisenia andrei* and *Perionyx excavatus. Soil Biol. Biochem.*

Barrett, T.J. (1949). Harnessing the Earthworm. Faber and Faber, London.

Bhadauria, T. and Ramakrishnan, P.S. (1989). Earthworm population dynamics and contribution to nutrient cycling during cropping and fallow phases of shifting agriculture (Jhum) in North-east India. *J. Appl. Ecol.,* 26: 505-520.

Bhadauria, T. and Ramakrishnan, P.S. (1996). Role of earthworms in nitrogen cycling during the cropping phase of shifting agriculture (Jhum) in North-east India, *Biol. Fertil. Soil.,* 22: 350-354.

Bohlen, P.J., Parmelee, R.W., McCartney, D.A. and Edwards, C.A. (1997). Earthworm effects on carbon and nitrogen dynamics. Applications, 4: 1341-1349.

Christensen, O. (1988). The direct effects of earthworms on nitrogen turnover in cultivated soil. *Ecol. Bull.* (*Copenhagen*), 39: 41-44.

Dash, M.C. and Patra, U.C. (1979). Worm cast production and nitrogen contribution to soil by a tropical earthworm population from a grassland site from Orissa, India. *Rev. Ecol. Biol. Soil.*, 16: 79-83.

Dash, M.C. and Senapati B.K. (1986). Vermitechnology, option for organic waste management in India. In: *Proc. Natl. Sem. Org. Waste Utilize.* Part B: Verms and vermicomposting (M.C. Dash, B.K. Senapati and P.C. Mishra, eds.), pp. 157-172. Five Star Printing Press, Burla, India.

Dominquez, J. and Edwards, C.A. (1991). Effect of stocking rate and moisture content on the growth and maturation of *Eisenia fetida* in pig manure. Paper presented at ISEE5.

Dowey, S.P. (1963). Effects of chemicals on earthworms. A review of literature, special scientific report. Wildlife, No. 74, Washington.

Ebing, W., Pflugmacher, J. and Haque, A. (1984). Der Regenwurm als Schlusselorganisms zur Messung der Bodenbelastung mit organischen Fremdchemi Kalien. *Berichte uber Landwirtschaft,* 62: 222-255.

Edwards, C.A. (1988). Breakdown of animal, vegetables and industrial organic wastes by earthworm. In: *Earthworms in Waste and Environmental Management* (C.A. Edwards and E.P. Neuhauser, Eds.), pp. 21-31. S.P.B. academic, The Hague.

Edwards, C.A., and Bholen, P. (1992). The effects of toxic chemicals upon earthworms. *Rev. Environ. Contam. Toxicol.*, 125: 23-99.

Edwards, C.A. and John Bater, E. (1992). The use of earthworms in environmental management. *Soil. Biol. Biochem.*, 24: 1683-1689.

Edwards, C.A. and Neuhauser, E.F. (1988) Earthworms in waste and environmental management, p. 388 S.P./B. Academic, The Hague.

Edwards, C.A. and Thompson, A.R. (1973). Pesticides and the soil fauna. *Residue Reviews,* 45: 1-79.

Eijsacker, H. (1998). Earthworms in environment research: still a promising tool. In: *Earthworm Ecology* (C.A. Edwards, ed.), pp. 295-323, CRC Press, The Netherlands.

Hoogerkamp, M., Rogaar, H. and Eijsackers, H.T.P. (1983). Effects of earthworms on grassland on recently reclaimed polder soils in the Netherlands. In: *Earthworms Ecology—from Darwin to Vermiculture* (J.E. Satchell, ed.), pp. 85-105. Chapman and Hall, London, England.

Kale, R.D. (2001). The functional role of Earthworms in agroecosystems. Workshop on Tropical Soil Biology and Fertility

Programme. School of Environmental Sciences, J.N.U., New Delhi, pp. 1-5.

Kale, R.D. and Krishnamoorthy, R.V. (1980). Enrichment of soil fertility by earthworm activity. In: *Progress in Soil Biology and Ecology in India* (Eds.) G.K. Veeresh, pp. 64-68.

Kale, R.D., Seenappa, S.N. and Rao, C.B.J. (1994). Sugar factory refuse for production of vermicompost and worm biomass. Paper presented at ISEE5.

Kreis, B., Edwards, P., Cenet, G. and Tarradellas, J. (1987). The dynamics of RCBS between earthworm populations and agricultural soil. *Pedobiologia.,* 30: 379-388.

Lavelle, P., Bignell, D., Lepage, M., Wolters, V., Roger, P., Ineson, P., Heal, O.W. and Dhillion, S. (1997). Soil function in a changing world: The role of invertebrate ecosystem engineers. *European J. Soil Biol.,* 33:151-193.

Lee, K.E. (1985) Earthworms: Their Ecology and Relationships with Soil and Land Use. Academic Press, Sydney.

Ma, W.C., Edelman, T., Van Beersum, I. and Jans, T. (1982). Uptake of cadmium, zinc, lead and cooper by earthworms near a zinc-smelting complex: influence of soil Ph and organic matter. *Bull. Enviorn. Contam. Toxdicol.,* 30:24-427.

Madan, M. and Sharma, N. (1986). Recycling of organic wastes through vermicomposting. *Bioenergy Newsletter,* 2:30-32.'

Marinissen, J.C.Y. and de Ruiter, P.C. (1993). Contribution of earthworms to carbon and nitrogen cycling in agroecosystems. *Agric. Ecosys. Environ.,* 47:59-74.

Martunucci, G.B., Crespi, P., Omodeo, P., Osella, G. and Tralds, G. (1983). Earthworms and TCDD (2, 3, 7, 8, tetra-chlorodibenzo-p-dioxin) in Seveso. I: *Earthworm Ecology* ((J.E. Satechell, ed.), pp. 275-283. Chapman & Hall, London.

Mitchell, M.J. and Horner, S.G. (1980). Decomposition processes in sewage sludge and sludge amended soil. In: *Soil Biology as related to land use practices* (D.L. Dindle, ed.), *Proc. 7th Int. Colloq. Soil Zool.,* New York: 129-138. Office of pesticides and toxic substances, EPA, Washington, D.C.

Muys, B. and Granval, P.H. (1997). Earthworms as bioindicators of forest site quality. *Soil Biol. Biochem.,* 29:763-766.

Peramaki, P., Itamies, J., Karttunen, V., Lajunen, L.H.J. and Pulliainen, E. (1992). Influence of pH on the accumulation of cadmium and lead in earthworms (*Aporrectodea caliginosa*) under controlled condition. *Ann. Zool. Fennici,* 29: 105-111.

Reddy M.V. (1983). Annual cast production by the megascolecid earthworm, Pheretima alexandri (Beddard). *Comp. Physiol. Ecol.,* 8:84-86.

Reinecke, A.J. and Reinecke, S.A. (1988). The use of Earthworms in ecotoxicological evaluation and risk assessment: New approaches. In: Earthworms Ecology (A.C. Edwards, ed.), pp. 273-293. CRC Press, The Netherlands.

Sabine, J.R. (1978). The nutritive value of earthworm meal. In: *Utilization of soil organisms in sludge management* (R. Hartenstein, ed.), pp. 122-130, National Institute of Technical Information Services, Springfield, Virginia, pp. 286-932.

Satchell, J.E. (1955). Some aspects of earthworm ecology. In: *Soil Zool.* (D.K. McEkevan, ed.), pp. 180-201.

Satchell, J.E. (1967). Lumbricidae. In: *Soil Biology* (A. Burges and F. Raw, eds.), pp. 259-322. Academic Press. London, England.

Satchell, J.E. (1983). Earthworms Ecology—from Darwin to Vermiculture. Chapman and Hall, London, England, p. 495.

Singh J. and Rai, S.N. (1998). Potential of earthworms in sustainable vermiculture. *Yojana* (November, 1998), pp. 10-12.

Stein, A. Bekker R.M., Blom, J.H.C. and Rogaar, H. (1992). Spatial variability of earthworm populations in a permanent polder grassland. *Boil. Fertil. Soil,* 14:262-266.

Stockdill, S.M.J. (1959). Earthworms improve pasture growth. *New Zealand J. Agricul.,* 98:227-233.

Stockdill, S.M.J. (1982). Effects of introduced earthworms on the productivity of New Zealand pastures. *Pedobiologia,* 24:29-35.

Stockdill, S.M.J. and Cossens, G.G. (1966). The role of earthworms in pasture production and moisture conservation. *Proc. New Zealand Grassland Association,* pp. 168-183.

Tarradellas, J., Diercxsens, P. and Boucha, M.D. (1982). Methods of extraction and analysis of PCB's from earthworms. *Int. J. Environ. Anal. Chem.,* 13:55-67.

Tomati, U., Galli, E., Grappeli, A. and Di Lena, G. (1986). Effect of earthworm cast on protein synthesis in Radish (*Raphamus sativum*) and Lettuce (Lactuca satoiva) Seedlings. *Biol. Fertil. Soils,* 9:288-299.

Tomati, U. and Grappeli, A. (1983). *Proceedings of Int. Symposium on Agricultural and Environmental Prospects in Earthworm Farming,* Rome. Tipografia Euromodema.

18

Potential Predators for Aphid Management

Omkar and *Geetanjali Mishra*

Aphids are tiny, pear shaped, siphunculi bearing insects belonging to the family Aphididae of the order Homoptera. These are serious pests of numerous agricultural and horticultural crops owing to their phytophagous and sap-sucking nature. They have been a source of great fascination and frustration to man for a very long time because of their intricate life cycles, close association with their host plants, polyphenism and their ability to reproduce both sexually and asexually.

There are about 7,757 species of aphids (Misra and Kurl, 1979) in comparison to 10,000 grasshopper and 60,000 weevil species. However, display of polyphenism by the aphids takes care of the lack in species diversity. The greatest number of aphid species occur in the temperate regions where one plant species in four is found infested (Dixon, 1998).

The damage causing abilities of the aphids can be assessed from the 80-96% damage that they are reported to cause to the mustard crop in India (Suri *et al.*, 1988). Reports of other severe losses also abound (Anonymous 2003a; Anonymous, 2003b). Other than the direct losses caused by the direct feeding by aphids and their sap sucking habit, aphids also cause indirect effects by acting as vectors of viruses and also lead to fungal attacks on the infested plants (Anonymous, 2003c).

As a result of the multifaceted damages caused by aphids, management of their populations is a matter of prime importance. Traditionally, synthetic pesticides were used for the purpose. However, due to the numerous drawbacks such as ecodegradation, health hazards, development of resistance among non-target species, resurgence of new species as pests, *etc*., the use of synthetic pesticides is now being frowned upon and the search for alternate pest management techniques is on, the world over.

During the search for an alternative pest management technique, attention of workers turned to the many natural enemies such as predators, parasitoids and pathogens that are commonly found in aphid colonies, predating or parasitising them. Though, a debate is on among biocontrol specialists on the efficacy of parasitoids versus pathogens versus predators, we have chosen to deal with predators owing to their high visibility, easy mass multiplication, easy releases and, of course, their high potential as biocontrol agents.

The predators most commonly found in aphid colonies are chrysopids, coccinellids, syrphids, anthocorid bugs, assassin bugs, chickadees, spiders and empid flies (Anonymous, 2003d; Dixon, 1998). Of these, we will be concentrating on chrysopids, coccinellids and syrphids as these are found most commonly and abundantly in aphid colonies. The present chapter will take into account their general biology, growth, development and reproduction, prey consumption, fitness in guild, *i.e.* intraguild predation and the releases of the three predatory groups. A comparative analysis on all the aspects followed by a conclusion would help identify the most suitable predatory group for the management of aphid populations.

CHRYSOPIDS

Chrysopids or lacewings belong to the family Chrysopidae of order Neuroptera, with the common name being an allusion to the diaphanous, lacy wings that they bear. In chrysopids, most adults are non-predatory while all larval instars are predatory (Figure-1). They attack and consume a wide variety of pests, including aphids, chinch bugs, mealybugs, scale insects, whiteflies, leafhoppers, lepidopterous eggs and larvae, and mites (Hydorn, 1971). Lacewing larvae are also known as *aphid lions* owing to their effective and deadly predatory approaches towards aphids.

(a)

(b)

(c)

Fig. 1. *(a)* Larva of chrysopid feeding on an aphid

(b) Adult coccinellid consuming an aphid

(c) Larva of a syrphid feeding on an aphid

General Biology

Adult chrysopids are medium-sized (7-15 mm), yellowish-green to grey insects with red, yellow or brown markings with cosmopolitan distribution. They have large copper coloured compound eyes, long filiform antennae and large lacy wings, which are held roof-like over their body (Omkar and Mishra, 2003). Most of the adults depend upon pollen, nectar and insect honeydew for sustenance (*Chrysopa carnea*), with only a few being predaceous (*Chrysopa commata, C. oculata* and *C. perla*).

Sixty-five species belonging to 21 genera have been recorded from various crops in India. Amongst them *C. carnea, Apertochrysa crassinervis, Mallada boninensis and Mallada astur* are the most common ones in India. *C. carnea* and *Chrysoperla rufilabris* are among the most common commercially available global species of chrysopids (Burke and Martin, 1956; Whitcomb and Bell, 1964). *C. carnea* adults are commonly known as the *green lacewings.*

Sexually mature adults contact each other by substrate-borne signals or "calls" produced by abdominal vibrations. These calls are species specific and induce reproductive isolation. After a melodious courtship display, the adults mate and deposit eggs on long, slender stalks, singly or in groups of 20-30 (Figure-2). Each female *C. carnea* lays several hundred eggs, which hatch in 4-12 days depending on the abiotic factors prevailing in the microcosm. There are three mottled, grayish-brown, alligator shaped larval instars (3/8 inch long before pupation) during a developmental period of about 11-12 days (20°C and 75% R.H.) (Figure-2), which are predatory in nature (Figure 1). Pupation takes place in silken cocoons, in concealed places and in the soil. The pupal stage lasts about 19 days after which the adults of *C. carnea* emerge out.

REPRODUCTION

Reproduction is one of the most important aspects of any organism's biology as it helps in the furtherance of generations. Reproduction in most of the organisms can be divided into courtship, mating and production of progeny. As the chapter

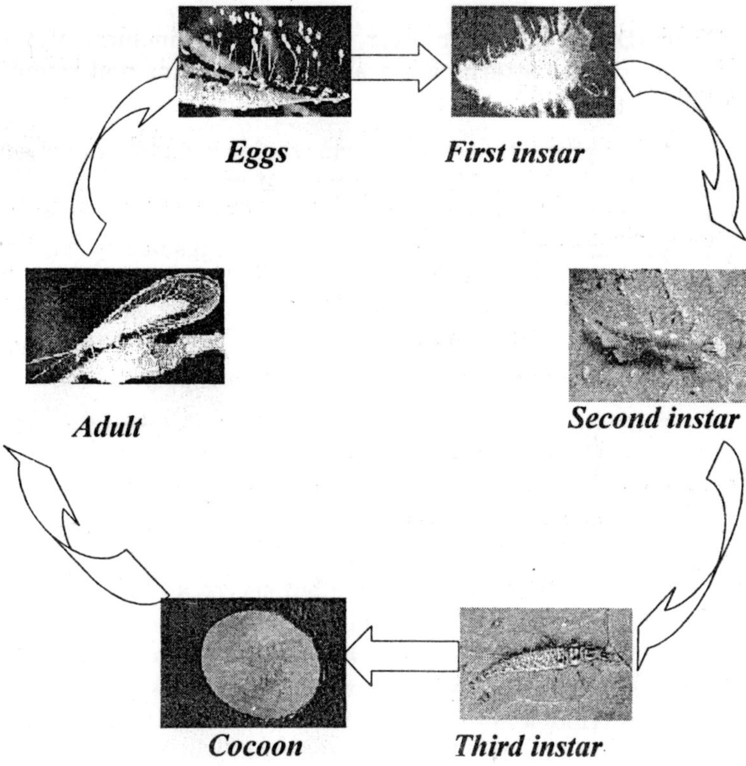

Fig. 2. Life cycle of a chrysopid

deals with the characteristics of predators relevant to their application in aphid management programmes, the behavioural aspect of courtship has not been dealt with despite its intrinsic interest value.

Reproductive Biology

Matings in chrysopids are usually preceded by courtship duets between the two sexes. Post-mating, the females deposit eggs on long slender stalks either singly or in groups of 20-30. Virgin females too lay eggs although infertile with most lacking pedicels/stalks (Ribeiro and Carvalho, 1991). Oviposition in most predators usually follows a triangular pathway, *i.e.* the daily oviposition increases and after reaching a peak begins to decline (Dixon, 2000). However, unlike the coccinellids, studies on oviposition patterns are scarce in chrysopids and require greater attention.

The reproductive output of chrysopids is highly influenced by a number of factors, *viz.* number of matings, the quality, the quantity of prey and abiotic factors such as temperature, photoperiods and humidity. The number of eggs laid by females that were paired with males for a lifetime were found to be higher (1139.01) than those laid by females that were isolated after a single mating (531.59) (Ribeiro and Carvalho, 1991) and is probably a result of the increased sperm supply.

Prey quality plays a major role in determining the reproductive output of chrysopids. Hodek (1959; 1962) classified prey on the basis of their suitability and specificity to the predators. The prey capable of sustaining a good level of larval development and oviposition were termed as *essential food* and the ones which were accepted but served as a mere source of survival and not of sustenance of generations were termed as *alternative food.* The term, *rejected prey,* was later added in context of prey that were unpalatable and unsuitable to the predators (Hodek and Honek, 1996). Malcolm (1992) used the term *included* for those food sources that can be successfully exploited, *peripheral* for those that are consumed but which result in a significant decline in the biological attributes and *excluded* for those that are not consumed as a result of their high toxicity and unpalatibility. Impinging on the essential and alternative food range are the borderline foods that ensure very good survival and minimal reproduction, these are known as *accepted foods.* There are certain alternative and rejected prey which are *toxic* in nature.

When adults of *C. carnea* were provided with pollen of different plants, the fecundity was found to be highest on castor (*Ricinus communis*) and was twice the least fecundity obtained (Deotale *et al.,* 1998). There was a many fold increase in the fecundity when females of *C. carnea* were fed on composite food comprising flowers of cotton, okra, ganja, maize and castor (781.60 eggs/female). This was followed by pollen grains of castor (269.60 eggs/female) as against honey (25.40 eggs/female)

(Gautam and Paul, 1988). Adults of *C. carnea* reared on aphids, *viz. Aphis gossypii* and *Myzus persicae* laid a higher number of eggs on *M. persicae* (103.00 eggs/fem) than on *A. gossypii* (84.70 eggs/fem) (Mannan *et al.*, 1997). Eggs of grain moth, *Sitotroga cerealella* resulted in an extremely high fecundity of 1082.58 eggs/female *Chrysoperla externa* (Costa *et al.*, 1999). The pre-oviposition period was also found to vary with the food supply. It was shorter on flowers of ganja (3.60 days) than on castor (6.80 days) (Gautam and Paul, 1988). Reproductive periods increase and non-reproductive periods, *viz.* pre-oviposition and post-oviposition periods decrease with the increase in nutritional quality of prey supplied (McEwen *et al.*, 1996; Mannan *et al.*, 1997; Macedo *et al.*, 1999).

In *C. externa* and *Chrysopa mediterranea*, females that fed on yeast plus honey (1:1) had a greater fecundity than those fed on the other diets. Females of *C. externa* laid an average of 2,304 eggs during 84.5 days, while females of *C. mediterranea* laid 2,160 eggs in 103.3 days at 25°C and 70-80% R.H. (Carvalho *et al.*, 1996). The prey dependent fecundity has been presented below (Table-1), which reveals that *Trialeurodes vaporariorum* is the most suitable prey for optimum reproductive output, while *A. gossypii* the least.

Table 1. Fecundity of *C. carnea* females on different prey

Food	*Fecundity (eggs / female)*	*References*
Trialeurodes vaporariorum	818.16	Yoldas, 1994
Macrosiphum euphorbiae	750.66	Yoldas, 1994
Aphis craccivora (cowpea)	318.40	Saminathan *et al.*, 1999
Aphis gossypii (okra)	271.20	Saminathan *et al.*, 1999
Aphis gossypii (guava)	266.80	Saminathan *et al.*, 1999
Aphis gossypii (cotton)	262.40	Saminathan *et al.*, 1999

Apart from prey quality, the onset of oviposition in lacewings was also influenced by abiotic factors, *viz.* photoperiod and temperature. The females laid eggs a week after emergence at LD (light: day) 16:8. It was extended up to 2 months at LD 12:12 or 8:16 *i.e.* medium and short days, respectively. Fourteen hours of light was the threshold for oviposition without diapause (Canard *et al.*, 1994) and reproductive rates were found to be relatively high and directly related to temperature between 15.6 and 26.7°C (Albuquerque *et al.*, 1994). Thus, the reproductive activity of the chrysopids is subject to a number of abiotic and biotic factors.

Oviposition Strategies

It is widely accepted that females, in general, prefer to oviposit at sites that are suitable and stable enough to ensure a secure future for their progeny (Kindlmann and Dixon, 1993). The experiments conducted to assess the validity of this hypothesis in lacewings have showed supportive results (Ruzicka, 1996; Ruzicka, 1998; Freitas and Scalloppi, 1996; Ballal and Singh, 1999; Bakthavatsalam *et al.*, 2000).

More eggs of *C. carnea* were found on maize plants sprayed with varying concentrations of molasses although the distribution of eggs was not affected and they were found at the base of plants and on the undersides of leaves (Freitas and Scalloppi, 1996). Wind tunnel studies indicated that males of *C. carnea* had a significantly higher preference for sunflower and females for both sunflower and cotton, while pigeonpea was least preferred. Females of *M. boninensis* responded to cotton while neither males nor females of chrysopid, *M. astur* responded to any of the host plants. Similar degree of ovipositional response of the three chrysopids to the three plants was also observed (Ballal and Singh, 1999). Adults of *C. carnea* responded well to volatiles from flowers and bolls of cotton infested by *Heliothis armigera,* while the response to infested leaves was very poor (Bakthavatsalam *et al.*, 2000).

Substrates contaminated with intra and interspecific abdominal secretions of first instars of *Chrysopa oculata* and *Chrysopa perla* deterred both the females from ovipositing. Also, it was noted that the number of larvae infesting a patch did not make a difference but rather the time spent on the patch caused a differential response (Ruzicka, 1996). Females of species predatory as adults (*C. commata, C. oculata* and

C. perla), and also species not predatory as adults (*C. carnea*) oviposited more on clean substrates than on those exposed previously to conspecific or heterospecific first-instar larvae (Ruzicka, 1998). These observations indicate the inbuilt adaptive strategies of the chrysopids in ensuring the future of their progeny.

FORAGING

The foraging behaviour of the chrysopids is an integral part of detailed studies, as it demands major consideration and attention while discussing their applicability as bioagents. Optimal foraging behaviour represents a compromise between benefits achieved (*e.g.*. finding of food, mates or shelter) and costs involved (*e.g.*. risk of mortality, energy consumption) (Bell, 1990). Theoretically, a forager should try to achieve a maximum of energy gain or reproductive success at a minimum of expense (Charnov, 1976).

Predators usually follow a well-defined path while searching for prey sites. These search paths are initially extensive in nature but turn more intensive (reduced mean walking speed, an increased mean turning frequency and an increased mean turning angle) in the presence of visual and chemical signals, indicating prey sites. These switchovers from extensive to intensive occur at three levels: search for habitat, search for plant and then search for prey patch.

Attraction to Prey Sites

The simultaneous action and intermingling of visual and chemical cues guides the predator towards the food source. There is evidence in support of attraction of chrysopids towards colours and odours. Yellow colour was found to attract the chrysopids most followed by green and red (Capinera and Walmsley, 1978). The preference for yellow and green indicates the attraction towards plant foliage while red probably indicates the role of flower colours in attracting chrysopids.

Chemical cues in the form of odours emanating from the prey and infested host plants play a major role in attracting the chrysopids to the prey site. Numbers of *C. carnea* on L-tryptophan sprayed tree canopies increased significantly

(McEwen *et al.,* 1994). Sprays of L-tryptophan on infested parts of cotton plants also induced response in adult chrysopids. Highest response of *C. carnea* was noticed for infested bolls. Larvae spent more time in zones where infested bolls were placed (Bakthavatsalam *et al.,* 2000).

First and third instars of *C. carnea* respond significantly to the kairomones of their aphid prey, *Acyrthosiphon pisum* and *M. persicae*. The third instars also responded to the odour of host plants, *Vicia faba*, *Brassica oleracea*, and *Phaseolus vulgaris* (Sengonca *et al.,* 1995a). Starvation of chrysopids leads to a more pronounced chemical attraction and a faster induction of intensive search in *C. carnea* (Sengonca *et al.,* 1995b; Mahabaleshwar and Kulkarni, 2000a).

Prey Consumption

The third instars consume a major portion (60-85%) of the total number of prey (Balasubramani and Swamiappan, 1994; Klingen *et al.,* 1996; Thite and Shivpuje, 1999). *C. carnea* larvae consumed 30.3 eggs of *Corcyra cephalonica*, 33.4 eggs of *H. armigera*, 0.54 egg masses of *Spodoptera litura*, 33.3 *A. gossypii* and 24.6 *P. citri* in a day (Geethlakshmi *et al.,* 2000). 3783.3 eggs, 6728.52 nymphs and 96.39 females were consumed in its lifetime (Malleshaiah *et al.,* 2000). The total larval consumption by chrysopids of different prey is given in Table 2.

Table 2. Total consumption by *C. carnea* during its larval development period

Prey	*Consumption*	*Reference*
Aphis gossypii at 25°C	952.4	Yuksel and Gocmen, 1992
Aphis gossypii at 30°C	655.6	Yuksel and Gocmen, 1992
Bemisia tabaci (pupae)	329.70	Balasubramani and Swamiappan, 1994
Corcyra cephalonica (eggs)	732.35	Balasubramani and Swamiappan, 1994
Heliothis armigera (neonates)	409.55	Balasubramani and Swamiappan, 1994
Lipaphis erysimi	164.00	Singh and Kumar, 2000

Prey consumption by the lacewings follows a predicted path in the face of increasing prey densities. The lacewings usually respond to the increasing prey densities in the manner predicted by Holling (1959). Holling (1959) proposed three functional responses, the initial response of which differed in all the three, though a saturation threshold was later obtained. The consumption is directly proportional to the prey density in the Type I response resulting in a linear trend. The Type II response is curvilinear in nature with an initial rapid increase in consumption followed by a subsequent slowing down of this increase. The Type III response has a sigmoidal pathway, which is obtained as a result of the increasing prey consumption over a limited range of prey densities. The mortality in prey was density independent in Type I, negatively density dependent in Type II and positively density dependent in Type III responses.

Third instar of *C. carnea* when exposed to different densities of aphids, *A. craccivora*, *Uroleucon cichorii*, *Pterochlorides persicae* and *A. gossypii* exhibited Type II response (Yuksel and Gocmen, 1992; Mirabzadeh, 1998; Fonseca *et al.*, 2000). The functional responses of *Chrysopa sinica* to the prey also fitted Holling's type II equation (Zhou *et al.*, 1991). Second instars exhibit the lowest searching efficiency, while first instar has the greatest searching and handling time. Handling time shows a progressive decrease with larval development (El- Kareim, 1998; Mirabzadeh, 1998; Fonseca *et al.*, 2000).

Plant texture and architecture have a major influence on the predation efficacy of most of the predators. The same is also evident in chrysopids. Plants differing in their architecture differ in providing suitable sheltered microenvironments for harbouring the prey. Thus, topography, architecture and chemical characteristics of the plant too play an important role in the predation by chrysopids.

Chrysoperla plorabunda spent more time walking and less scrambling on glossy type leaves than on normal wax type leaves, since a reduction in wax bloom substantially affects the mobility and effectiveness of the generalist predator (Eigenbrode *et al.*, 1999). In greenhouse and field experiments in Utah (U.S.A) that assessed the ability of *C. carnea* as a generalist predator, aphid colonies were reduced to near zero on slender leaved grasses but were maintained at moderate densities on intermediate wheatgrass (Messina *et al.,* 1995). This indicates the variation in predation efficacy in relation to differential architecture of plants. The area of residence of the prey is influenced by the plant structure, thus modifying the degree of accessibility afforded to the predator (Clark and Messina, 1998).

GROWTH AND DEVELOPMENT

Growth and developmental parameters of lacewings like other predators are highly influenced by biotic and abiotic factors. Prey quality and quantity have a major influence on the developmental parameters along with the role of other abiotic factors such as temperature, light and humidity.

Different prey influence the growth and developmental parameters of the chrysopids with better and faster development on suitable aphids as evident in Table 3. While one prey may be suitable to a particular chrysopid, its nutrients may not fulfill the nutritional requirements of another. For example, the total larval period of chrysopid, *Mallada desjardinsi* on *A. gossypii* was 14.9 days, while that of *C. congrua* was 13.5 days on the same prey (Kabissa *et al.,* 1993).

C. carnea that fed on poinsettia reared whitefly, *Bemisia argentifolii* did not survive to the pupal stage, while those fed on cantaloupes and cucumber reared *B. argentifolii* developed into adults (Legaspi *et al.,* 1996). The effect of host plant on the suitability of the prey is evident since the chemical constituents of prey are dependent on the host plant.

Developmental and reproductive rates of *C. externa* were relatively high and directly related to temperature between 15.6 and 26.7°C. Survival was also high throughout this temperature regime. Lower thermal thresholds for development of immature stages fell between 11 and 12.5°C and K value for total development was 320 days-degrees above 11.8°C (Alburqueque *et al.,* 1994).

Temperature significantly influences the biological attributes of lacewings. When *C. carnea* was tested at 14, 18,

Table 3. Total development period of *C. carnea* on different foods

Food	*Total Developmental period (days)*	*References*
Eggs of *Heteropsylla cubana* (psyllid)	28.2	Jalali and Singh, 1992
Bemisia tabaci	19.15	Yoldas, 1994
Eggs of *Corcyra cephalonica*	19.35	Yoldas, 1994
Eggs of *Mamestra brassicae*	27.4	Klingen *et al.*, 1996
Larvae of *Mamestra brassicae*	21.5	Klingen *et al.*, 1996
Eggs of *Heliothis armigera*	19.95	Balasubramani and Swamiappan, 1994
Aphis gossypii	20.15	Balasubramani and Swamiappan, 1994
Neonates of *Heliothis armigera*	22.50	Balasubramani and Swamiappan, 1994
	27.4	Saminathan *et al.*, 1999
Aphis craccivora (groundnut)	18.59	Saminathan *et al.*, 1999

22 and 27°C temperatures and 6-22 and –2-22°C thermo periods with pyralid, *Ephestia kuehniella* as prey, it was found that the survival of embryo and first instar larvae was lowest (42 and 14%, respectively) at the lowest temperature and pupation did not occur. The highest survival (80-92%) was for the two thermoperiods (Nicoli *et al.,* 1991). In *C. externa* reared on noctuid, *Albama argillacea* eggs, growth rate increased with increase in temperature. An increase in temperature from 6 to 15°C reduced the length of maturation from larvae to adult by 54 days. The temperature most suitable for the development of *C. carnea* was 24°C. The range of the low temperature threshold was from 9.5 to 11.5°C. The thermal constants were 58.1, 50.8, 36.0, 55.7, 102.6 and 290.2 degree-days for the egg, first, second and third instars, pupa and the period from larva to adult, respectively (Figueira *et al.,* 2000).

INTRAGUILD PREDATION

Prey abundance leads to the advent of a number of predators belonging to different species, different genera and even different orders. These heterospecifics exploit the prey and constitute a guild (Root, 1967). In their joint exploitation of the prey patch, they tend to affect the foraging success of others by competing for the food source, by interference and by active predation of the other guild members. Intraguild predation (IGP) is the consumption of guild members belonging to different species or genera having a common food source. The active aggressor is known as the intraguild predator, the vulnerable victim as intraguild prey and the primary food source as the extraguild prey (Omkar *et al.,* 2002). The intricacies and complexities of intraguild predation have been extensively reviewed (Polis *et al.*, 1989; Polis and Holt, 1992; Omkar *et al.*, 2002).

Lacewing populations are specific to each particular habitat, with one species being dominant. The dominant member of the guild suppresses the population of the other members, which are weaker and more vulnerable to its predation attempts. Ideally, a combination of natural enemies with a common food source can be an asset to applied biocontrol (Turnbull and Chant, 1961; Ehler, 1990; Benrey and Lamp, 1994) as long as there is no or little mutual interference and competition among the predators. For effective biocontrol with a combination of predators, it is imperative that the results obtained are either synergistic or additive but not non-additive.

Of the six ant species occurring in olive orchards infested with yponomeutid, *Prays oleae*, *Crematogaster scutellaris* and *Tapinoma nigerrimum* were most effective predators of chrysopid eggs with the latter being probably more damaging due to both large nest size and its high activity. An ant, *Formica subrufa* was intermediate in its activity and ability to consume *C. carnea* eggs (Morris *et al.*, 1998).

Feeding specificity, size, mobility and aggressiveness of the protagonists as well as the extraguild prey, influence the magnitude and direction of IGP. Cecidomyiid, *Aphidoletes aphidimyza* was more vulnerable to IGP than *C. rufilabris* and ladybeetle, *Coleomegilla maculata*. Among *C. maculata* and *C. rufilabris* the interaction was symmetric while in the interaction with *A. aphidimyza,* it was in favour of the ladybeetle and lacewing. Sessile and low mobility stages of all species were extremely vulnerable to IGP (Lucas *et al.,* 1998).

Interspecific predation of *C. carnea* and ladybeetle, *Hippodamia variegata* was investigated in the absence and presence of prey, *A. gossypii*. In both the cases, *i.e.* the presence or absence of prey the larvae and adults of lacewings emerged superior to those of ladybeetle. The lacewing larvae fed on different life stages of ladybeetle (eggs, larvae, pupae and adults) while the larvae and adults of ladybeetle only fed on the eggs and larvae of lacewing. Presence of prey reduced IGP significantly (Hatami and Zibai, 2000).

Nabid, *Nabis* sp. and assassin bug, *Zelus renardii* added to experimental cotton fields, attacked and predated on the non-target, *C. carnea* and thereby caused hindrance in the aphid biocontrol by the latter (Rosenheim *et al.,* 1993).

Effect of interactions between erigonid spiders and larvae of lacewing, *C. carnea* on start populations of 15 alatiforme L4 instars of aphid, *Sitobion avenae* were investigated in 7-day microcosm experiments. Unidirectional IGP of erigonid females on lacewing larvae were found. The effects on the aphids were additive, *i.e.* both *Erigone atra* and lacewing larvae survived.

Non-additivity of effects on aphids was found for *Oedothorax apicatus* females plus surviving lacewing larvae (Dinter, 1998). Eggs of three lepidopterans, namely, *H. armigera, Earias vittella* and *S. litura* were allowed to be parasitized by *Trichogramma chilonis, Trichogramma acaeae* and *Telonomus remus*. The parasitized eggs when offered to the larvae of *Chrysoperla scelestes* were readily attacked. The incidence of attacks was directly proportionate to the aging in eggs. The predators even damage the eggs, while the parasites were emerging, killing the parasites. Non-additive effects are clearly visible in this case and, thus, release of either egg parasite or *C. scelestes* alone is suggested rather than a combination (Krishnamoorthy and Mani, 1985).

Thus, intraguild predation carries the potential to undermine some of the most well thought of releases if the local population and combination of natural enemies released have not been studied thoroughly. It has been revealed that chrysopids usually dominate over other members of guilds with the exceptions of a few ant species and spiders. They are especially dominant over coccinellids, the other major group of aphidophaga.

EXPERIMENTAL TRIALS (RELEASES)

Laboratory studies reveal chrysopids to be efficient predators of a number of insect pests, especially aphids. Test releases have been widely conducted to assess their effectiveness as bioagents in the fields, the size of initial releases, kind of release (eggs, larvae or adults) and release rates. *Chrysoperla* sp. are ranked as some of the most commonly used and commercially available natural enemies. For many years, *C. carnea* and *C. rufilabris* have been mass-reared and marketed commercially in North America and Europe. Two additional species, *C. externa* and *C. nipponensis*, are marketed in Latin America and Asia. In response to a questionnaire in 1992, members of American Association of Applied Ecologists ranked *Chrysoperla* sp. as unrivaled on the list of commonly applied and commercially available predators (Anonymous, 2002a).

In studies carried out in Braunschweig, Germany, during 1992, the efficiency of *C. carnea* in controlling aphids,

Aulacorthum solani (75%), *Macrosiphum euphorbiae* (12%), *Nasonovia ribisnigri* (2%), and *M. persicae* (11%), on lettuces in greenhouses was tested. Of the three predatory species tested, only the application of *C. carnea* (eggs) resulted in reasonable aphid control. The best effects from this predator were achieved when eggs were first added to the young lettuce plants before transplanting, subsequently followed by 3 sprays of *C. carnea* eggs at weekly intervals (25-30 eggs/m^2) (Quentin *et al.*, 1995).

The effective stage and rate of releases of *C. carnea* against safflower aphid, *Dactynotus carthami* were evaluated at the College of Agriculture Farm, Nagpur, in the Rabi season of 1998-1999. The results revealed the highest aphid population reduction in the plots treated with dimethoate 0.03%. Larval releases of *C. carnea* at 2 or 3 larvae at the population level of 10 or 20 aphids per shoot reduced the population between 40 to 50%. Egg releases of *C. carnea* did not help in satisfactory reduction of aphid population. On the basis of yield, promising treatments were dimethoate 0.03% (15.8 q/ha), 3 larvae of *C. carnea* released at 10 aphids (13.43 q/ ha) and 20 aphids (13.20 q/ ha) population levels (Badgujar *et al.*, 2000).

Green lacewings are highly predaceous on cardamom aphid, *Pentalonia nigronervosa*. Second instars and adults of *C. carnea* were released at ratios of 1:5, 1:10, 1:15, 1:120, 1:25 and 1: 50. Predator prey ratio of 1:5 was superior with 78.33% population reduction after 3 days (Mathew *et al.*, 1999).

The continuous rearing of a coccinellid predator on a synthetic diet leads to a decrease in its predation efficacy (Dixon, 2000). This does not seem to be applicable in chrysopids. *C. carnea* reared for 10 successive generations on a semi synthetic diet containing soybean hydrolyzed powder (1.3%), water (38.7%), egg yolk (32.3%), honey (16.1%), petroleum jelly (0.7%) and paraffin wax (9.6%). Semi synthetic diet reared larvae readily attacked and consumed cotton aphid, *A. gossypii* and eggs of *H. armigera* (Venkatesan *et al.*, 2000).

The test releases support the high efficacy of the chrysopids as biocontrol agents. There is need to properly mass produce the chrysopids and then to release them in appropriate quantities at the appropriate time by using the most effective application techniques so as to obtain best results.

REARING TECHNIQUES

The mass production and commercialization of effective biocontrol agents depends upon the ability of insectaries to produce and market highly reliable and relatively inexpensive supply of natural enemies. Rearing of larvae constitutes the most costly aspect in mass-production of *Chrysoperla* sp. largely because all three instars are predaceous. Most of the insectaries depend on mass-produced insect prey as food, which is relatively expensive compared with artificial diets.

In the traditional and conventional methods of rearing, 200 pairs of adults are kept in oviposition units measuring 75 x 30 cms, with sides lined by nylon wire mesh. The sliding top cover is slided over a comb and is fitted with a black cloth as a substrate for oviposition. The sliding top cover is replaced everyday starting from the fourth day. In the meantime, adults are daily provided cotton swabs dipped in water, 50% honey, castor pollen and equal quantities of protinex, fructose, honey and powdered yeast dissolved and mixed together in little water. One day after being laid, the eggs are removed by gently working with sponge and then stored for future use (Anonymous, 1995).

Newer and less space consuming units are being developed the world over. Two cylindrical and cubical adult feeding and oviposition units for *Chrysoperla* sp. have been developed that contain 5.4 and 9.3 times the internal surface area (with higher internal surface: volume ratios) of the traditional 11.30 litres cardboard units (Nordlund and Correa, 1995a). A few more effective rearing and harvesting cages such as the AFOU and hot wire system have been designed (Nordlund and Correa. 1995b).

An even more compact method of mass production is the Biosec Bioassay tray. Each tray (44 × 21 cms) comprises 8 blocks, each block containing 16 wells with the dimensions of each well being 2 × 1.5 cms. The eggs of *C. carnea* are transferred along with 0.045 cc of *Corcyra* eggs. The Pull-N-Peel tab is pulled firmly over the wells and the trays stacked one over the other. It takes 12-16 days for the adults to emerge. The recovery is approximately 95%. These trays give better adult recovery, prevent the larvae from escaping, occupy less space and are also relatively cheaper. This method is recently being tried out in the Project Directorate of Biological Control, Bangalore, India.

The filler material used in the units and the degree of filling also influence the development of the chrysopids. The percent pupation was significantly higher in most of the filler materials when used at 75% of the container's capacity. Maximum pupation (88%) was obtained in larvae reared in wood shavings, which was at par with those reared in injection vials (93%) and plastic louvres (91%) (Mahabaleshwar and Kulkarni, 2000b).

RELEASE TECHNIQUES

A major decision to be taken, while planning field releases, is the form in which the releases are to take place. Whether eggs, larvae or adults should be released? How many and how often should the releases be made? The question of how many and how often is answered by extrapolating the results of experimental releases. Eggs and larvae are usually the recommended stages for biocontrol application since adults are liable to fly away. Eggs, however, delay the start of biocontrol by a week because that is the time required to reach the voracious second and third instar. The application of larvae avoids this delay. Also more eggs are needed than larvae to compensate for the delay in onset of biocontrol mechanisms.

What are the mechanisms used for the application of chrysopids? Historically, chrysopid eggs were dispensed manually, mixed with a solid medium such as rice hulls or vermiculite to practice uniform field distribution. One disadvantage of these solid carriers is the poor retention of eggs on the plants; eggs fall off the leaves whereas liquid carriers help attach eggs to the targeted plants.

Recently, there have been notable advances in the development of liquid carriers and commercial sprayers. For example, distributing *C. carnea* eggs in an agar solution rather than sucrose-based carriers has the advantage of lowered attractiveness to ants and other predators (Anonymous, 2002b).

Special spray techniques for applying mass reared eggs of *C. carnea* in fields need to be developed. Water is a suitable liquid medium in which the eggs could be suspended for up to 12 hours. A pressure of up to 4.5 bar liquid pressure was found to be suitable for application of eggs as the hatching percent was not affected. A flat jet producing Flood Jet TK 1.5 nozzle is suitable for the application of eggs at a pressure of about 3 bar (Lochte and Sengonca, 1995).

A new solid application method is directly related to the means of transportation. Chrysopids are shipped as eggs on cards, eggs on tapes or as pre-fed larvae in hexcel units. There are 5,000 or 10,000 lacewings eggs per card with each one-inch square holding an average of approximately 170 and 330 eggs, respectively. Lacewing tapes are made up in units of 20 tapes, each 6 inches long and 3/4 inches wide with waxed paper on each end so that the ends can be conveniently attached to form a bracelet around an infested limb or vine. Each tape holds an average of approximately 250 lacewing eggs with equal volume of moth egg diet and corn grit to cover the exposed sticky areas on the tapes. These tapes and cards can be directly placed firmly on the foliage as soon as larvae start crawling (Anonymous, 2002a).

After the release of bioagents there is need to increase the attractiveness of the fields by spraying food substances and attractants to augment the local populations of chrysopids. Augmentation of *C. externa* in soybeans and corn did not affect the resulting number of lacewing larvae but applying wheast and sugar at the time of the release gave a 2- to 6-fold increase in the densities of adults and eggs of *C. externa* in cornfields (Anonymous, 2002b). Commercial yeast hydrolysates, like wheast; 4:7:10 brewer's or baker's yeast hydrolysate (or whey substitute): sugar: water, in weekly spray onto alfalfa attract more lacewing adults and eggs. 4.5 kg of fermented whey plus 4.5 kg sugar in 25-40 litres of water can be sprayed each 1-2 weeks to attract and induce oviposition in *C. carnea* (Anonymous, 2002a).

However, the effectiveness of food sprays in manipulating field populations varies and recent studies emphasize the need of further basic and applied research in this area. In most trials, protein-sprays without sugar fail to increase the number of lacewings and sugar-sprays without protein attract lacewing adults but do not stimulate egg laying. By using food sprays to attract and induce chrysopids to lay eggs before natural honeydew becomes abundant, it is also possible to suppress honeydew-producing or other pests before their numbers become large (Anonymous, 2002b). Proper and timely releases are essential for the success of a biocontrol programme. A lot of basic research is needed before the mass rearing, storage and release techniques get standardized enough to make each biocontrol release a success.

COCCINELLIDS

Coccinellids or ladybird beetles or ladybeetles belong to the family Coccinellidae of order Coleoptera. They constitute one of the most colourful predators in aphid colonies. With their bright colouration they form a rather loved group of insects and have since long been considered as harbingers of good luck and bountiful crops. Their adults as well larval stages are predatory in nature and have a wide prey range, which includes aphids, coccids, mealybugs, whiteflies, leafhoppers, lepidopterous eggs and larvae, mites and other soft-bodied insects (Figure 1) (Hodek and Honek, 1996; Dixon, 2000).

General Biology

Ladybeetles are oval or spherical in shape, ranging from 0.8-18 mm. The dorsal surface is convex and the ventral flat. They occur in a variety of shades of cream, brown, black and red. Some may sport species-specific patterns in dots, stripes or both on their elytra. The elytra may be smooth and shiny or may be covered with short recumbent hairs, while the hindwings are membranous (Omkar and Mishra, 2002a; 2002b). Females are usually larger than males. Ladybeetles mate after a short but pronounced courtship display of approach, watch, examine, mount and copulatory attempt by the male. After mating, they lay light yellow to deep orange eggs, which are oval or spindle shaped and are laid either singly or in clusters with the narrow ends attached to substrata (Figure 3). The first instar larvae usually hatch out of the eggs after 2-3 days in the tropical regions of the world. There are four instar stages, followed by prepupa and pupa. After a few days, a ladybird with light colouration and soft delicate wings bearing no pattern emerges from pupa. With the passage of time, the elytra harden and deepen in colour, to assume the commonly seen form (Omkar and Mishra, 2002a; 2002b).

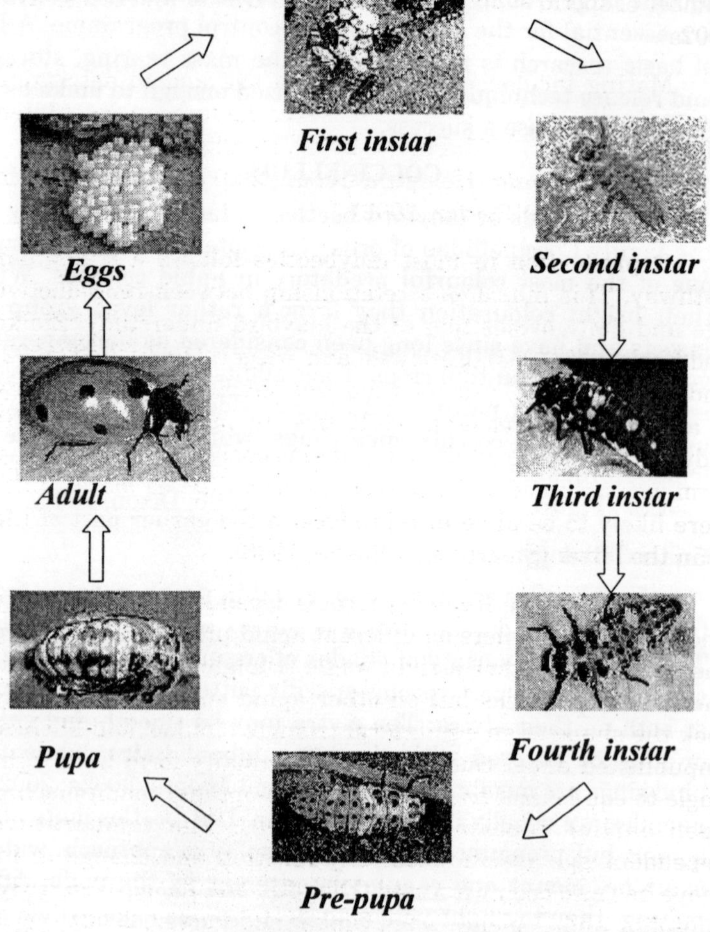

Fig. 3. Life cycle of a coccinellid

REPRODUCTION

Reproductive Biology

Post-emergence the ladybeetles display a small pre-mating period, which may range from an hour to 10-12 days (Hodek and Honek, 1996; Dixon, 2000). After the pre-mating period, mating takes place, with multiple matings a common occurrence. Increase in the number of matings increases the fecundity and percent viability of eggs in ladybeetles. The increase in egg viability can be attributed to the increased number of sperm supply (Majerus, 1994; Omkar and Srivastava, 2002a).

Mating incidences in relation to age and to determine the maturity period of the two sexes have been calculated (Hemptinne *et al.*, 2001; Pervez, 2002). Results reveal protogyny in *Adalia bipunctata* (Hemptinne *et al.*, 2001) and protandry in *Propylea dissecta* (Pervez. 2002).

Reproduction in most ladybeetles follows a triangular pathway. This indicates a relationship between reproductive age and daily oviposition of the ladybird understudy. Ziolko and Kozlowski (1983), Kozlowski and Wiegert (1986), Roff (1992) and Stearns (1992) assumed that the shaping of the fecundity is a result of increase in adult size. In insects, such as the ladybeetles, where adult size remains constant, the fecundity form was assumed to be shaped by mortality, as more adults were likely to be alive to reproduce in the earlier part of life than the latter (Stearns and Koella, 1986).

The triangular fecundity form is dependent on food quality and quantity. It differs on different aphid preys: in *P. dissecta*, the shape was rather a right angle triangle on *A. craccivora*, the favoured species but on other aphid species it gradually took the shape of an equilateral triangle (Omkar and Mishra, unpublished data). Such a shift in fecundity form from right angle to equilateral triangle due to sub-optimal conditions has been earlier reported (Dixon, 2000). The temperature dependent age specific fecundity function and change in its shape has also been witnessed (Ding-Xin and Zhong-Wen, 1987; Ponsonby and Copland, 1998; Omkar and Pervez, 2002).

The mean number of eggs laid/female/day also increases curvilinearly with increase in aphid density (Agarwala and Bardhanroy, 1997; Omkar and Pervez, 2003a). Usually females lay eggs in batches, whose sizes may be anatomically determined by the number of ovarioles, as number of eggs per batch is equal to half the number of ovarioles (Stewart *et al.*, 1991). Thus, half the number of ovarioles participates in the egg output while the other half prepares the next batch (Stewart *et al.*, 1991). The optimum ovariole range found in *Coccinella septempunctata* was 41-50 ovarioles per ovary (Rhamhalinghan, 1986).

The reproductive output of a ladybeetle also depends upon the quality as well as quantity of prey. Not all prey are equally suitable as has been discussed in the case of chrysopids. *C. septempunctata* laid maximum number of eggs on mustard aphid, *Lipaphis erysimi* and minimum on aphid, *Aphis nerii* (Omkar and Srivastava, 2003a) while *Cheilomenes sexmaculata* has been reported to respond best to *A. nerii* (Choudhury *et al.*, 1983). In a contrary report, *A. nerii* has been considered worst prey resulting in less reproductive output of *C. sexmaculata* and *A. craccivora* was considered the best (Bind, 1998). *C. transversalis* reproduced best on *A. gossypii* and also showed the fastest development on it (Omkar and James, 2003). Pale morph of ladybeetle, *P. dissecta* reproduced best on *A. gossypii* (Pervez, 2002) while the typical morph was found to lay maximum number of eggs on *A. craccivora* (Omkar and Mishra, unpubl. data). On the other hand *A. craccivora* has been found unsuitable to a number of ladybeetles (Hodek, 1960; Hukusima and Kamei, 1970; Obatake and Suzuki, 1985). Fecundity of a few ladybeetles in relation to prey have been presented in Table 4.

The host plant too plays a major role in determining the fecundity of the ladybeetle. The palatability and suitability of *A. craccivora* was found to be highly dependent on the host plant from which it was obtained (Hodek, 1960; Hukusima and Kamei, 1970; Obatake and Suzuki, 1985). The predatory coccinellids *C. sexmaculata* and *Micraspis vincta* were reared on *A. craccivora* (on cowpeas and beans), *L. erysimi* (on mustard and cabbages), and *A. gossypii* (on brinjal). Marked

Table 4. Fecundity of a few ladybeetles in relation to different prey

Ladybeetle	*Aphid*	*Fecundity*	*Reference*
Exochomus quadripustulatus	*Dysaphis devecta*	172.8±18.4	Radwan and Lovei, 1983
E. quadripustulatus	*A. pisum*	91.3±16.0	Radwan and Lovei, 1983
Lemnia biplagiata	*A. craccivora*	1064.0±96.0	Semyanov, 2001
L. biplagiata	*M. persicae*	1651.0±84.0	Semyanov, 2001
C. transversalis	*A. gossypii*	1399.60±22.58	Omkar and James, 2003
C. transversalis	*A. craccivora*	1155.90±33.18	Omkar and James, 2003
C. septempunctata	*L. erysimi*	1764.10± 8.46	Omkar and Srivastava, 2003a
C. septempunctata	*A. nerii*	203.20±11.83	Omkar and Srivastava, 2003a

differences in the growth and development of predators were observed when reared on different aphids and the same aphid on different food plants. *C. sexmaculata* developed faster when fed on *A. craccivora* (on cowpeas), and *M. vincta* developed faster when fed on *L. erysimi* (on cabbages) (Rao *et al.*, 1997).

When aphids are scarce, ovarioles in *C. septempunctata* ranged between 31-40 and in contrast high ranges of 51-60 and 61-70 ovarioles per ovary have been observed at adequate and high prey supply (Rhamhalinghan, 1986). Thus, fecundity is positively correlated with food consumption (Rhamhalinghan, 1987a; Veeravel and Baskaran, 1997; Omkar and Srivastava, 2003a). A supporting result was obtained in the case of pale morph of *P. dissecta* where there was a reduction in the fecundity as well as oviposition period on being subjected to prey deprivation. However, an interesting observation was that the reproductive rate did not vary even with the increase in duration of prey deprivation, *i.e.* the female continued to reproduce vigourously even after having been prey deprived. This observation led to the formulation of a hypothesis that *females live to lay eggs* (Omkar and Pervez, 2003b).

Ovarioles are known to show a seasonal variation with their numbers increasing during winter and decreasing during summer and monsoon (Rhamhalinghan, 1987b). This indicates the change in reproductive output with change in temperature. Among the adults, the females of *C. sexmaculata* lived longer and produced more eggs at 30 than at 20°C (Veeravel and Baskaran, 1995). The fecundity of females of ladybeetle, *Scymnus mediterrinus* reared at different temperatures varied between 1.7 eggs at 15°C and 601.86 eggs at 30°C (Ba M'Hamed and Chemseddine, 2001).

Oviposition Strategies

Some ladybeetles provide a defense shield to their eggs. *Pseudoscymnus kurohime* (Miyatake) protects its eggs by covering them with faeces like secretion (Arakaki, 1988), while *Megalocaria dillata* (Fabricius) lays eggs on a spine or needle like plant structure and surrounds them with a colourless sticky substance secreted from the abdomen (Liu, 1933).

The theory of optimal foraging by Kindlmann and Dixon (1993) deals with the considerations of the female in the search for suitable sites, the examination of the sites and her subsequent decision of oviposition. Dixon (1997) suggested the presence of a window in the aphid population graph, with parameters favourable for oviposition. The requirement for a minimum aphid density for oviposition indicates the possible presence of a nutritional threshold below which oviposition cannot take place (Dixon, 1959, 2000). Though, it is purported that females in captivity with mature eggs will oviposit regardless of the presence of aphids (Hodek and Honek, 1996).

The larval tracks of con- as well as hetero-specifics also determines the oviposition (Ruzicka, 1997; Doumbia *et al.*, 1998; Yasuda *et al.*, 2000). Chloroform extracts of the larval tracks reveal the presence of an oviposition deterring pheromone (ODP) consisting of a mixture of alkanes in which methyl branched (C_{23} to C_{33}) and straight chain alkanes (C_{20}-C_{31}) predominate. These alkanes spread easily on the hydrophobic cuticle of the plants and their effects last for a minimum of ten days since saturated hydrocarbons are not quickly oxidised (Hemptinne *et al.,* 1992; Merlin *et al.,* 1996; Ruzicka, 1997; Doumbia *et al.,* 1998; Hemptinne *et al.,* 2000a).

Ladybeetles, *Adonia variegata, C. septempunctata* and *Semiadalia undecimnotata* prefer ovipositing on host plant while *A. bipunctata* and *Propylea quatuordecimpunctata* prefer the lid (Iperti and Quilici, 1986). The preference for host plant twig by *C. septempunctata* has been previously recorded (Omkar and Srivastava, 2001a). While red colour was preferred by *A. bipunctata* (Iperti and Prudent, 1986), green and yellow colours were preferred by *C. septempunctata* (Omkar and Srivastava, 2001a).

FORAGING

Attraction to Prey Sites

Ladybeetles have an instinctive negative geotropic and positive phototropic association, which puts them in the vicinity of the prey. The tendency of the larvae and the adults to walk along the leaf edges, interveinal and intraveinal areas also serves the same purpose. Ladybeetles have been known to search in daylight with active searching put on the backseat in the night though prey venturing into the vicinity are consumed (Nakamuta, 1987).

The search for prey usually follows an extensive mode with a switchover to an intensive one in the presence of active signals, which indicate the desired and suitable sites. The initial signals help locate the habitat, the following cues the prey location and the final cues, a particular prey patch. Each successive cue leads to sublimation in the extensive and random searching mode, with the most extensive search involved in habitat location and the least extensive in locating the prey patch. What are the cues that elicit this switchover?

Visual cues may help the predator in orientating towards the plant. Vertical lines on the horizon, silhouettes of trees and ovate shape of leaves seem to direct the ladybeetle, *Chilocorus nigritus* towards the host plant (Hattingh and Samways, 1995). After the visual cues, the chemical cues probably play a major role in ensuring that the predator reaches the infested plants. *C. septempunctata* was found to orientate towards crushed *Berberis vulgaris* leaves and towards the methanol extracts of the leaves inside an olfactometer, indicating the role of olfaction in habitat location (Schaller and Nentwig, 2000). Aphid body fluid has been considered to be the final cue in the prey recognition (Nakamuta, 1984), thus, pointing towards chemical detection. Honeydew acts as an arrestant stimulus for the coccinellids and induces intensive and localized search in the area (Carter and Dixon, 1984). The wax of mealy bugs acts as an arrestant stimulus for *Diomus* sp., *Exochomus* sp. and *Cryptolaemus montrouzieri* (Merlin *et al.,* 1996). The ladybeetles can detect and orientate towards the prey prior to physical contact, which has been proved by a number of works on *C. septempunctata* (Stubbs, 1980; Nakamuta, 1984). Change in the searching behaviour in the presence of aqueous aphid extract points towards the presence and role of kairomones in the signaling process (Shonouda, 1999). If unsuccessful in locating more prey after the initial switchover inducing encounter, the ladybeetles tend to switch back to an extensive search. Ladybird larvae are known to chase the prey that eludes capture even after an encounter.

The pursuit is fast and is attributed to the larval motivation to forage (Carter and Dixon, 1982). Larvae avoid areas bearing their tracks, so as to save precious time and energy (Marks, 1977; Dixon, 2000).

Prey Consumption

Ladybeetles are polyphagous having wide food-range, which include a number of soft bodied arthropods, fungi, *etc.* along with certain natural products, such as pollen and honeydew. Female ladybeetles tend to consume more than males. Mean prey consumption by male and female, *C. septempunctata* was 62.24 and 69.42 individuals of *L. erysimi* per day, respectively (Sharma *et al.*, 1997) Lifetime consumption of *L. erysimi* by adult male and female *C. septempunctata* was 5180 and 6151, respectively (Omkar and Srivastava, 2003a). Male consumes about 88.70% of the female's diet (Lucas *et al.*, 1997).

The first meal of the larval life has been deemed critically important for its survival (Dixon, 1959). Since first instars have limited abilities of search and capture, they require a high aphid density, so that it is easier for them to capture their first prey. Social feeding has evolved in the first instar to make the first meal easier to obtain. Kairomones released by a captured aphid attracts other first instars in the vicinity to site of capture and leads to its sharing (Hemptinne *et al.*, 2000b). The first three instars were more efficient in converting the food to body mass than the last one (Ferran and Larroque, 1977). This may have an adaptive significance; the young larvae are less efficient in prey capture and hence consume less and need to make out the best of whatever resources they have at their disposal. About 22.4% of total weight was gained during the duration of the three larval instars, which consumed only 17.7% of the total consumption by the immature stages (Ferran and Larroque, 1977). The reduced conversion of prey consumed into body mass by the fourth instar may be due to the increased energy expenditure during the process of pupation (Ferran and Larroque, 1977). Also, the older the larvae gets, the wider becomes its feeding niche (Samways and Wilson, 1988; Hattingh and Samways, 1992).

Coccinellids exhibit Type II functional response (Holling 1959), in which, initially there is an increase in the number of prey consumed, but the rate of prey consumption decreased with increase in prey density. The prey consumption of *C. sexmaculata* (Agarwala and Bardhanroy, 1997) and *Coelophora inaequalis* (Mora *et al.*, 1995) increased up to a certain point of increase in prey density, thereafter it became constant. Functional response of the fourth instar of *C. septempunctata* on *M. persicae, A. craccivora, A. gossypii, U. compositae* and *A. nerii* confirmed a curvilinear relationship between prey consumption and prey density (Omkar and James 2001; Omkar and Srivastava, 2001b; 2002b; 2003b; c). A study on functional response of adults and larval instars of a ladybeetle, *Stethorus gilvifrons* (Mulsant) on a phytophagous mite, *Panonychus ulmi* Koch revealed that adults exhibited Holling's Type III, whereas the larvae Type II response (Haji-Zadeh *et al.*, 1994). Holling's Type III response is scarcely reported in predaceous coccinellids.

Several workers have noticed a high degree of specificity of some ladybeetles for certain prey. *C. septempunctata* prefers *L. erysimi* (Atwal and Sethi, 1963; Srivastava, 2000). Biochemical analysis of this aphid ascribed the preference to its high protein content (Atwal and Sethi, 1963). Aphids, *A. gossypii* (Singh and Devi, 1996) and *Hydaphis coriandri* (Omkar *et al.*, 1997) were also preferred by *C. septempunctata*. *C. sexmaculata* preferred *A. gossypii* over aphids, *Uroleucon compositae, A. craccivora, Brevicoryne brassicae, L. erysimi* and *A. nerii* (Omkar and Bind, 1998). *C. transversalis* prefers *A. craccivora* and *A. gossypii* (Omkar *et al.*, 1999; James, 2001). Ladybeetle, *Anegleis cardoni* preferred *B. brassicae* (Afroze, 1999) while *Micraspis discolor* preferred *A. gossypii* and *A. craccivora* (Omkar and Pervez, 2001). Adults along with third and fourth instars larvae of *P. japonica* preferred older instars of aphids, whereas first and second instars preferred young aphid instars (Zhou *et al.*, 1994).

Aphids, such as *A. nerii* possess high alkaloid contents, making it unsuitable prey for *S. undecimpunctata, P. quatuordecimpunctata* and *A. bipunctata* with exception of *A. variegata* (Iperti, 1966). It is also toxic to *Harmonia axyridis* along with *C. septempunctata* and *C. transversalis* (Tao and Chiu, 1971; Omkar and James, 2003; Omkar and Srivastava, 2003a). High content of alkaloid, quenlizedine in aphid, *Macrosiphum albifrons* appears to be a reason for it being an excluded food for *C. septempunctata* (Hodek and Honek, 1996).

Conspecific aphids existing on different host plants also affect the growth, development and survival of coccinellid predators (Kumar *et al.*, 1999). A few prey are unacceptable to certain ladybeetles and their stages due to their inherent or accumulated toxic effects. Aphid, *Aphis sambuci* is reported to have a toxic effect on individuals of *C. septempunctata* with no survivors (Hodek, 1957a) or very few extremely small sized survivors (Blackman, 1967). *A. craccivora* has been reported as toxic to *H. axyridis* (Pallas) when it is obtained from the plants, *Robinia pseudoaccacia* and *Vicia faba* (Hukusima and Kameii, 1970).

The minimal food requirement for a developing larva could be as low as one-seventh of its normal requirement. Immature stages of *C. septempunctata* may even develop on one large aphid, *A. pisum* per day (Obrycki *et al.*, 1997). Larvae of *M. discolor* required a minimum of 35.2 aphids (*i.e.* 6,4,11.2, 14 aphids for I, II, III, IV instars, respectively) to undergo pupation (Agarwala *et al.*, 1988). The mean daily feeding of larvae was 1.99 aphids for 17.6 days (Agarwala *et al.*, 1988). Prey scarcity may induce premature moulting and the adults may switch on to the secondary prey or even to honeydews, pollens and nectars (Mills, 1981).

GROWTH AND DEVELOPMENT

The amount and quality of food in addition to a number of abiotic factors, chiefly temperature, substantially influence the rate of larval growth and development.

Several studies reveal that decrease in the amount of food increases the development time and slows down the larval growth. Larvae of *A. bipunctata* when fed with excess food develop faster than when they were provided with minimum food (Wratten, 1973). Similar effects of starvation were seen on a number of ladybeetles (Kaddou, 1960; Hukusima and

Ohwaki, 1972; Dimetry 1976; Kawauchi, 1979). Prey scarcity may induce premature moulting and the adults may switch on to the secondary prey or even to honeydew, pollens and nectars (Mills, 1981). Prey scarcity reduces the fitness of coccinellid larvae, which may develop even when the prey supply is reduced to 40% or even less (Baumgaertner *et al.*, 1981).

Smith (1965) and Obrycki and Orr (1990) studied the effect of two aphids, *A. pisum* and *Rhopalosiphum maidis* on the development of seven Palearctic and Nearctic ladybeetles and found that the development time increased with the decrease in prey suitability. Omkar and Srivastava (2003) reported highest weight of immature stages of *C. septempunctata* and lowest developmental period on *L. erysimi* while the reverse was the case when the ladybeetle was fed *A. nerii*. *A. nerii* was however suitable to *C. sexmaculata* (Choudhury *et al.*, 1983).

The daily mean temperature ranges of 20-27^{0}C favours the activity of coccinellids (Freier and Triltsch, 1996). High temperature shortens the developmental period of coccinellids (Padmaja *et al.*, 1995; Jalali *et al.*, 1999; Jarosik *et al.*, 2000). Developmental rate was found to have a linear relationship with the increase in temperature from 15°C to 30°C (Baumgartner *et al.*, 1986; Ponsonby and Copland, 1996). The incubation period of *C. septempunctata* was 8.5 and 1.8 days at 15°C and 30°C (Singh and Malhotra, 1979). The average larval duration of *C. septempunctata* varied from 13-14 days to 10-11 days at 18.35°C to 23.60°C (Singh *et al.*, 1994). The total developmental periods of a few ladybeetles at different temperatures are given in Table 5.

Temperature below 17°C is also not suitable for growth and immature survival of certain ladybeetles (Jean *et al.*, 1990). The mortality of *Hippodamia convergens* (83%) (Miller, 1992) and *Hippodamia parenthesis* (67%) (Orr and Obrycki, 1990) was maximum at the lowest temperature tested, *i.e.* 17°C. Similarly, mortality was maximum (74%) at 15°C (Naranjo *et al.*, 1990). At all the tested temperatures, first instars suffered the highest mortality, while pupae had the lowest (Ponsonby and Copland,

Table 5. Total developmental period of ladybeetles at different temperatures

Ladybeetle	*Temperature (in °C)*	*Total Developmental Period (in days)*	*Reference*
Pseudoscymnus tsugae	18.7	45.0	Carole *et al.*, 2000
P. tsugae	22.7	31.5	Carole *et al.*, 2000
Coelophora inaequalis	20	24.7	Wang and Tsai, 2001
C. inaequalis	25	15.9	Wang and Tsai, 2001
Harmonia convergens	18	57.0	Rodriguez-Saona and Miller, 1999
H. convergens	30	14.0	Rodriguez-Saona and Miller, 1999
Scymnus levaillanti	15	63.9	Uygun and Atlihan, 2000
S. levaillanti	35	11.1	Uygun and Atlihan, 2000

1996). In comparison to melanics, the typicals of *C. septempunctata* were more tolerant to high temperatures, *i.e.* 32 and 35°C probably due to their lower internal temperatures than melanics (Rhamhalinghan, 1987a). *C. septempunctata* responds positively to long day photoperiod for its growth and development, whereas *C. septempunctata brucki* prefers short day photoperiods (Sakurai *et al.*, 1991).

INTRAGUILD PREDATION

This results in instant energy gain to the starving predators and also reduces the potential competitors (Phoofolo and Obrycki, 1998). Simple experiments have proved the presence of intraguild predation, with the larger predator consuming the smaller ones and this is aggravated in the absence of the aphids (Sengonca and Frings, 1985; Lucas *et al.,* 1997; 1998; Phoofolo and Obrycki, 1998). The experiments conducted by Evans (1991) and Chang (1996) to test the efficacy of ladybird combinations in biocontrol, however, did not reveal the presence of intraguild predation, possibly owing to the relatively similar size of the ladybeetles used in the experiments and also due to the abundance of extraguild prey.

Most of the reports obtained from numerous cage and field studies point towards non-additive results. The efficacy of combined efforts was found to be less than the sum of individual efficiencies. Intraguild predators were found to avoid immature stages of fellow members, though parasitised extra guild prey were readily consumed (Prakasan and Bhat, 1985; Guiterrez *et al.,* 1988; Samways and Wilson, 1988; Colfer and Rosenheim, 1995; Ferguson and Stiling, 1996; Taylor *et al.,* 1998, Takizawa *et al.,* 2000), thus reducing the overall efficacy of biocontrol. Synergistic effects in biocontrol of pest populations have also been reported. This synergistic effect occurs when the foraging activity of the predatory stages of ladybeetle causes alteration in the prey habitats and thus makes it more susceptible to another predatory species that inhabits its new habitat (Losey and Denno, 1998).

It is the larger ladybeetles that constitute a hazard to the smaller and less abundant ladybeetles, thus the smaller ones invest more in defence and are toxic to the other guild members, *e.g.* the eggs of *A. bipunctata* and *C. septempunctata* are toxic to heterospecifics (Agarwala and Dixon, 1992; Agarwala *et al.*, 1998; Hemptinne *et al.*, 2000c).

The disappearance or reduction in the numbers of the native ladybeetles on introduction of an exotic ladybird (Subramanyam 1953; Watt, 1986; Calunga and Gage, 1999) is probably a result of intraguild predation by the introduced *top predator* ladybeetle. Top predators, in this context, are those ladybeetles that are also voracious feeders with a good developmental rate on immature stages of other ladybeetles. *H. axyridis* has been reported to be a top predator with an excellent development rate on immature stages of *C. septempunctata* (Takahashi, 1989; Dixon, 2000). The establishment of *C. septempunctata* in North America is probably a result of its superiority in the guild (Wheeler and Hoebeke, 1995). It has, however, now been dominated by *H. axyridis*, the second most established exotic predator in North America (Yasuda and Katsuhiro, 1997). The larvae of *H. axyridis* appeared to be a poor food to *C. septempunctata*, as the survival of latter reduced after feeding on the former (Yasuda and Ohnuma, 1999). The failure of establishment of exotic predator, *A. bipunctata* in Japan may be attributed to its vulnerability to the other guild members *viz. H. axyridis* and *C. septempunctata* (Kajita *et al.*, 2000).

RELEASES

The facts given above reveal that ladybeetles have great potential as biocontrol agents. The history of their success begins with the Vedalia beetle, *Rodolia cardinalis*, which was released in 1889 after importation from Australia, into the citrus orchards of California for the control of cottony cushion scale, *Icerya purchasi*. Since then a number of releases of ladybeetles have been conducted the world over. But, however, the successes are few and far between. The success stories that ladybeetles can boast of, are mostly against scale insects (Hodek and Honek, 1996; Dixon, 2000; Omkar and Pervez, 2003c).

Aphidophagous ladybeetles have unfortunately not been so successful. The earliest release of an aphidophagous ladybeetle dates back to the year 1874, when adults of *Coccinella undecimnotata* were released into New Zealand for the control of aphids. Though, the species established itself, it failed to make an impact due to the inadvertent introduction of its parasite (Dumbleton, 1936).

C. septempunctata (released in USA), *P. quatuordecimpunctata* (released in USA), *H axyridis* (released in USA) and *A. variegata* (released in USA) are the recent successful examples of introduction of an exotic species in terms of their successful establishment into the native guild. However, neither of these introductions could result in the successful management of aphid populations (McClure, 1987). Of the inundative releases of aphidophagous ladybeetles, few have been marginally successful, while most have failed (Hagen, 1962; Cooke, 1963; DeBach, 1964; Savoiskya, 1966; Hodek and Honek, 1996; Dixon, 2000).

The failure of aphidophagous ladybeetles in comparison to coccidophagous ones can be attributed to the greatly fluctuating population dynamics of the aphids. Also, the lack of synchronization among the aphids and the ladybeetles because of difference in their temperature thresholds is responsible for the failures. The unplanned introduction of species without deep study of its suitability in the new habitat in terms of abiotic factors and also interactions with the guild members have responsible for the failures befalling aphidophagous ladybeetles.

However, further detailed investigations into the biology as well as the population dynamics of the ladybeetles, their rearing and release techniques may smoothen the path of the future.

SYRPHIDS

Syrphids belong to the family Syrphidae of order Diptera. They are commonly known as hoverflies because of their hovering motion over flowers. In America they are known as flower flies. While the adults are active pollinators, most of the larvae are predatory on a number of soft bodied arthropods (Figure-1), though a few may be phytophagous, saprophagous, coprophagous or even aquatic filter feeders.

General Biology

Syrphids are true flies that are 1/4-3/4 inch in length with large eyes that take up most of their head area. They have a single pair of wings, a characteristic of Diptera. Many have distinct black and yellow patterns on their abdomens (Figure-4). They are Batesian mimics of bees and wasps. The adults hover over flowers and the most interesting thing is that they can fly sideways. Adults sip nectars and are very active pollinators of our gardens.

The syrphids do not display any courtship behaviour and lay long white solitary eggs after having mated. The eggs hatch within two to three days, releasing small legless maggots

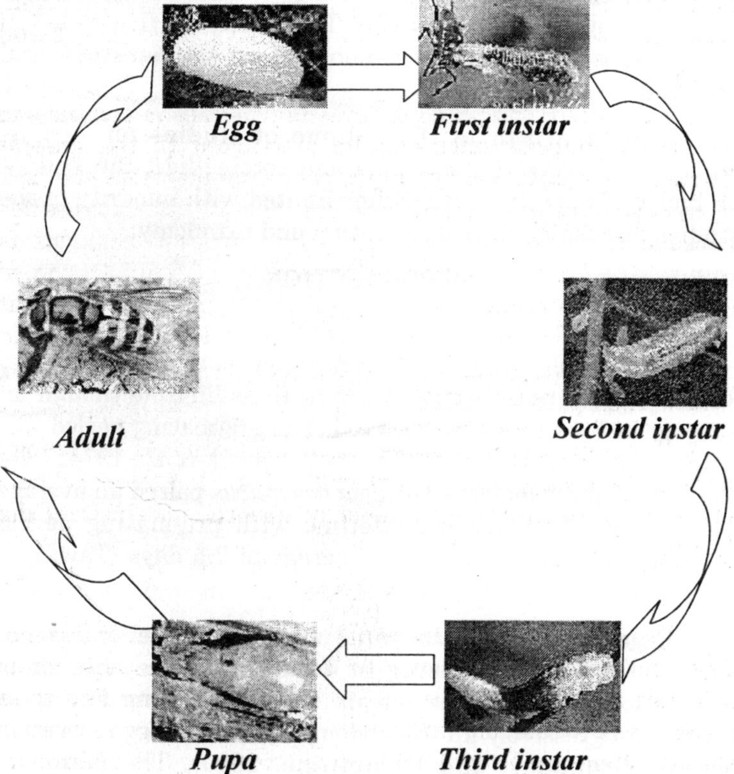

Fig. 4. Life cycle of a syrphid

about ¼″ long and usually transparent and in various hues of light green and grey (Figure 4). They look somewhat slug-like, and taper towards the head (Anonymous, 2003a; b). The larvae feed on aphids, small caterpillars, scale, thrips and other small larvae such as corn borer and corn earworm. They move by sticking their rear end to a substrate and striking around with their head in all directions. On coming contact with the aphid, the larva grabs the aphid and lifts it into the air to suck out the body fluids. After having fed for two to three weeks and having consumed up to 400 aphids, the larvae pupate either on the foliage or in the soil. The pupa is enclosed in a hardened, teardrop-shaped casing, which is the hardened skin of the last larval stage. During the growing season, adult hover flies emerge from the casing in one to two weeks. Otherwise, hoverflies overwinter as pupae. There may be five to seven generations each season, depending on temperature and availability of food.

In comparison to the volume of studies on various biological aspects of chrysopids and coccinellids, the studies on biology of syrphids are rather limited with majority of the studies discussing their distribution and taxonomy.

REPRODUCTION

Reproductive Biology

The females of syrphid, *Xanthogramma aegypticus* are known to pair upto twenty times in their lifetime though an average of 5.4 times was observed. The pre-mating period was found to range from 2-5 days at 26°C (Tawfik *et al.*, 1974a). On the other hand females of *Paragus aegypticus* paired an average number of 9.2 times in a lifetime with pre-mating period ranging from 1-13 days or an average of 7.5 days (Tawfik *et al.*, 1974b).

Eggs of *Episyrphus balteatus* only mature after emergence with the ovaries developing in 4 stages. In the absence of oviposition sites, females refrain from ovipositing and their ovaries progressively fill the abdomen and then egg resorption occurs (Branquart and Hemptinne, 2000). The potential fecundity, which is expressed by the ovariole number, the reproductive biomass and the abdomen volume, varies isometrically with the size of females. Egg size is much less variable and does not rise proportionally to body size. In laboratory conditions, females of *E. balteatus* might lay between 2000 and 4500 eggs during their life-time at a rate of 1 to 2 eggs per ovariole per day (Branquart and Hemptinne, 2000).

E. balteatus laid about 55-80 eggs when fed on *L. erysimi* (Khan and Yunus, 1970). The total number of eggs laid by *X. aegypticus* was an average of 67.9 on *A. gossypii* and 105.5 on *M. persicae*. *P. aegypticus*, however, had a low average fecundity of 28.2 eggs (Tawfik *et al.*, 1974b). *Metasyrphus corollae* laid an average number of 57.3 eggs in a lifetime and paired for an average of 4.6 times in a lifetime (Tawfik *et al.*, 1974c). Fecundity of *Eristalis cerealis* on different main diets was 3225, 2774, 553, and 1119 eggs on oak leaves, mud soils, fermented rice straw, and sawdust, respectively (Kim *et al.*, 1994). *Ischiodon scutellaris* laid 608 eggs, followed by *Paragus serratus* with 40 eggs when fed on *A. craccivora* (Joshi *et al.*, 1999). The mean fecundity per female *Sphaerophoria macrogaster* was 20.8 eggs (She *et al.*, 1998).

Unlike coccinellids where a trade-off is witnessed between fecundity and longevity, i.e. oviposition takes its toll through the reduction of longevity; the same could not be seen in the syrphids. Females of *Sphaerophoria. flavicauda* that mated and oviposited, lived longer than those that did not mate and oviposit. Mated males however had a shorter life span than the unmated ones (Tawfik *et al.*, 1974d; Makhmoor and Verma, 1987).

Oviposition Cues

Foraging theory indicates that aphid predators should lay their eggs early in the development of an aphid colony. Syrphids appear to respond to cues associated with the age of an aphid colony while assessing its quality for oviposition (Hemptinne *et al.*, 1993). The addition of artificial honeydew to potato fields increased the number of predators including syrphid predator, *Sphaerophoria cylindrica* but did not induce oviposition since the presence of aphids in the vicinity seems to be a prerequisite for oviposition (Ben Saad and Bishop, 1976). Females of *M. corollae*, a syrphid predator of *A. fabae*, responded to aqueous extract containing kairomones of the prey when topically applied on to clean leaves of *Vicia faba*. The aqueous extract of aphids stimulated females of *M. corollae* to lay eggs at all aphid densities, the number of eggs being directly proportional to aphid concentration (Shonouda, 1996). The oviposition behaviour of *E. balteatus* (Diptera: Syrphidae) was investigated and it was observed that females laid hardly any eggs in the absence of aphids and preferred larger aphid colonies to small ones. Honeydew attracted the females via olfaction and elicited oviposition. Females were able to perceive the presence of conspecific eggs and oviposited less often in aphid colonies in which eggs were present. The oviposition deterring stimulus was also active when the eggs were removed, hence, the deterrent did not act visually, but probably olfactorily (Scholz and Poehling, 2000).

FORAGING

Attraction to Prey Sites

Increased plant size reduces foraging efficiency of a syrphid, *Eupeodes corollae* at a given prey density. In addition, increased prey number was expected to increase production efficiency. As the number of aphids offered was increased, larvae of *E. corollae* killed more of them but left larger carcasses. Foraging efficiency (percentage of available aphids eaten) declined as aphid abundance and plant size increased. Foraging costs were not decreased by increasing prey density, and therefore production efficiency was not increased (Scott and Barlow, 1986).

Foraging rates for syrphids were slower in comparison to coccinellids, and did not vary with temperature (Brodsky, and Barlow, 1986). Syrphids like coccinellids are known to follow the optimal forging theory as has been demonstrated by Hemptinne *et al*., (1993). In choice experiments, *E. balteatus* demonstrated attraction using visual as well as olfactory cues by increasing their foraging in the presence of visual and chemical cues. *E. balteatus* exhibited enhanced behavioural responses to flowers painted yellow. Smaller artificial flowers were preferred over bigger flowers. *E. balteatus* also showed preference for the greatest nectar concentrations, whilst increasing pollen did not affect foraging behaviour (Sutherland *et al.*, 1999). Studies on the feeding behaviour and efficacy of the larvae of three aphidophagous syrphid species against four aphid species were conducted in the laboratory during 1985-86 which showed that *E. balteatus* was the most efficient in prey searching (69.8 sec.), followed by *Metasyrphus confrater*, while *I. scutellaris* was the least efficient (120.4 sec.) (Kumar *et al.*, 1996).

Prey-location behaviour of larvae of *E. balteatus* was investigated in two different experimental set-ups. First-instar larvae exhibited directed search over short distances, guided by olfactory cues from aphids, but not from honeydew. However, second and third instars did not respond to aphid-plant-complex odours in a 4-arm-olfactometer. Aphid extracts, honeydew and sucrose were also found to be feeding stimulants for the larvae (Bargen *et al.*, 1998).

Prey Consumption

Larvae of syrphids, *viz. E. confrater, Scaeva pyrastri, Macrosyrphus sp., E. balteatus, Eupeodes corollae and I. scutellaris* consumed 418.8 ±8.36, 384.9 ±8.80, 268.2 ±8.46, 252.0 ±11.89, 235.1 ±5.99 and 222.8 ±3.90 aphids, respectively in their lifetime. There were significant differences in the number of aphids consumed per larva between S. *pyrastri* and *Macrosyrphus* sp., *E. confrater* and *Macrosyrphus sp., S. pyrastri* and *E. balteatus,* and *E. balteatus* and *Eupeodes confrater* (Sharma and Bhalla, 1988). Larvae of *E. balteatus* consumed only 21 aphids/day when fed on *R. maidis* or *Schizaphis graminum*, but 52-56 aphids/day when fed on *Myzus obtusirostris* (Hamid, 1986). The 1st-, 2nd- and 3rd-instar larvae of syrphid *Paragus serratus* consumed 9.2, 33.0 and 34.0 individuals of *A. craccivora*, respectively. A single larva of the predator consumed an average of 258.3 aphids during its development (Patro and Behera, 1993). The predatory potential of the larvae of the hoverfly, *B. serarius*, on *Aphis fabae solanella* under laboratory conditions was investigated. To become a full-fed larvae, it moulted twice and consumed 187 to 263 aphids with an average of 232. The mean daily consumption of the first, second and third instar larva was 2.8, 11.7 and 30 aphids, respectively (Dayal and Gupta, 2001).

The last instar of *Betasyrphus fletcheri*, *Betasyrphus linga*, *Dideopsis aegrota*, *I. scutellaris*, *P. serratus* and *Paragus yerburiensis* was the most voracious on *A. craccivora* (Joshi *et al.*, 1999). During the larval stage of *S. macrogaster* the total number of prey consumed by each larva was 286.7 aphids, of which 98.2% was consumed by the 2nd- and 3rd-instar larvae (She *et al.*, 1998).

Eupeodes sp. consumed 40 individuals of *L. erysimi*, 44 individuals of *B. brassicae* and 52 individuals of *M. persicae* per day. Larvae consumed a total of 215.42 ± 15.2, 247.32 ± 16.2 and 416.26 ± 14.28 aphids in the three larval instars (Lakhanpal, and Desh Raj, 1998). The total number of prey consumed by larvae of *E. corollae* were 97.63, 73.25 and 90.38/larva for *Dysaphis devecta, Myzus persicae* and *Hyalopterus pruni*, respectively, while for *E. balteatus* these values were 86.22, 75.00 and 68.67 (Zeki and Kilincer, 1994). *I. aegyptius* appeared to be more voracious (consuming an average of 36 ± 3 aphids/day) than ladybeetle, *Scymnus* sp. (26 ± 5 aphids/day) (Ekukole and Misari 1993). The predation efficiency of larvae *of E. balteatus* was studied in the laboratory and in field cages on winter wheat. In the laboratory, 3-day-old syrphid larvae consumed 1140 and 660 3rd-instar nymphs of *Metopolophium dirhodum* at 14 and 22°C, respectively (Tenhumberg, and Poehling, 1994).

Laboratory observations of the predatory behaviour of *I. aegyptius* and *Scymnus* sp. at different densities of *A. gossypii* showed that prey consumption increased with increasing prey density. Both predators exhibited a Type II functional response to varying aphid densities (Ekukole and Misari 1993). The feeding rate of syrphid larvae corresponded to a type III functional response (Tenhumberg and Poehling, 1994). The type of functional response was found to depend on the instar of *I. scutellaris* larvae; type II for 1st-instar larvae, type III for 2nd-instar larvae and type I for 3rd-instar larvae (Singh and Mishra, 1988).

Five to seven-day-old larvae of *M. corollae* ingested the contents of the aphid *A. pisum at* a decreasing rate over time because the contents of the aphid became increasingly difficult to obtain as the body of the prey was emptied. Larvae that had been starved for 24 hours usually handled prey for a longer period than well-fed larvae. Younger starved larvae fed for a longer time than older starved larvae. Larvae that had been starved for 24 hours ate more of a prey item than well-fed larvae or larvae that had been starved for 36 hours. On average, larvae held a prey item until 71% of its dry mass had been extracted, after which it was discarded. The prey-handling time and amount of each prey item consumed were thus determined by the size and hunger of the syrphid larvae and the degree of depletion of the contents of the prey (Barlow and Whittingham, 1986).

GROWTH AND DEVELOPMENT

M. corollae completed normal development on aphids, *viz. Macrosiphum avenae, Rhopalosiphum padi, Acyrthosiphon dirhodum* and *A. pisum*. Larval development on the 4 species required 11.0, 10.2, 10.1 and 10.6 days, respectively. Complete development required 29.0, 30.6, 27.4 and 26.1 days, respectively. The pupae weighed 33.9-40.7gm. No significant differences were shown among the 4 prey species. The predator did not, however, complete its development on *B. brassicae* (Du and Chen, 1993). In a laboratory experiment in which one larva of *M. corollae* was supplied with 5, 10, 20, 40 and 80 2nd- to 3rd-instar nymphs of *M. dirhodum* daily, the larval development period of the predator was significantly longer when 5 aphids were supplied each day than when more prey were supplied. The larval survival rate, weight of pupae and the adult emergence rate increased as the prey density increased. Eighty aphid nymphs were sufficient for an *M. corollae* larva to complete normal development (Du and Chen, 1993).

The developmental period of *Betasyrphus fletcheri, B. linga, D. aegrota, I. scutellaris, P. serratus* and *P. yerburiensis* feeding on the cowpea aphid, *A. craccivora* from egg to adult emergence ranged from 20.10 to 25.80 days at 26 ±2°C and 75±2% RH. *I. scutellaris* showed the greatest longevity (36 days) and fecundity. Due to its short developmental time, high fecundity, longevity and relatively large consumption, *I. scutellaris* was concluded to be the most suitable species for mass rearing (Joshi *et al.*, 1999). *Eupeodes* sp. completed larval development in 23.30 ± 1.89, 19.50 ± 1.72 and 22.50 ± 4.74 days on *L. erysimi, B. brassicae* and

M. persicae, respectively (Lakhanpal and Desh Raj, 1998). Development of eggs, larvae and pupae of *Melangyna viridiceps* and *Symosyrphus grandicornis* fed on *Macrosiphum rosae*, was studied at constant temperatures of 4, 8, 10, 15, 20, 25 or 30°C. Although eggs hatched at 4°C, first-instar larvae died at this temperature. The developmental rates were directly proportional to the temperatures (Soleyman-Nezhadiyan and Laughlin, 1998).

INTRAGUILD PREDATION

Larval interaction between *Scymnus* and syrphid predators was investigated in relation to effectiveness of wax cover of *Scymnus* against predation from syrphids. Waxless first instar larvae were not protected but wax-covered larvae of second, third and fourth instars were protected from predation by syrphid larvae. The protection was lower in the second instar which has a thin wax cover and significantly higher in the third and fourth instars having a thick wax cover. In addition, larvae from which the wax was removed were significantly more vulnerable to predation. Vulnerability of *Scymnus* larvae to predation from syrphids was directly related to the thickness of wax cover (Agarwala and Yasuda, 2001).

Eggs and first- and second-instar larvae of *E. balteatus* were highly susceptible to predation by the other predators, while only the larvae of *C. carnea* preyed upon the pupae of *E. balteatus*. Eggs and first- and second-instar larvae of *E. balteatus* sustained intraguild predation, irrespective of arena size and extraguild prey, while predation on third-instar larvae of *E. balteatus* was significantly reduced (Hindayana *et al*., 2001).

In an interesting result when individuals of *A. bipunctata* and *E. balteatus* were released together in field experiments in the year 1997, for the management of *D. plantaginea*, additive results were obtained. The two species did not significantly interact and thus an additive model best explains their joint effect. Combined releases of the two predator species reduced aphid densities to 5% of the control (Wyss *et al*., 1999).

CONCLUSIONS

Chrysopids, coccinellids and syrphids are major predators of aphids and commonly occur in aphidophagous guilds. They individually or in combination play a major role in the management of aphid populations in nature. Despite the occurrence of these three major predatory groups, aphid outbreaks are still quite common. There is, therefore, a need of lending these three groups further assistance by either enhancing the number of existing predators through enhancement of natural populations or through the introduction of new exotic species. The introduction of new exotic species, however, remains a risk without detailed and extensive research. At times, the introduction of a new species has been known to backfire with danger of extinction of native species becoming imminent.

The above comparative account reveals coccinellids to be more voracious, with both adults and larvae predatory in nature. Also, coccinellids are more fecund than both chrysopids and syrphids and have a longer oviposition period. They also lend themselves easily for mass multiplication, though the same is also true for the other two predatory groups. The coccinellids require lesser infrastructure for mass multiplication. But, it is for the chrysopids that extensive researches on rearing techniques have been conducted. Despite these advantages, when it comes to competitive ability, coccinellids lag behind the chrysopids. Syrphids on the other hand are comparatively sluggish, less voracious and easily outclassed in the fight for supremacy in an aphidophagous guild. Their commercial supply is still not available, though chrysopids and coccinellids can be purchased in plenty.

On the basis of comparisons, we would suggest the use of chrysopids or coccinellids (depending on the prey and habitat) followed by the syrphids for the management of aphid populations. A combination of compatible predators of all the three groups would be a boon but if left to introduction of a single species, chrysopids and coccinellids are an easy choice over the syrphids.

REFERENCES

Afroze, S. (1999). Preferential predation by *Anegleis cardoni* (Weise) (Coccinellidae: Coleoptera) on some aphid species, *J. Aphidol.* 13: 95-97.

Agarwala, B.K. and Bardhanroy, P. (1997). Oviposition behaviour and reproduction efficiency in ladybeetles (Coccinellidae: Coleoptera). A case study of *Menochilus sexmaculatus* (Fabr.) *J. Aphidol.* 11(1): 49-59.

Agarwala, B.K., Das, S. and Chaudhuri, M. S. (1988). Biology and food relations of *Micraspis discolor* (F.): An aphidophagous coccinellid of India. *J. Aphidol.* 2 (1&2): 7-17.

Agarwala, B.K. and Dixon, A.F.G. (1992). Laboratory study of cannibalism and interspecific predation of ladybeetles. *Ecol. Entomol.* 17: 303-309

Agarwala, B.K. and Yasuda, H. (2001). Larval interactions in aphidophagous predators: effectiveness of wax cover as defence shield of *Scymnus* larvae against predation from syrphids. *Ent. Exp. Appl.* 100(1):101-107

Albuquerque, G.S., Tauber, C.A. and Tauber, M.J. (1994). *Chrysoperla externa* (Neuroptera: Chrysopidae): life history and potential for biological control in Central and South America. *Biol. Cont.* 4 (1): 8-13.

Anonymous (1995). Technology for production of natural enemies. *Technical Bulletin No. 4* (ICAR). Ed. S.P.Singh.

Anonymous (2002 a). www.rain.org/~sals/lace.html.

Anonymous (2002b). www.entomology.wicu/mbcn/fea702.html

Anonymous (2003a). http://www.agric.wa.gov.au/cropupdates/1998/lupins/aphid.htm; http://www.bugwood.org/sl96/pecan.html

Anonymous (2003b). ttp://www.bugwood.org/sl96/pecan.html

Anonymous (2003c). http://www.agric.wa.gov.au/results.pasp?pid=1 & search=aphid

Anonymous (2003d). http://www.cals.wisc.edu/sciencereport/02SRstories/Uninvited-2.html

Arakaki, N. (1988). Egg protection with faeces in the ladybeetle, *Pseudoscymnus kurohime* (Miyatake) (Coleoptera: Coccinellidae). *Appl. Entomol. Zool.* 23: 495-497.

Atwal, A.S. and Sethi, SL. (1963). Biochemical basis for food preference of a predator beetle. *Curr. Sci.* 32:511-512.

Ba M'Hamed, T. and Chemseddine, M. (2001). Assessment of temperature effects on the development and fecundity of *Pullus mediterraneus* (Col., Coccinellidae) and consumption of *Saissetia oleae* eggs (Hom., Coccoida). *J. Appl. Entomol.* 125(9/10):527-531.

Badgujar, M.P., Deotale, V.Y., Sharnagat, B.K. and Nandanwar, V.N. (2000). Performance of *Chrysoperla carnea* against safflower aphid, *Dactynotus carthami* (HRL). *J. Soils Crops.* 10(1): 125-127.

Bakthavatsalam, N., Singh, S.P., Tandon, P.L., Chaudhary, M. and Preethi, S. (2000). Synomone mediated behavioural responses of *Chrysoperla carnea* (Stephens)(Neuroptera: Chrysopidae) to cotton infested by *Helicoverpa armigera* (Hübner) (Lepidoptera: Noctuidae). *J. Biol. Cont.* 14 (2):1-6.

Balasubramani, V. and Swamiappan, M. (1994). Development and feeding potential of the green lacewing *Chrysoperla carnea* Steph. (Neur., Chrysopidae) on different insect pests of cotton. *Anzeiger für Schädlingskunde, Pflanzenschutz, Umweltschutz.* 67 (8): 165-167.

Ballal, C.R. and Singh, S.P. (1999). Host plant-mediated orientational and ovipositional behavior of three species of chrysopids (Neuroptera: Chrysopidae). *Biol. Cont.* 16 (1): 47-53.

Bargen, H., Saudhof, K. and Poehling, H. M. (1998). Prey finding by larvae and adult females of *Episyrphus balteatus. Ent. Exp. Appl.* 87(3): 245-254.

Barlow, C.A.and Whittingham, J.A. (1986). Feeding economy of larvae of a flower fly, *Metasyrphus corollae* (Dip.: Syrphidae): partial consumption of prey. *Entomophaga.* 31(1): 49-57.

Baumgaertner, J.U., Guitierrez, A.P. and Summers, C.G. (1981). The influence of aphid prey consumption on searching behaviour, weight increase, developmental time, and mortality of *Chrysopa carnea* (Neuroptera: Chrysopidae) and *Hippodamia convergens* (Coleoptera: Coccinellidae) larvae. *Can. Ent.* 114: 1007-1014.

Bell, W J. (1990). Searching behaviour patterns in insects. *Ann. Rev. Entomol.* 35: 447-467.

Ben Saad, A.A. and Bishop, G.W. (1976). Effect of artificial honeydews on insect communities in potato fields. *Environ. Entomol.* 5(3): 453-457

Benrey, B. and Lamp, W.O. (1994). Biological control in the management of planthopper populations. In: *Planthoppers: Their Ecology and Management* (ed.R.F.Denno & T.J. Perfect) pp. 519-550. New York: Chapman and Hall.

Bind, R.B. 1998. *Bioecology and behaviour of a ladybird beetle,* Cheilomenes (=Menochilus) sexmaculata *(Fabricius) (Coleoptera: Coccinellidae).* Ph.D. Thesis. University of Lucknow. p. 164.

Blackman R L (1967). The effects of different aphid foods on *Adalia bipunctata* L. and *Coccinella 7-punctata* L. *Ann. Appl. Biol.* 59: 207-219.

Branquart, E. and Hemptinne, J.L. (2000). Development of ovaries, allometry of reproductive traits and fecundity of *Episyrphus balteatus* (Diptera: Syrphidae). *Eur. J. Entomol.* 97(2): 165-170.

Brodsky, L.M. and Barlow, C. A. (1986). Escape responses of the pea aphid, *Acyrthosiphon pisum* (Harris) (Homoptera: Aphididae): influence of predator type and temperature. *Can. J. Zool.* 64(4): 937-939.

Burke, H. R. and Martin, D. F. (1956). The biology of three chrysopid predators of the cotton aphid. *J. Econ. Entomol.* 49: 498-500.

Butler, G.D. Jr. (1982). Development time of *Coccinella septempunctata* in relation to constant temperatures (Coleoptera: Coccinellidae). *Entomophaga*. 27(3): 349-353.

Calunga, G.M. and Gage, S.H. (1999). Arrival, establishment, and habitat use of the multicoloured Asian ladybeetle (Coleoptera: Coccinellidae) in a Michigan landscape *Environ. Entomol.* 27: 1574-1580.

Canard, M., Carvalho, C. F. and Sissoko, F. (1994). Diapause in *Chrysoperla mediterranea* (Hölzel, 1972): influence of photoperiod on the duration of the pre-ovipositional period (Neuroptera, Chrysopidae). *Bull. Soc.Entomol.France*. 99 (5): 455-461.

Capinera, J.L. and Walmsley, M.R. (1978). Visual response of some sugar beet insects to sticky traps and water pan raps of various colors. *J. Econ. Entomol.* 71: 926-927.

Carole, A.S., Cheah, J. and McClure, M.S. (2000). Seasonal synchrony of life cycles between the exotic predator, *Pseudoscymnus tsugae* (Coleoptera: Coccinellidae) and its prey, the hemlock woolly adelgid *Adelges tsugae* (Homoptera: Adelgidae). *Agric. Forest Ent.* 2(4): 241.

Carter, M.C. and Dixon, A. F.G (1982). Habitat quality and foraging behaviour of coccinellids. *J. Anim. Ecol.* 51: 865-878.

Carter, M.C. and Dixon, A.F.G. (1984). Honeydew an arrestant stimulus for coccinellids. *Ecol. Entomol.* 9: 383-387.

Carvalho, C. F., Canard, M., and Lauzet, C. (1996). Comparison of the fecundities of the neotropical green lacewing *Chrysoperla externa* (Hagen) and the West Palaearctic *Chrysoperla mediterranea* (Hölzel) (Insecta: Neuroptera: Chrysopidae). In: *Pure and applied research in neuropterology*. Proceedings of the

Fifth International Symposium on Neuropterology Cairo, Egypt, 2-6 May 1994 (Insecta: Neuroptera, Megaloptera, Raphidioptera). Eds Canard, M., Aspock, H.and Mansell, M. W. 103-107.

Chang, G.C. (1996). Comparison of single versus multiple species of generalist predators for biological control. *Environ. Entomol.* 25: 207-212

Charnov, E.L. (1976). Optimal foraging: the marginal value theorem. *Theo. Pop. Biol.* 9: 126-136.

Choudhury, D., David, B.A. and Singh, D.R. (1983). Studies on the host preference of *Cheilomenes sexmaculata* (Fabr.) (Coleoptera: Coccinellidae). *Comp. Physiol. Ecol.* 8(4): 289-290.

Clark, T. L. and Messina, F. J. (1998). Foraging behavior of lacewing larvae (Neuroptera: Chrysopidae) on plants with divergent architectures. *J. Insect Behavi* 11(3): 303-317.

Colfer, R.G. and Rosenheim, J.A. (1995). Intraguild predation by coccinellid beetles on an aphid parasitoid, *Lysiphelbus testaceipes;* In *1995 Proc. Beltwide Cot. Conf.* 2:1033-1036.

Cooke, W.C. (1963). Ecology of the pea aphid in the Blue mountain area of Eastern Washington and Oregon. *Tech. Bull. U.S. Dep. Agric.* 1287:48 pp.

Costa, R. I. F. da, Macedo, L. P. M. de, Almeida, S. A. de and Soares, J. J. (1999). Oviposition potential and longevity of *Chrysoperla externa* (Hagen, 1861) (Neuroptera: Chrysopidae) in the laboratory. *Anais II Congresso Brasileiro de Algodão: O algodão no século XX, perspectivar para o século XXI, Ribeirão Preto, SP, Brasil, 5-10 Setembro 1999*. 253-255.

Dayal, K. and Gupta, P. R. (2001). Predatary potential of the larvae of the hoverfly, *Betasyrphus serarius* (Wiedemann) on *Aphis fabae solanella* (Theobald). *Insect Environment*. 6(4):150-151

DeBach, P. (Ed.) (1964). *Biological Control of Insect Pests and Weeds*. Chapman & Hall, London. 844 pp.

Deotale, V.Y., Nimbalkar, S.A., Shamagat, B.K. and Nandanwar, V.N. (1998). Effect of feeding different pollen grains on fecundity of *Chrysoperla carnea* Stephens. *Journal of Soils and Crops.* 8 (2): 223.

Dimetry, N.Z. (1976). The role of predator and prey density as factors affecting behavioural and biological aspects of *Adalia bipunctata* (L.) larvae. *Z. Angew. Entomol*. 81:386-392.

Ding-Xin, Z., and Zhong-Wen, W. (1987). Influence of temperature on the development of the coccinellid beetle, *Scymnus hofmanni* Weise. *Acta Entomol.Sinica* 30: 47-54.

Dinter, A (1998). Interactions between spider and lacewing predators (Araneae: Erigonidae and Neuroptera: Chrysopidae) and their effects on the grain aphid *Sitobion avenae* Fab. (Homoptera: Aphididae). *Bulletin OILB/SROP*. 21 (8): 91-101.

Dixon, A.F.G. (1959). An experimental study in the searching behaviour of the predatory coccinellid beetle *Adalia decempunctata* (L.). *J. Anim. Ecol.* 28 259-281.

Dixon, A.F.G. (1997). Patch quality and fitness in predatory ladybeetles *Ecol. Stud.* 130: 205- 223.

Dixon, A.F.G. (1998). *Aphid Ecology*. Chapman & Hall. pp. 300.

Dixon, A.F.G. (2000). *Insect Predator- Prey Dynamics: Ladybird Beetles and Biological Control.* Cambridge University Press. pp. 257.

Doumbia, M., Hemptinne, J.L. and Dixon, A.F.G. (1998). Assessment of patch quality by ladybird: role of larval tracks. *Oecologia.* 113(2): 197-202.

Du, Y. Z. and Chen, X. Z. (1993). Influence of different aphid prey on the development of *Metasyrphus corollae* (Dip.: Syrphidae). *Chinese J. Biol. Cont.* 9(3):111-113.

Dumbleton, L.D. (1936). The biological control of fruit pests in New Zealand. *N.Z. JL.Sci. Technol.* 18:588-595.

Ehler, L.E. (1990). Introduction strategies in biological control of insects. In: *Critical Issues in Biological Control* (Eds M Macauer, L.E. Ehler and J Roland), Andover: Intercept. 111-134.

Eigenbrode, S.D., Kabalo, N.N. and Stoner, K.A. (1999). Predation, behavior, and attachment by *Chrysoperla plorabunda* larvae on *Brassica oleracea* with different surface waxblooms. *Ent. Exp. Appl.* 90(3): 225-235.

Ekukole, G. and Misari, S. M. (1993). Effect of prey density on the predatory behaviour of *Ischiodon aegyptius* Weid. and *Scymnus* sp. on the cotton aphid, *Aphis gossypii* Glover in Cameroon. *Samaru J. Agric. Res.* 10:55-62.

El-Kareim, A.I.A. (1998). Searching rate and potential of some natural enemies as bioagent against the cotton whitefly, *Bemisia tabaci* Genn. (Hom., Aleyrodidae). *J. Appl. Entomol.* 122 (8): 487-492.

Evans, E.W. (1991). Intra versus interspecific interactions of ladybeetles (Coleoptera: Coccinellidae) attacking aphids. *Oecologia* 87: 401-408.

Ferguson, K.I. and Stiling, P. (1996). Non-additive effects of multiple natural enemies on aphid populations. *Oecologia* 108: 375-379.

Ferran, A. and Larroque, M.M. (1977). Etude des relations hote predateur: la consommaion et l'utilisationd'unpuceron, *Myzus persicae* Sulz. Par les differents stades larvaires de la Coccinelle *Semiadalia undecimnotata* Schn. (Col. Coccinellidae). *Ann. Zool-Ecol. Anim.* 9: 665-691.

Figueira, L.K., Carvalho, C.F. and Souza, B. (2000). Biology and thermal requirements of *Chrysoperla externa* (Hagen, 1861) (Neuroptera: Chrysopidae) fed on *Alabama argillacea* (Hübner, 1818) (Lepidoptera: Noctuidae) eggs. *Ciência e Agrotecnologia.* 24 (2): 319-326.

Fonseca, A.R., Carvalho, C.F. and Souza, B. (2000). Functional response of *Chrysoperla externa* (Hagen) (Neuroptera: Chrysopidae) fed on *Schizaphis graminum* (Rondani) (Hemiptera: Aphididae). *Anais da Sociedade Entomológica do Brasil.* 29 (2): 309-317.

Freier, B. and Triltsh, H. (1996). Climate chamber experiments and computer simulations on the influence of increasing temperature on wheat-aphid-predator interactions. *Implications of 'Global environmental change' for crops in Europe, 1-3 April 1996, Churchill College, Cambridge, UK. Aspects-of-Applied-Biology.* 45: 293-298.

Freitas, S. de, and Scaloppi, E.A.G. (1996). Effect of spraying molasses on maize on a population of *Chrysoperla externa* (Hagen) and the distribution of eggs on plants. *Revista de Agricultura* 71(2):251-258.

Gautam R.D. and Narvarajan Paul, A.V. (1988). Influence of adult food supplements on *Chrysopa scelestes* Banks (Chrysopidae: Neuroptera). *J. Entomol. Res.* 12(1): 125-127.

Geethalakshmi, L., Muthukrishnan, N., Chandrasekaran, M. and Raghuraman, M. (2000). Chrysopids biology on *Corcyra cephalonica* and feeding potential on different host insects. *Ann. Plant Pro. Sci.* 8(2):132-135.

Gutierrez, A.P., Neuenschwander, P., Schulthess, F., Herren, H.R, Baumgaertner, J.U., Wermelinger, B., Lohr, B. and Ellis, C.K. (1988). Analysis of biological control of cassava pests in Africa II. Cassava mealybug, *Phenacoccus manihoti. J. Appl. Ecol.* 25 921-940.

Hagen, K.S (1962). Biology and ecology of predaceous coccinellidae. *Ann. Rev. Entomol.* 7:289-326.

Haji-Zadeh, J., Kamali, K. and Assadi, H.B. (1994). Studies on functional response and population dynamics of *Stethorus gilvifrons* (Mulsant) to *Panonychus ulmi* in Karaj vicinity. *Appl. Entomol. Phytopathol.* 61(1&2): 32-34.

Hamid, S. (1986). Aphid control potential of *Episyrphus balteatus* (DeGeer). Proceedings of the 5th Pakistan Congress of Zoology, University of Karachi, Karachi, January 8-11, 1987, 153-158.

Hatami, B. and Zibai, K. (2000). Laboratory study of interspecific predation of green lacewing, *Chrysoperla carnea* (Steph.) and ladybeetle, *Hippodamia variegata* (Goeze). *J. Sci. Tech. Agric. Nat. Res.* 4 (3): 99-110.

Hattingh, V. and Samways M.J. (1992). Prey choice and substitution in *Chilocorus* spp. (Coleoptera: Coccinellidae). *Bull. Ent. Res.* 82: 327-334.

Hattingh, V. and Samways, M.J. (1995). Visual and olfactory location of biotopes, prey patches, and individual prey by the ladybeetle *Chilocorus nigritus*. *Ent. Exp. Appl.* 75: 87-98.

Hemptinne, J.L., Dixon, A.F.G. and Coffin, J. (1992). Attack strategy of ladybird beetles (Coccinellidae): factors shaping their numerical response. *Oecologia* 90: 238-245.

Hemptinne, J. L., Dixon, F. G., Doucet, J.L. and Petersen, J.E. (1993). Optimal foraging by hoverflies (Diptera: Syrphidae) and lady birds (Coleoptera: Coccinellidae): mechanisms. *Eur. J. Entomol.* 90(4):451-455.

Hemptinne, J.L., Dixon, A.F.G. and Gauthier, C. (2000a). Nutritive cost of intraguild predation on eggs of *Coccinella septempunctata* and *Adalia bipunctata* (Coleoptera: Coccinellidae). *Eur. J. Entomol.* 97(4): 559-562.

Hemptinne, J.L., Lognay, J., Gauthier, C., and Dixon, A.F.G. (2000b). Role of surface chemical signals in egg cannibalism and intraguild predation in Ladybeetles (Coleoptera : Coccinellidae). *Chemoecology* 10: 123-128.

Hemptinne, J.L., Gaudin, M., Dixon, A.F.G., and Lognay, J. (2000c). Social feeding in ladybird beetles: adaptive significance and mechanism. *Chemoecology* 10: 149-152.

Hemptinne, J.-L., Dixon, A.F.G. and Adam, B. (2001). Do males and females of the two-spot ladybird, *Adalia bipunctata* (L.), differ in when they mature sexually? *J. Insect Behav.* 14(3): 411-419.

Hindayana, D., Meyhofer, R., Scholz, D. and Poehling, H. M. (2001). Intraguild predation among the hoverfly *Episyrphus balteatus* deGeer (Diptera: Syrphidae) and other aphidophagous predators. *Biol. Cont.* 20(3): 236-246.

Hodek, I. (1957). The influence of *Aphis sambuci* L. as food for *Coccinella 7-punctata* L. II. *Cas. Cs. Spol. Entomol.* 54: 10-17.

Hodek, I. (1960). The influence of various aphid species as food for two lady-birds *Coccinella 7-punctata* L. and *Adalia bipunctata* L. In: *The Ontogeny of Insects*. (Proc. Symp. Praha 1959). Academia, Prague. pp. 314-316.

Hodek, I. and Honek, A. (1996). *Ecology of Coccinellidae*. Kluwer Academic Publishers Dordrecht Boston London. p. 464.

Holling, C.S. (1959). Some characteristics of simple types of predation and parasitism. *Can. Ent.* 91: 385-398.

Hukusima, S. and Kamei, M. (1970). Effects of various species of aphids as food on development, fecundity and longevity of *Harmonia axyridis* Pallas (Coleoptera: Coccinellidae). *Res. Bull. Fac. Agric. Gifu. Univ*. 29: 53-66.

Hukusima, S. and Ohwaki, T. (1972). Further notes on feeding biology of *Harmonia axyridis* Pallas (Coleoptera: Coccinellidae). *Res. Bull. Fac. Agric. Gifu. Univ*. 33: 75-82.

Hydorn, S.B. (1971). Food preferences of *Chrysopa rufilabris* (Burmeister) in north central Florida. M. S. Thesis, Univ. Florida., Tallahassee. Florida p. 79.

Iperti, G. (1966). The choice of oviposition sites in aphidophagous Coccinellidae. In: Hodek I. (ed.): Ecology of Aphidophagous Insects. *Academia, Prague and Dr. W. Junk, The Hague*. pp. 121-122.

Iperti, G. and Prudent, P. (1986). Effect of the substrate properties on the choice of oviposition sites by *Adalia bipunctata* In: Hodek I. (ed.): Ecology of Aphidophaga. *Academia, Prague and Dr. W. Junk, Dordrecht*. pp. 143-149.

Iperti, G. and Quilici, S. (1986). Some factors influencing the selection of oviposition site by *Propylea quatuordecimpunctata*. In: Hodek, I., (ed), *Ecology of Aphidophaga*. Academia, Prague & Dr. W. Junk, Dordrecht, pp.137-142.

Jalali, S.K. and Singh, S.P. (1992). Biology and feeding potential of *Curinus coeruleus* (Mulsant) and *Chrysoperla carnea* (Stephens) on subabul psyllid, *Heteropsylla cubana* Crawford. *J. Insect Sci.*. 5 (1): 89-90.

Jalali, S.K., Singh, S.P. and Biswas, S.R. (1999). Effect of temperature and female age on the development and progeny production of *Cryptolaemus montrouzieri* Mulsant (Coleoptera: Coccinellidae). *Entomon* 24(3): 293-296.

James, B.E. (2001). *Contribution on certain aspects of bioecology and behaviour of a ladybeetle,* Coccinella transversalis Fabricius *(Coccinellidae: Coleoptera).* Ph. D. Thesis. University of Lucknow, Lucknow, India. p. 190.

Jarosik, V., Honek, A. and Dixon, A.F.G. (2000). Rate isomorphy in insects and mites. In: A.F.G. Dixon (ed.). *Insect Predator- Prey Dynamics, Ladybird Beetles and Biological Control*. Cambridge University Press.

Jean, C., Coderre, D. and Tourneur, J.C. (1990). Effect of temperature and substrate on survival and lipid consumption of hibernating *Coleomegilla maculata lengi* (Coleoptera: Coccinellidae). *Environ. Entomol.* 19(6): 1657-1662.

Joshi, S. Venkatesan, T. Ballal, C.R. and Rao, N.S. (1999). Comparative biology and predatory efficiency of six syrphids on *Aphis craccivora* Koch. *Pest Management in Horticultural Ecosystems*. 5(1): 1-6

Kabissa, J.C.B., Kayumbo, H.Y. and Yarro, J.G. (1993). Comparative biology of *Mallada desjardinsi* (Navas) and *Chrysoperla congrua* (Walker) (Neuroptera: Chrysopidae), predators of *Helicoverpa armigera* (Hubner) (Lepidoptera: Noctuidae) and *Aphis gossypii* (Glover) (Homoptera: Aphididae) on cotton in eastern Tanzania. *Int. J. Pest Man.* 41(4):214-218.

Kaddou, I. (1960). The feeding behaviour of *Hippodamia quinquesignata* (Kirby). *Univ. Calif. Publ. Ent.* 16, 181-232.

Kajita, Y., Takano, F., Yasuda, H. and Agarwala, B. K. (2000). Effects of indigenous ladybird species (Coleoptera: Coccinellidae) on the survival of an exotic species in relation to prey abundance. *Appl. Entomol. Zool*. 35(4): 473-479.

Kawauchi, S. (1979). Effects of temperature on the aphidophagous coccinellids. *Kurume. Univ .J.* 28(1): 47-52.

Khan, S. and Yunus, M. (1970). Biology of *Syrphus balteatus* DeGeer. *Pakistan J. Zool.* 2(2):215-217.

Kim InSoo, Uhm KiBaik, Goh HyunGwan and Choi KuiMoon (1994) Selection of artificial diet for mass rearing of the pollinator, *Eristalis cerealis* Fabricius. *RDA J. Agric. Sci., Crop Protec.* 36(2):369-372.

Kindlmann, P. and Dixon, A.F.G. (1993). Optimal foraging in ladybird beetles (Coleoptera: Coccinellidae) and its consequences for their use in biological control. *Eur. J. Entomol.* 90(4): 443-450.

Klingen, I., Johansen, N.S. and Hofsvang, T. (1996). The predation of *Chrysoperla carnea* (Neurop., Chrysopidae) on eggs and larvae of *Mamestra brassicae* (Lep., Noctuidae). *J. Appl. Entomol.* 120(6):363-367.

Kozlowski, J., and Wiegert, R.G. (1986). Optimal allocation of energy to growth and reproduction. *Theor. Pop. Biol.* 29: 16-37.

Krishnamoorthy, A. and Mani, M. (1989). Records of green lacewings preying on mealybugs in India. *Curr. Sci.* 58 (3):155-156.

Kumar, A. Kapoor, V. C. and Mahal, M. S. (1996). Feeding behaviour and efficacy of three aphidophagous syrphids. *J. Insect Sci.*, 9(1):15-18.

Kumar, A., Kumar, N., Siddiqui, A. and Tripathi, C.P.M. (1999). Prey-predator relationship between *Lipaphis erysimi* Kalt. (Hom., Aphididae) and *Coccinella septempunctata* L. (Col., Coccinellidae). II. Effect of host plants on the functional response of the predator. *J. Appl.Ent.* 123(10): 591-596.

Lakhanpal, G.C. and Desh Raj (1998). Predation potential of coccinellid and syrphid on important aphid species infesting rapeseed in Himachal Pradesh. *J. Entomol. Res.* 22(2):181-190.

Legaspi, J.C., Nordlund, D.A. and Legaspi, B.C.Jr. (1996). Tri-trophic interactions and predation rates in *Chrysoperla* spp. attacking the silverleaf whitefly. *Southwest. Entomol.* 21(1): 33-42.

Liu C Y (1933). Notes on the biology of two giant coccinellids in Kwangsi (*Caria dilatata* Fabr. and *Synonchia grandis* Thunbg.) with special reference to the morphology of *Caria dilatata*. *Y. B. Bur. Entomol. Hangchow,* 1:205-250.

Löchte, C. and Sengonca, C. (1995). Applying the eggs of *Chrysoperla carnea* (Stephens) using a spray technique for biological control of aphids in the field. *Mitteilungen der Deutschen Gesellschaft für Allgemeine und Angewandte*. 10 (1-6): 243-246.

Losey, J.E. and Denno, R.F. (1998). Positive predator-predator interactions: enhanced predation rates and synergistic suppression of aphid populations. *Ecology* 79:2143-2152.

Lucas, E., Coderre, D. and Brodeur, J. (1997). Instar specific defence of *Coleomegilla maculata lengi* (Col.: Coccinellidae): influence on attack success of the intraguild predator *Chrysoperla rufilabris* (Neur.: Chrysopidae). *Entomophaga* 42:3-12

Lucas, E., Coderre, D. and Brodeur, J. (1998). Intraguild predation among aphid predators : characterization and influence of extraguild prey density. *Ecology*. 79:1084-1092.

Macedo, L.P.M.de, Costa, R.I.F.da, Guerra, C.L. and Soares, J.J. (1999). Study of reproductive variation in *Chrysoperla externa*

(Hagen, 1861). (Neuroptera: Chrysopidae) in the laboratory. *Anais II Congresso Brasileiro de Algodão: O algodão no século XX, perspectivar para o século XXI, Ribeirão Preto, SP, Brasil, 5-10 Setembro 1999*. 250-252.

Mahabaleshwar, H. and Kulkarni, K.A. (2000a). Response of *Chrysoperla carnea* (Stephens) to kairomonal substances. *Karnataka J. Agric. Sci.* 13 (2):445-447.

Mahabaleshwar, H. and Kulkarni, K.A. (2000b). Mass rearing of *Chrysoperla carnea* (Stephens) by using wood shavings as filler material. *Karnataka J. Agric. Sci.*13(1): 184-186.

Majerus, M.E.N. (1994). Female promiscuity maintains high fertility in ladybirds (Coleoptera: Coccinellidae). *Entomologist's Mon. Mag.* 130(1564-67): 205-209.

Makhmoor, H.D. and Verma, A.K. (1987). Bionomics of major aphidophagous syrphids occurring in mid-hill regions of Himachal Pradesh. *J. Biol. Cont.* 1(1):23-31.

Malcolm, S.B. (1992). Prey defence and predator foraging. In: *Natural Enemies: The population biology of predators, parasites and diseases*, M. J. Crawley. Oxford: Blackwell Scientific Publication pp. 458-475.

Malleshaiah, Rajagopal, B.K. and Gowda, K.N.M. (2000). Feeding potential of *Chrysoperla carnea* (Neuroptera: Chrysopidae) of different stages of citrus mealybug, *Planococcus citri* (Hemiptera: Pseudococcidae). *Crop Research*. 20(1): 126-129.

Mannan, V.D., Varma, G.C. and Brar, K.S. (1997). Biology of *Chrysoperla carnea* (Stephens) on *Aphis gossypii* Glover and *Myzus persicae* (Sulzer). *J. Insect Sci.*10 (2): 143-145.

Marks R.J. (1977). Laboratory studies of plant searching by *Coccinella septempunctata* L. larvae. *Bull. Ent. Res.* 67:235-241.

Mathew, M.J., Venugopal, M.N. and Saju, K.A. (1999). Predatory potential of green lacewing on cardamom aphid. *Insect Environment*. 4(4): 152-153.

McClure, M.S. (1987). Potential of Asian predators, *Harmonia axyridis* Pallas (Coleoptera: Coccinellidae), to control *Matsucoccus resinosea* Bean and Godwin (Homoptera: Margarodidae) in the United States. *Environ. Entomol.*16: 224-230.

McEwen, P.K. (1996). Viability of green lacewing *Chrysoperla carnea* Steph. (Neuropt., Chrysopidae) eggs stored in potential spray media, and subsequent effects on survival of first instar larvae. *J. Appl. Ent.* . 120 (3):171- 173.

McEwen, P.K., Jervis, M.A. and Kidd, N.A.C. (1994). Use of a sprayed L-tryptophan solution to concentrate numbers of the green lacewing *Chrysoperla carnea* in olive tree canopy. *Ent. Exp. Appl.* 70 (1):97-99.

Merlin, J., Lemaitre, O. and Gregoire, J.C. (1996). Chemical cues produced by conspecific larvae deter oviposition by the coccidophagous ladybird beetle, *Cryptolaemus montrouzieri. Ent. Exp. Appl.* 79: 147-151.

Messina, F.J., Jones, T.A. and Nielson, D.C. (1995). Host plant affects the interaction between the Russian wheat aphid and a generalist predator, *Chrysoperla carnea. J. Kansas Entomol. Soc.* 68 (3):313-319.

Miller, J.C. (1992). Temperature-dependent development of the convergent lady beetle (Coleoptera: Coccinellidae). *Environ. Entomol.* 21: 1139-1142.

Mills, N.J. 1981. Essential and alternative foods for some British Coccinellidae (Coleoptera). *Entomol. Gaz.*, 32: 197-202.

Mirabzadeh, A., Sahragard, A. and Azma, M. (1998). Functional response of *Chrysoperla carnea* Steph. larvae to *Pterochloroides persicae* , *Uroleucon (Urol.) cichorii* and *Aphis craccivora* . *Seed and Plant.* 14(3):29-34.

Misra, S.D. and Kurl, S.P. (1979). The life of aphids. *Science Reporter.* 16 (9): 608-613.

Mora, J.G., Gapud, V.P. and Velasco, L.R.I. (1995). Life history and voracity of *Coelophora inaequalis* (Fabricius) (Coleoptera: Coccinellidae) on *Aphis craccivora* Koch. (Hemiptera: Aphididae) *Philippine Entomol.* 9(5): 523-553.

Morris, T.I., Campos, M., Jervis, M.A., McEwen, P.K. and Kidd, N.A.C. (1998). Potential effects of various ant species on green lacewing, *Chrysoperla carnea* (Stephens) (Neuropt., Chrysopidae) egg numbers. *J. Appl. Ent.* 122 (7): 401-403.

Nakamuta, K. (1984). Visual orientation of a ladybeetle, *Coccinella septempunctata* L. (Coleoptera: Coccinellidae), towards its prey *Appl. Entomol. Zool.* 19: 82-86.

Nakamuta, K. (1987). Diel rhythmicity of prey search activity and its predominance over starvation in the ladybeetle, *Coccinella septempunctata brucki. Physiol. Entomol.* 12: 91-98.

Naranjo, S.E., Gibson, R.L. and David, D.W. (1990). Development, survival and reproduction of *Scymnus frontalis* (Coleoptera: Coccinellidae), an important predator of Russian wheat aphid at four fluctuating temperature. *Ann. Entomol. Soc. Amer.* 83(3): 527-531.

Nicoli, G., Galazzi, D., Mosti, M., Burgio, G (1991). Embryonic and larval development of *Chrysoperla carnea* (Steph.) (Neur., Chrysopidae) at different temperature regimes. *Bulletin SROP.* 14(5):43-49.

Nordlund, D.A. and Correa, J.A (1995a). Description of green lacewing adult feeding and oviposition units and a sodium hypochlorite-based egg harvesting system. *Southwestern Entomol.* 20 (3): 293-301.

Nordlund, D.A. and Correa, J.A. (1995b). Improvements in the production system for green lacewings: an adult feeding and oviposition unit and hot wire egg harvesting system. *Biol. Cont.* 5(2): 179-188.

Obatake, H. and Suzuki, H. (1985). On the isolation and identification of canavanine and ethanolamine contained in the young leaves of black locus, *Robinia pseudoaccacia*, lethal for the ladybeetle, *Harmonia axyridis. Tech. Bull. Fac. Agri. Kagawa Univ.* 36: 107-115.

Obrycki, J.J. and Orr, C.J. (1990). Suitability of three prey species for Nearctic populations of *Coccinella septempunctata*, *Hippodamia variegata* and *Propylea quatuordecimpunctata* (Coleoptera: Coccinellidae). *J. Econ. Entomol.* 83: 1292-1297.

Omkar and Bind R.B. (1998). Prey preference of a lady beetle, *Cheilomenes (=Menochilus) sexmaculata* (Fabricius) (Coleoptera: Coccinellidae). *J. Aphidol.* 12(1&2): 63-66.

Omkar and James, B.E. (2001). Predation efficacy of a ladybeetle, *Coccinella transversalis* Fabricius in relation to certain aphids. *Biol. Mem.* 27(1): 20-26.

Omkar and James, B.E. (2003). Influence of prey species on immature survival, development, predation and reproduction of *Coccinella transversalis* Fabricius (Coleoptera: Coccinellidae). *Journal of Applied Entomology*, Germany (Accepted).

Omkar, James B. E. and Srivastava S. (1999). Prey preference of ladybeetle, *Coccinella transversalis* Fabricius (Coleoptera: Coccinellidae). *Adv. Bios.* 18(2): 117-122.

Omkar and Mishra, G. (2002a). Lovable Ladybirds. *Science Reporter*, 39(12): 56-59.

Omkar and Mishra, G. (2002b). Ladybirds: The Ecofriendly Option. *Everyman's Science*, 37(3): 162-167.

Omkar and Mishra, G. (2003). Chrysopids: An Effective Pest Management Tool. In: *Biopesticides and Bioagents in Insect Pest Management of Agricultural Crops* (ed. Dr. R.P. Srivastava) Publ. International Book Distributing Co., Lucknow pp. 671-719.

Omkar, Mishra, G. and Pervez A. (2002). Intraguild predation by ladybeetles: An ultimate survival strategy or an aid to advanced aphid biocontrol? *Prof. S.B. Singh Comm. Vol. Zool. Soc. India*, 77-90.

Omkar and Pervez, A. (2001). Prey preference of a ladybeetle, *Micraspis discolor* (Fabricius). *Entomon* 26(2): 195-197.

Omkar and Pervez, A. (2002). Influence of temperature on age specific fecundity of a ladybeetle, *Micraspis discolor* (Fabricius). *Insect Sci. Applic.* 22(1): 61-65.

Omkar and Pervez, A. (2003a). Functional and numerical responses of pale morph of *Propylea dissecta* (Mulsant) (Coccinellidae: Coleoptera). *Eur. J. Entomol.* (MS under revsion).

Omkar and Pervez, A. (2003b). Influence of prey deprivation on the biological attributes of pale morph of *Propylea dissecta* (Mulsant). *Insect Sci. Applic.* 23(2): 143-148.

Omkar and Pervez, A. (2003c). Ecology and Biocontrol Potential of a Successful Coccidophagous scale-predator, *Chilocorus nigritus*. *Biocontrol Sci. Tech.* 13(4): 379-390.

Omkar and Srivastava, S. (2001a). Ovipositional preference of *Coccinella septempunctata* Linnaeus (Coccinellidae: Coleoptera). *J. Aphidol.* 15(1&2), 5-8.

Omkar and Srivastava, S. (2001b). Comparative Predatory Potential of a ladybird beetle, *Coccinella septempunctata* Linn. on six prey species. *Biol. Memoirs*, 27(2): 59-63.

Omkar and Srivastava, S. (2002a). The reproductive behaviour of an aphidophagous ladybeetle, *Coccinella septempunctata* Linnaeus. *Eur. J. Entomol.* 99 (4) : 465-470.

Omkar and Srivastava, S. (2002b). Functional response of the ladybeetle, *Coccinella septempunctata* Linnaeus on *Uroleucon compositae* (Theobald) and *Aphis nerii* Boyer de Fonscolombe. *Prof. S.B. Singh Comm. Vol. Zool. Soc. India,* 115-128.

Omkar and Srivastava, S. (2003a). Influence of six aphid prey species on reproduction development and of a ladybird beetle, *Coccinella septempunctata* Linnaeus. *Biocontrol* 48(4): 379-393.

Omkar and Srivastava, S. (2003b). Functional response of seven-spotted ladybeetle, *Coccinella septempunctata* Linnaeus for mustard aphid, *Lipaphis erysimi* (Kaltenbach). *Insect Sci. Applic.* 23(2): 149-152.

Omkar and Srivastava, S. (2003c). Comparative prey consumption and searching efficiency of ladybeetles, *Coccinella septempunctata* Linnaeus and *Coccinella transversalis* Fabricius for different aphid species. *J. Biol. Control.* (Accepted).

Omkar, Srivastava S. and B. E. James (1997). Prey preference of a lady beetle, *Coccinella septempunctata* Linnaeus (Coleoptera : Coccinellidae). *J. Adv. Zool.*, 18(2), 96-97.

Orr, C.J. and Obrycki, J.J. (1990) Thermal and dietary requirements for development of *Hippodamia parenthesis* (Coleoptera: Coccinellidae). *Environ. Entomol.* 19(5):1523-1527.

Padmaja, Ch., Ramesh Babu, T., Reddy, D.D.R. and Sriramulu, M. (1995) Biology and predation of *Scymnus coccivora* Aiyar (Coleoptera: Coccinellidae) on mealy bugs. *J. Ent. Res.* 19(1): 79-81.

Pervez, A. 2002. *Contribution on prey-predator relationship and reproductive biology of a colour morph of* Propylea dissecta *(Mulsant) (Coccinellidae: Coleoptera.* Ph.D. Thesis. University of Lucknow, India, p. 197.

Phoofolo, M.W. and Obrycki, J.J. (1998). Potential for intraguild predation and competition among predatory Coccinellidae and Chrysopidae. *Ent. Exp. Appl.* 89: 47-55.

Polis, G.A,, Myers, C.A. and Holt, R.D. (1989). The ecology and evolution of intraguild predation: potential competitors that eat each other. *Ann. Rev. Ecol. Sys.* 20: 297-330.

Polis, G.A. and Holt, R.D. (1992). Intraguild predation: the dynamics of complex trophic interactions. *Trends in Ecology and Evolution.* 7: 151-154.

Ponsonby, D. J. and Copland, M. J. W. (1996). Effect of temperature on development and immature survival in the scale insect predator, *Chilocorus nigritus* (F.) (Coleoptera: Coccinellidae). *Biocontrol Sci. Technol.* 6(1): 101-109.

Ponsonby, D.J. and Copland, M.J.W. (1998). Environmental influences on fecundity, egg viability and egg cannibalism in the scale insect predator, *Chilocorus nigritus*. *BioControl*, 43: 39-52.

Prakasan, C. B. and Bhat, K. P. (1985). Interference of the predator, *Cryptolaemus montrouzieri* with the performance of a newly introduced parasite, *Leptomastix dactylopii; J. Cof. Res.* 15: 29-32.

Quentin, U., Hommes, M. and Basedow, T. (1995). Studies on the biological control of aphids (Hom., Aphididae) on lettuce in greenhouses. *J. Appl. Ent.* 119 (3): 227-232.

Radwan, Z. and Lõveii, G.L. (1983). Aphids as prey for the coccinellid, *Exochomus quadripustulatus*and *Ent. Exp. Appl.* 34:283-286.

Rao, B. S., Rao, A. S., Nagalingam, B. (1997). Biology of two coccinellid predators on three species of aphids infesting different crops. *Madras Agric. J.* 84(2): 53-56.

Rhamhalinghan, M. (1986). Seasonal variation in ovariole numbers/ ovary in *Coccinella septempunctata* L. (Coleoptera: Coccinellidae). *Proc. Indian Natn. Sci. Acad.* 52:619-623.

Rhamhalinghan, M. (1987a). The behaviour of *Coccinella septempunctata* L. associated with the exit of *Coccinellimermis Rubtzov* (Nematoda: Mermithidae): 1. Prior to the emergence of the parasite. *Indian Zool.* 10: 95-97.

Rhamhalinghan, M. (1987b). Feeding behaviour of *Coccinella septempunctata* L. *var. confusa* Wiedemann (Coleoptera: Coccinellidae) in relation to temperature- I. Pre-oviposition period. *J. Ent. Res.* 11(2): 178-183.

Ribeiro, M.J. and Carvalho, C.F.de (1991). Aspects of the biology of *Chrysoperla externa* (Hagen, 1861) (Neuroptera, Chrysopidae) in different mating conditions. *Revista Brasileira de Entomologia.* 35(2):423-427.

Rodriguez-Saona, C. and Miller, J.C. (1999). Temperature-dependent effects on development, mortality, and growth of *Hippodamia convergens* (Coleoptera: Coccinellidae). *Environ. Entomol.* 28(3): 518-522.

Roff, D.A. (1992). *The Evolution of Life Histories*. Chapman and Hall, New York.

Root, R.B. (1967). The niche exploitation patterns of the blue-gray gnatcatcher. *Ecol. Mono.* 37: 317-350.

Rosenheim, J.A., Wilhoit, L.R. and Armer, C.A. (1993). Influence of intraguild predation among generalist insect predators on the suppression of an herbivore population. *Oecologia.* 96(3): 439-449.

Ruzicka, Z. (1996). Oviposition-deterring pheromone in Chrysopidae (Neuroptera): intra- and interspecific effects. *Eur. J. Entomol.* 93 (2): 161-166.

Ruzicka, Z. (1997). Recognition of oviposition-deterring allomones by aphidophagous predators (Neuroptera: Chrysopidae, Coleoptera: Coccinellidae). *Eur. J. Entomol.* 94: 431-434.

Ruzicka, Z. (1998). Further evidence of oviposition-deterring allomone in chrysopids (Neuroptera: Chrysopidae. *Eur. J. Entomol.* 95(1):35-39.

Sakurai, H., Yoshida, N., Kobayashi, C and Takeda, S. (1991). Effect of temperature and day length on oviposition and growth of ladybeetle *Coccinella septempunctata bruckii. Res. Bull. Fac. Agric. Gifu. Univ*. 56: 45-50.

Saminathan, V.R., Baskaran, R.K.M. and Mahadevan, N.R. (1999). Biology and predatory potential of green lacewing (*Chrysoperla carnea*) (Neuroptera: Chrysopidae) on different insect hosts. *Indian J. Agric. Sci.* 69 (7):502-505.

Samways, M. J. and Wilson, S.J. (1988). Aspects of the feeding behaviour of *Chilocorus nigritus* (F.) (Col., Coccinellidae) relative to its effectiveness as a biocontrol agent *J. Appl. Entomol.* 106:177-182.

Schaller, M. and Nentwig, W. (2000). Olfactory orientation of the seven-spot ladybird beetle, *Coccinella septempunctata* (Coleoptera: Coccinellidae): attraction of adults to plants and conspecific females *Eur. J. Entomol.* 97(2):155-159.

Scholz, D. and Poehling, H. M. (2000). Oviposition site selection of *Episyrphus balteatus*. *Ent. Exp. Appl*. 94(2):149-158.

Scott, S. M. and Barlow, C. A. (1986). Effect of prey availability on foraging and production efficiencies of larval *Metasyrphus corollae* (Dipt.: Syrphidae). *Entomophaga*. 31(3):243-250.

Semyanov, V. P (2001). The biology of coccinellids (Coleoptera, Coccinellidae) from South-East Asia. III. *Lemnia biplagiata* (Swartz). *Entomologicheskoe Obozrenie*. 80(3):578-584.

Sengonca, C. and Fringe, B. (1985). Interference and competitive behaviour of the aphid predators, *Chrysoperla carnea* and *Coccinella septempunctata* in the laboratory. *Entomophaga* 30:245-251.

Sengonca, C., Griesbach, M. and Löchte, C. (1995). Suitable predator-prey ratios for the use of *Chrysoperla carnea* (Stephens) eggs against aphids on sugar beet under laboratory and field conditions. *Zeitschrift für Pflanzenkrankheiten und Pflanzenschutz*. 102(2):113-120.

Sengonca, C., Kotikal, Y.K. and Schade, M (1995a). Olfactory reactions of the larvae of *Chrysoperla carnea* (Stephens) (Neuroptera, Chrysopidae) to various odour sources in an eight-armed olfactometer. *Mitteilungen der Deutschen Gesellschaft für Allgemeine und Angewandte Entomologie*. 10(1/6):605-608.

Sharma, D.K., Varma, G.C. and Kishore, L. (1997). Feeding capacity of predators of mustard aphid, *Lipaphis erysimi*. *J. Aphidol.* 11(2): 171-174.

Sharma, K. C. and Bhalla, O. P. (1988). Biology of six syrphid predators of cabbage aphid (*Brevicoryne brassicae*) on seed crop of cauliflower (*Brassica oleracea var. botrytis*). *Indian J. Agric Sci.* 58(8): 652-654.

She ChunRen, Pan RongYing, Yang ShengHua, Lin LiLing and Guo Chen (1998). Studies on the biology of Sphaerophoria macrogaster (Diptera: Syrphidae). *Wuyi Science Journal.* **14**: 73-77.

Shonouda, M. L (1996). Crude aqueous-extract (kairomone) from *Aphis fabae* Scop. (Hom., Aphidae) and its effect on the behaviour of the predator *Metasyrphus corollae* Fabr. (Dipt., Syrphidae) female. *J. Appl. Ent.*. 120(8):489-492.

Shonouda, M. L. (1999). Aphid aqueous-extract as a source of host searching kairomones for the aphidophagous predator *Coccinella septempunctata* L. (Col., Coccinellidae). *Anzeiger für Schädlingskunde* 72(5): 126-128.

Singh, M. P. and Devi, C.S. (1996). Food preference of *C. septempunctata* (Linn.) on aphids. *Insect Environ.* 2(1): 5-6.

Singh, N.N. and Kumar, M. (2000). Potentiality of *Chrysoperla carnea* (Stephens) in suppression of mustard aphid population. *Indian J. Ent.* 62(4):323-326.

Singh, R. and Malhotra, R.K. (1979). Bionomics of *Coccinella septempunctata* Linn. *Indian J. Entomol.* 41(3), 244-249.

Singh, R. and Mishra, S. (1988). Development of syrphid fly, *Ischiodon scutellaris* (Fabricius) on *Rhopalosiphum maidis* (Fitch). *J. Aphidol.* 2(1-2):28-34.

Singh, V.S., Yadav, R.R. and Singh, R. (1994) Post embryonic development, survival rate and predation potential of *Coccinella septempunctata* Linn. in relation to the mustard aphid (*Lipaphis erysimi* Kalt.) *J. Entomol. Res.* 18(1): 5-10.

Smith, B. C. (1965). Effects of food on the longevity, fecundity and development of adult coccinellids. *Can. Ent.* 97: 910-919.

Soleyman-Nezhadiyan, E. and Laughlin, R.(1998). Voracity of larvae, rate of development in eggs, larvae and pupae, and flight seasons of adults of the hoverflies *Melangyna viridiceps* Macquart and *Symosyrphus grandicornis* Macquart (Diptera: Syrphidae). *Australian J. Entomol.* 37(3):.243- 248.

Srivastava, S. (2000). *Certain aspects of Bioecology and ethology of a ladybeetle, Coccinella septempunctata Linnaeus (Coccinellidae: Coleoptera).* University of Lucknow, Lucknow. p. 160.

Stearns, S.C. (1992). *The Evolution of Life Histories.* Oxford University Press, Oxford.

Stearns, S.C. and Koella, J.C. (1986). The evolution of phenotopic plasticity in life-history trait: Prediction of reaction norm for age and size at maturity. *Evolution* 40:893-913.

Stewart, L.A., Hemptinne, J.L. and Dixon, A.F.G. (1991). Reproductive tactics of lady beetles: relationships between egg size, ovariole number and developmental time. *Fun.Ecol.* 5: 380-385.

Stubbs, M. (1980). Another look at prey detection by coccinellid *Ecol. Entomol.* 5:179-182.

Subramanyam, V.K. (1953) Control of the fluted scale in peninsular India. *Indian J. Ent.* 17:103-120.

Suri, S. M., Singh, D. and Brar, K. S. (1988). Estimation of losses in yield of brown Sarson due to aphids in Kangra valley. I, effect of crop growth stage and aphid feeding exposure. *J. Insect Sci.,* 1(2): 162-167.

Sutherland, J. P., Sullivan, M. S. and Poppy, G. M. (1999). The influence of floral character on the foraging behaviour of the hoverfly, *Episyrphus balteatus*. *Ent. Exp. Appl.* 93(2):157-164.

Takahashi, K. (1989). Intra- and interspecific predations of ladybeetles in spring alfalfa fields. *Japanese J. Entomol.* 57:199-203.

Takizawa, T., Yasuda, H. and Agarwala, B.K. (2000). Effects of parasitized aphids (Homoptera: Aphididae) as food on larval performance of three predatory ladybeetles (Coleoptera: Coccinellidae). *Appl. Entomol. Zool.* 35 (4):467-472.

Tao, C.C. and Chui, S.C. (1971). Biological control of citrus, vegetables and tobacco aphids. *Spec. Publ. Taiwan Agric. Res. Inst.* 10: 1-110.

Tawfik, M. F. S., Azab, A. K. and Awadallah, K. T. (1974a). Studies on the life-history and descriptions of the immature forms of the Egyptian aphidophagous Syrphids. III - *Xanthogramma aegyptium* Wied. (Diptera:Syrphidae). *Bulletin de la Societe Entomologique d'Egypte*.58 : 73-83.

Tawfik, M. F. S., Azab, A. K .and Awadallah, K. T.(1974b). Studies on the life-history and description of the immature forms of the Egyptian aphidophagous Syrphids. II - *Paragus aegyptius* Macq.

(Diptera:Syrphidae). *Bulletin de la Societe Entomologique d'Egypte*. 58: 35-44.

Tawfik, M. F. S., Azab, A. K. and Awadallah, K. T. (1974c). Studies on the life-history and description of the immature forms of the Egyptian aphidophagous Syrphids. I - *Syrphus corollae* Fabre (Diptera:Syrphidae). *Bulletin de la Societe Entomologique d'Egypte*.58:1-16.

Tawfik, M. F. S., Azab, A. K. and Awadallah, K. T.(1974d.) Studies on the life-history and description of the immature form of the Egyptian aphidophagous Syrphids. IV - *Sphaerophoria flavicauda* Zett. (Diptera: Syrphidae). *Bulletin de la Societe Entomologique d'Egypte*.58:103-115.

Taylor A.J., Muller C.B. and Godfray H.C. J. (1998). Effect of aphid predators on oviposition behaviour of aphid parasitoids *J. Insect Behav.* 11:297-302.

Tenhumberg, B. and Poehling, H. M. (1994). Quantification of the predation efficacy of *E. balteatus* (Diptera: Syrphidae) with the help of traditional models. *Bulletin OILB / SROP.* 17(4):112-126.

Thite, N.R. and Shivpuje, P.R. (1999). Biology, feeding potential and development of *Chrysoperla carnea* (Stephens) on *Aphis gossypii* (Glover). *J. Maharashtra Agric. Univ.* 24(3):240-241.

Turnbull A.L. and Chant, D.A. (1961). The practice and theory of biological control of insects in Canada. *Can. J. Zool.* 36: 697-753.

Uygun, N. and Atlihan, R. (2000). The effect of temperature on development and fecundity of *Scymnus levaillanti*. *BioControl*, 45(4): 453-462.

Veeravel, R. and Baskaran, P. (1995) Cannibalistic behaviour and survival rate of two coccinellid predators, *Coccinella transversalis* Fabricius *and Menochilus sexmaculatus* (Fabricius). *Indian J. Ecol.* 22(2): 113-117.

Veeravel, R. and Baskaran, P. (1997). Functional and numerical responses of *Coccinella transversalis* and *Cheilomenes sexmaculata* Fabr. Feeding on the melon aphid, *Aphis gossypii* Glov. *Insect Sci Appl.* 17(3/4): 335-339.

Venkatesan, T., Singh, S.P. and Jalali, S.K. (2000). Rearing of *Chrysoperla carnea* (Stephens) (Neuroptera: Chrysopidae) on semi-synthetic diet and its predatory efficiency against cotton pests. *Entomon*. 25 (2): 81-89.

Wang J. and Tsai, J.H. (2001). Development and functional response of *Coelophora inaequalis* (Coleoptera: Coccinellidae) feeding on brown citrus aphid, *Toxoptera citricida* (Homoptera: Aphididae). *Agric. Forest Entomol.*, 3(1): 65-69.

Watt, J.C., (1986). Beetles (Coleoptera) of the offshore islands of northern New Zealand In: The *Offshore Islands of Northern New Zealand* pp 221-228 (eds) A.E. Wright and R.E. Beever (Wellington: Department of Lands and Survey).

Wheeler, A.G. Jr. and Hoebeke, E.R. (1995). *Coccinella novemnotata* in Northeastern North America: Historical occurrence and current status (Coleoptera: Coccinellidae). *Proc. Entomol. Soc. Wash.* 97(3): 701-716.

Whitcomb, W.H. and Bell, K. (1964). Predaceous insects, spiders, and mites of Arkansas cotton fields. *Univ. Arkansas Agric. Exp. Stn. Bull.* 690: 1-84.

Wratten, S.D. (1973). The effectiveness of the coccinellid beetle, *Adalia bipunctata* (L.) as a predator of the lime aphid, *Eucallipterus tiliae* L.. *J. Anim. Ecol.* 42: 785-802.

Wyss, E. (1999). Enhancement and release of predaceous arthropods to control aphids in organic apple orchards. *Bulletin OILB/SROP.* 22(7):47-51.

Yasuda, H. and Katsuhiro, S. (1997). Cannibalism and interspecific predation in two predatory ladybirds in relation to prey abundance in the field. *Entomophaga* 42(1/2): 153-163.

Yasuda, H. and Ohnuma, N. (1999). Effect of cannibalism and predation on the larval performance of two ladybird beetles. *Ent. Exp. Appl.* 93(1): 63-67.

Yasuda, H., Takagi, T. and Kogi, K. (2000). Effects of conspecific and heterospecific larval tracks on the oviposition behaviour of the predatory ladybird, *Harmonia axyridis* (Coleoptera: Coccinellidae). *Eur. J. Entomol.* 97: 551-553.

Yoldas, Z. (1994). Studies on the biology of *Chrysoperla carnea* (Stephens) (Neuroptera: Chrysopidae) feeding on two different prey. *Turkiye III. Biyolojik Mucadele Kongresi Bildirileeri, 25-28 Ocak 1994, Ege Universitesi Ziraat Fakultesi, Bitki Koruma Bolumu, Izmir.* 375-380.

Yüksel, S. and Gocmen, H. (1992). The effectiveness of *Chrysoperla carnea* (Stephens) (Neuroptera, Chrysopidae) as a predator on cotton aphid *Aphis gossypii* Glov. (Homoptera, Aphididae). *Proceedings of the Second Turkish National Congress of Entomology*. 209-216.

Zeki, C.and Kilincer, N. (1994). Studies on the starvation period and prey selection of larvae of *Metasyrphus corollae* (F.) and

Episyrphus balteatus (De Geer) (Diptera: Syrphidae). *Turkiye III. Biyolojik Mucadele Kongresi Bildirileeri, 25-28 Ocak 1994, Ege Universitesi Ziraat Fakultesi, Bitki Koruma Bolumu, Izmir.*, pp. 59-66.

Zhou X, Carter, N. and Powell W. (1994). Seasonal distribution and aerial movement of adult coccinellids on farmland. *Biocontrol Sci. & Tech.* 4(2): 167-175.

Zhou, C. A., Zou, J.J., Peng, J.C., Ouyang, Z.Y., Hu, L.C., Yang, Z.L. and Wang, X. B. (1991). Predation of major natural enemies on *Panonychus citri* and its comprehensive evaluation in citrus orchards in Hunan, China. *Acta Phytophylacica Sinica* .18 (3): 225-229.

Ziolko, M., and Kozlowski, J. (1983). Evolution of body size: an optimization model. *Math. Biosci.* 64: 127-143.

19

Soil Fauna in Sustainable Management of Agroecosystem

G. Tripathi, S. Ram and *B.M. Sharma*

Biological diversity encompasses all species of plant, animals and microorganisms in the ecosystems, and it is the basis of livelihood and ecological security in a true sense. India with 2.5% of the world area has 8% of the world's biodiversity with species count of 0.126 million and two of the 18 hot spots of the world. Biological diversity refers to the structural and functional variety of life forms at genetic, population, species, community and ecosystem levels. It may be used for food, medicine, domestication, wood, research, education, monitoring, recreation and cultural values. The ecosystem services of biodiversity are related to pollination and seed dispersal, protection of water resources, breakdown of pollutants, nutrient cycling and maintenance of soil quality. The landmark United Nations Conference on Environmental and Development held in 1992 at Rio de Janeiro, Brazil, popularized the importance of biological diversity at the global scenario. Most of the biological issues brought to world attention were dealing with the problems of visual biodiversity and below-ground biodiversity was not given due consideration.

India is the seventh largest country in the world with an area of 32,87,263 square kilometers and with a population of more than one billion. About one lakh species of animals have been identified from India and many more are to be reported. In the next three decades our country which is rich in biodiversity may lose a considerable number of species even without identification. In the beginning of the present millennium, the Ministry of Environment and Forest has launched a nation-wide program to survey and identify biological species. In 2001, Department of Biotechnology of the Ministry of Science and Technology has also initiated program for exploring potentials of Indian biodiversity. These two programs mainly focus on the above-ground biodiversity. In contrast Indian Council of Agricultural Research has not introduced any separate program for exploring the potentials of soil faunal biodiversity. However, the fact is that the greatest concentration of life occurs in soil and below-ground biodiversity is the basis for the above-ground biological diversity.

Soil animals act as buffer energy source for soil and plant both. It is well established that such animals are important contributors to soil fertility and humification processes. The degradation of soil biological community is the main reason for rapid loss of sustainability in tropical India. The activities of soil fauna play a vital role in minimizing the rate of soil degradation. The direct contributions of soil invertebrates to mineral fluxes are through heterotroph community metabolism and the indirect effects involve feedbacks to the physicochemical controls over the activities of organisms of lower tropic levels. The comminution of uningested organic matter increases the surface area for microbial colonization and defaecation produces resources of changed chemical quality. The indirect effects of invertebrates on soil physicochemical processes modify expression of the direct effects by altering the soil as an environment for microbial activities and plant roots. Hence the importance of below-ground invertebrates cannot be ignored at this stage.

Faunastic Composition

The greatest concentration of life occurs in soil. There is a fascinating world of soil invertebrates about which very little is known today. Soil organisms, based on their body size, have been categorized into the following three groups.

1. Microfauna (20µ to 200µ): Mostly protozoans belong to this group. Mites, nematodes, rotifers, tardigrades may also be included near the upper limit of this group.

2. Mesofauna or meiofauna (>200μ to 1 cm): Microarthropods, Acari (mites) and Collembola (spring tail) are important members of this group. It also includes larger nematodes, potworms and some insect larvae.
3. Macrofauna (> 1cm): This group is represented by larger animals like millipedes, centipedes, wood lice, sluge and snails, earthworms, and some vertebrates etc.

Singh and Pillai (1973) reported that soil-inhabiting organisms play a vital role in humus formation and maintaining a control over plant-damaging species. Animals in the soil are diverse in their habitats. Their feeding and breeding habitats, protective and destructive devices, are all best suited for the subterranean life. They utilize soil, as their hide out. It is for the biologist to bring this hide and seek game played by these animals to light.

Since the establishment of Japan international Biological Programme-Productivity of Terrestrial communities (JIBP-PT), in 1996, a remarkable advance was achieved concerning the integral analysis of soil fauna by a teamwork of many research workers in various localities in Japan (Kitazawa *et al.,* 1968). Some preliminary attempts were made in relation to soil fauna of the grassland for amelioration in Hokkaido of Japan (Hokkaido Development Bureau, 1965, 1966). Tamura *et al.,* (1969) made a preliminary survey of soil fauna in Hidaka–Mombetsu of Southern Hokkaido in Japan. They surveyed macrofauna, ants and mesofauna with special reference to collembolans and oribatid mites in three different habitats such as Quercus forest, Larix forest and grassland. Faunal composition in these habitats differed both qualitatively and quantitatively. Van der drift (1969) investigated quantitative composition of soil fauna of macroarthropods, oribatids, collemboles and enchytraeids in the Netherlands. Kumar *et al.,* (1999) investigated soil microarthropods and annelids at different altitudes in natural reed growing areas in Vazhachal Reserved Forest, Kerala. The reed growing soil had a high diversity and density of soil fauna, and was very high in organic carbon content. The high fertility and organic carbon content of the soil are attributed to the presence of the diverse soil fauna which assist in humus formation.

Population Distribution

The animal populations have patchy distribution in soil (Anderson, 1977; Swift *et al.,* 1979; Usher *et al.,* 1982). Patchy distribution is a consequence of various abiotic and biotic properties of the soil, such as porosity, acidity, humidity and resources. Moreover, organisms show themselves the discontinuous distribution patterns by their mode of life such as grouping behavior, limited dispersal and nonuniform resource consumption (Kareiva, 1986). Specialization of soil organisms to particular microhabitats may be due to high species diversity in soil (Anderson, 1975, 1978a; Atkinson and Shorrocks, 1981). Many anthropogenic activities can create secondary discontinuity in soil. It has been shown that along heavy metal gradients not only the abundance of individuals but also the structure of the whole below-ground community is changed (Bengtsson and Tranvik, 1989). Species-specific distribution patterns in discontinuously contaminated soil can be a consequence of general tendency of organisms to disperse, adaptation to toxicants, interspecies interaction, and the capability of organisms to perceive different concentrations of toxicants in their environment (Bengtsson *et al.*, 1994). Soil animals may localize, colonize and utilize clean patches in discontinuously contaminated soil (Tranvik and Eijsaskers, 1989; Bengtsson *et al.*, 1994). Colonization patterns and successional development of soil animal communities have been studied in many soil types and environmental conditions (Coleman and Macfadyen, 1966; Anderson, 1975b; Huhta *et al.,* 1979; Stanton, 1979; Hagvar and Abrahamsen, 1980; Hudson, 1980; Streit *et al.,* 1985). These studies have proved that environmental factors, such as vegetation cover and soil pH, are the main forces driving the community development. However, biotic interactions among the fauna and between the fauna and microorganisms can also affect the structure of a developing community (Setala *et al.*, 1995), although only few studies suggest the presence of the biotic interactions, such as competition, to be common in soils (Anderson, 1978b; Strong, 1983; Usher, 1985).

Faunal Aggregation

Several studies on the horizontal distribution of Collembola and other soil microarthropods have shown that their populations are distributed non-randomly. The animals tend to aggregate in soil (Paclt, 1956; Haarlov, 1960; Poole, 1961) and no suggestions have been given on the biological significance of aggregations in Collembola. Two possible explanations of the formation of aggregations have been proposed (Hale, 1966). Firstly, aggregation may result from the slow dispersion of individuals from an egg batch, and secondly, aggregations may result from individuals coming together at either a food source or a suitable microhabitat.

Joosse (1970) studied the formation and biological significance of aggregations of six species of surface dwelling Collembola of a coniferous forest soil. These species are *Tomocerus minor*, *Lepidocyrtus cyaneus*, *Orchesella cincta*, *Entomobrya nivalis*, *Isotomurus palustris* and *Isotoma viridis*. They demonstrate a marked aggregation, but to a different degree and in different places, since the degree of aggregation of total Collembola appeared to be less than that of the individual species.

In natural situation the presence of other organisms of different species is an unavoidable and a necessary part of the environment. The existence of other species may be crucially important in the provision of food, shelter or other necessities. Choudhuri and Roy (1970) documented interactions between soil Collembola and other subterranean arthropods. It is well known that many of the arthropods are predators of Collembola and therefore act as a biotic factor limiting the population of Collembola. Collembola are adopted to terrestrial life and thus exposed to a more or less dry environment. Their survival is greatly affected by the humidity of the surrounding air. Most species lack a tracheal or other internal respiratory system but respire through the cuticle, especially of the ventral tube (Nutman, 1941; Ruppel, 1953). Thus the cuticle is very permeable to water, and the animals lose considerable amounts of water by transpiration. The loss is proportional to the vapour pressure difference between the evaporating surface of the organism and the immediately surrounding air layer. Drought stimulates Collembola to start locomotory activity which may lead them to aggregate in optimal humidity condition. Thus this reaction may play a part in increasing the survival rate of these animals (Joosse and Groen, 1970). Ecological aspects of aggregation in Collembola has been reported by Joosse (1971). He said that a relation exists between the humidity requirements of the Collembola species and their distribution and locomotory activity in the field. Species with high humidity requirements are forced in aggregation and show a lower general activity level in the field. This general activity level appears to be determined by the temperature saturation deficient conditions. This survival mechanism, by which the animals are forced into aggregation may play an important role in increasing the meeting chance between individuals in a population, which may be important in reproduction. If this is the case, one could expect that species with lower aggregational habits should need other mechanisms to bring the individuals together. Christiansen (1970) studied in detail the aggregation and dispersion of Collembola. Looking at all the data, it appears that the single epigeic form (*S. curviseta)* shows a slight decrease of total movement and speed of movement under cave as compared to room conditions but with no striking change in pattern of movement.

Gut Digestion and Microorganism

Some member's of the soil fauna are able to assist directly in the breakdown of dead plant material by digesting certain plant structural polysaccharides either through the possession of the necessary enzymes, or a symbiotic gut flora (Nielsen, 1962). Such animals include sluge and snails, tipulid larvae, termites, nematodes and earthworms. Soil oribatids, on the other hand, have generally been presumed to play a less direct role in the breakdown of decaying higher plant material, and even the macrophytophages were considered to utilize only the microbial flora for organic matter digestion (Nielsen, 1962; Wallwork, 1967). This hypothesis was given credence by the experiments of Hartenstein (1962) who showed that *Protoribates lophotrichus* (Berlese) fed vigorously on dead leaf tissue until the microorganisms within it were destroyed either by heating or by treatment with fungicide. Littlewood (1969) showed that when sterile mites were provided with food free from bacteria and fungi no feeding occurred. Littlewood (1969) suggested that food is not recognized as such by the oribatids unless microorganisms are present on it, or alternately that the microorganisms might affect the food in some way (perhaps by beginning its breakdown) which makes it palatable to the mites.

It has been stated that oribatid mites in woodland soil must have an important effect on the organic layers within which they mostly live. There have been several general reports of observations on the feeding habits of oribatids (Berthet, 1964; Hartenstein, 1962a-g; Luxton, 1966; Pauly, 1956; Rajski, 1966; Riha, 1951; Schuster,1956; Van der Drift, 1951; Wallwork, 1967; Woodring, 1963; Woolley, 1960). The influence of oribatid mites on forest litter layers may also be assessed in terms of their digestive capabilities. The study of carbohydrase enzymes in oribatid mites suggests that their digestive abilities are as wide ranging as their ingestive abilities. The smallness of most oribatid mites is a great hindrance to the detailed study of food quantities processed by them.

An entirely different approach at the problem of chemical attack of the litter by the fauna is the examination of the enzymes occurring in these animals. Neilsen (1962) studied thirty soil animals and suggested that only snails are capable of attacking cell wall materials. Nevertheless, the droppings of these animals also contain many fragments. It is probable that only the part absorbed in the hepatopancreas is subject to a sufficiently strong cellulose action. Its black colour and amorphous structure clearly distinguish it from the rest of the excrement and it is obviously far more highly digested than the other food. The exclusive presence of pectinase and xylanase is Tipulidae larvae may possibly be related to the caeca of these animals, in which, part of the food is absorbed and severely attacked. In this case, the enzymes in question might have been derived from the symbiotic microorganisms (Buchner, 1953). According to Nielsen (1962), the food assimilated by litter fauna chiefly consists of the bacteria and fungi ingested with the vegetable material. The soluble carbohydrates and nitrogenous compounds of the cell contents remaining in the litter are also components in the diet. The decomposition of litter is almost entirely due to microorganisms, but for these to thrive, animal activity is essential. Experiments in which animals were excluded have shown this (Kurceva, 1960). The interrelation between litter fauna and microorganisms a fascinating field of study for soil zoologists and soil microbiologists.

Spain and Luxton (1971) have shown that certain oribatid can digest cellulose and other plant polysaccharides. Information has been provided about the carbohydrase enzymes possessed by some twelve species of oribatid (Luxton, 1972). It is clear that the litter-eaters possess enzymes capable of digesting certain plant structural polysaccharides such as cellulose, xylan and pectin. There must, however, be some degree of digestion as the food passes through the gut. Careful weighing of food, remnants and excrement of milliped and isopods showed retention percentages of from 4 to 20 per cent. Bocock (1963) found that *Glomeris* had fairly high assimilation percentages for crude fats, soluble carbohydrates and even holocellulose. Priesner (1969) was able to demonstrate by means of paper chromatography the assimilation of nitrogenous compounds by *Tipula maximapoda*.

Energy metabolism of soil fauna should be studied. Only by this is it possible to evaluate the relative importance of various faunrl population in the energy flow through the soil. By measuring ingestion and egestion, assimilation may be estimated by subtraction. The percentage of ingested food which is assimilated varies widely, depending on animal and food species and on the rate of food intake. In most litter feeding macroarthropod species, not more than 5-10% of the ingested food is assimilated. Soil animals act as an important agent in diminution of the litter, rendering it more accessible for attack by microorganisms. Assimilated food is partly respired, partly fixed in body tissue as growth. The proportion of these two parts also varies widely. In animals with a long lifetime, the energy content of the body tissue will be very low compared with the energy dissipated by respiration. In small terrestrial animals, it is assumed that respiration is equivalent to 75-90% of assimilation leaving only 10-25% for growth. In microorganisms with a quick turnover the proportion of energy stored up in body tissue may be much greater than this. Conversion of the dead body tissue may be greatly promoted by the activity of soil fauna (Van der drift, 1966).

Faunal Effect on Soil Nutrient

Termites are dominant soil animals in many tropical and sub-tropical terrestrial ecosystems and play important roles in both nutrient cycling (decomposition) and soil-forming processes (Wood and Sands, 1978). They feed on plant materials at every stage of the decomposition process, ranging from fresh litter to wood to humified materials. Their selected food materials are digested partly through the enzymatic capacities of the termites themselves (Breznak and Brune, 1994; Slayter, 1992) and partly through symbiotic relationships with micro-organisms (Higashi and Abe, 1997). The tropical savannas of north-eastern Australia support large populations of litter feeding and grass-harvesting termites. Spain *et al.* (1983) reported mean populations of 230 mounds per hectare in the Townsville area. *Amitermes laurensis* builds conical nest (mean heights 0.4-0.6 m in the Townsville area; Spain *et al.*, 1983). While those of *Drepanotermes rubriceps* are low, compact and irregularly dome-shaped; both dominantly consists of earthen materials. *A. laurensis* is recorded as consuming grass liter (Lee and Wood, 1971) and the dung of cattle and other animals (Ferrar and Watson, 1970). It is considered to feed on more decomposed materials than *D. rubriceps,* which predominantly stores harvested fragments of the dead leaves and stems of perennial grasses in its nests (Hill, 1942). Tayosu *et al.* (2002) studied N values of termite tissues and the stored food materials of the litter-feeding species *Amitermes laurensis*. N value in the species were significantly greater than their equivalents in the grass-harvesting species *Drepanotermes rubriceps*. In addition, the carbohydrate/lignin and C/N ratio were higher and chitin/carbohydrate ratio was lower in food materials stored in the mounds of *D. rubriceps*. Both sets of analysis confirmed that *D. rubriceps* assimilates less decomposed food materials than *A. laurensis*.

Termites play an important role in decomposition processes in savanna and tropical forest ecosystems. In the savanna soils of South America termite mounds are abundant and their establishment is strongly related to the soil physical properties determining water balance during critical periods (San Jose *et al.*, 1989). It is well documented that, in the process of mound building, termites modify the physical, chemical and biochemical characteristics of the soil (Lee and Wood, 1971a; Lopez-Hernandez and Febres, 1984; Arshad *et al.*, 1988; Lavelle *et al.*, 1984). The modification of the soil material in epigeous termites is a consequence of the use of organic compounds (salivary or faecal originated) in their mound construction (Lee and Wood, 1971b). Therefore, organic compounds, particularly C and N, in general, are more abundant in termite nests than in surrounding soil. The phosphorus content of termite mounds suggests a role of feeding habits in phosphorus accumulation (Lopez-Hernandez *et al.*, 1989). The available P has been reported to be more abundant in mounds of soil-feeding termites (Wood *et al.*, 1983) than in those of plant debris feeders. However, Lopez-Hernandez *et al.* (1989a, b) reported a high level of both total and available phosphorus in termite mounds of *Nasutitermes ephratae*, a common plant debris feeding termite from savanna.

A large body of literature discussing the accumulation of nutrients (C, N, P and available cations) in termite mounds and surrounding soils has already been reviewed (Lee and Wood,1971; Okello-Oloya *et al.,*1985; Lopez-Hernandez and Febres,1984; Abbadie and Lepage,1989; Lavelle *et al.,*1994). Evidence suggests that nutrients accumulate in the mound relative to nearby areas, so that the abundant termite populations could play an important role in controlling nutrient cycling in tropical savannas, which are typically nutrient deficient. Few studies have, however, looked at the dynamics of biogenic elements (C, N, P) in termite mounds compared with their associated soil, although some evidence exists that mounds act as sinks for some elements and protect them from seasonal fires (Wood and Sands, 1978; Lopez-Hernandez and Febres, 1984). Lopez-Hernandez (2001) showed the mounds of *Nasutitermes ephratae* contain more C, N and P than adjacent top soils due to the use of faecal material to build gallery walls. Mounds accumulate 2.25 times more total P than adjacent soil. Available and potentially mineralizable organic P (Po) were significantly higher in *Nasutitermes* mounds than in adjacent soils. The mound materials significantly stimulate soil microbial metabolism, due to the accumulation of organic carbon. The results for nitrogen dynamics emphasize the low ammonification rate in savanna soil. Ammonium production in the savanna soil is very low, whereas in the mound the value increases several folds. About 20% of the total nitrogen is inorganic nitrogen within the termitaria. Nitrate production on the other hand was negligible in both soil and mound, so that the nitrification rate seemed to be not limited by ammonium production. The nutrient depleted conditions of this environment together with the important concentrations of C, N and P found within the mounds of *Nasutitermes ephratae* corroborated the hypothesis that termite mounds can act as 'sinks' in the nutrient economy of well-drained savanna with low productivity.

Population Dynamics

Knowledge of desert-inhabiting arthropods is largely confined to the larger forms such as scorpions, spiders, myriapods and beetles that can be collected by hand and observed in the field (Edney, 1957; Lawrence, 1959; Cloudsley-Thompson, 1964). Microarthropods of desert have received little attention. Wood (1972) described the distribution and abundance of *Folsomides deserticola* (Collembola:Isotomidae) and other microarthropods in arid and semi-arid soils in Southern Australia. Although Rapoport and Tschapek (1956) have reported the occurrence of an Isotomid Collembolan (*Isotoma* sp.) in the desert of Northern Chile, an unexpected feature of the microarthropod population of arid areas in Southern Australia is the dominance of Collembola small prostigmatid mites, small cladoceran and copepod crustaceans which belong to groups which are generally regarded as being susceptible to desiccation. One might have expected the dominant species in the desert to be those with apparent morphological adaptations for survival in dry habitats such as the several genera of scale-covered Collembola belonging to the family Entomobryidae and the groups of xerophyllous oribatid mites with a well-sclerotised cuticle covered by an epicuticular wax of high (50-60□C) melting point (Madge, 1964). Moreover, microarthropods are incapable of burrowing to a great extent and thus diurnal vertical movements in the soil, such as that exhibited by many larger desert-inhabiting arthropods (Cloudsley-Thompson, 1964), to escape heat and desiccation during the day and to utilize surface dew during the night are unlikely to contribute to their survival. However, Rose (1968) working near Aliee Springs in Central Australia showed that as the soil surface cools after sunset the downward vapour flux in the soil decreases in magnitude and eventually reverses during the night with the result that moisture condenses within the soil in the cool 0-3 cm layer. The maximum condensation rate was calculated to be 6-10 cm second^{-1}. If this is a regular phenomenon, microarthropods living near the surface of the soil would be assured of a regular supply of moisture even in the hottest and driest regions of the desert.

Among the microarthropod fauna of the arid areas there are many species which occur more frequently and at greater densities than in sub-humid regions (Wood, 1972). In case of *F. deserticola,* evidence has been presented that suggests a positive correlation between its abundance and seasonal and regional aridity. A superficial examination of the summer and winter populations in different sites indicates a negative correlation between the density of *F. deserticola* and that of other microarthropods. However, when populations in the sites in arid and semi-arid areas are considered individually no such relationship is evident. In fact there is a tendency for a positive correlation between the density of *F. deserticola* and the density of the microarthropods, largely due to the fact that sites with high densities of microarthropods tend to have high densities of Folsomides (Wood, 1972).

Sanyal *et al.* (1996) studied the composition and seasonal abundance of soil inhabiting arthropods and their relationship with the temperature and moisture content of soil in banana and guava plantations at Krishnanagar (Nadia) in West Bengal of India. They found Acari population not abundant which followed by Collembola and Hymenoptera. The fauna of the banana site was rich compared to guava site. The arthropod population peaked in August at both the sites. Edaphic factors directly influence faunal abundance. Reddy (1995) described association between soil organisms and litter decomposition in the tropics. Shweta (1997) studied apterygotes comprising seven Collemboles (*Xenylla, Folsomia, Isotomurus, hypogastrura, Proisotoma, Bourletiella* and *Sminthurus*), three diplurans (*Campodea, Heterojapyx* and *Anajapyx*) and some Proturans in tubewell and sewage irrigated fields for two years. Intergeneric relationships and their relations with Acari, nematodes, protozoans and physicochemical parameters were computed. Spring-tails were dominant and proturans the least. Collembolans were maximum during September-December, diplurans during September-October and proturans during June-July, with distinct generic variations. Collembolans show differences in microhabitat utilization, tolerance potential to become resistant and to have different food niches to reduce competition. Proturans appeared to be utilized as food resources by Collembolans and diplurans.

Jain *et al.* (1990) studied the seasonal composition of the Acarine fauna in the leaf litter and top soil of a *Dalbergia sisoo* plantation in Haryana. Quantification of mites at two sites with high moisture soil and low temperature revealed the predominance of cryptostigmatids (oribatids) (58.8%) in the former site and of mesostigmatids (40.9%) in the latter site. In total, 25 species and genera, belonging to 22 families, were identified. Of these, 13 genera belonged to the Cryptostigmata and 8 genera to the Mesostigmata. *Epilohmannia cylindrical, Oppia* sp., *Scheloribates* sp. and *Galumna* sp. were the most abundant mite fauna. Astigmatids represented by the genera of *Tyrophagus* which were the lowest in number in both soil types. In the beech wood, population was at its maximum from December to April, and minimum in June-August. The older swamp shows a maximum in October, a minimum in the spring, when the subsoil water was high and the soil wet and cold (Hammer, 1972). Depending on their age or on their seasonally varying physiological condition, the animals variable behaviour could obstruct a quantitative explanation of seasonal changes in age structure of field populations (Vannier, 1970; Nef, 1971).

Litter Decomposition

The effects of animal activity in woodland litter can often be more readily determined than the animals themselves.

Holes made in leaf tissue frequently accompanied by accumulation of excrement can be easily seen with the naked eye. A ten or twenty-fold magnification clearly shows that smaller animals also attack the dead leaf and leaves behind their feeding patterns and droppings. In general, it can be stated that the animals living in and on litter influence the surrounding environment by their movement and feeding habitats. As they move they shift both organic and inorganic material (transportation), and they also affect the distribution of microorganisms (inoculation). As the fauna feed, it may give rise to a physical and chemical attack on the litter (fragmentation and digestion), but feeding may also result in removal of microorganisms (sanitation). Microscopic and microchemical examination of the faecal pellets of animals feeding on a particular type of litter clearly shows that there is only a very slight degree of digestion and that the most important effect is the reduction of the litter to fragments which are combined into porous agglomerates. The faecal pellets of Enoicyla larvae contain an average of 3000 particles of 60m diameter. The surface exposed to attack is multiplied fifteen times, and there is a considerable increase in aeration and moisture retentivity. The pH is also slightly increased, possibly owing to ammonia secretion, and this also increases bacterial development, as reflected in a marked increase in the carbon dioxide production (Van der drift and Witkamp, 1959).

The arthropod fauna associated with the decomposition of herbage of various species on the surface of the soil was studied by Curry (1973). There are quantitative and qualitative differences between the microarthropod communities associated with various litters. Highest populations of microarthropods have been recorded from *Dactylis*. *Dactylis* and *Lolium* were the most similar in the species composition of their faunas. Changes occur in the relative abundance of various arthropod groups and species during the course of decomposition in all litters. In a grazed pasture, decomposition of organic materials, both of the plant and animal origin, is thought to be an important stage in the recirculation of plant nutrients via the soil-plant-animal cycle. Cattle dung may be a main component of the organic material originated from animals. It is decomposed primarily by a result of microbiological activity leading to the production of carbon dioxide and nutrients (Burges, 1967; Marsh and Campling, 1970). Small animals, mainly invertebrates, may also play important roles in the decomposition of cattle dung. However, little information is available on the role of small animals in the decomposition of cattle dung (Mohr, 1993; Laurence, 1953; Nakamura, 1973). Nakamura (1959) measured the rates of decrease in dry weight and water content of cow dung. During the first month, the water contents of the cow dung of a pasture and forest set in spring (May) decrease at greater rates than that of a pasture set in autumn (October).

Both the numbers and the faunal composition of soil animals obtained from the dung and the soil with or without dung are different between vegetation and between seasons. As soon as the dung is placed on the ground of the pasture in spring, various invertebrates, mainly Coleoptera and Diptera, visit it, but such a phenomenon was not observed in the forest station, nor in the pasture station set out in autumn. Coleptera and Diptera seem to be the important soil animal groups which are concerned with the disappearance of cow dung. The soil with or without dung maintain a different fauna from that in dung. Numbers were often larger in the soil with dung than in the soil without dung (Nakamura, 1975b).

Anti-nutritional Effect

The rate of colonization and succession of soil microarthropod groups were compared with respect to antinutritional chemical parameters in the leaf litter of 12 different tree species under controlled field conditions (Maity and Joy, 1999). The capacity of arthropods to migrate and utilize litter resources was evident within one month of decomposition. The initial colonization was high in *Tectona grandis*, *Casuarina equisetifalia*, *Anthocephalus chinensis*, *Cassia siamea* and *Acacia auriculiformis*. However, leaf litter of *Anacardium occidentale*, *Shorea robusta*, *Mangifera indica*, *Syzygium cumini* and *Eucalyptus citriodora* could harbour only a very poor population. In *Dalbergia sissoo* and *Psidium guajava,* the microarthropod population gradually increases with the microbial conditioning of leaf litter. In the majority of litter types, the higher initial level of total polyphenols and tannins undergoes a marked decline, however, total lipid and lignins register very slow rates of degradation. Regarding the impact of anti-nutritional factors, the initial colonization was high in leaf litter with less initial amounts of polyphenols and tannins. Similarly, leaf litter with minimum initial amount of lipids and lignins show a gradual increase in the colonization rate of microarthropods. Among different groups, Collembola predominates in all litter types which is also the pioneer colonizer. The importance of Acari gradually increases and is evident in the case of *Anthocephalus chinensis*, *T. grandis*, *S. cumini* and *Casuarina equisetifolia*. The colonization of microarthropods demonstrates their significance in trapping energy and nutrients from fast decomposing litter and suitability of some liter types in enhancing biological activity in soil (Maity and Joy, 1999).

Impact of Pollution

Sulkava *et al.* (1996) studied the impact of pollution on soil biological community. Soil materials were defaunated and soil organisms were reinoculated into a mesh basket in the center of each microcosm. Part of the humus layer was contaminated with three concentrations of sodium pentachlorophenate (0, 50 or 500 mg kg^{-1} of dry humus) and put into mesh baskets around the central patch. No differences in dispersion ability from the reinoculated patch were found between microarthropod species. Sodiumpentachlorophenate (PCP) decreases microbial biomass in the humus. Numbers of collembolans were significantly lower in the contaminated patches than in the clean ones. The same trend was observed in the numbers of other microarthropods (mainly oribatid mites). Microarthropods have been found to move intensively in the microcosm and taxa composition in the patches are different. Although PCP reduces numbers of individuals it did not affect diversity of the microarthropod community (Sulkava, 1996). Rising atmospheric carbondioxide changed the abundance and species composition of Collembola and fungal-feeding arthropods (Jones *et al.*, 1998).

Field experiment using insecticides show that excluding microarthropods reduces rates of forest litter decomposition (Seastedt and Crossley, 1983; Blair *et al.*, 1992). Knoepp *et al.* (2000) studied the biological indices of soil quality and compared with four common groups of soil biological indicators including soil microarthropods. Soil organisms, in close interaction with each other and their environment, are responsible for decomposition of organic matter in ecosystems. Many other studies have also demonstrated the importance of the soil fauna as a regulator in nutrient cycling (Anderson *et al.*, 1985; Persson, 1989; Setala, 1990).

Impact of Agricultural Practices

Agroecosystems which are under different management practices often experience vastly differing disturbance regimes, and this has potentially important effects upon the structure of the detritus food-web (Paustian *et al.*, 1990). These impacts have generated considerable interest because of the involvement of the soil biota in regulating nutrient cycling, energy flow and decomposition processes (Coleman *et al.*, 1983). It is becoming increasingly evident that those systems which bear the closest resemblance to natural ecosystems may require fewer inputs because greater reliance can be placed upon ecosystem self-regulating (Altieri, 1991; House *et al.*, 1984). The increasing use of cover crops and mulches for weed management, especially in intensive agricultural systems, also appears to modify the soil-associated biota, and may substantially enhance both microfloral and faunal groups (Favretto *et al.*, 1992; Mulongoy, 1986).

Wardle (1993a) studied soil fauna in an annual cropping system (asparagus: *Asparagus afficinale* L.), to determine the response of the below-ground ecosystem to different forms of agricultural management, particularly those associated with weed control. The impact of agricultural practices on soil microflora and litter decomposition (Wardle *et al.*,1993a), and primary productivity and meso- and macro-fauna (Wardle *et al.*,1993b) have been reported. Further, Wardle (1995) investigated detritus food web and diversity of soil fauna in perennial (asparagus) cropping system. Biota in the perennial system is the most responsive to disturbance. Soil-associated beetle show relation with other organisms in the mulched plots. The soil-associated macro-fauna is usually linked to high weed and surface organic residue levels. The soil-associated beetles was strongly enhanced by mulching and sometimes high weed levels. Approaches based on either functional group or species composition demonstrate reasonably sensitive indicators for assessing the response of the soil biota to disturbance (Wardle, 1995).

The primary goal of the land management practices adopted is to enhance productivity of the desired crop or livestock species and this is achieved by manipulating the habitat so as to make it favourable for the growth of biological species. This frequently involves such practices as the addition of synthetic compounds (eg., pesticides, fertilizers), manipulation of residue and the disturbance of the soil itself (eg., through cultivation, seedbed preparation). Intensification of agriculture is associated with the increasing use of these practices (Ruthenberg, 1980; Giller *et al.*, 1997). While agricultural intensification is generally aimed at maximizing agricultural productivity in the short term maintenance of environmental quality is often perceived to be necessary for sustaining the long-term performance of the system (Philips, 1984; Altieri, 1991).

Disturbance associated with tillage has been shown to alter the structure of soil-associated food-webs (Hendrix *et al.*, 1986; Andren *et al.*, 1990), as well as the composition and diversity of soil arthropod communities (Emmunuel *et al.*, 1985; Lagerlof and Andren, 1991; Robertson *et al.*, 1994). Although the direction of these effects appears to be idiosyncrate and case-specific (Wardle, 1995). Herbicide application can also exert demonstrable effects showing involvement of arthropods responding to indirect effects (eg., loss of weed cover following herbicide spraying) rather than direct effects of the chemical itself (Mahn and Kastner, 1985; Wardle *et al.*, 1998). Above-ground manipulation, such as ground cover vegetation or residue management, also have potentially important effects on soil arthropods, manifest largely through alteration of microclimatic conditions and resource availability (Reichert and Bishop, 1990; Paolelli *et al.*, 1991).

Wardle *et al.* (1999) investigated the effects of three component of agricultural intensification i.e., cultivation (disturbance), herbicide addition (modification of floristic composition) and mulching (resource addition) on soil-associated arthropods in an annual (maize) and a perennial (*asparagus*) cropping system over a 7-year period. An additional treatment (hand-hoeing of weeds during the crop growing season) was used to represent minimal intensification. Many taxa of arthropods responded positively to mulching and to treatments which allowed high weed biomass in the non-crop production period eg., the hand-hoeing and cultivation treatments in the perennial crop. Herbicide treatments also facilitated high numbers of many taxa in the annual crop when this coincided with plot invasion by herbicide-tolerant weeds. Generally, arthropod taxa are positively correlated with weed biomass and negatively with crop plant biomass, probably because of the superior resource (litter) quality produced by the former. Mulching and allowing high weed biomass also promote a high species richness of soil associated Coleoptera but its diversity is not related to weed species diversity (Wardle *et al.*, 1999). They suggested that agricultural intensification is not consistently harmful to the soil fauna. In fact the soil-associated arthropods are most responsive to management practices which affect the nature and quality of resource input. Long-term experiments are essential for answering questions about how agricultural practices affect soil organisms.

Decomposition processes in the soil are mainly biologically mediated phenomena, climatic and other physico-chemical conditions at the site do determine general patterns in decomposition. Reddy and Reddy (1999) reviewed the effects of tillage practices, organic amendments, pesticides, and cover crops, on the population structure of meso- and macro-fauna. Soil inhabiting microarthropods and termites were very sensitive to annual treatment with pesticides. Annual tillage decreased the population of earthworm and microarthropods and increased the abundance of Hymenoptera, Cryptostigmata and Astigmata. In Alfisols, zero tillage and mulching with rice straw were favourable to the endogeic fauna, and farmyard manure increased earthworms abundance and biomass. Experimental studies investigating interactions between soil food web component and their implications for decomposition, nutrient mineralization and plant growth have been reviewed (Scheu *et al.,* 1999). It has been concluded that perturbation experiments in ecosystems which are little disturbed, like forests, are vital for a better understanding of structuring forces and the functioning of soil decomposer communities.

Microbial Activity and Mineralization

Various scientists reported decomposition similar for forest litters (Melillo *et al.*, 1982; Agren and Bosatta, 1996; Heal *et al.*, 1997; Couteaux *et al.*, 1998; Sanger *et al.*, 1998), and crop residue (Amato *et al.*, 1984; Angers and Recous, 1997; Mary *et al.*, 1996; Quemada and Cabrera, 1995). The carbon mineralization of added organic materials (AOM) in soil was assessed by combining laboratory and modeling approaches (Thuries *et al.,* 2002). The AOM used in the organic fertilizer industry included plant residue from agri-food origin, animal wastes, manures, compost and organic fertilizers. They were fractionated by sequential analysis of fibers and analyzed for C, N and ash contents. The effect of fire-induced soil heating on carbon (C) and nitrogen (N) dynamics in forest mineral soils has received only limited scientific attention. It is known that fire enhances chemical oxidation of soil organic matter thus altering its chemical composition (Fernandez *et al.*, 1997) but the degree of organic matter oxidation depends on temperature generated by fire, fire duration, and heat penetration (Hungerford *et al.*, 1991).

Temperatures higher than 50°C result in death of heat-sensitive microbes and temperatures higher than 70°C can directly affect vegetation (Hernandez *et al.*, 1997; Neary *et al.,* 1999). Subsequently, dead plant and microbial biomass can be rapidly oxidized contributing to the pulse of inorganic N that typically follow burning (Diaz-Ravina *et al.*, 1996). However, the combustion of organic matter at exceedingly high temperatures can result in volatilization and potential loss of N previously found in complex organic forms (Giovannini *et al.,* 1990). Soil moisture content may become a confounding factor in assessing the impact of fire on soil biological and chemical properties (Albini and Reinhardt, 1995). The recovery of microbial biomass during 14 days of post-heating incubation is in important factor. Soil demonstrate slow rates of recovery of nitrifying organisms despite high rates of net NH_4^+ -N accumulation. (Choromanska, 2002). It appears that low soil water potential at the time of heat exposure reduces losses of mineralizable N.

Soil respiration is a common and simple measure of biological activity in soil. In particular, C mineralization has been thought of as a convenient surrogate for N dynamics because C mineralization is thought to be coupled to N mineralization the C: N ratio of detrital resources and their consumers (Smith,1994; Mary *et al.*,1996; Myrold,1998). Mineralized C is released as CO_2, while mineralized N may be immobilized by plants and microbes, resulting in measures of N mineralization which reflect net but not gross N mineralization. If C and N mineralization is coupled, measurements of C mineralization could be used to estimate N mineralization. Temporal variation of CO_2 evaluation can be predicted by models that consider soil moisture and temperature (Edwards, 1975; Wilding,*et al.*, 1998). However, spatial and temporal heterogeneity of the soil environment plus the influence of biotic factors (Burgess and Webster, 1980; Rao and Wagenet, 1985; Graffman and Tiedje, 1988; Robertson *et al.*, 1994; Gorres *et al.*,1998) and the effect of root respiration make it difficult to measure or predict the variation of C mineralization at the field scale. In particular, microbial community structure and population dynamics are affected by microbivory and may thus affect mineralization (Coleman *et al.*,1977; Anderson *et al.*,1981; Ingham *et al.*,1985; Hunt *et al.*,1987; Verhoef and Brussaard,1990; Wardle and Yeates,1993; Griffiths,1994).

Microbivorous grazers are thought to enhance nutrient mineralization. The predicted effect of microbivory on nutrient cycling depends on the pore habitat model used. Savin *et al.* (2001) evaluated CO_2 evolution and mineral N content of an old field soil. They investigated the effect of soil moisture on the relationship between microbial biomass, microbivorous and predaceous nematodes. Soil respiration and mineral N concentrations in an old field remain five times during the course of a year. Press and Folster (2001) hypothesized that this accumulation is caused by low microbial activity.

The presence and activity of the microbial component of the surface and subsoil can be detected and measured in many ways, including microbial biomass as well as functional and enzyme activities (Kastner *et al.,* 1994; Lovell *et al.,* 1995, Murphy *et al.,* 1998, Bandick and Dick, 1999). In addition, newer molecular approaches including probing for specific genotypes (Sayler *et al.*, 1995) and monitorıng mRNA expression (Wilson *et al.*, 1999) have an increasingly important part to play in advancing our understanding. Knowledge of microbial numbers and activity in subsoils is essential for understanding the transformation and downward movement of natural and synthetic organics. Bacterial numbers decreased with depth as indicated by viable counts and calculations based upon biomass carbon and extracted DNA. Soil enzyme activities (arylsulphatase, β-glucosidase, phosphomonoesterases urease, dehydrogenase) are also decreased with depth (Taylor *et.al.*, 2002).

ACKNOWLEDGEMENT

The authors are grateful to Indian Council of Agricultural Research (ICAR), New Delhi, for providing financial assistance in the form of a research project.

REFERENCES

Agren, G.I. and Bosatta, E. (1996). Quality: a bridge between theory and experiment in soil organic matter studies. *Oikos*, 76: 522-528.

Albini, F.A., Reinhardt, E.D. (1995). Modeling ignition and burning rate of large woody natural fuels. *Inter. J. Wildland Fire.*, 5: 81-91.

Altieri, M.G. (1991). How can we best use biodiversity in agrosystems? *Outlook on Agriculture*, 20: 15-23.

Amato, M., Jackson, R.B., Butler, J.H.A. and Ladd, J.N. (1984). Decomposition of plant material in Australian soils. II. Residual organic ^{14}C and ^{15}N from legume plant parts decomposing under field and laboratory conditions. *Aust. J. Soil Resea.*, 22: 331-341.

Anderson, J.M. (1975). The enigma of soil animal species diversity. In: Progress in Soil Zoology, Proceedings of the Eight

International Calloquium on Soil Zoology, (ed. J. Vanek). *Academia Prague*, 51-58.

Anderson, J.M. (1977). The organization of soil animal communities. *Ecol. Bulletins* (Stockholm) 25: 15-22.

Anderson, J.M. (1978a). Inter- and intra-habitat relationship between woodland *Cryptostigmata* species diversity and the diversity of soil and litter microhabitats. *Oecologia*, 32: 341-348.

Anderson, J.M. (1978b). Competition between two unrelated species of soil *Cryptostigmata* (Acari) in experimental microcosms. *J. Anim. Ecol.*, 47: 801-803.

Anderson, J.M., Leonard, M.A., Ineson, P. and Huish, S.A. (1985). Faunal biomass; a key component of a general model of N mineralization. *Soil. Biol. Biochem.*, 17; 735-737.

Anderson, R.V., Coleman, D.C., Cole, C.V. and Elliott, E.T. (1981). Effect of the nematodes *Acrobeloides* sp. and *Mesodiplogaster lheritieri* on substrate utilization and nitrogen and phosphorus mineralization in soil. *Ecology*, 62: 549-555.

Andren, O., Lindberg, T., Paustian, K. and Rosswall, T. (eds.), (1990). Ecology of Arable Land. Ecological Bulletins, Copenhagen.

Angers, D.A. and Recous, S. (1997). Decomposition of wheat straw and rye residues as affected by particle size. *Plant Soil*, 189: 197-203.

Arshad, M.A., Schnitzer, M. and Preston, C.M. (1988). Characterization of humic acids from termite mounds and surrounding soils. Kenya. *Geoderma*, 42: 213-225.

Atkinson, W.D. and Shorrocks, B. (1981). Competition on divided and ephemeral resource: a simulation model. *J. Ani. Ecol.*, 50: 461-471.

Bandick, A.K. and Dick, R.P. (1999). Field management effects on soil enzyme activities. *Soil Biol. Biochem.*, 31: 1471-1479.

Bengtsson, G., Rundgren, S. and Sjogren, M. (1994). Modeling dispersal distances in a soil gradient: the influence of metal tolerance, competition and experience. *Oikos*, 71: 13-23.

Bengtsson, G. and Tranvik, L. (1989). Critical metal concentration for forest soil invertebrates. In: *Water, Air, Soil Pollution*, 47: 381-417.

Berthet, P. (1964). Lactivite des oribatides d'une chenaie. Mem. Inst. Roy. *Sci. Nat. Belg.*, 152: 1-152.

Blair, J.M., Crossley, Jr., D.A. and Callaham, L.A. (1992). Effects of litter quality and microarthropods on N-dynamics and retention of exogenous N-15 in decomposing litter. *Boil. Fertil. Soil*, 12: 241-252.

Bocock, K.L. (1963). The digestion and assimilation of food by Glomeris. In: Soil Organisms (ed. Doeksen and Van der Drift), 85p. North-Holland Publ. Co., Amsterdam.

Breznak, J.A. and Brune, A. (1994). Role of microorganisms in the digestion of lignocellulose by termites. *Ann. Rev. Entomo.*, 39: 453-487.

Buchner, P. (1953). Endosymbiose der Tiere mit Pflanzlichena Mikroorganismen, Birkhauser Basel/Stuttgart.

Burges, A. (1967). The decomposition of organic matter in the soil. In: Soil Biology (ed. A. Burges and F. Raw) Academic Press, 479-492.

Burgess, T.M. and Webster, R. (1980). Optimal interpolation and isarithmic mapping of soil properties. I. The semivariogram and punctual kriging. *J. Soil Sci.*, 31: 315-331.

Choromanska, U. and Deluca, T.H. (2002). Microbial activity and nitrogen mineralization in forest mineral soil following heating: evaluation of post-fire effects. *Soil Biol. Biochem.*, 34: 263-271.

Christiansen, K. (1970). Experimental studies on the aggregation and dispersion of Collembola. *Pedobiol.*, Bd. 10.s.: 180-198.

Cloudsley-Thompson, J.L. (1964). Terrestrial animals in dry heat: arthropods. In: Handbook of Physiology. 4. Adaptation to the Environment, American Physiological Society, Washington D.C., 451-465.

Coleman, D.C., Cole,C.V., Anderson, R.V., Blaha, M., Campion, M.K., Clarholm, M., Elliott, E.T., Hunt, H.W., Shaefer, B. and Sinclair, J. (1977). An analysis of rhizosphere-saprophage interactions in terrestrial ecosystems. In: *Soil Organisms as Components of Ecosystems. Proceedings of the 6th International Colloquium of Soil Zoology* (Eds. Lohm, U. and Persson, T.). 25: 299-309. Stockholm.

Coleman, D. and Macfadyen (1966). The recolonization of gamma-irradiated soil by small arthropods. *Oikos* 17; 62-70.

Coleman, D.C., Reid, C.P. and Cole, V.C. (1983). Biological strategies of nutrient cycling in soil systems. *Adv. Ecol. Res.*, 13: 1-55.

Couteaux, M.M., Mc Tiernan, K.B., Berg, B., Szuberla, D., Dardenne, P. and Bottner, P. (1998). Chemical composition and carbon mineralization potential of scots pine needles at different stages of decomposition. *Soil Biol. Biochem.*, 30: 583-595.

Curry, J.P. (1973). The arthropods associated with the decomposition of some common grass and weed species in the soil. *Soil Biol. Biochem.*, 5: 645-657.

Diaz-Ravina, M., Prieto, A. and Baath, E. (1996). Bacterial activity in a forest soil after soil heating and organic amendments measured by the thymidine and leucine incorporations technique. *Soil Biol. Biochem.*, 28: 419-426.

Edney, E.B. (1957a). The winter relations of terrestrial arthropods. Cambridge Univ. Press.

Edwards, N.T. (1975). Effects of temperature and moisture on carbon dioxide evolution in a mixed deciduous forest floor. *Soil Sci. A. Proceedings*, 39: 361-365.

Emmanuel, N., Curry, J.P. and Evans,G.O. (1985). The soil Acari of barley plots with different cultural treatments. *Experimental and Applied Acarology*, 1: 101-113.

Fernandez, I., Cabaneiro, A. and Carballas, T. (1997). Organic matter changes immediately after a wildfire in an Atlantic forest soil and comparison with laboratory soil heating. *Soil Biol. Biochem.*, 29: 1-11.

Ferrar, P. and Watson, J.A.L. (1970). Termites associated with dung in Austalia. *J. Aust. Entomo. Soc.*, 9: 100-102.

Giller, K.E., Beare, M.H., Lavelle, P., Izac, A.M.N. and Swift, M.J. (1997). Agricultural intensification, soil biodiversity and ecosystem function. *Appl. Soil Ecd.*, 6: 3-16.

Giovannini, G., Lucchesi, L. and Giachetti, M. (1990) Effects of soil heating on some chemical parameters related to soil fertility and plant growth. *Soil Sci.*, 149: 344-350.

Gorres, J.H., Dichiaro, M.J., Lyons, J.B. and Amador, J.A. (1998). Spatial and temporal patterns of soil biological activity in a forest and an old field. *Soil Biol. Biochem.*, 30: 219-230.

Graffman, P.M. and Tiedje, J.M. (1988). Denitrification hysteresis during wetting and drying cycles in soil. *Soil Sci*. Soc. A. J., 52: 1626-1629.

Griffiths, B.S. (1994). Soil nutrient flow. In: *Soil protozoa*, (ed. Darbyshire J.F.), CAB International, Oxford. pp. 65-91.

Haarlov, N. (1960). Microarthropods from Danish soils. *Oikos Suppl.*, 3: 1-176.

Hagvar, S. and Abrahamsen, G. (1980). Colonization by Enchytraeidae, Collembola and Acari in sterile soil samples with adjusted pH levels. *Oikos*, 34: 245-258.

Hale, W.G. (1966). A population study of moorland Collembola. *Pedobiol.*, 6: 65-99.

Hammer, M. (1972). Microhabitats of oribatid mites on a Danish woodland floor. ***Pedobiologia***, 12: 412-423.

Hartenstein, R. (1962). Soil Oribatei I. Feeding specificity among forest soil Oribatei. *Ann. Entomo. Soc. America*, 55: 2, 202-206.

Heal, O.W., Anderson, J.M. and Swift, M.J. (1997). Plant litter quality and decomposition: an historical overview. In: *Driven by Nature, Plant Litter Quality and Decomposition*. (eds. Cadisch, G. and Giller, K.E.), 47-66 pp., CAB International, Wallingford.

Hendrix, P.F., Parmelee, R.W., Crossley, D.A., Coleman, D.C., Odum, E.P. and Groffman, P.M. (1986). Detritus food-webs in conventional and no-tillage agroecosystems. *Biosci.*, 36: 374-380.

Hernandez, T., Garcia, C. and Reinhardt, I. (1997). Short-term effects of wildfire on the chemical, biochemical and microbiological properties of Mediterranean pine forest soils. *Boil. Fertil. Soil*, 25: 109-116.

Higashi, M. and Abe, T. (1997). Global diversification of termites driven by the evolution of symbiosis and Sociality. In: *Biodiversity: An Ecological Perspective* (eds. Abe. T., Levin, S.A. and Higashi, M.), 83-112 pp., Springer, New York.

House, W. (1984). Nitrogen cycling in conventional and no-tillage agroecosystems in the southern piedmont. *J. Soil Water Conserv.*, 39: 194-200.

Hudson, B. (1980). Colonization of industrial reclamation site by Acari, Collembola and other invertebrates. *J. Appl. Ecol.*, 17: 255-275.

Huhta, V., Ikonen, E. and Vilkamaa, P. (1979). Succession of invertebrate populations in artificial soil made of sewage sludge and crushed bark. *Annals Zoologici Fennici*, 16: 223-270.

Hungerford, R.D., Harrington, M.G., Frandsen, W.H., Ryan, K.C. and Niehoff, G.J. (1991). Influence of fire factors that affect site productivity. Intermountain Research Station General Technical Report INT-280, USDA-FS. 50p.

Hunt, H.W., Coleman, D.C., Ingham, E.R., Ingham, R.E., Elliott, E.T., Moore, J.C., Rose, S.L., Reid, C.P.P. and Morley, C.R. (1987). The detrital food web in a short-grass prairie. *Boil. Fertil. Soil*, 3: 57-68.

Ingham, R.E., Trofymow, J.A., Ingham, E.R. and Coleman, D.C. (1985). Interactions of bacteria, fungi and their nematode grazers: effect on nutrient cycling and plant growth. *Ecological Monographs*, 55: 119-140.

Jain, K.L., Singh, M., Dogra, D., Singh, M. and Dogra, D. (1990). Composition of Acarine fauna inhabiting soil and litter in *Dalbergia sissoo* plantation under semiarid condition in India. *J. Entomo. Res.*, 22: 147-152.

Jones, T.H., Thompson, L.J., Lawton, J.H., Bezemer, T.M., Bardgett, R.D., Blackburn, T.M., Bruce, K.D., Cannon, P.F., Hall, G.S., Hartley, S.E., Howson, G., Jones, C.G.C., Kampichler,+ Kandeler, E. Ritchin, D.A. (1998). Impacts of Rising Atmospheric Carbon Dioxide on Model Terrestrial Ecosystems. *Science*, 280: 441-442.

Joosse, E.N.G. and Groen, J.B. (1970). Relationship between suturation deficit and the survival and locomotory activity of surface dwelling Collembola. *Entomol. Exp. Appl.,* 13: 229-235.

Joosse, E.N.G. and Groen, J.B. (1970). Relationship between saturation deficit and the survival and locomotory activity of surface dwelling Collembola. *Entomo. Exp. Appl.*, 13: 229-235.

Joosse, E.N.G. (1971). The formation and biological significance of aggregations in the distribution of Collembola. *Netherlands J. of Zoology,* 20(3): 299-314.

Kareiva, P. (1986). Patchiness, dispersal and species interactions: Consequences for communities of herbivorous insects. In: Community Ecology (eds. J. Diamond and T.J. Case), 192-206 pp. Harper and Row, New York.

Kastner, M., Breuer-Jammali, M. and Mahro, B. (1994). Enumeration and characterization of the soil microflora aromatic hydrocarbons. *Appl. Microbiol. Biotech.*, 41: 267-273.

Kitazawa, Y. (1968). Measurement of secondary production of terrestrial animal community. In: JIBP Progress Report 1968 (ed. Kitazawa), 1-206.

Knoepp[a], D.J., Coleman, D.C., Crossley Jr[b], D.A. and Clark[c], J.S. (2000). Biological indices of soil quality: an ecosystem case study of their use. *Forest Ecol. And Manage.,* 138: 357-368.

Kumar, M.G.S., Sujatha, M.P. and Sankar, S. (1999). Population density and diversity of microarthropods and annelids in the reed growing. *J. Trop. Fort.,* 15: 135-143.

Kurceva, G.F. (1960). The food assimilated by litter fauna chiefly consists of the bacteria and fungi ingested with the vegetable material. *Pedology, Leningr,* 4: 16.

Lagerlof, J. and Andren, O. (1991). Abundance and activity of Collembola, Protura and Diplura (Insecta, Apterygota) in four cropping system. *Pedobiol.*, 35: 337-350.

Lawrence, R.R. (1959). The sand-dune fauna of the Namib desert. *Sth. Afr. J. Sci.*, 55: 233-239.

Lee, K.E. and Wood, T.G. (1971a). Physical and chemical effects of Australian termites and their pedological significance. *Pedobiol.*, 11: 376-409.

Lee, K.E. and Wood, T.G. (1971b). Termites and soils. Academic Press, London. 251 p.

Littlewood, C.F. (1969). A surface sterilization technique used in feeding algae to Oribatei. In Proc. 2nd Int. Congr. Acarology, Sutton Bonington, 1967, (ed G.O. Evans). Hungarian Academy of Sciences, Budapest, 53-56.

Lopez-Hernandez, D., Nino, M., Fardeau, J.C., Nannipieri, P. and Chacon, P. (1989). Phosphorus accumulation in savanna termite mound in Venezuela. *Sci.*, 40: 635-640.

Lopez-Hernandez, D. and Febres, A. (1984). Changements chimiques et granulometriques produits dans le sols de Cote d'Ivoire par la presence de trois especes de termites. *Revue d'Ecologie et de Biologie. du Sol*, 21: 477-489.

Lovell, R.D., Jarvis,S.C. and Bardgett, R.D. (1995). Soil microbial biomass and activity in long-term grassland: effects of management changes. *Soil Biol. Biochem.*, 27: 969-975.

Luxton, M. (1972). Studies on the oribatid mites of a Danish beech wood soil. *Pedobiol.*, 12: 434-463.

Madge, D.S. (1964). The water relations of *Belbageniculosa* Oud., and other species of Oribatid mites. *Acarologia,* 6: 199-223.

Maity, S.K. and Joy, V.C. (1999). Impact of antinutritional chemical compounds of leaf litter on detritivore soil arthropod fauna. *J. Ecobiol.*, 11: 193-202.

Marsh, R.R. and Campling, C. (1970). Fouling of pasture by dung. *Harb. Abst.,* 40: 123-130.

Mary, B., Recous, S., Darwis, D. and Robin, D. (1996). Interactions between decomposition of plant residue and nitrogen cycling in soil. *Plant Soil*, 181: 71-82.

Melillo, J.M., Aber, J.D. and Muratore, J.F. (1982). Nitrogen and lignin control of hardwood leaf litter decomposition dynamics. *Ecol.*, 63: 621-626.

Mulongoy, K. (1986). Microbial biomass and maize nitrogen uptake under a Posphocarpus palustris live-mulch grown on a tropical alfisol. *Boil. Biochem.*, 18: 395-398.

Murphy, D.V., Sparling, G.P. and Fillery, I.R.P. (1998). Elemental analysis for biomass C and N and gross N mineralization with soil depth in two contrasting Western Australian agricultural soil. *Aust. J. Soil Res.*, 36: 45-55.

Myrold, D.D. (1998). Transformations of nitrogen. In: *Principles and Applications of Soil Microbiology* (eds. Sylvia, D.M., Fuhrmann, J.J., Hartel, P.G. and Zuberer, D.A.), 259-294 pp. Prentice Hall, Upper Saddle River, NJ.

Nakamura, Y. (1959). Decomposition of organic materials and soil fauna in pasture. 2. Disappearance of cow dung. *Pedobiol.*, 15: 129-132.

Nakamura, Y. (1975). Decomposition of organic materials and soil fauna in pasture. 2. Disappearance of cow dung. *Pedobiol.*, 15: 129-132.

Neary, D.G., Klopatek, C.C., De Bano, L.F. and Ffolliott, P.F. (1999). Fire effects on belowground sustainability: a review and synthesis. *Fort. Ecol. Manag.*, 122: 51-71.

Nef, L.(1971). Influence de I'Humidite' sur le geotactisme des Oribates (Acarina) dans I extracteur de Berlese-Tullgren. *Pedobiol.*, 11: 433-445.

Nielsen, C.O. (1962). Carbohydrases in soil and litter invertebrates. *Oikos*, 13:(2): 200-215.

Nutman, S.R. (1941) Function of the ventral tube in *Onychiurus armatus* (Collembola). *Nature*, Lond., 148: 168-169.

Okello-Oloya, T., Spain, A.V. and John, R.D. (1985). Selected chemical characteristics of the mounds of two species of Amitermes (Isoptera Termitinae) and their adjacent surface soil from North-eastern Australia. *Revue d'Ecologie et de Biologie du Sol*, 22: 291-311.

Paclt, J. (1956). Biologie der priar flugellosen Insekten. Gustav Fischer Verlag, Jena. 258 p.

Paoletti, M.G., Favretto, M.R., Stinner, B.R., Purrington, F.F. and Bater, G.E. (1991). Invertebrates as indicators of soil use. *Agricul. Ecosys. Environ.*, 34: 341-362.

Paustian, K., Andren, O., Clarhalm, M., Hansson, A-C., Johansson, G., Lagerlof, J., Lindberg, T., Pettersson, R. and Sohlenius, B.

(1990). Carbon and nitrogen budgets of four agro-ecosystems with annual and perennial crops, with and without N fertilization. *J. Appl. Ecol.,* 27: 60-84.

Phillips, S.H. (1984). Introduction. In: *No Tillage Agriculture: Principles and Practices.* (eds. Phillips, R.E. and Phillips, S.H.), 1-10 pp. Van Nostrand Reinhold Company, New York.

Poole, T.B. (1961). An ecological study of the Collembola in a coniferous forest soil. *Pedobiol.,* 1: 113-137.

Priess, J.A. and Folster, H. (2001). Microbial properties and soil respiration in submountane forests of Venezuelian Guyana: characteristics and response to fertilizer treatments. *Soil Biol. Biochem.,* 33:503-509.

Quemada, M. and Cabrera, M.L. (1995). Carbon and nitrogen mineralized from leaves and stems of four cover crops. *Soil Sci. Soc. Am. J.,* 59: 471-477.

Rao, P.S.C. and Wagenet, R.J., (1985). Spatial variability of pesticides in the field soils: methods for data analysis and consequences. *Weed Sci.,* 33: 18-24.

Reddy, M.V. (1995) Soil organisms and litter decomposition in the tropics. p. 274.

Reddy, M.V., and Reddy, M.V. (1999). Soil management and beneficial soil meso and macrofauna with particular reference to semi-arid tropical Alfisol. 223-271pp. Sci. Pub. Inc. Enfield, USA.

Riechert, S.E. and Bishop, L. (1990). Prey control by an assemblage of generalist predators: spiders in test garden system. *Ecol.,* 71: 1441-1450.

Robertson, L.N., Kettle, B.A. and Simpson, G.B. (1994). The influence of tillage practices on soil macrofauna in a semi-arid ecosystem in northeastern Australia. *Agricul. Ecosys. Environ.,* 48:149-156.

Rose, C.W. (1968). Evaporation from bare soil under high radiation condition. *Trans. 9th Cong. Int. Soil Sci.* 1: 57-66.

Ruthenberg, H. (1980). Farming systems in the Tropics. Clarendon Press, Oxford.

Sanger, L.J. Cox, P., Splatt, P., Whelan. M. and Anderson, J.M. (1998). Variability in the quality and potential decomposability of pinus sylvestris litter from sites with different soil characteristics: acid detergent fiber (ADF) and carbohydrate signatures. *Soil Biol. Biochem.,* 30:455-461.

San Jose, J., Montes, R., Stansly, P.A. and Bentley, B. (1989). Environmental factors related to the occurrence of mound-building Nasute termites in *Trachypogon* savannas of the Orinoco Llanos. *Biotropica*, 21: 353-358.

Sanyal, A.K. (1996). Soil arthropod population in two contrasting sites at Nadia, West Bengal. *Environ. Ecol.*, 14: 346-350.

Savin, M.C., Gorres, J.H., Neher, D.A. and Amador, J.A. (2001). Biogeophysical factors influencing soil respiration and mineral nitrogen content in an old field soil. *Soil Biol. Biochem.*, 33: 429-438.

Sayler, G.S., Layton, A., Lajoie, C.,Bowman, J., Tschantz, M. and Fleming, J.T. (1995). Molecular site assessment and process monitoring in bioremediation and natural attenuation. *Appl. Biochem. Biotech.*, 54: 277-290.

Schuster, R. (1956). Der anteil der oribatiden an den zersetzvngsvorga – ngen im B boden. Z. Morph. *Okal. Tiere* 45: 1-33.

Seastedt, T.R. and Crossley, Jr. D.A. (1983). Nutrients in forest litter treated with naphthalene and simulated througfall: a field microcosm study. *Soil Bio. Biochem.*, 15: 159-165.

Setala, H. (1990). Effects of soil fauna on decomposition and nutrient dynamics in coniferous forest soil. Dissertation, Uni. Jyvaskyla.

Setala, H., Marshall, V. and Trofymow, (1995). Influence of micro- and macro-habitat factors on Collembolan communities in Douglas-fir stumps during forest succession. *Appl. Soil Ecol.*, 2: 227-242.

Shweta and Gupta, B.B. (1997). Apterygote ecology at Aligarh (India). *J. Ecobiol.*, 9:17-30.

Singh, J. and Pillai, K.S. (1973). The world Beneath us. *Science Reporter.* 318-321 pp.

Slaytor, M.(1992). Cellulose digestion in termites and cockroaches: What role do symbionts play? *Comp. Biochem.Physiol.*, 103B: 775-784.

Smith, J.L.(1994). Cycling of nitrogen through microbial activity. In: *Soil Biology: Effects of Soil Quality* (eds. Hatfield, J. L. and Stewart, B.A.), pp. 91-120. Lewis Publishers, Boca Raton, FL.

Spain, A. and Luxton, M. (1971). Catelog and bibliography of the acari of the New Zealand sub-region. *Pacific Insects Monographs.*, 25: 179-226.

Spain, A.V., Okello-Oloya, T. and Brown, A.J. (1983). Abundance, aboveground masses and basal areas of termites mounds at six location in tropical north- eastern Australia. *Revued' Ecologie et de Biologie du sol,* 20: 547-566.

Stanton, N. (1979). Patterns of species of diversity in temperate and tropic litter mites. *Ecol.,* 60:295-304.

Streit, B., Buehlmann, A. and Reutimann, P. (1985). Mites succession in compost communities: Studies with Oribtei. Gamasina and Uropodina. *Pedobiol.*, 28: 1-12.

Strong, D.R. Jr (1983). Natural variability and the manifold mechanisms of ecological communities. *Am. Nat.*, 122: 636-660.

Sulkava, P., Huhta, V. and Laakso, J. (1996). Impact of soil faunal structure on decomposition and N-mineralisation in relation to temperature and soil moisture in forest soil. *Pedobiol.,* 40: 505-513.

Swift, M.J., Heal, J.M. and Anderson, J.M. (1979). Decomposition in Terrestrial Ecosystems. Blackwell Scientific Publications, Oxford.

Tamura, H., Nakamura, Y., Yamauchi, K. and Fujikawa, T. (1969). An ecological survey of soil fauna in Hidaka-Mombetsu, Southern Hokkaido. *J. Fac. Sci.* Hokkaido Univ. Ser. VI, Zool. p. 17.

Tayosu, I, Hyodo, F., Abe, T., Inoue, T. and Spain, Alister, V. (2002). Nitrogen and carbon stable isotope rations in the sympatric Australian termites, Amitermes laurensis and Drepanotermes rubriceps (Isoptera: Termitidae) in relation to their feeding habits and the quality of their food material. *Soil Biol. Biochem.*, 34: 297-301.

Taylor, J.P., Wilson, B., Mills, M.S. and Burns, R.G. (2002). Comparison of microbial numbers and enzymatic activities in surface soils and subsoils using various techniques. *Soil Biol. Biochem.*, 34: 387-401.

Thuries, L., Pansu, M., Larre-Larrouy, M.C. and Feller, C. (2002). Biochemical composition and mineralization kinetics of organic inputs in a sandy soil. *Soil Biol. Biochem.*, 34: 239-250.

Tranvik, L. and Eijsackers, H. (1989). On the advantage of *Folsomia fumetarioides* over *Isotomiella minor* (Collembola) in a metal polluted soil. *Oecologia*, 80: 195-200.

Usher, M.S., Booth, R.G. and Sparkes, K.E. (1982). A review of progress in understanding the organization of communities of soil arthropods. *Pedobiol.,* 23: 126-144.

Usher, M.S. (1985).. Population and community dynamics in the soil ecosystem. In *Ecological Interactions in Soil* (eds. A. Fitter, D. Atkinson, D. Read and M. Usher), 243-265 pp. Blackwell Scientific Publications, Oxford.

Vander drift, J. (1966). The effects of animal activity in the little layer. *Experi., Pedology* 227-235.

Vander drift, J. (1969). Production and decomposition of organic matter in an oakwood in the Netherlands. *Proc. Brussels Symp.,* 4: 631-634.

Vannier, G. (1970). Reactions des Micro-arthropods auch variation de l'etat hydrique du sol. Techniques relatives a l'extraction des Arthropods du Sol. Editions du Centre National de la Recherche Scientifique, Paris.

Verhoef, H.A. and Brussaard, L. (1990). Decomposition and nitrogen mineralization in natural and agroecosystems: the contribution of soil animals. *Biogeochem.*, 11: 175-211.

Wardle, D.A., Nicholson, K.S. and Yeates, G.W. (1993). The effects of weed management strategies on some soil-associated arthropods in maize and asparagus ecosystems. *Pedobiol.*, 37: 257-269.

Wardle, D.A., Verhoef, H.A. and Clarhalm, M. (1998) Tropic relationships in the soil microfood-web: predicting the responses to changing global environment. *Global Change Biol.*, 4: 713-722.

Wardle, D.A., Yeates, G.W., Watson, R.N. and Nicholson, K.S. (1995) The detritus food-web and diversity of soil fauna as indication disturbance regimes in agro-ecosystems. *Plant Soil Zool.*, 70: 35-43.

Wardle, D.A., Yeates, G.W., Watson, R.N. and Nicholson, K.S. (1993b). Response of soil microbial biomass and plant litter decomposition to weed-management strategies in maize and asparagus cropping systems. *Soil Biol. Biochem.*, 25: 857-868.

Wardle, D.A., Nicholson, K.S., Bonner, and Yeates, G.W. (1999). Effects of agricultural intensification on soil-associated arthropods population dynamics, community, structure, diversity and temporal variability over a seven-year period. *Soil Biol. Boichem.,* 31: 1691-1706.

Wilson, M.S., Bakermans, C. and Madsen, E.L. (1999). In situ real-lime catabolic gene expression: extraction and characterization of naphthalene dioxygenase mRNA transcripts from ground water. *Appl. Environ. Microbiol.*, 65: 80-87.

Woodring, J.P. (1963). The nutrition and biology of saprophytic sarcoptiformes. In: *Advances in Acarology I* (ed. J. Naegele). Comstock, Cornell, 89-111.

Wood, T.G. (1972). The distribution and abundance of *Folsomides deserticola* (Collembola: Isotomidae) and other micro-arthropods in arid and semi-arid soils in Southern Australia, with a note on nematode populations. *Pedobiol.,* Bd. 11: 446-468.

Wood, T.G., Johnson, R.A. and Anderson, J.M. (1983). Modification of soils in Nigerian savanna by soil-feeding Cubitermes, Termitidae. *Soil Biol. Biochem.*, 15: 575-579.

Wood, T.G. and Sands, W.A. (1978). The role of termites in ecosystems. In: *Production Ecology of Ants and Termites*. (ed. Brain, M.V.), 245-292 pp., University Press, Cambridge.

